HUMAN RETROVIROLOGY: HTLV

Human Retrovirology: HTLV

Editor

William A. Blattner, M.D.

Viral Epidemiology Section
Environmental Epidemiology Branch
National Cancer Institute
National Institutes of Health
Bethesda, Maryland

RAVEN PRESS ✍ NEW YORK

Raven Press, 1185 Avenue of the Americas, New York, New York 10036

Made in the United States of America

Library of Congress Cataloging-in-Publication Data

Human retrovirology: HTLV/edited by William A. Blattner.
      p. cm.
    Based on a 3 day conference held in Port of Spain, Trinidad and Tobago in Feb., 1989.
    ISBN 0-88167-661-6
    1. HTLV infections—Congresses.   2. HTLV (Viruses)—Congresses.
I. Blattner, William A.
    [DNLM: 1. HTLV Viruses—congresses.   2. Retrovirus Infections—congresses.   QW 166 H9183 1989]
   QR201.H86H85   1990
   616.9′25—dc20
   DNLM/DLC
   for Library of Congress   90-8204
                                                      CIP

9 8 7 6 5 4 3 2 1

Preface

The isolation and characterization of the first human retrovirus, human T-cell lymphotrophic virus type I (HTLV-I), in 1980, culminated a search for such an agent that was begun in the early decades of this century. Propelled by this sentinel discovery, the decade of the 1980s has been one of rapid discovery, including the recognition that the epidemic of acquired immunodeficiency is caused by a retrovirus, human immunodeficiency virus, HIV-1, a discovery directly dependent on the techniques perfected in the isolation of HTLV-I. Because of the pleotrophic manifestations of HTLV-I, research on this agent has brought together scientists and researchers from many disciplines. Since no single traditional scientific forum provides an adequate venue for exploring the range of insights provided by this virus and its close cousin HTLV-II, a three-day meeting involving scientists and researchers from all over the world with expertise in many areas of science was convened in Port of Spain, Trinidad, and Tobago to provide an update and synthesis of current knowledge on these viruses. In this volume are distilled emerging new insights from multiple disciplines and a comprehensive summary of the state of the art surrounding HTLV-I/II.

The section concerning pathogenesis presents a series of chapters that provide a framework for considering how viruses of this class cause the diversity of diseases that have now been recognized. A major issue in considering the pathogenesis of HTLV-associated diseases is the question of latency from exposure to disease. HTLV has the unique property of entering the host years to decades before disease subsequently emerges.

In chapters in this section and the subsequent section on virology, the unusual properties of the regulatory genes of HTLV provide a framework for addressing the pathogenic mechanisms of viral diseases of long latency. Other manifestations of this interaction between virus and host include immunologic effects such as virally induced lymphocyte proliferation and cytokine induction by viral regulatory genes. Recent discoveries reported in this section bolster suggestions that other newly discovered or suspected viruses of this class may be linked to other diseases such as mycosis fungoides and multiple sclerosis.

The clinical manifestations of HTLV infection range from aggressive forms of T-cell leukemia and lymphoma to chronic demyelinating neurologic syndromes. The new recognition that HTLV also may be associated with polymyositis reinforces the concept that a broad spectrum of organ systems are affected by these viruses.

The section on epidemiology provides an in-depth presentation of issues surrounding the transmission and natural history of HTLV infection as well as broad information on the geographic patterns of virus and disease occurrence.

The detection of these viruses by serologic techniques has been a major challenge not only because of their low level of expression but also because of significant cross reaction between HTLV-I and HTLV-II; this is of concern regarding the safety of the blood supply and in regard to etiologic studies evaluating patterns of immune response in relationship to disease.

The publication of this book at the end of the first decade of discovery of the HTLV family of viruses documents the tremendous progress made and the new insights gained concerning disease causation that have resulted from their discovery. The decade of the 1990s promises to bring forth even greater knowledge as the tools of modern biology are interfaced with the best of clinical and epidemiologic research to gain new understanding about the role of currently known and yet-to-be discovered human retroviruses in disease.

William A. Blattner, M.D.

Acknowledgments

I would like to acknowledge the many contributions of Professor Courtenay Bartholomew, University of the West Indies, Port of Spain, his staff, and many friends and colleagues who provided a special atmosphere of hospitality unique in the experience of conference participants. Thanks as well to Dr. Howard Streicher and Dr. Arwin Diwan who contributed to the organization of the meeting and to Mrs. Judy Lichaa who organized many details in producing the conference. Thanks as well to Angela Mayes and Susan Redfield, and to the staff at Raven Press, especially Erin Thomas, Marta Núñez, and Mary Martin Rogers who were critical to producing the book. The sessions were expertly moderated by the following chairpersons: Dr. Myron Essex, Dr. Elaine Jaffe, Professor Guy de Thé, Dr. Pamela Rodgers-Johnson, Dr. Arwin Diwan, Dr. Howard Streicher, Dr. Dani Bolognesi, Dr. Michael Lairmore, and Dr. Carl Saxinger. Rapporteurs whose syntheses of the discussion following each session are recorded at the end of many chapters were Dr. Dean Mann, Dr. Estela Matutes, Dr. William A. Sheremata, Dr. Alexander Krämer, Dr. Jonathan Kaplan, Dr. David Waters, Dr. Jeffrey Nakamura, and Dr. Anne Bodner. Mr. Roger Morris, External Affairs Department, E. I. Dupont de Nemours, and his staff provided logistical support, and publication of this book was underwritten by a grant from E. I. Dupont de Nemours. Finally, thanks to the participants who shared openly of often unpublished data and new insights, which are the fuel for new scientific discovery.

Contents

Section I—Pathogenesis

Section II—Virology

Section III—Clinical Manifestations

Section IV—Epidemiology

Section V—Serology and Dx

Contributors

J. Todd Abrams
The Wistar Institute
Thirty-sixth Street at Spruce
Philadelphia, Pennsylvania 19104 4268

David W. Anderson
Division of Blood and Blood Products
Center for Biologics Evaluation and
    Research
Food and Drug Administration
8800 Rockville Pike
Bethesda, Maryland 20892

Cesar Arango
Department of Internal Medicine
Universidad Del Valle
Carrera 38A #5A-108
Cali, Colombia

Dean W. Ballard
Howard Hughes Medical Institute
Duke University School of Medicine
Box 3037
Durham, North Carolina 27710

Courtenay F. Bartholomew
Department of Medicine
University of the West Indies
General Hospital
14 Trinidad Crescent, Federation Park
Port of Spain
Trinidad, West Indies

R. Bellance
Service de Neurologie
Hôpital la Meynard
97200-Fort de France, Martinique

Gerald A. Beltz
Cambridge BioScience Corporation
365 Plantation Street
Biotechnology Research Park
Worcester, Massachusetts 01605

Robin Biojo
General Surgery Department
San Andres Hospital
Tumaco, Columbia

William A. Blattner
Viral Epidemiology Section
National Cancer Institute
National Institutes of Health
Bethesda, Maryland 20892

Ernst Böhnlein
Sandoz Research Institute
1235 Vienna, Austria

Virginia M. Braman
Cambridge BioScience Corporation
365 Plantation Street
Biotechnology Research Park
Worcester, Massachusetts 01605

G. G. Buisson
Service de Neurologie
Hôpital la Meynard
97200-Fort de France, Martinique

M. Campbell
Department of Pathology
University of the West Indies
Kingston 7, Jamaica

Christiane Caudie
Laboratoire d'Immuno-Biologie
Hôpital P. Whertheimer
69, Lyon Cedex, France

W. Charles
Department of Haematology
General Hospital
Port of Spain, Trinidad

Yi-ming A. Chen
Department of Cancer Biology
Harvard School of Public Health
665 Huntington Ave.
Boston, Massachusetts 02115

Farley R. Cleghorn
Caribbean Epidemiology Center
P.O. Box 164
Port of Spain, Trinidad

Enrico Collalti
I University
Department of Experimental Medicine
Via Regina Elena
00161 Roma, Italy

Mauricio Concha
Tropical Spastic Paraparesis
Research Unit
Tumaco, Columbia

Mark C. Connelly
E. I. DuPont de Nemours & Co. Inc.
Medical Products Department
Glasgow Research Laboratory
Box 713
Glasgow, Delaware 19702

Maria Coronesi
Division of Infectious Disease
North Shore University Hospital
300 Community Drive
Manhasset, New York 11030

Beverly Cranston
Department of Pathology
University of the West Indies
Kingston 7, Jamaica

J. K. Cruickshank
Department of Epidemiology
Northwick Park Hospital
Watford Road
Harrow, Middlesex, HA 1 3UJ
United Kingdom

Mabel Cruz
Karolinska Institute
Department of Neurology
Huddinge University Hospital
Stockholm, Sweden

Bryan R. Cullen
Howard Hughes Medical Institute
Duke University School of Medicine
Box 3037
Durham, North Carolina 27710

A. G. Dalgleish
Retrovirus Research Group
Clinical Research Centre
Watford Road
Harrow, Middlesex, HA 1 3UJ
United Kingdom

Barun De
Retrovirus Diseases Branch
Division of Viral Diseases
Center for Infectious Diseases
Centers for Disease Control
Atlanta, Georgia 30333

Guy de Thé
CNRS Laboratory of Epidemiology &
Immunovirology of Tumors
Faculty of Medicine Alexis Carrel
Rue Guillaume Paradin
69372 Lyon
Cedex 08, France

Elaine DeFreitas
The Wistar Institute
Thirty-sixth Street at Spruce
Philadelphia, Pennsylvania 19104-4268

Jay S. Epstein
Division of Blood and Blood Products
Center for Biologics Evaluation and
Research
Food and Drug Administration
8800 Rockville Pike
Bethesda, Maryland 20892

Violet Esquenazi
Histocompatibility Laboratory
University of Miami/Jackson Memorial
  Hospital and *Veterans*
  Administration Medical Center
P.O. Box 016960
Miami, Florida 33101

Myron Essex
Department of Cancer Biology
Harvard School of Public Health
665 Huntington Avenue
Building 1, Room 811
Boston, Massachusetts 02115

Anthony Evangelista
University of Minnesota, Microbiology
420 Delaware Street SE
1460 Mayo Bldg.
Minneapolis, Minnesota 55455

Chyang T. Fang
Transmissible Diseases Laboratory and
  National Reference Laboratories for
  Infectious Diseases
American National Red Cross
Jerome H. Holland Laboratory
15601 Crabbs Branch Way
Rockville, Maryland 20855

J. Peter Figueroa
Jamaican Ministry of Health
Kingston, Jamaica

Steven K. H. Foung
Department of Pathology
Stanford University Blood Bank
800 Welch Road
Palo Alto, California 94304

Luigi Frati
I University
Department of Experimental Medicine
Via Regina Elena
00161 Roma, Italy

Genjiro Futami
Blood Transfusion Service and the
  Second Department of Internal
  Medicine
Kumamoto University Medical School
Honjo 1-1-1
Kumamoto 860, Japan

D. Carleton Gajdusek
Laboratory of Central Nervous System
  Studies
National Institute of Neurological
  Disorders and Stroke
National Institutes of Health
Building 36, Room 5B21
Bethesda, Maryland 20892

Robert C. Gallo
Laboratory of Tumor Cell Biology
National Cancer Institute
National Institutes of Health
Building 37, Room 6A11
Bethesda, Maryland 20892

Murray B. Gardner
Department of Medical Pathology, 2047
University of California, Davis
Davis, California 95616

Antoine Gessain
Laboratoire d'Hématologie Moléculaire
et Laboratoire Central d'Hématologie
Hôpital Saint-Louis
75010 Paris, France

Clarence J. Gibbs, Jr.
Laboratory of Central Nervous System
  Studies
National Institute of Neurological
  Disorders and Stroke
National Institutes of Health
Building 36, Room 5B21
Bethesda, Maryland 20892

W. N. Gibbs
World Health Organization
Geneva, Switzerland

Olivier Gout
Clinique de Neurologie et de
Neuropsychologie
Hôpital de la Salpetrière
75013 Paris, France

Steven J. Greenberg
Metabolism Branch
National Cancer Institute
National Institutes of Health
Building 10, Room 4N109
Bethesda, Maryland 20892

Warner C. Greene
Professor of Medicine
Howard Hughes Medical Institute
Duke University School of Medicine
Box 3037
Durham, North Carolina 27710

Ajay Gupta
Neuroimmunology Branch
National Institutes of Health
Building 10, Room 5B-16
Bethesda, Maryland 20892

Curado Gurgo
Laboratory of Tumor Cell Biology
National Cancer Institute
National Institutes of Health
Bethesda, Maryland 20892

Ashley Haase
Department of Microbiology
University of Minnesota
420 Delaware Street SE
Minneapolis, Minnesota 55455

William W. Hall
Division of Infectious Disease
North Shore University Hospital
300 Community Drive
Manhasset, New York 11030

Barrie Hanchard
Department of Pathology
University of the West Indies
Kingston 7, Jamaica

Sarah M. Hanly
Howard Hughes Medical Institute
Duke University School of Medicine
Box 3037
Durham, North Carolina 27710

Jonathan Harris
Neuroimmunology Branch
National Institutes of Health
Building 10, Room 5B-16
Bethesda, Maryland 20892

Trudie M. Hartley
Retrovirus Diseases Branch
Division of Viral Diseases
Center for Infectious Diseases
Centers for Disease Control
Atlanta, Georgia 30333

S. Havard
Service de Neurologie
Hôpital la Meynard
97200-Fort de France, Martinique

Shigeo Hino
Department of Bacteriology
Nagasaki University School of Medicine
Nagasaki 852, Japan

Peter Horal
Department of Clinical Virology
University of Göteborg
Guldhedsgatan 10B
S-413 46 Göteborg, Sweden

Chung-Ho Hung
Cambridge BioScience Corporation
365 Plantation Street
Biotechnology Research Park
Worcester, Massachusetts 01605

Toshinori Ishii
Blood Transfusion Service and the
Second Department of Internal
Medicine
Kumamoto University Medical School
Honjo 1-1-1
Kumamoto 860, Japan

Junzo Ishizaki

Second Department of Internal Medicine
Miyazaki Medical College
Miyazaki 889-16, Japan

Shin-Ichiro Ito

Director, Tsushima
Izuhara Hospital
Shimoagata-gun
Nagasaki 817, Japan

Steven Jacobson

National Institute of Neurological and
*   Communicative Disorders and Stroke*
National Institutes of Health (NIH)
Building 10, Room 5B-16
Bethesda, Maryland 20892

Elaine S. Jaffe

Laboratory of Pathology
National Cancer Institute, NIH
Bethesda, Maryland 20892

Stig Jeansson

Department of Clinical Virology
University of Göteborg
Guldhedsgatan 10B
S-413 46 Göteborg, Sweden

William Ju

The Wistar Institute
Department of Dermatology
University of Pennsylvania
Philadelphia, Pennsylvania 19104 4268

Marshall Kadin

Beth Israel Hospital
Boston, Massachusetts 02215

V. S. Kalyanaraman

Advanced Bioscience Laboratories, Inc.
5510 Nicholson Lane
Kensington, Maryland 20895

Jonathan E. Kaplan

Retrovirus Diseases Branch
Division of Viral Diseases
Center for Infectious Diseases
Centers for Disease Control
Atlanta, Georgia 30333

Mark H. Kaplan

Division of Infectious Disease
North Shore University Hospital
300 Community Drive
Manhasset, New York 11030

Rima F. Khabbaz

Retrovirus Diseases Branch
Division of Viral Diseases
Center for Infectious Diseases
Centers for Disease Control
Atlanta, Georgia 30333

Jerome H. Kim

Howard Hughes Medical Institute
Duke University School of Medicine
Box 3037
Durham, North Carolina 27710

Tetsuyuki Kiyokawa

Blood Transfusion Service and the
*   Second Department of Internal*
*   Medicine*
Kumamoto University Medical School
Honjo 1-1-1
Kumamoto 860, Japan

Alexander Krämer

National Cancer Institute
National Institutes of Health
Bethesda, Maryland 20892

L. LaGrenade

Department of Pathology
University of the West Indies
Kingston 7, Jamaica

Michael D. Lairmore

Retrovirus Disease Branch
Division of Viral Diseases
Center for Infectious Diseases
Centers for Disease Control
Atlanta, Georgia 30333

Andrew Larson

University of Minnesota, Microbiology
420 Delaware Street SE
1460 Mayo Bldg.
Minneapolis, Minnesota 55455

Tun-Hou Lee
Department of Cancer Cell Biology
Harvard School of Public Health
Boston, Massachusetts 02115

Stuart Lessin
The Wistar Institute
Department of Dermatology
University of Pennsylvania
Philadelphia, Pennsylvania 19104 4268

Paul H. Levine
Viral Epidemiology Section
Environmental Epidemiology Branch
Division of Cancer Etiology
National Cancer Institute
National Institutes of Health
Bethesda, Maryland 20892

James J. Lipka
Department of Pathology
Stanford University Blood Bank
800 Welch Road
Palo Alto, California 94304

W. Lofters
Department of Pathology
University of the West Indies
Kingston 7, Jamaica

Olivier Lyon-Caen
CNRS Laboratory of Epidemiology &
 Immunovirology of Tumors
Faculty of Medicine Alexis Carrel
Rue Guillaume Paradin
69372 Lyon
Cedex 08, France

Michael H. Malim
Howard Hughes Medical Institute
Duke University School of Medicine
Box 3037
Durham, North Carolina 27710

K. S. Mani
Neurological Clinical
1 Old Veterinary Hospital Road
Basavanagudi, Bangalore
 India

Angela Manns
Environmental Epidemiology Branch
National Cancer Institute
National Institutes of Health
Rockville, Maryland 20892

Vittorio Manzari
II University
Department of Experimental Medicine
 and Biochemical Sciences
Via Orazio Raimondo
00173 Roma, Italy

Dante J. Marciani
Cambridge BioScience Corporation
365 Plantation Street
Biotechnology Research Park
Worcester, Massachusetts 01605

Stacene Maroushek
Department of Microbiology
University of Minnesota
420 Delaware Street SE
Minneapolis, Minnesota 55455

David Mattson
Neuroimmunology Branch
National Institutes of Health
Building 10, Room 5B-16
Bethesda, Maryland 20892

Dale E. McFarlin
National Institute of Neurological and
 Communicative Disorders and Stroke
National Institutes of Health
Building 10, Room 5B-16
Bethesda, Maryland 20892

Jean Michel Miclea
Hôpital Saint-Louis
1 Avenue Claude Vellefaux
75010 Paris, France

J. Mikol
Service de Neuropathologie
Hôpital Lariboisière
2 rue Amboise Paré
75010 Paris, France

Hal Minnigan
Department of Microbiology
University of Minnesota
420 Delaware Street SE
Minneapolis, Minnesota 55455

Masanao Miwa
Virology Division
National Cancer Center Research
Institute
Tsukiji 5-chome
Chuo-ku, Tokyo, Japan

Andrea Modesti
I University
Department of Experimental Medicine
Via Regina Elena
00161 Roma, Italy

Joseph G. Montes
Department of Biophysics
School of Medicine
University of Maryland
660 Redwood Street
Baltimore, Maryland 21201

Owen St. Claire Morgan
Department of Medicine
University of West Indies
Mona, Kingston 7, Jamaica,
West Indies

Shigehisa Mori
Virology Division
National Cancer Center Research
Institute
Tsukiji 5-chome
Chuo-ku, Tokyo, Japan

Nancy Mueller
Harvard School of Public Health
Department of Epidemiology
677 Huntington Avenue
Boston, Massachusetts 02115

Koichi Murai
Second Department of Medicine
Miyazaki Medical School
Miyazaki, Kyushu, Japan

Edward L. Murphy
Departments of Laboratory Medicine
and Medicine
University of California at San
Francisco
San Francisco, California 94143

Kazuo Nagashima
Department of Pathology
University of Hokkaido
Sapporo 060, Japan

A. L. Newell
Retrovirus Research Group
Clinical Research Centre
Watford Road
Harrow, Middlesex, HA 1 3UJ
United Kingdom

P. K. Newman
Department of Neurology
General Hospital
Middlesbrough, Cleveland
United Kingdom

Peter Nowell
Department of Pathology
University of Pennsylvania
Philadelphia, Pennsylvania 19104

Yuko Ohno
Epidemiology Division
National Cancer Center Research
Institute
Tsukiji 5-chome
Chuo-ku, Tokyo, Japan

Takashi Okamoto
Section Head, Virology Division
National Cancer Center Research
Institute
Tsukiji 5-chome
Chuo-ku, Tokyo, Japan

Akihiko Okayama
Second Department of Medicine
Miyazaki Medical School
Miyazaki 889-16 Japan and
Department of Cancer Biology
Harvard School of Public Health
665 Huntington Avenue
Boston, Massachusetts 02115

Michael B. A. Oldstone
Department of Immunology
Research Institute of Scripps Clinic
10666 North Torrey Pines Road
La Jolla, California 92037

Steven Ono
Laboratory of Central Nervous System
Studies
National Institute of Neurological
Disorders and Stroke
National Institutes of Health (NIH)
Building 36, Room 5B21
Bethesda, Maryland 20892

Mitsuhiro Osame
The Third Department of Internal
Medicine
Faculty of Medicine
Kagoshima University
1208-1 Usuki-cho
Kagoshima 890 Japan

Naoki Oyaizu
Division of Infectious Disease
North Shore University Hospital
300 Community Drive
Manhasset, New York 11030

Thomas J. Palker
Division of Rheumatology and
Immunology
Duke University Medical Center
Box 3307
Durham, North Carolina 27710

L. D. Panchoosingh
Department of Pathology
University of the West Indies
Kingston 7, Jamaica

Judy L. Parker
Department of Pathology
Stanford University Blood Bank
800 Welch Road
Palo Alto, California 94304

Wade Parks
Department of Pediatrics
New York University Medical Center
550 First Avenue
New York, New York 10016

Donald W. Paty
Division of Neurology
The University of British Columbia
Department of Medicine
Vancouver General Hospital
#222-2775 Heather Street
Vancouver, BC, Canada V5Z 3J5

Jorge Peries
Hôpital Saint-Louis
1 avenue Claude Vellefaux
75010 Paris, France

Lauren T. Pierik
Division of Blood and Blood Products
Center for Biologics Evaluation and
Research
Food and Drug Administration
8800 Rockville Pike
Bethesda, Maryland 20892

Bernard J. Poiesz
Department of Medicine and
Microbiology
Division of Hematology/Oncology
State University of New York, Syracuse
750 E. Adams Street
Syracuse, New York 13210

G. Rangan
Neurological Clinical
1 Old Veterinary Hospital Road
Basavanagudi, Bangalore
India

Lee Ratner
Departments of Medicine and Molecular
*   Microbiology*
Washington University
Box 8125, 660 S. Euclid
St. Louis, Missouri 63110

William C. Reeves
Division of Epidemiology
Gorgas Memorial Laboratory
P.O. Box 6991
Panama 5, Republic of Panama

Marvin S. Reitz, Jr.
Laboratory of Tumor Cell Biology
National Cancer Institute
National Institutes of Health
Building 37, Room 6A11
Bethesda, Maryland 20892

Ernest Retzel
Department of Microbiology
University of Minnesota
420 Delaware Street SE
Minneapolis, Minnesota 55455

Jayne S. Reuben
National Institute of Neurological and
*   Communicative Disorders and Stroke*
National Institutes of Health
Building 10, Room 5B-16
Bethesda, Maryland 20892

J. H. Richardson
Retrovirus Research Group
Clinical Research Centre
Watford Road
Harrow, Middlesex, HA 1 3UJ
*   United Kingdom*

Laurence T. Rimsky
Howard Hughes Medical Institute
Duke University School of Medicine
Box 3037
Durham, North Carolina 27710

Pamela E. B. Rodgers-Johnson
Laboratory of Central Nervous System
*   Studies*
National Institute of Neurological
*   Disorders and Stroke*
National Institutes of Health
Building 36, Room 5B21
Bethesda, Maryland 20892

G. Roman
Neuroepidemiology Branch
National Institutes of Health
Bethesda, Maryland 20892

Alain Rook
Department of Dermatology
University of Pennsylvania
Philadelphia, Pennsylvania 19104

P. Rudge
Department of Neurology
Northwick Park Hospital
Watford Road
Harrow, Middlesex, HA 1 3UJ
*   United Kingdom*

Lars Rymo
Medical Biochemistry
University of Göteborg
Guldhedsgatan 10B
S-413 46 Göteborg, Sweden

Fortuna Saal
Hôpital Saint-Louis
1 avenue Claude Vellefaux
75010 Paris, France

S. Zaki Salahuddin
Laboratory of Tumor Cell Biology
National Cancer Institute
National Institutes of Health
Bethesda, Maryland 20892

S. Gerald Sandler
Transmissible Diseases Laboratory and
*   Blood Services*
American National Red Cross
Jerome H. Holland Laboratory
15601 Crabbs Branch Way
Rockville, Maryland 20855

Angela Santoni
I University
Department of Experimental Medicine
Via Regina Elena
00161 Roma, Italy

Carl Saxinger
Laboratory of Tumor Cell Biology
National Cancer Institute
National Institutes of Health
Bethesda, Maryland 20892

Jonathan Seals
Cambridge BioScience Corporation
365 Plantation Street
Biotechnology Research Park
Worcester, Massachusetts 01605

William A. Sheremata
Multiple Sclerosis Center
Department of Neurology (D4-5)
Jackson Memorial Hospital
P.O. Box 016960
Miami, Florida 33101

Kunitada Shimotohno
Virology Division
National Cancer Center Research
  Institute
Tsukiji 5-chome
Chuo-ku, Tokyo, Japan

Masanori Shimoyama
Hematology-Oncology and Clinical
  Cancer Chemotherapy Division
National Cancer Center Research
  Institute
Tsukiji 5-chome
Chuo-ku, Tokyo, Japan

Shigemasa Shioiri
Second Department of Medicine
Miyazaki Medical School
Miyazaki, Kyushu, Japan

Eiichi Shishime
Second Department of Medicine
Miyazaki Medical School
Miyazaki, Kyushu, Japan

Gerald Shulman
American Red Cross Blood Services
Atlanta Region
Atlanta, Georgia 30324

François Sigaux
Hôpital Saint-Louis
1 avenue Claude Vellefaux
75010 Paris, France

Ida Silvestri
I University
Department of Experimental Medicine
Via Regina Elena
00161 Roma, Italy

Dennis Slamon
Division of Hematology and Oncology
UCLA School of Medicine
Los Angeles, California

Shunro Sonoda
Department of Virology
Faculty of Medicine
Kagoshima University
1208-1 Usuki-cho
Kagoshima 890 Japan

Sherri O. Stuver
Harvard School of Public Health
Department of Epidemiology
677 Huntington Avenue
Boston, Massachusetts 02115

Bo Svennerholm
Department of Clinical Virology
University of Göteborg
Guldhedsgatan 10B
S-413 46 Göteborg, Sweden

Nobuyoshi Tachibana
Second Department of Internal Medicine
Miyazaki Medical College
Miyazaki 889-16, Japan

Kazuo Tajima
Division of Epidemiology
Aichi Cancer Center
Research Institute
1-1 Kanokoden
Chikusa-ku, Nagoya 464, Japan

Kiyoshi Takatsuki
Blood Transfusion Service and the
  Second Department of Internal
  Medicine
Kumamoto University Medical School
Honjo 1-1-1
Kumamoto 860, Japan

Craig L. Tendler
Metabolism Branch
National Cancer Institute
National Institutes of Health
Building 10, Room 4N109
Bethesda, Maryland 20892

Richard M. Thorn
Cambridge BioScience Corporation
365 Plantation Street
Biotechnology Research Park
Worcester, Massachusetts 01605

Jorge M. Trujillo
Tropical Spastic Paraparesis
Research Unit
Tumaco, Columbia

Erwin Tschachler
Laboratory of Tumor Cell Biology
National Cancer Institute
National Institutes of Health
Building 37, Room 6A11
Bethesda, Maryland 20892

Kazunori Tsuda
Second Department of Internal Medicine
Miyazaki Medical College
Miyazaki 889-16, Japan

Shoichiro Tsugane
Epidemiology Division
National Cancer Center Research
  Institute
Tsukiji 5-chome
Chuo-ku, Tokyo, Japan

Anders Vahlne
Department of Clinical Virology
University of Göteborg
Guldhedsgatan 10B
S-413 46 Göteborg, Sweden

J. C. Vernant
Service de Neurologie
Hôpital la Meynard
97200-Fort de France, Martinique

Eric Vonderheid
Department of Medicine
Division of Dermatology
Hahnemann University
Philadelphia, Pennsylvania 19102

Thomas A. Waldmann
Metabolism Branch
National Cancer Institute
National Institutes of Health
Building 10, Room 4N109
Bethesda, Maryland 20892

Shaw Watanabe
Epidemiology Division
National Cancer Center Research
  Institute
Tsukiji 5-chome
Chuo-ku, Tokyo, Japan

Stefan Z. Wiktor
Viral Epidemiology Section
National Cancer Institute
National Institutes of Health
Bethesda, Maryland 20892

Annelie Wilde
Cambridge BioScience Corporation
365 Plantation Street
Biotechnology Research Park
Worcester, Massachusetts 01605

R. Wilks
Department of Pathology
University of the West Indies
Kingston 7, Jamaica

Alan E. Williams
Transmissible Diseases Laboratory
American National Red Cross
Jerome H. Holland Laboratory
15601 Crabbs Branch Way
Rockville, Maryland 20855

E. Williams
Department of Pathology
University of the West Indies
Kingston 7, Jamaica

N. Williams
Department of Pathology
University of the West Indies
Kingston 7, Jamaica

Kazunari Yamaguchi
Blood Transfusion Service and the
   Second Department of Internal
   Medicine
Kumamoto University Medical School
Honjo 1-1-1
Kumamoto 860, Japan

HUMAN RETROVIROLOGY: HTLV

Human Retrovirology: HTLV,
edited by William A. Blattner.
Raven Press, Ltd., New York © 1990.

Models of Retroviral Leukemogenesis

Murray B. Gardner

Department of Medical Pathology, University of California, Davis, Davis, California 95616

INTRODUCTION

The last decade has seen the dramatic emergence of human retroviruses as the cause of adult T-cell leukemia (ATL), a chronic progressive myelopathy called HTLV-associated myelopathy (HAM) in Japan or tropical spastic paraparesis (TSP) in equatorial regions, and the acquired immunodeficiency syndrome (AIDS). With more sensitive assays now at hand, yet other diseases will undoubtedly be linked to human retroviruses. Research done on animal retroviruses during the first eight decades of this century (1) set the precedence for the current era of human retrovirology by establishing the dual tropism of these viruses for both the immune and central nervous systems (CNS), by showing their ability to cause both malignant and nonmalignant diseases under natural conditions, by uncovering mechanisms of pathogenesis, and by demonstrating the efficacy of various antiviral therapeutic and preventive measures. Because this conference concerns mainly HTLV, a type C oncovirus, I will give a brief historical perspective and summarize several mechanisms of leukemogenesis exhibited by the animal type C oncoviruses; highlight the contribution of animal oncoviruses to our understanding of human leukemogenesis; and call special attention to the relevance of the bovine, simian, and wild mouse type C oncovirus models to HTLV.

HISTORICAL PERSPECTIVE

Retroviruses of animals and humans are classified in three subfamilies—the oncoviruses, the lentiviruses, and the spumaviruses (2). The type C oncoviruses are naturally occurring causes of leukemia in chickens, mice, cows, cats, gibbon apes, and humans; and type B oncoviruses cause breast cancer in mice. Less commonly, type C oncoviruses cause sarcomas in chickens, mice, and cats. A search for oncoviruses in human leukemia and sarcomas was a major effort of the National Cancer Institute (NCI) Virus Cancer Program initiated in 1970 (3). From the animal studies it also became apparent that leukemia viruses could cause, in addition to cancer, various cytopathic diseases in their natural hosts,

1

such as spleen necrosis in birds, lower motor neuron disease in wild mice, and aplastic anemia and immunosuppression in domestic cats (4). More recently, nononcogenic type D retroviruses were found to cause fatal simian AIDS (SAIDS) in various species of macaques (5).

Naturally occurring lentivirus infections of farm animals, such as visna virus in sheep, caprine arthritis encephalitis virus in goats, and equine infectious anemia virus in horses, cause chronic inflammatory diseases of the brain, lung, joints, and hematopoietic system (4). Because they, apparently, do not directly cause cancer, the lentiviruses received relatively less scientific attention than did oncoviruses in past years. However, visna virus was the prototype "slow infection" of the brain discovered in the 1950s (6). Lentivirus disease models more recently discovered include the feline immunodeficiency virus in domestic cats (7) and simian immunodeficiency viruses in captive macaques (8). Each of these animal models is highly relevant to our understanding of the general biology of lentiviruses, and the feline and simian models are particularly pertinent to the pathogenesis of HIV and AIDS. Features common to each lentivirus model include persistent infection in face of a strong immune response, macrophage tropism, restricted expression, and marked antigenic variation of the envelope (9). The human foamy viruses are spumaviruses; they are cytopathic in vitro, but, apparently, are not pathogenic in vitro (10). They may prove of interest from the evolutionary standpoint and for understanding mechanisms of cytopathology in vitro, and they may be useful as vectors for genetic engineering.

Study of animal retroviruses during these former years was for many investigators a "ticket of admission" to cell biology; much of the knowledge gained, along with reagents and assay systems developed, have made possible the rapid progress with detection and characterization of human retroviruses in the 1980's. For the chicken, mouse, and cat type C oncoviruses, it was found that the major determinants of tissue tropism, leukemogenesis, and cytopathology are localized within the *env* gene and long terminal repeat (LTR) sequences (4). The envelope interaction with cell surface receptors determines which specific cell types are susceptible to virus entry (11), and filling of these receptors with endogenous or exogenous envelope glycoprotein accounts for the phenomenon of viral interference (12). The interaction of cell-specific factors with the viral LTR, especially in the enhancer region, governs the extent of viral expression and disease specificity (13). The discovery of reverse transcriptase enzyme in avian and murine leukemia viruses in 1970 helped to explain how the RNA genomes of exogenous oncoviruses become integrated as complete DNA proviral copies in the host chromosomes (14,15) and led to a practical assay still used today for detection of replicating retroviruses. Ready availability of this enzyme also now makes possible the cloning of cDNA from mRNA, an important technique of modern DNA biotechnology.

The ability of the type C leukemia viruses in chickens, mice, and cats to occasionally recombine with and transduce host cellular genes to form sarcoma viruses led, fortuitously, to the identification of oncogenes, which are themselves now a growth industry with great relevance to human biology (16). Because viral genes are deleted to make room for the oncogenes, the animal sarcoma viruses are replication defective and require helper leukemia viruses for their "rescue" and transmission. This important retroviral property was first shown 25 years ago with the Bryan strain of Rous sarcoma virus (17). Similarly, by recombining with endogenous retrovirogenes, the leukemia viruses occasionally gain an altered host cell range and enhanced virulence. An example is the spleen focus forming virus (SFFV) component of Friend erythroid leukemia virus in which the defective SFFV encodes an incomplete *env* gene derived from the recombination of exogenous and endogenous sequences (18). Defective oncoviruses have recently been shown capable of inducing severe immunosuppression in cats and mice (19,20). The phenomena of defectivity, rescue with helper virus, and transduction of cellular oncogenes have not as yet been described with lentiviruses or spumaviruses, nor is there any evidence of genetic recombination or phenotypic mixing between lentiviruses, spumaviruses, and oncoviruses. The human oncovirus HTLV-I is not defective and has not been shown to transduce or recombine with cellular oncogenes or retrovirogenes in infected cells. Furthermore, the question of whether defective oncoviral genomes are responsible for the immunosuppression that often occurs in infected animals—or whether defective HIV plays a role in human AIDS—is still unresolved (21).

All of the disease-producing oncoviruses and lentiviruses are acquired exogenously, either congenitally or perinatally. However, an exception occurs in certain inbred mouse strains, e.g., AKR mice, in which infectious (ecotropic) leukemia viruses are primarily inherited as two separate endogenous proviral genes (22). Following a convoluted course, envelope sequences recombine between the infectious and other noninfectious (xenotropic) provirus genes and form virulent thymotropic virus, which transforms mink spleen cells (MCF virus) in vitro and accelerates development of thymic lymphomas (23). Knowledge of inherited virogenes—their origin and evolution, their activation by physical and chemical carcinogens and aging, and their interaction with cellular oncogenes in inbred chickens and mice—formed the basis of the "viral oncogene" and "protoviral" hypotheses that guided cancer virus research in the early 1970's (24,25). It was hypothesized that the generality of human cancers might even be explained, at least in part, by the activation of similar endogenous virogenes and oncogenes in the human genome. Because bona fide human retroviruses could not be isolated during these years, the Virus Cancer Program was dismantled in the late 1970's. However, the isolation of T-cell growth factor (IL-2) in 1976 at the NCI (26) made possible the long-term growth of antigen-specific T-cell lines, and led ultimately to the discovery of HTLV-I in 1978, which was reported in 1980 (27–29). However, HTLV proved not endogenous to humans, but exogenously acquired as an infectious virus. In this respect, HTLV resembled all other pathogenic retroviruses of outbred species, including wild mice and domestic cats and was quite different from the inherited oncoviruses of inbred mice. Endogenous proviral DNA sequences related to, but not identical to, HTLV and various animal oncovirus sequences are, indeed, inherited in the genome of all humans and other mammalian species, but these sequences are mostly defective and thus unable to synthesize complete, infectious virions (30–32). The biological significance of the endogenous oncoviral related DNA sequences in the human genome remains largely unknown.

MECHANISMS OF ANIMAL RETROVIRAL LEUKEMOGENESIS

Type C leukemia viruses of animals lead toward leukemogenesis by at least four general mechanisms: 1) immunosuppression, 2) chronic antigenic stimulation, 3) transduction of cellular oncogenes, and 4) proviral activation of cellular oncogenes. These processes may occur in various combinations and do not exclude other mechanisms yet to be discovered.

Immunosuppression and Antigenic Stimulation

Immune dysregulation and clinical immunosuppression accompany infection by some strains of avian, murine, and feline leukemia viruses. The mechanisms differ and have not been precisely defined. Infection of chickens with avian leukosis virus may cause atrophy of the bursa and thymus and inhibit the cellular immune response (4). The replication competent helper viruses associated with avian acute reticuloendotheliosis virus (REV) severely suppress the cellular immune response of infected birds by lysing lymphoid cells and activating a suppressor cell population. The cytopathic effect on immune cells may result from accumulation of large quantities of viral proteins and/or unintegrated viral DNA molecules in the cytoplasm, the consequence, perhaps, of repeated cycles of infection (34,35). The suppressor cells inhibit the proliferation of cytotoxic T cells capable of lysing rel-oncogene–bearing REV tumor cells, and lymphomas result (35). In certain strains of laboratory mice MuLV induces immunosuppression indirectly by serving as an antigenic stimulus for polyclonal B-cell activation and proliferation, which in turn, leads, via autoantibodies and perhaps T-suppressor cells, to a decrease in T–helper cell activity. An example is the severe immunodeficiency disease called murine AIDS (MAIDS) induced in C57B1/6 mice with the Duplan defective strain of MuLV (20,36). Antigenic stimulation by MuLV-receptor interaction may serve as a mitogenic signal and trigger lymphoid proliferation and eventual leukemogenesis, possibly even in the absence of concomitant immunosuppression (37). Antigenic stimulation by MuLV in donor lymphocytes appears to underlie, at least in part, the immunosuppression associated with the host-versus-graft (HVG) disease in susceptible strains of inbred mice (38). Mice that survive acute autoimmune disease may develop lymphomas following formation of MuLV recombinants (39).

Cats infected with feline leukemia virus are often immunosuppressed (40). Lymphoid mitogenesis is inhibited by inactivated FeLV, by the purified transmembrane *env* polypeptide (p15E), and by a synthetic peptide representing a 14-amino-acid sequence in FeLV p15E (41,42). The 14-amino-acid immuno-suppressive peptide of FeLV p15E is contained within a highly conserved stretch of 26 amino acids in the C-terminal *env* domains of other type C and D on-coviruses. A synthetic peptide representing this 26-amino-acid sequence also has immunosuppressive effects on cultured lymphoid cells. This envelope peptide has thus been implicated in the pathogenesis of retrovirus immunosuppression. However, the significance of these in vitro observations with respect to patho-genesis of oncovirus induced immunosuppression in vivo remains to be de-termined.

An immunosuppressive strain of FeLV was molecularly cloned and shown to contain a defective *env* variant which is formed in the cytoplasm of target bone marrow cells at the onset of the clinical immunodeficiency (19). Immune dys-function may result in this instance from the intracellular accumulation of altered *env* glycoprotein, possibly via a failure to interfere with superinfection at the receptor level.

The mechanism whereby nononcogenic type D retroviruses (SRV) induce fatal immunosuppression in macaques is unknown (5). The virus has a broad host range, readily infecting T and B cells and other cell types, independent of the CD4 receptor (43). The receptor for this virus has not been identified. SRV causes an asymptomatic infection of the macaque CNS (44). The virus induces syncytia in human B cells (Raji) but little or no cytopathology in vitro. Yet, a profound depletion of T and B cells occurs from in vivo infection with this virus. This finding suggests that the host immune response to the virus may contribute to the immune depletion, but this mechanism has not as yet been documented. These are but a few examples of the complex host-oncovirus interactions that result in immunosuppression, chronic antigenic stimulation, and leukemogenesis in animals.

Transduction of Cellular Oncogenes

As is now well known, retroviral oncogenes derive from the normal cellular homologs, called protooncogenes, that are transduced, mutated, and/or activated by naturally occurring type C oncovirus infection of chickens, mice, and cats with single examples in turkey and wooly monkey (16). The prototype oncogene is the *src* gene transduced by avian leukosis virus 80 yr ago to form the Rous sarcoma virus (45). About 20 such retroviral oncogenes of animal origin have been isolated and shown to induce target cell transformation rapidly in vitro and acute leukemia or sarcoma of short latent period in vivo (46). Because these viruses are defective, they are not infectious in nature. Of the 20 retroviral on-cogenes of animal origin, at least 10 of the homologous cellular protooncogenes (*c-abl, c-erb, c-ets, c-mos, c-myb, c-fos, c-myc, c-H-ras, c-K-ras, c-sis*) have been incriminated in human cancer. Four of these protooncogenes (*c-myc, c-fos, c-*

abl, c-ets), in particular, have been specifically incriminated in human leukemogenesis (47). In several instances homologous oncogenes have been transduced
by different helper viruses from different animal species, indicating, perhaps, a
limited number of protooncogenes in nature. An example is the *fes* oncogene
present in both avian and feline sarcoma viruses (48). Several animal retroviruses
even include two separately transduced oncogenes or an oncogene and another
cellular gene in the same genome. Examples include the *myb* and *ets* oncogenes
in an avian leukemia virus (49) and the *fgr* oncogene and actin gene in the GR
strain of feline sarcoma virus (50).

Oncogenes function at different steps in the signal pathways controlling cell
growth and differentiation (51,52). These steps include 1) activation of growth
factors, e.g., *sis*-encoded, platelet-derived growth factors, 2) activation of growth
factor receptors, e.g., *erb-B*-encoded epithelial growth factor receptor, 3) activation
of intracellular growth signals, e.g., *src* and *ras* stimulation of second-messenger
pathways, and 4) DNA transcriptional activation, e.g., *myc* and *fos* stimulation
of genes involved in cell division. Retroviral oncogenes have indeed provided
insight into the mechanisms responsible for conveying growth stimuli at the cell
surface to genes in the nucleus, and they have illuminated a variety of ways that
normal cells are converted into tumor cells.

Proviral Activation of Cellular Oncogenes

cis-*Activation*

The great majority of animal oncoviruses in nature do not contain transduced
oncogenes. These viruses are nontransforming in vitro and cause chronic leukemia of long latent period and low incidence. They initiate leukemogenesis, at
least in some instances, by insertional mutagenesis mediated by integration of
the provirus in the domain of cellular protooncogenes, whose expression is then
altered. The integrated proviruses appear to stimulate transcription of neighboring
cellular genes, i.e., *cis*-activation, via the promoter and enhancer elements in
the viral LTR and/or truncation of the normal gene. An example of this phenomenon is the induction of T-cell lymphomas in chickens, mice, and cats by
the insertional activation of the *C-myc* locus by exogenous type C oncoviruses
(53–55). A number of cellular genes other than those transduced by oncoviruses
are commonly *cis*-activated among the retroviral-induced tumors of chickens,
mice, and rats. Such genes include the int-1 and int-2 loci activated by the type
B oncovirus, which leads to breast cancer in mice (56), and the Mlvi-1 locus
activated by proviral insertion in Moloney-MuLV–induced murine thymomas
(57). At least 15 additional common regions of integration have been identified
in MuLV-induced murine hematopoietic tumors, but the genes deregulated by
proviral integration in most of these loci have not been identified yet (58).

trans-*Activation*

Another means by which some retroviruses, including HTLV, may initiate leukemia of long latency period and low incidence is by the production of viral regulatory proteins which stimulate cell genes at a distance, i.e., *trans*-activation, within the chromosomes of the infected cell. Viruses of this type include bovine leukemia virus (BLV), HTLV, and HIV (59–61). In contrast to *cis*-activation, these viruses integrate essentially at random into the host genome and they do not contain cell-derived oncogenes. The 3′ ends of the genomes of BLV, HTLV, and HIV contain X sequences (*Tax* and *Tat,* respectively, for HTLV and BLV or HIV) encoding a set of novel proteins that are necessary for efficient transcription from the viral promoter (62,63). In addition, these transactivating proteins may "turn on" cellular growth promoting genes such as *fos,* IL-2, and the IL-2 receptor, and thus set in motion the stepwise process of clonal evolution leading to eventual malignancy (64,65). Once the process has been initiated, continued viral transcription is apparently not required, because viral-specific RNAs are not detected in mononuclear cells from the blood or tumor cells of BLV-infected cows or sheep (66,67) or HTLV-infected humans (68). However, virus expression increases within a few hours after the blood cells are placed in culture, and mitogenic or immune stimuli may increase virus expression even more. Moreover, immune responses to unrelated antigens may similarly increase virus expression in vivo and accelerate the course of disease (69). Exposure to various infectious agents, including DNA viruses, apparently accelerates the development of AIDS in HIV-infected individuals by this mechanism (70). The HTLV-I *Tax* gene stimulates the HIV-1 enhancer by the action of such inducible cellular proteins (71) and may thereby accelerate the progression to AIDS (72). At the molecular level the activation of HTLV or HIV expression and cell gene expression in lately infected T helper cells may occur through participation of host cell factors, e.g., NFkB, which are normally induced during the immune response. The NFkB protein binds to similar 5′ enhancer elements present in the viral LTR, k-light chain, B interferon, and IL-2 receptor genes. This transcription factor thus plays a broad role in gene regulation similar to a second-messenger system in that it mediates extracellular signals into specific patterns of gene expression (73).

Mice that are transgenic for the HTLV-I *Tax* gene under the control of the HTLV-I LTR develop multiple nerve sheath tumors in which the *Tax* protein is expressed in significant amounts (74). These findings suggest that the HTLV LTR gene has neurotropic properties and that the *Tax* gene may activate cell oncogenes, resulting in cellular proliferation. Interestingly, as yet none of these transgenic mice has developed lymphomas, nor has HTLV-I infection been linked to neural or other soft-tissue tumors in humans.

Mice transgenic for the HIV *Tat* gene under control of its own LTR develop skin lesions resembling Kaposi's sarcoma (75). The *Tat* protein is expressed in the epidermis but not in the adjacent tumors, suggesting an indirect tumorigenic effect—perhaps via release of cytokines with autocrine and paracrine growth effects, as postulated for the pathogenesis of Kaposi's sarcoma in humans (76).

ONCOGENES AND HUMAN LEUKEMIA

Humans appear totally resistant to infection with animal retroviruses (77), and HTLV causes <1% of all human leukemias (28). Obviously, other means of oncogene activation, apparently unrelated to retrovirus infection, apply to the generality of human leukemias. These other activating mechanisms include 1) chromosome gross rearrangements, 2) gene amplification, and 3) subtle point mutations (46,47). Prime examples include Burkitt's lymphoma, in which the *C-myc* gene on chromosome 8 is translocated to one of three immunoglobulin loci on chromosome 14, 2, or 22; and chronic myelogenous leukemia, where the *c-abl* gene on chromosome 9 is translocated to chromosome 22. In some T-cell leukemias, *C-myc* may be activated by downstream translocation of the alpha locus of the T cell receptor. *C-myc* or *C-myb* amplification may also contribute to some human leukemias. Activation of members of the *ras* gene family by somatic point mutations around codons 12, 13, or 59 may trigger leukemogenesis in some individuals. Other cellular oncogenes, unrelated to animal retroviruses and also involved in human leukemogenesis, are commonly located at or near the site of specific chromosomal breakpoints (e.g., bcl 1,2) or are detected only by the DNA transfection technique (e.g., B-lym. T-lym, dbl). Study of the animal type C oncoviruses has clearly provided key knowledge to our understanding of cellular oncogenes and their activation in leukemogenesis and other types of cancer.

ANIMAL MODELS FOR HTLV

Among the animal type C oncoviruses, the most directly applicable models to HTLV are the simian T-lymphotropic viruses (STLV) and BLV. From the historical perspective, the wild mouse MuLV model also provided an excellent forerunner of events to come with HTLV.

STLV

Simian viruses, closely related (90–95%) to, but distinct from, HTLV-I, are highly prevalent in nonhuman primates from Africa and Asia (78). Little is known about the transmission and natural history of this virus. An age-dependent increase in STLV seropositivity suggests that the natural history of STLV infection in monkeys may be similar to that of HTLV infection in humans. ATL-like disease has been observed in one African Green monkey naturally infected with STLV (79), and STLV antibodies are quite prevalent in lymphomatous baboons in the Sukhumi Primate Center in southern Russia (80). However, like HTLV, the virulence of this virus appears low and the latent period long, indicating that it is not a practical model for pathogenesis or vaccine studies. Biological features common to STLV, HTLV, and BLV are the strong degree of cell association, absence of viremia, and powerful restriction of virus expression in the infected cells in vivo.

BLV

BLV infection is also quite prevalent in cattle throughout the world (59). Infection by BLV occurs early in life and is followed by a long latency period and only a low incidence of leukemia. BLV and HTLV share numerous features of infection and pathology, have a similar genomic structure (81), and use common strategies for gene expression. With both viruses, a polyclonal proliferation of infected T cells precedes development of a monoclonal leukemia. BLV is more distantly related to HTLV than is STLV, and BLV causes B-cell leukemia, whereas HTLV transforms T cells. The ability to readily infect sheep with BLV and induce B-cell leukemia after a shortened latent period makes this a more suitable model system in which to study mechanisms of viral latency, activation, and tumorigenesis. Indeed, the same immune stimuli that activate latent HTLV and HIV appear to activate expression of latent BLV in experimentally infected sheep (67).

Wild Mouse MuLV

The natural history of MuLV in feral mice was investigated in the early 1970s as a model for the yet-to-be-discovered human retroviruses (82,83). The biology of MuLV in wild mice (WM-MuLV) revealed many similarities with HTLV, and the wild mouse was certainly more accurate than the inbred mouse in predicting later events associated with HTLV infection and ATL (84). Remarkable similarities are 1) the solely exogenous mode of virus transmission, 2) the regional and familial clustering of virus infection in association with a specific type of lymphoma, 3) the relatively low pathogenicity and long latent period for tumor development, 4) the neurotropism, 5) the virus stability, and 6) the random viral integration in resultant lymphomas. At the time these studies were initiated, the major model for cancer was the lymphoma-prone AKR inbred mouse, and the conceptual emphasis was on the activation of endogenous, inherited retrovirus genes and oncogenes (22,24). The revelation that, in feral mice, infectious MuLV was transmitted mainly by mother's milk (85) was thus contrary to the then-current dogma, but it proved later to be correct in pointing the way to a major route of HTLV transmission. At about this time, the evidence that FeLV was horizontally spread among cats by saliva was also met with some disbelief but now is well established. The incidental finding that WM-MuLV could also be spread by sexual contact (and by blood) gave further precedence for the ominous transmission, a decade later, of HTLV and HIV. Of further relevance to HTLV was the discovery that the infectious, ecotropic MuLV of feral mice, even the same molecular clone, could induce both pre–B-cell lymphoma and a fatal lower motor neuron disease with hind leg paresis (86,87). The amphotropic MuLV that are apparently also unique to wild mice in southern California (88) have now found a very practical application as wide host range vectors for genetic engineering purposes.

Not unexpectedly, differences between the wild mouse MuLV and HTLV biologies were also apparent. HTLV is more T-cell restricted, directly transforms T-lymphoid cells in vitro and induces T-cell leukemia, is more latent with lower levels of viremia, and contains a *Tat* gene. However, both WM-MuLV and HTLV integrate monoclonally in leukemia cells but lack preferred integration sites. Thus, the genetic alterations that follow integration of the provirus and result in cancer are essentially unknown in both wild mouse and humans. Understanding the nature of these steps in the wild mouse model might help to explain the pathogenicity of HTLV and ATL. The mechanism of oncoviral neuropathology probably differs in the mouse and human in that TSP apparently represents a host inflammatory, immune response to HTLV-infected mononuclear cells in the CNS and peripheral blood, whereas neuroparalysis is caused in mice by a direct viral noninflammatory, nonimmunogenic injury to motor neurons in the spinal cord (89). The cytolytic effect on neurons appears caused by the intracytoplasmic accumulation of aberrantly replicating Type C oncovirus particles. Type C particles have not as yet been demonstrated in neural cells of patients with TSP.

In feral mice the major determinant of individual susceptibility or resistance to the neurologic disease is a dominant gene called *FV-4* (formerly *AKvr-1*), that segregates in this population (90). *FV-4* represents a defective endogenous provirus that encodes an envelope glycoprotein on the cell surface that blocks entry of the closely related infectious MuLV (91). Virus can therefore not replicate to the high level required to infect and destroy neurons. *FV-4* is found only in wild mice in southern California, Japan, and southeast Asia (92,93). Whether a similar mechanism of receptor interference pertains to humans and accounts for genetic variations in susceptibility to HTLV (or HIV) infection and disease remains to be determined. However, in view of the many parallels already found between the leukemia viruses of wild mice and humans, this possibility deserves serious consideration.

CONCLUSION

Animal models of retrovirus infection have provided a wealth of information about mechanisms of pathogenesis and natural history features which have pointed the way to the current era of human retrovirology. Further study of animal retrovirus models and transgenic mice will help answer some of the important unresolved questions concerning the pathogenesis of the human retroviruses.

REFERENCES

1. Gardner MB. Historical background: In: Stephenson JR, ed. *Molecular biology of RNA tumor viruses.* New York: Academic Press, 1980;1–46.
2. Teich N. Taxonomy of retroviruses. In: Weiss R, Teich N, Varmus H, Coffin J, eds. *RNA tumor viruses.* New York: Cold Spring Harbor, 1985;25–207.
3. Rettig RA. *Cancer crusade. The story of the National Cancer Act of 1971.* Princeton, New Jersey: Princeton University Press, 1971.
4. Teich N, Wyke J, Mak T, et al. Pathogenesis of retrovirus-induced disease. In: Weiss R, Teich N, Varmus H, Coffin J, eds. *RNA tumor viruses.* New York: Cold Spring Harbor, 1985;785–998.
5. Gardner MB, Marx P. Simian acquired immunodeficiency syndrome. In: Klein G, ed. *Advances in viral oncology,* vol. 5. New York: Raven Press, 1985;57–81.
6. Sigurdsson B, Palsson PA, Grimmson H. Visna, a demyelinating transmissible disease of sheep. *J Neuropathol Exp Neurol* 1957;16:389–403.
7. Pedersen HC, Ho E, Brown MJ, Yamamoto K. Isolation of a T-lymphotropic virus from domestic cats with an immunodeficiency-like syndrome. *Science* 1987;235:790–793.
8. Daniel MD, Letvin NL, King NW, et al. Isolation of T-cell tropic HTLV-III-like retrovirus from macaques. *Science* 1985;228:1201–1204.
9. Haase AT. Pathogenesis of lentivirus infections. *Nature* 1986;322:130–136.
10. Hooks JJ, Detrick-Hooks B. Spumaviriniae: foamy virus group infections: comparative aspects of diagnosis. In: Kurstak E, Kurstak C, eds. *Comparative diagnosis of viral diseases,* vol. 4. New York: Academic Press, 1981;599–618.
11. Brown DW, Robinson HL. Influence of *env* and long terminal repeat sequences on the tissue tropism of avian leukosis viruses. *J Virol* 1988;62:4828–4831.
12. Vogt PK, Ishizaki R. Patterns of viral interference in the avian leukosis and sarcoma complex. *Virology* 1986;30:368–374.
13. Golemis E, Li Y, Fredrickson TN, Hartley JW, Hopkins N. Distinct segments within the enhancer region collaborate to specify the type of leukemia induced by nondefective Friend and Moloney viruses. *J Virol* 1989;63:328–337.
14. Temin HM, Mizutani S. RNA-dependent DNA polymerase in virions of Rous sarcoma cells. *Nature* 1970;226:1211–1213.
15. Baltimore D. RNA-dependent DNA polymerase in virions of RNA tumor viruses. *Nature* 1970;226:1209–1211.
16. Bishop JM. Cellular oncogenes and retroviruses. *Annu Rev Biochem* 1983;52:301–354.
17. Hanafusa H, Hanafusa T, Rubin H. The defectiveness of Rous sarcoma virus. *Proc Natl Acad Sci USA* 1963;49:572–580.
18. Li JP, Bestwick RK, Spiro C, Kabat D. The membrane glycoprotein of Friend spleen focus-forming virus: evidence that the cell surface component is required for pathogenesis and that it binds to a receptor. *J Virol* 1987;61:2782–2792.
19. Mullins JI, Chen CS, Hoover EA. Disease-specific and tissue-specific production of unintegrated feline leukaemia virus variant DNA in feline AIDS. *Nature* 1986;319:333–335.
20. Aziz DC, Hanna Z, Jolicoeur P. Severe immunodeficiency disease induced by a defective murine leukaemia virus. *Nature* 1989;338:505–508.
21. Weiss RA. (Editorial) Defective viruses to blame? *Nature* 1989;338:458.
22. Rowe WP. Genetic factors in the natural history of murine leukemia virus infection. *Cancer Res* 1973;33:3061–3068.
23. Hartley JW, Wolford NK, Old LJ, Rowe WP. A new class of murine leukemia virus associated with development of spontaneous lymphoma. *Proc Natl Acad Sci USA* 1977;74:789–792.
24. Huebner RJ, Todaro GT. Oncogenes of RNA tumor viruses as determinants of cancer. *Proc Natl Acad Sci USA* 1969;64:1087–1094.
25. Temin HM. On the origin of the genes for neoplasia: GHA Clowes Memorial Lecture. *Cancer Res* 1974;34:2835–2841.

26. Morgan DA, Ruscetti FW, Gallo R. Selective in vitro growth of T lymphocytes from normal human bone marrows. *Science* 1976;193:1007–1010.
27. Poiesz BJ, Ruscetti FW, Gazdar AF, Bunn PA, Minna JD, Gallo RC. Detection and isolation of type C retrovirus particles from fresh and cultured lymphocytes of a patient with cutaneous T-cell lymphoma. *Proc Natl Acad Sci USA* 1980;77:7415–7419.
28. Hinuma Y, Nagata K, Hanaoka M et al. Adult T-cell leukemia: antigen in a ATL cell line and detection of antibodies to the antigen in human sera. *Proc Natl Acad Sci USA* 1981;78:6476–6480.
29. Blattner WA, Takatsuki K, Gallo RC. Human T-cell leukemia/lymphoma virus and adult T-cell leukemia. *JAMA* 1983;250:1074–1080.
30. Rabson AB, Hamagishi Y, Steele PE, Tykocins M, Martin MA. Characterization of human endogenous retrovirus envelope RNA transcripts. *J Virol* 1985;56:176–182.
31. Steele PE, Martin MA, Rabson AB, Bryan T, O'Brien SJ. Amplification and chromosomal dispersion of human endogenous retroviral sequences. *J Virol* 1986;59:545–550.
32. Mager DL, Freeman JD. Human endogenous retroviruslike genome with type C *pol* sequences and *gag* sequences related to human T-cell lymphotropic viruses. *J Virol* 1987;61:4060–4066.
33. Temin HM. Mechanisms of cell killing/cytopathic effects by retroviruses. *Rev Infect Dis* 1986;10:399–405.
34. Dorner AJ, Coffin JM. Determinants for receptor interaction and cell killing on the avian retrovirus glycoprotein gp 85. *Cell* 1986;45:365–374.
35. Bose HR. Reticuloendotheliosis virus and disturbance in immune regulation. *Microbiol Sci* 1984;1:107–112.
36. Mosier DE, Yetter RA, Morse HC III. Retroviral induction of acute lymphoproliferative disease and profound immunosuppression in adult C57BL/6 mice. *J Exp Med* 1985;161:766–784.
37. McGrath MS, Weissman IL. A receptor-mediated model of viral leukemogenesis: hypothesis and experiments. *Cold Spring Harbor Conference Cell Proliferation* 1979;5:547–589.
38. Hard RC, Cross SS. Pathology, immunology and virology of the host versus graft syndrome. *Surv Immunol Res* 1983;2:1–11.
39. Cross SS, Brede G, Tucker HSG III, Maloney M, Montour JL, Hard RC Jr. Expression of murine leukemia viruses in RFM mice with host versus graft disease after perinatal inoculation of (T6 × RFM)F1 lymphohemopoietic cells. *Infect Immun* 1983;41:570–577.
40. Hardy WD, Hess PW, MacEwen EG, et al. Biology of feline leukemia virus in the natural environment. *Cancer Res* 1976;36:582–588.
41. Mathes LE, Olsen RG, Hebebrand LC, et al. Immunosuppressive properties of a virion polypeptide, a 15,000 dalton protein, from feline leukemia virus. *Cancer Res* 1979;39:950–955.
42. Cianciolo GJ, Copeland TD, Oroszlan S, Synderman R. Inhibition of lymphocyte proliferation by a synthetic peptide homologous to retroviral envelope protein. *Science* 1985;230:453–455.
43. Maul DH, Zaiss CP, Mackenzie MR, Shiigi SM, Marx PA, Gardner MB. Simian retrovirus D serogroup has a broad cellular tropism for lymphoid and non-lymphoid cells. *J Virol* 1988;62:1768–1773.
44. Lackner AA, Marx PA, Lerche W, et al. Asymptomatic infection of the central nervous system by the macaque immunosuppressive type D retrovirus, SRV-1. *J Gen Virol* 1989;70:1641–1651.
45. Swanstrom R, Parker RC, Varmus HE, Bishop JM. Transduction of a cellular oncogene: the genesis of Rous sarcoma virus. *Proc Natl Acad Sci USA* 1983;80:2519–2523.
46. Varmus HE. The molecular genetics of cellular oncogenes. *Annu Rev Genet* 1984;18:553–612.
47. Gardner MB. Oncogenes and acute leukemia. In: Stass SA, ed. *The acute leukemias.* New York: Marcel Dekker, 1987;327–359.
48. Shibuya M, Hanafusa T, Hanafusa H, Stephenson R. Homology exists among the transforming sequences of avian and feline sarcoma viruses. *Proc Natl Acad Sci USA* 1980;77:6536–6540.
49. Duesberg PH, Bister K, Moscovici C. Genetic structure of avian myeloblastosis virus released from transformed myeloblasts as a defective virus particle. *Proc Natl Acad Sci USA* 1980;77:5120–5124.
50. Naharro G, Robbins KC, Reddy EP. Gene product of v-fgr onc: hybrid protein containing a portion of actin and tyrosine-specific protein kinase. *Science* 1984;223:63–66.
51. Heldin C-H, Westermark B. Growth factors: mechanism of action and relation to oncogenes. *Cell* 1984;37:9–20.
52. Weinberg RA. The action of oncogenes in the cytoplasm and nucleus. *Science* 1985;230:770–776.

53. Hayward WS, Neel BG, Astrin SM. Activation of a cellular onc gene by promoter insertion in ALV-induced lymphoid leukosis. *Nature* 1981;209:475–449.
54. Steffen D. Proviruses are adjacent to c-myc in some murine leukemia virus–induced lymphomas. *Proc Natl Acad Sci USA* 1984;81:2097–3001.
55. Neil JC, Hughes D, McFarlane R, et al. Transduction and rearrangement of the myc gene by feline leukemia virus in naturally occurring T-cell lymphomas. *Nature* 1984;308:814–820.
56. Nusse R, Varmus HE. Many tumors induced by the mouse mammary tumor virus contain a provirus integrated in the same region of the host genome. *Cell* 1982;31:99–109.
57. Tsichlis PN, Strauss PG, Hu LF. A common region for proviral DNA integration in MoMuLV-induced rat thymic lymphomas. *Nature* 1983;302:445–449.
58. Neil JC, Forrest D. Mechanisms of retrovirus-induced leukemia: selected aspects. *Biochim Biophys Acta* 1987;907:71–91.
59. Burny A, Cleuter R, Kettmann M, et al. Bovine leukemia: facts and hypothesis from the study of an infectious cancer. *Cancer Surveys* 1987;6:139–159.
60. Yoshida M, Seiki M. Recent advances in the molecular biology of HTLV-1: trans-activation of viral and cellular genes. *Annu Rev Immunol* 1987;5:541–559.
61. Wong-Staal F, Gallo RC. Human T-lymphotropic retroviruses. *Nature* 1985;317:395–403.
62. Rosen CA, Park R, Sodroski JG, Haseltine WA. Multiple sequence elements are required for regulation of human T-cell leukemia virus gene expression. *Proc Natl Acad Sci USA* 1987;84: 4919–4923.
63. Seiki M, Inoue J-I, Hidaka M, Yoshida M. Two cis-acting elements responsible for posttranscriptional trans-regulation of gene expression of human T-cell leukemia virus type I. *Proc Natl Acad Sci USA* 1988;85:7124–7128.
64. Leung K, Nabel GJ. HTLV-1 transactivator induces interleukin-2 receptor expression through an NF-kB-like factor. *Nature* 1988;333:776–778.
65. Wano Y, Feinberg M, Hosking JB, Bogerd H, Greene WC. Stable expression of the *tax* gene of type I human T-cell leukemia virus in human T cells activates specific cellular genes involved in growth. *Proc Natl Acad Sci USA* 1988;85:9733–9737.
66. Van den Broke A, Cleuter Y, Chen G, et al. Even transcriptionally competent proviruses are silent in bovine leukemia virus–induced sheep tumor cells. *Proc Natl Acad Sci USA* 1988;85: 9263–9267.
67. Lagarias DM, Radke K. Transcriptional activation of bovine leukemia virus in blood cells from experimentally infected, asymptomatic sheep with latent infections. *J Virol* 1989;63:2099–2107.
68. Desgranges C, Duc Dodon M, Gazzolo L. Virological aspects of HTLV-1 and related retroviruses. In: *HTLV-I and the nervous system.* New York: Alan R. Liss, 1989;9–18.
69. Fauci AS. The human immunodeficiency virus: infectivity and mechanisms of pathogenesis. *Science* 1988;129:617–622.
70. Tong-Starksen SE, Luciw PA, Peterlin BM. Human immunodeficiency virus long terminal repeat responds to T-cell activation signals. *Proc Natl Acad Sci USA* 1987;84:6845–6849.
71. Bohnlein E, Siekevitz M, Ballard DW, et al. Stimulation of the human immunodeficiency virus type 1 enhancer by the human T-cell leukemia virus type I *tax* gene product involves the action of inducible cellular proteins. *J Virol* 1989;63:1578–1586.
72. Bartholomew C, Blattner W, Cleghorn F. Progression of AIDS in homosexual men co-infected with HIV and HTLV-I in Trinidad. *Lancet* 1987;2:1469.
73. Lenardo MJ, Fan C-M, Maniatis T, Baltimore D. The involvement of NK-kB in B-interferon gene regulation reveals its role as widely induceable mediator of signal transduction. *Cell* 1989;57: 287–294.
74. Hinrichs SH, Nerenberg M, Reynolds RK, Khoury G, Jay G. A transgenic mouse model for human neurofibromatosis. *Science* 1987;237:1340–1343.
75. Vogel J, Hinrichs SH, Reynolds RK, Luciw PA, Jay G. The HIV *tat* gene induces dermal lesions resembling Kaposi's sarcoma in transgenic mice. *Nature* 1988;335:606–611.
76. Ensoli B, Nakamura S, Salahuddin SZ, et al. AIDS-Kaposi's sarcoma-derived cells express cytokines with autocrine and paracrine growth effects. *Science* 1989;2343:223–226.
77. Gardner MB, Rasheed S, Shimizu S, et al. Search for RNA tumor virus in humans. In: Hiatt HH, Watson JD, Winsten JA, eds. *Origins of human cancer.* New York: Cold Spring Harbor, 1977;1235–1251.
78. Ishikawa K, Fukasawa M, Tsujimoto H, et al. Serological survey and virus isolation of simian T-cell leukemia/T-lymphotropic virus type I (STLV-I) in non-human primates in their native countries. *Int J Cancer* 1987;40:233–239.

79. Tsujimoto H, Seiko M, Nakamura H, et al. Adult T cell leukemia-like disease in monkeys naturally infected with simian retrovirus related to human T cell leukemia virus type 1. *Gann* 1985;76:911–914.
80. Voevodin AF, Lapin BA, Yakovleva LA, et al. Antibodies reacting with human T lymphotropic retrovirus (HTLV-1) or related antigens in lymphomatous and healthy hamadyas baboons. *Int J Cancer* 1985;36:579–584.
81. Sagata N, Yasunaga T, Tsuzuku-Kawamura J, et al. Complete nucleotide sequence of the genome of bovine leukemia virus: its evolutionary relationship to other retroviruses. *Proc Natl Acad Sci USA* 1984;81:4741–4745.
82. Gardner MB. Type C viruses of wild mice: characterization and natural history of amphotropic, ecotropic, and xenotropic MuLV. *Curr Top Microbiol Immunol* 1978;79:215–259.
83. Gardner MB, Rasheed S. Retroviruses in feral mice. *Intern Rev Exper Path* 1982;23:209–267.
84. Gardner MB. Naturally occurring leukaemia viruses in wild mice: how good a model for humans? *Cancer Surveys* 1987;6:55–71.
85. Gardner MB, Chiri A, Dougherty MF, Casagrande J, Estes JD. Congenital transmission of murine leukemia virus from wild mice prone to the development of lymphoma and paralysis. *J Natl Cancer Inst* 1979;62:63–70.
86. Gardner MB, Henderson BE, Officer JE, et al. A spontaneous lower motor neuron disease apparently caused by indigenous type C RNA virus in wild mice. *J Natl Cancer Inst* 1973;51:1242–1254.
87. Jolicoeur P, Nicolaiew N, DesGroseillers L, Rassart E. Molecular cloning of infectious viral DNA from ecotropic neurotropic wild mouse retrovirus. *J Virol* 1983;45:1159–1163.
88. Rasheed S, Gardner M, Chan E. Amphotropic host range of naturally occurring wild mouse leukemia viruses. *J Virol* 1976;19:13–18.
89. Gardner MB. Retroviral infection of the nervous system in animals and man. In: *Neuroimmune networks: physiology and disease.* New York: Alan R. Liss, 1989:179–192.
90. Gardner MB, Rasheed A, Pal BK, Estes JD, O'Brien SJ. Akvr-1, a dominant murine leukemia virus restriction gene, in polymorphic leukemia-prone wild mice. *Proc Natl Acad Sci USA* 1980;77:531–535.
91. Dandekar S, Rossitto P, Pickett S, et al. Molecular characterization of the Akvr-1 restriction gene: a defective endogenous retrovirus identical to Fv-4. *J Virol* 1987;61:308–314.
92. Odaka TH, Ikeda H, Yoshikura H, Moriawaka K, Suzuki S. Fv-4: gene controlling resistance to NB-tropic Friend murine leukemia virus. Distribution in wild mice. Introduction into genetic background of BALB/c mice, and mapping of chromosomes. *J Natl Cancer Inst* 1981;67:1123–1127.
93. Kozak CA, O'Neill RR. Diverse wild mouse origins of xenotropic, mink cell focus-forming, and two types of ecotropic proviral genes. *J Virol* 1987;61:3082–3088.

DISCUSSION

The first discussant pointed out the possible relevance of retrovirus transmission in humans by mother's milk and asked if milk transmission of murine retrovirus was cell-free or cell-associated. Another discussant stated that the virus is largely cell-free and thus different from the human retrovirus, which is more cell-associated.

Human Retrovirology: HTLV,
edited by William A. Blattner.
Raven Press, Ltd., New York © 1990.

The Issues of Causation and Neurotropism in Neurological Diseases Associated with Infections by Retroviruses

Ashley Haase, Anthony Evangelista, Hal Minnigan,
Stacene Maroushek, Andrew Larson, Ernest Retzel,
Dale McFarlin, Steve Jacobson, and Courtenay Bartholomew

Department of Microbiology, University of Minnesota, Minneapolis, Minnesota

INTRODUCTION

Retroviruses have been classified into major subfamilies whose members share characteristic behavior in their natural hosts. The oncogenic retroviruses transform cells in culture and cause tumors in vivo; the lentiviruses fuse and kill cells in culture and cause slow infections in vivo. However, when we look at retroviral behavior from the perspective of neurotropism, these distinctions become blurred (Fig. 1): oncogenic murine (1,2) and avian retroviruses (3)—and, in all likelihood, HTLV-I—join the animal lentiviruses and HIV as agents of chronic neurological diseases. Indeed, there are particularly striking parallels between HTLV-I–associated (4–11) neurological conditions and visna, a paralytic disease of sheep caused by the prototype of the lentivirus subfamily (12). Taking these similarities as evidence perhaps of other meaningful relationships, this chapter begins with an account in the experimental approaches of visna that have proved particularly useful in understanding this slow infection as a prelude to comparable analyses of HTLV-I.

PARALLELS BETWEEN VISNA AND THE HTLV-I–ASSOCIATED CHRONIC PROGRESSIVE MYELOPATHIES

Following the introduction of Karakul sheep into Iceland in the 1930s, new diseases emerged that reached epidemic proportions over the next decade. The most prominent of these diseases was a progressive interstitial pneumonia called maedi ("shortness of breath"). Some of the animals also developed a neurological condition which, like tropical spastic paraparesis (TSP) and the HTLV-I–associated myelopathy (HAM) in Japan (5–11), begins with paralysis of the extremities. This paralytic disease was progressive and the afflicted sheep eventually

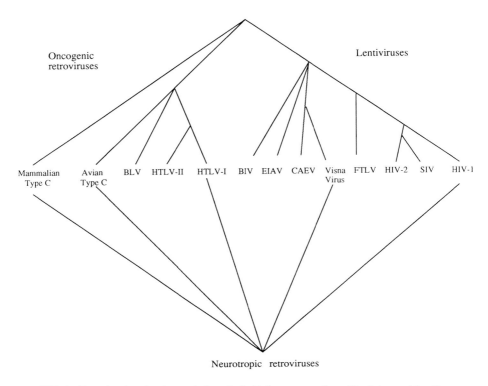

FIG. 1. Neurotropic retroviruses belong to both the oncogenic and lentivirus subfamilies.

succumbed to inanition and intercurrent infection. The farmers called the disease visna ("wasting") because of the striking loss of weight.

Sigurdsson investigated visna, maedi, and other infections introduced by the imported sheep, showed that they were transmissible conditions caused by viruses with long incubation periods to manifest illness, and based on that experience, he introduced the concept of slow infections (12). Sigurdsson also provided the first descriptions of the neuropathology of visna, in which he emphasized the demyelinating nature of the lesions which developed around foci of inflammation (13). Nathanson et al. later found that immunosuppressive therapy diminished the infiltration of mononuclear cells in the central nervous system (CNS) of infected animals and, pari passu, histopathological changes and development of symptoms. Their observations strongly suggested an immunopathological process in which infection incites an inflammatory response that indiscriminately kills infected cells and innocent bystanders (14).

In many respects the clinical presentation, neuropathology, and possibly the pathogenesis of TSP and HAM resemble visna. The slowly progressive paralysis of the extremities reflects the common underlying demyelination, particularly in the corticospinal tracts; this, in turn, is in all cases associated with foci of inflammation in which lymphocytes and monocytes accumulate as cuffs around blood vessels and in collections in the neuroparenchyma called glial nodules (Fig. 2) (15). Osame and his colleagues have also demonstrated a response to steroid treatment in the majority of individuals with HAM, further sustaining the analogy to visna (16).

ISSUES AND DEEPER IDENTITIES

Are these similarities between HAM and visna the result of still deeper identities and, if so, what lessons can be derived from the visna paradigm? This chapter is written on the premise that this is indeed the case and that there are two issues where the visna model should prove especially instructive. These issues are causation and neurotropism.

Causation: Koch's Postulates Revisited with In Situ Hybridization

In visna, Sigurdsson was able to satisfy Koch's postulates, slightly recast for viruses (12,18): (a) a filterable agent was consistently isolated from cases of visna and maedi; (b) the agent could be propagated in tissue culture, where it produced characteristic cytopathic effects (rounding up and degeneration of cells and formation of multinucleated giant cells); (c) reinoculation of the infected tissue culture fluids reproduced the disease; and (d) prior admixing of the tissue culture supernatant with sera from infected animals prevented transmission.

In the past decade we have used in situ hybridization to fulfill Koch's postulates in molecular terms: (a) the virus genome should be constantly associated with disease; and (b) the virus genome should be in a plausible site to produce the pathological changes. Thus, both the genomic RNA of visna virus and its cellular DNA intermediate have been consistently demonstrated in experimentally in-

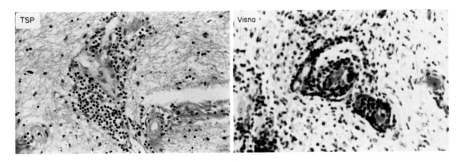

FIG. 2. Comparative neuropathology of TSP and visna. Demyelination occurs in areas in which mononuclear inflammatory cells accumulate, often arranged as a cuff surrounding blood vessels, as shown in this figure.

fected animals (19,20); and the viral genome, genes, and products bear a plausible relationship to the disease process (21).

Plausibility Argument

This notion of plausibility is of sufficient importance to warrant further discussion of the crucial contribution that single-cell technologies of in situ hybridization and immunocytochemistry make to understanding the pathogenesis of infectious diseases. Because lentiviruses like visna and retroviruses like HTLV-I are harbored in lymphocytes and monocytes (4,22), the isolation of virus from blood or cerebrospinal fluid (11) or detection of viral nucleic acids by the polymerase chain reaction (PCR) (23–25) does not necessarily define the role of the virus in the pathological process. In the CNS infiltrated by inflammatory mononuclear cells, the interpretation of these kinds of analyses will be confounded to

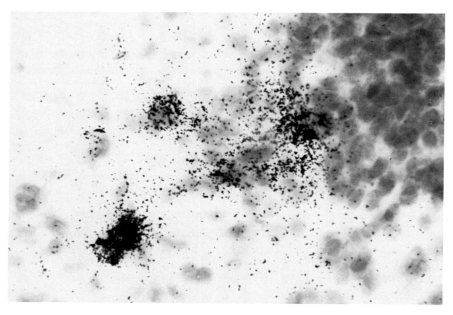

FIG. 3. Detection of visna virus RNA by in situ hybridization. In the developed radioautograph large numbers of silver grains are evident in several cells at the periphery of a perivascular cuff of inflammatory cells (upper right). Because the number of copies of viral RNA per cell is proportional to the number of grains, the densely labeled cells indicate concentrations of viral RNA similar to those in productive infections in tissue culture. In subjacent sections viral antigens can be detected. This population of infected cells is always found in close proximity to areas with the greatest inflammatory and demyelinating pathological alterations.

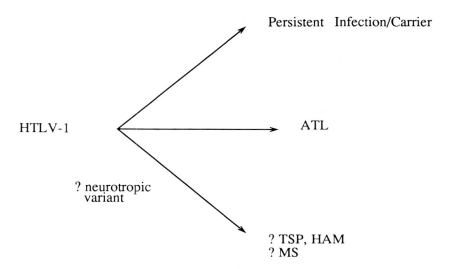

FIG. 4. Outcomes of HTLV-I infection. See text for explanation.

an even greater degree. On the other hand, with in situ hybridization and immunocytochemistry, visna RNA has been detected in neural cellular elements, such as oligodendrocytes, which can be directly linked to the demyelination process (21). Quantitative analyses too are congruent with characteristics that define the infection: persistence of virus despite host immunity, and the slow evolution of an immunopathological process predominantly affecting white matter. With in situ hybridization and immunofluorescence or immunocytochemistry, we have shown that most cells infected with visna virus harbor viral DNA in an immunologically silent state where viral transcripts and antigens are reduced by orders of magnitude vis-a-vis productive infections. The restricted nature of gene expression in this population provides a mechanism for viral persistence (12). The explanation for the slowly evolving paralytic disease is also derived from these single-cell analyses, which have documented a spectrum of states of gene expression tightly correlated with the extent of inflammation. Thus, there is a gradient of gene expression in which the number of copies of viral RNA will occasionally approach that seen in productive infections in tissue culture (Fig. 3). These cells also contain viral antigens and are associated with the most extensive inflammation and tissue destruction in white matter. These findings are in accord with the concept that the antigen positive cells provoke and sustain the immunopathological process. The tempo is set by the small fraction of infected cells at any time during infection at the high end of the spectrum of gene expression. The explanatory power of this approach in visna has encouraged us to embark on a similar undertaking in chronic progressive myelopathy (CPM).

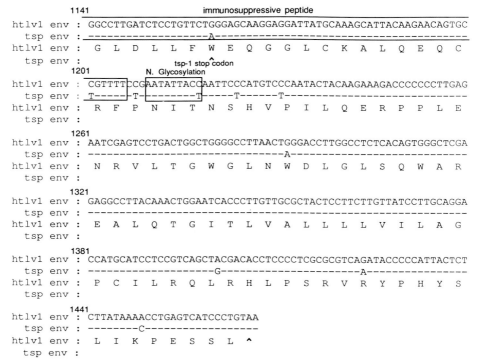

FIG. 5. Nucleotide (nt) and deduced amino acid sequences in the region of the *env* gene (gp21) containing two stop codons. Subclones from a full-length clone derived from the DNA in the first cell line established from a TSP patient (11) was sequenced by the Sanger dideoxy chain termination method (28). The upper two lines show the nucleotide sequence and the lower the deduced amino sequence of HTLV-I and TSP-1. Only differences are indicated for TSP-1. Lowercase, nonaligned bases; uppercase, aligned nonidentical bases. · · · ·, gap. Putative sites for N glycosylation, first stop codon, and amino acid sequence homologous to the immunosuppressive peptide region of other retroviruses are indicated.

Neurotropism Issue: Visna Precedent

The second major issue which needs to be addressed in retrovirus-associated immunological disease is whether the responsible agent is a neurotropic variant. Here again visna may offer useful precedents. Visna in Iceland occurred as a subset of chronically infected sheep which for the most part developed progressive pneumonia. One explanation for this pattern of disease is that a variant virus was responsible for neurological disease. To investigate this possibility, Petursson and his colleagues (unpublished) sought neurotropic variants by selecting viruses that produced neurological disease in a high proportion of infected animals. After three passages in vivo, they succeeded in isolating a virus from an animal that in Icelandic sheep does produce a high incidence of neurological disease

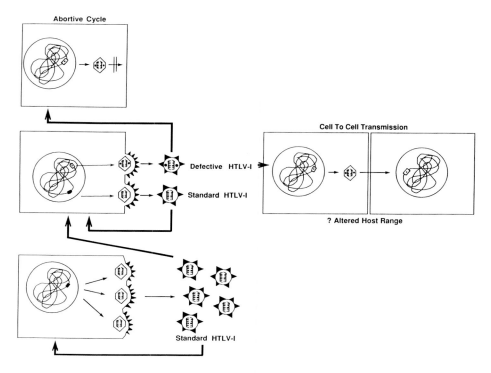

FIG. 6. Speculative diagram of how a defective HTLV-I provirus (⊶⊐⊷) could give rise to a variant ( ) RNA encapsidated by viral proteins (▲) provided in coinfected cells by standard HTLV-I ( ). On rare occasions the defective virus might enter the nervous system, infect neural cellular element, and slowly spread by cell-to-cell transmission.

with a shorter incubation period. Recent evidence points to more efficient replication in the CNS, as judged by the RNA levels determined by in situ hybridization (C. Zink, O. Narayan, unpublished), as the explanation for the neurovirulence of this strain of visna virus. These findings lend support to the argument that there may be neurotropic variants that cause visna rather than maedi.

Neurotropism and HTLV-I Infections

HTLV-I primarily establishes chronic infections in which ~5% of carriers ultimately develop adult T-cell leukemia/lymphoma and others perhaps have neurological disease (Fig. 4). With the visna experience in mind, we have begun to look for HTLV-I variants in the HTLV-I–associated neurological diseases. We initiated these studies by cloning and sequencing a provirus integrated in the DNA of cells from an individual with TSP (11) to compare it with the published sequences of HTLV-I from a patient with HAM (25) and with those

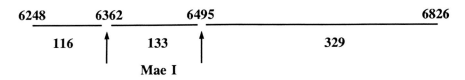

FIG. 7. Physical map (line diagram) of the region from nt6248 to 6826 with two stop codons at the positions indicated by the arrows where the G-to-A mutation introduces restriction sites for MAE1 $\left(\begin{smallmatrix} C{\downarrow}TAG \\ GAT{\uparrow}C \end{smallmatrix}\right)$. The numbers below the line indicate the number of base pairs in the fragments.

of a patient with ATL (26). The impelling motive and rationale underlying these efforts is the hope that sequence changes will be identified that distinguish the proviruses of HTLV-I in HAM and TSP from those in ATL.

Thus far, in addition to sequence microheterogeneity of indeterminate significance, there is one mutation in the external glycosylated polypeptide gp46 (a C → T transition that changes threonine to isoleucine) common to the TSP and HAM proviruses and two point mutations (G to A) that introduce stop codons in the *env* gene encoding the transmembrane anchor gp21. The first of these stop codons occurs in a region encoding an immunosuppressive amino acid sequence of retroviruses (27) that would result in the loss of ∼100 amino acids from the carboxy terminus of gp21 (Fig. 5).

Assessing the Possible Role of a Defective Virus in CPM

We think it likely that the TSP provirus we sequenced is defective, as it is likely that the loss of the carboxyterminus of the transmembrane anchor would seriously impair assembly of infectious virions. In Fig. 6 we present a speculative model for infection of the CNS by defective HTLV-I, where the altered host range is construed as a consequence of a new mode of spread from cell to cell imposed by the inability to assemble viral particles. In the TSP cell line from which the defective provirus was cloned (11), the model predicts a predominant population of standard HTLV-I, the product of singly infected cells, and a minor population of dually infected cells where standard HTLV-I provides gp21 to perpetuate infection by the defective variant. The model accounts for the detection of viral particles in the TSP cell line and the detection of glycosylated *env* gene products of gp63, gp21, and gp40 from the dominant population of standard HTLV-I (the slight decrease in molecular weight of gp40, compared with gp46, in the HUT102 cell line infected with HTLV-I may reflect differences in glycosylation). We did not detect the lower-molecular-weight species of the defective virus, but this may be simply the result of the relative infrequency of singly infected cells producing the truncated *env* gene products and the relatively small quantities of virion proteins overall in the TSP cell line compared with HUT102 (unpublished data).

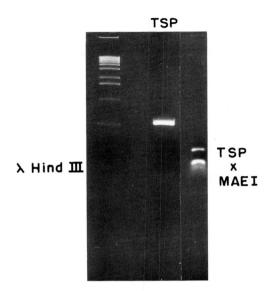

FIG. 8. Analysis of MAEI restriction site polymorphism. DNA from the plasmid containing the TSP provirus (1 ng) was amplified in the region of the genome containing the two stop codons using the polymerase chain reaction. The products were isolated and electrophoresed in agarose gels containing ethidium bromide. The amplified product lane (marked TSP) was of the expected size (587 nt) compared with molecular weight markers (λ *Hind*III digest). After digestion with MAEI the expected fragments of 329 nt and two species of 116 and 133 nt were generated (lane marked TSP × MAEI).

It should be possible to test the critical prediction of the model, that TSP patients harbor the defective provirus, by looking for *env* gene polymorphisms introduced by the G-to-A mutations in genomic DNA of cells from TSP and HAM patients. [The G-to-A mutations introduce two new restriction sites for MAEI (Fig. 7).] To illustrate this approach, the flanking region of the cloned proviral DNA from the TSP patient was amplified with the PCR and digested with MAEI. The expected products are species of 329, 133, and 166 bp; DNA lacking these sites should yield a fragment of 587 bp. As shown in Fig. 8, the expected products were amplified from the TSP DNA (the two smaller species were not resolved). We are currently applying this analysis to DNA from TSP, HAM, and ATL patients.

CONCLUSIONS

There are similarities in the clinical presentation and neuropathology of the slow infection of sheep visna and HTLV-I myelopathies that may reflect common mechanisms of disease. For this reason the experience with visna may well define approaches that will prove useful in understanding CPMs. In this chapter we examine two issues in some detail where we think the parallels will prove informative: the role of HTLV-I in the neurological conditions associated with infection, and the possibility that neurotropic variants of HTLV-I are responsible for CPM. We advance arguments that the issues of causation can best be resolved by systematic analyses of tissues by in situ hybridization, which provides information critical to determining whether HTLV-I in the CNS bears a plausible relationship to the neuropathological changes characteristic of HTLV-I. We also describe work in progress on a defective HTLV-I provirus in TSP cells that conceivably could have different cellular tropisms imposed by cell-to-cell spread. The search for the HTLV-I genome in the CNS and the role of defective virus in CPM constitute important objectives for ongoing investigations of human retroviral infections.

REFERENCES

1. Gardner MB. Retroviral spongiform polioencephalomyelopathy. *Rev Infect Dis* 1985;7:99–110.
2. Wong PKY, Soong MM, MacLeod R, Gallick G, Yuen PH. A group of temperature-sensitive mutants of Moloney leukemia virus which is defective in cleavage of *env* precursor polypeptide in infected cells also induces hindlimb paralysis in newborn CFW/D mice. *Virology* 1983;125: 513–518.
3. Whalen LR, Wheeler DW, Gould DH, Fiscus SA, Boggie LC, Smith RE. Functional and structural alterations of the nervous system induced by avian retrovirus RAV-7. *Microbial Pathogen* 1988;4: 401–416.
4. Wong-Staal F, Gallo RC. Human T-lymphotropic retroviruses. *Nature* 1985;317:395–403.
5. Gessain A, Barin F, Vernant JC, et al. Antibodies to human T-lymphotropic virus type-1 in patients with tropical spastic paraparesis. *Lancet* 1985;2:407–410.
6. Rodgers-Johnson P, Gajdusek DC, Mortan DC, et al. HTLV-I and HTLV-III antibodies and tropical spastic paraparesis. *Lancet* 1985;2:1247–1248.
7. Bartholomew C, Cleghorn F, Charles W, et al. HTLV-1 and tropical spastic paraparesis in Trinidad. *Lancet* 1986;2:99–100.
8. Roman G, Schoenberg B, Madden D, et al. HTLV-1 antibodies in the serum of patients with tropical spastic paraparesis in the Syechelles. *Arch Neurol* 1987;44:605–607.
9. Osame M, Matsumoto M, Usuku K, et al. Chronic progressive myelopathy associated with elevated antibodies to human T-lymphotropic virus type 1 and adult T-cell leukemia-like cells. *Ann Neurol* 1987;21:117–122.
10. Koprowski H, DeFreitas EC, Harper ME, et al. Multiple sclerosis and human T-cell lymphotropic retroviruses. *Nature* 1985;318:154–160.
11. Jacobson S, Raine CS, Mingioli ES, McFarlin DE. Isolation of an HTLV-1–like retrovirus from patients with tropical spastic paraparesis. *Nature* 1988;331:540–543.
12. Haase AT. Pathogenesis of lentivirus infections. *Nature* 1986;322:130–136.
13. Sigurdsson B, Pálsson PA, Van Bogaert L. Pathology of visna. *Acta Neuropathol* 1962;1:343–362.
14. Nathanson N, Panitch H, Pálsson P, Petursson G, Georgsson G. Pathogenesis of visna. II. Effect of immunosuppression upon early central nervous system lesions. *Lab Invest* 1976;35:444–451.
15. Montgomery RD, Cruickshank EK, Robertson WB, McMenemey WH. Clinical and pathological observations on Jamaican neuropathy. *Brain* 1964;87:427–462.
16. Osame M, Matsumoto M, Usuku K, et al. Chronic progressive myelopathy associated with elevated antibodies to human T-lymphotropic virus type 1 and adult T-cell leukemialike cells. *Ann Neurol* 1987;21:117–122.
17. Sigurdsson B, Pálsson PA. Visna of sheep. A slow, demyelinating infection. *Br J Exp Pathol* 1958;39:519–528.
18. Sigurdsson B, Thormar H, Pálsson PA. Cultivation of visna virus in tissue culture. *Arch Gesamte Virusforsch* 1960;10:368.

19. Haase AT, Stowring L, Narayan O, Griffin D, Price D. Slow persistent infection caused by visna virus: role of host restriction. *Science* 1977;195:175–177.
20. Brahic M, Stowring L, Ventura P, Haase AT. Gene expression in visna virus infection in sheep. *Nature* 1981;292:240–242.
21. Stowring L, Haase AT, Petursson G, et al. Detection of visna virus antigens and RNA in glial cells in foci of demyelination. *Virology* 1985;141:311–318.
22. Peluso R, Haase A, Stowring L, Edwards M, Ventura P. A trojan horse mechanism for the spread of visna virus in monocytes. *Virology* 1985;147:231–236.
23. Kowk S, Kellogg D, Erlich G, Poiesz B, Bhagavati S, Sninsky JJ. Characterization of a sequence of human T cell leukemia virus type 1 from a patient with chronic progressive myelopathy. *J Infect Dis* 1988;158:1193–1197.
24. Bhagavati S, Ehrlich G, Kula RW, et al. Detection of human T-cell lymphoma/leukemia virus type 1 DNA and antigen in spinal fluid and blood of patients with chronic progressive myelopathy. *N Engl J Med* 1988;318:1141–1147.
25. Tsujimoto A, Teruuchi T, Imamura J, Shimotohno K, Miyoshi I, Miwa M. Nucleotide sequence analysis of a provirus derived from HTLV-1-associated myelopathy (HAM). *Mol Biol Med* 1988;5: 29–42.
26. Seiki M, Hattori S, Yoshida M. Human adult T-cell leukemia virus: molecular cloning of the provirus DNA and the unique terminal structure. *Proc Natl Acad Sci USA* 1983;79:6899–6902.

DISCUSSION

In reference to one of the papers under discussion, one participant asked if any cross-reactivity with myelin basic protein (MBP) (and retroviral antigens) had been found, pointing out that there is limited sequence homology with MBP. The paper's presenter replied that no crossreactivity with MBP has been observed.

Another participant wondered what the frequency of defective provirus is. The presenter responded that microheterogeneity in vivo may be irrelevant for HIV. With HTLV-I, however, the findings of nucleotide sequence differences may be more significant.

A discussant wondered what the role of monocyte in HTLV-I infection might be. The presenter didn't know.

Another discussant offered that Gp 21E encoded by a defective provirus (by analogy with animal retroviruses) may be important. The discussant indicated that this has been found a variable region by PCR. Forty to fifty percent of ATL patients and none of the TSP patients generated an amplifiable signal. These data were interpreted as evidence that stability in region may not be the same in TSP.

Human Retrovirology: HTLV,
edited by William A. Blattner.
Raven Press, Ltd., New York © 1990.

Molecular Basis of Viral Persistence

Michael B. A. Oldstone

*Departments of Neuropharmacology and Immunology, Research Institute
of Scripps Clinic, La Jolla, California 92037*

MECHANISMS OF VIRAL PERSISTENCE

For a virus to persist in a host cell, it must accomplish two feats. First, its replicative cycle must change from the usual lytic mode to a nonlytic one, which provides the virus with a selective advantage of continuous housing. The alternative to this continuity is lysis of the cell and the virus' subsequent necessity to shop for a new cell or host in which to reside. Second, the virus must avoid immunologic surveillance. This can be accomplished by any of the several mechanisms listed in Table 1. The net result of these two achievements is a stable existence for the virus without the battering and eventual destruction by the host's immune system.

VIRUS-INDUCED IMMUNOSUPPRESSION

All the mechanisms cataloged in Table 1 are proven routes of establishing persistent infection for a variety of viruses; however, this presentation focuses on just one pathway: the ability of a virus to infect and inactivate specific lymphocytes that are ordinarily destined to react against that virus. In this scenario, the virus removes the policeman responsible for maintaining control in its neighborhood and thereby survives with little or no danger of eviction. Interestingly, this virus-induced immunosuppression can be directed selectively against only the infecting virus and no other infectious agents. For example, in the model of lymphocytic choriomeningitis virus (LCMV) infection of mice, a persistent infection is associated with the host's inability to mount an effective cytotoxic T-lymphocyte (CTL) response against LCMV—yet normal and efficient responses are made against other infectious agents such as Pichinde virus, a relative of LCMV in the arenavirus family, and the unrelated vaccinia and influenza viruses (Table 2). Further, persistently LCMV-infected mice respond to such nonviral antigens as bovine serum albumin, keyhole limpet hemocyanin, human immunoglobulin, and sheep red blood cells as efficiently as do their uninfected, age- and sex-matched brethren of similar haplotypes.

Because the defect is in the mounting of a LCMV-specific major histocom-

TABLE 1. *Mechanisms to establish persistent infection*

For persistent infection a virus must:
  1. Alter replication and transcription
      A. Assume an incomplete or defective form
      B. Generate mutants or variants
      C. Diminish expression of viral gene product
  2. Avoid immunologic surveillance
      Remove recognition molecules on infected cells
      A. Alter viral protein expression: e.g., antiviral antibody-induced capping and modulation
          Alter MHC expression
              Directly: adenovirus, etc.
              Indirectly: infection of release by lymphocytes
              Nonexpression: e.g., neurons
      B. Abrogate lymphocyte/macrophage function
          Generalized immunosuppression
          Selected immunosuppression
          Lymphokines alter transcription of host gene(s)

Viruses use a variety of mechanisms to escape immunologic surveillance and thereby establish persistent infection. These are active functions of the virus. In addition, the virus must alter its replicative strategy to one of nonlytic-host cell interaction.

patibility complex–restricted CTL response, the adoptive transfer of such effector CTL leads to the cleansing of infectious viral nucleic acid sequences and viral antigens from tissues of persistently infected mice, as shown in Fig. 1. These experiments demonstrate that persistent infections can be successfully treated and cured. The implication for numerous persistent viral infections of man is significant and obvious.

MOLECULAR MECHANISMS

How do the persistent LCMV state and the inactivation of LCMV-specific CTL develop? A few years ago my colleague R Ahmed and I noted that lymphoid cells obtained from persistently infected mice contain a unique viral variant. This variant can abort CTL activity in immunocompetent animals, thereby allowing the initiation and maintenance of viral persistence by so-called CTL^{nil} variants. Further, the effect of such CTL^{nil} viral variants is specific in that once persistence is achieved, it causes only the failure to generate CTL to LCMV, whereas the ability to generate CTL to either Pichinde or vaccinia virus remains intact. Figure 2 details several of these original observations, during which >85% of viral isolates obtained from lymphoid tissues of persistently infected mice prevented the generation of LCMV-specific CTL, accounting for the ongoing persistence. In contrast, both the parental virus and isolates from a variety of tissues, including brain and liver, almost uniformly induced LCMV CTL responses, thereby terminating the acute infection and preventing the viral persistence. These variants are termed CTL^{+}.

Recent studies of this LCMV-induced immunosuppression that accounts for

TABLE 2. *LCMV-specific H-2–restricted CTL killing is selectively diminished during persistent LCMV infection*

STATUS OF H-2d BALB/W MOUSE		PERCENT SPECIFIC ^{51}CR RELEASE						
		VIRUS INFECTED H-2d				LCMV INFECTED		
		LCMV		PICHIND	UNINF	H-2d		H-2b
	#	50:1	5:1	50:1	50:1	50:1	5:1	50:1
MOCK INFECTED	1	8	2	8	5	8	2	4
	2	3	0	9	7	3	0	2
	3	5	3	2	6	5	3	3
	4	6	2	5	1	6	2	5
ACUTE LCMV INFECTION (DAY 7 P° SPL)	1	81	28	7	6	81	28	6
	2	76	26	8	4	76	26	7
	3	76	12	5	3	76	12	4
	4	61	18	8	0	61	18	1
PERSISTENT LCMV INFECTION	1	6	3	4	5	6	3	2
	2	16	5	9	0	16	5	4
	3	8	4	3	1	8	4	6
	4	12	5	4	2	12	5	0
PERSISTENT LCMV INFECTION (PICHIND DAY 7 P° SPL)	1	8	3	54	4	8	3	4
	2	7	2	63	3	7	2	2
	3	14	5	58	5	14	5	1
	4	12	4	46	1	12	4	1

During acute infection, cytotoxic T lymphocytes (CTL) are generated (to LCMV: day 7 [primary splenic] CTL) that are virus specific (react against LCMV infected but not against Pichinde [Pichind] infected targets) and are restricted by the major histocompatibility complex (H-2 in the mouse). That is, virus-specific CTL generated by H-2^d mice lyse H-2^d but not H-2^b targets and vice versa (data not shown). In contrast, during persistent infection with LCMV, mice fail to generate a significant CTL response to LCMV (spontaneously or when challenged with new dose of LCMV). However, such persistently infected mice are capable of generating CTL specific for other viruses (Pichinde shown here). They also make vigorous immune responses to many nonviral antigens and produce high titers of antibodies to LCMV (data not shown). Hence, the virus-induced immunosuppression is specific and directed against the generation of LCMV-specific CTL.

virus persistence have focused on obtaining and characterizing the phenotype and genotype of CTL^{nil} variants and the revertants. The LCMV genome is composed of two RNA strands, a large one (L) of ~7.1 kb and a smaller (S) one of 3.7 kb. Ahmed and associates made reassortants between CTL^{nil} and CTL$^+$ LCMV viruses and showed that the immunosuppressive event mapped to the L RNA segment. Concurrently and independently, M Salvato showed by direct cloning and sequencing that the mutation for this immunosuppressive phenotype resides in the L RNA and not the S RNA. Presently a large number of CTL^{nil}

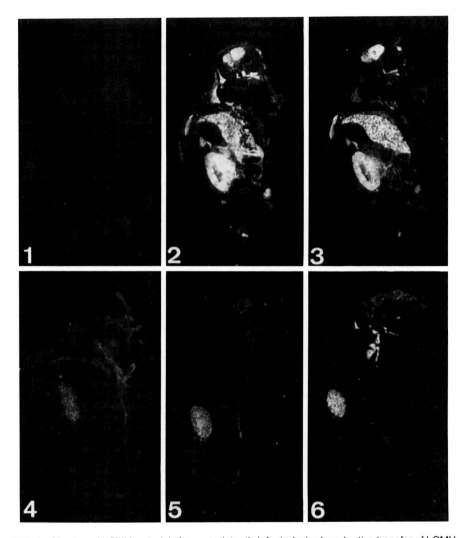

FIG. 1. Clearing of LCMV materials from persistently infected mice by adoptive transfer of LCMV-specific, H-2-restricted CTL. Viral materials, in this case LCMV nucleoprotein, are detectable in whole-body sections by the use of mouse monoclonal antibody to LCMV nucleoprotein, rabbit antibody to mouse IgG, and ^{125}I Staphylococcus protein A. Viral protein is prominently displayed in brain, liver, adrenal, spleen, kidney, tissues, etc. of 10- to 12-wk-old mice that have been persistently infected with LCMV since birth (panels 2, 3). A whole-body section of an age-matched uninfected mouse (negative control) is shown in panel 1 for comparison. Within 15 d of adoptive transfer of LCMV-specific, H-2-restricted CTL, virus titers in the blood decrease from 10^4–10^5 PFU/ml of sera to <50 PFU, and viral material clears from most tissues, except the kidney and brain. Panels 4, 5, and 6 show whole-body sections from such persistently infected mice approximately 30 d after CTL transfer. Viral proteins in the kidney are restricted to the glomeruli, where they reside as virus-antibody immune complexes. By ≥60 d after the transfers, the kidneys are virtually free of viral materials. Similarly, neuronal cells in the brain harbor viral materials for ~60 d after the transfer, followed by clearance. The kinetics and mechanism of clearance

variants and revertants (CTL$^+$) are being sequenced to determine whether one or both of the hot spots (mutations) originally found is responsible for the virus-induced immunosuppression. One hot spot occurs at the 5' end of the L RNA in the noncoding region of the Z gene. The other is found at the 3' end open reading frame in the viral polymerase. Once these data are complete and confirmed by analysis of several variants, cDNA and site-specific mutagenesis will

Lymphocyte Isolate	#10	#17	Brain Isolate
CTL(^{51}Cr release) 50:1 25:1	9 3	7 6	71 69
PFU (plasma)	1×10^6	9×10^5	nil
Whole Body Section Probed with ^{32}P LCMV GP	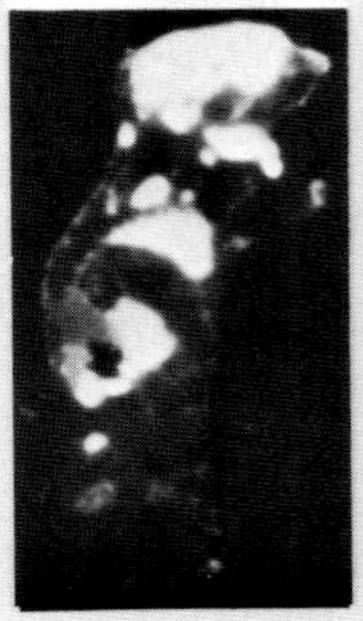		

FIG. 2. LCMV isolated from lymphocytes (plaque-purified isolates #10, #17) actively suppress the generation of LCMV-specific CTL and allow the establishment and maintenance of viral persistence [shown as infectious virus (PFU) in the plasma, and as viral nucleic acid sequences in whole body sections (LCMV specific ^{32}P cDNA probe)]. These are termed CTL^{nil} variants. In contrast, the parental virus, and isolates from the liver or brain (data from one brain isolate shown), uniformly generate LCMV-specific CTL termed CTL$^+$. These CTL$^+$ types eliminate the virus and prevent the establishment of persistence.

from neurons differ from that in other infected cells (Oldstone et al. 1986), primarily because neurons lack (fail to express) H-2 glycoproteins. The effectiveness of lymphocyte transfers illustrated here indicates the potential usefulness of immunocytotherapy in treating and curing persistent viral infections.

provide a precise biochemical answer. The phenotype of two CTL^{nil} variants (#10 and #17) being analyzed is displayed in Fig. 2. Several other laboratories are actively engaged in similar research on the replication, or restriction of replication, of CTL^{nil} and CTL^{+} variants in lymphocytes, as well as recording the effect of such viruses on lymphocyte biology and survival.

IMPLICATIONS TO VIRAL PATHOGENESIS

The lessons learned from studying the animal model of LCMV persistent infection, as well as the biochemical and in vitro analysis of LCMV-lymphocyte interactions, are likely to provide a sound foundation on which to build an understanding of virus-induced immunosuppression. It has not escaped our attention, nor that of others, that viruses and their genetic products are frequently

TABLE 3. *Viruses that infect lymphocytes and monocytes*

Virus	Host	Infected cells
Double-strand DNA viruses		
Hepatitis B virus	Human, monkey	PBMC, T and B lymph
Papovavirus	Human, monkey	PBMC
Group C adenoviruses	Human	T, B and null lymph
Herpes simplex virus	Human	T lymph
Epstein-Barr virus	Human	B lymph
Cytomegalovirus	Human	Lymph, mono
Pox virus	Rabbit	Spleen cells
Single-strand DNA viruses		
Porcine parvovirus	Pig	Spleen cells
Minute virus of mice	Mouse	Lymph
Positive-strand RNA viruses		
Poliovirus	Human	Lymph, mono
Rubella	Human	T and B lymph
Negative-strand RNA viruses		
Measles	Human	T and B lymph, mono
Mumps	Human	T and B lymph
Respiratory syncytial virus	Human	Lymph, mono
Vesicular stomatitis virus	Human, mouse	T lymph
Influenza A	Human	Lymph, mono
Parainfluenza	Human	Lymph, mono
Ambisense RNA viruses		
Lymphocytic choriomeningitis virus	Mouse	T and B lymph, mono
Junin virus	Human	PBMC
Retroviruses		
Murine leukemia virus	Mouse	B lymph
Feline leukemia virus	Cat	T and B lymph, mono
HTLV-I, -II	Human	T, B and null lymph
HIV-1, -2	Human	T and B lymph, mono
Endogenous C-type virus	Mouse	Spleen cells

Viruses that infect monocytes but not lymphocytes are not listed. Abbreviations: PBMC—peripheral blood mononuclear cells, lymph—lymphocyte, mono—monocyte or macrophage. (From McChesney and Oldstone. Ann. Rev. Immunol. 1987; 5:279–304).

found in the lymphocytes of animals and humans with persistent infections (Table 3). Hence, although their strategies of replication and transcription may differ, conceptually those viruses tropic for cells of the immune system enjoy a selective advantage for enabling their own persistence. The outcome of such virus-lymphocyte interactions can be the heightened or hyperimmune response or partially to completely suppressed immune responses. It follows that 1) any unexplained defect in immunoresponsiveness (either elevated or depressed) should be studied with an infectious etiology in mind, and 2) the analysis of virus-lymphocyte interactions and experimental replacement of defective lymphocytes should lead to the understanding of persistent infection for the eventual treatment and elimination of virus that persist, thereby preventing the accompanying chronic, degenerative, and demyelinating diseases.

REFERENCES

1. McChesney MB, Oldstone MBA. Viruses perturb lymphocyte functions: selected principles characterizing virus induced immunosuppression. *Annu Rev Immunol* 1987;5:279–304.
2. Doyle MV, Oldstone MBA. Interactions between viruses and lymphocytes. I. *In vivo* replication of lymphocytic choriomeningitis virus in mononuclear cells during both chronic and acute viral infections. *J Immunol* 1978;121:1262–1269.
3. Casali P, Rice GPA, Oldstone MBA. Viruses disrupt functions of human lymphocytes: effects of measles virus and influenza virus on lymphocyte-mediated killing and antibody production. *J Exp Med* 1984;159:1322–1337.
4. Ahmed R, Salmi A, Butler LD, Chiller JM, Oldstone MBA. Selection of genetic variants of lymphocytic choriomeningitis virus in spleens of persistently infected mice: role in suppression of cytotoxic T lymphocyte response and viral persistence. *J Exp Med* 1984;160:521–540.
5. Southern PJ, Blount P, Oldstone MBA. Analysis of persistent virus infections by *in situ* hybridization to whole-mouse sections. *Nature* 1984;312:555–558.
6. Oldstone MBA, Blount P, Southern PJ, Lampert PW. Cytoimmunotherapy for persistent virus infection: unique clearance pattern from the central nervous system. *Nature* 1986;321:239–243.
7. Schrier RD, Oldstone MBA. Recent clinical isolates of cytomegalovirus suppress human cytomegalovirus-specific human leukocyte antigen-restricted cytotoxic T-lymphocyte activity. *J Virol* 1986;59:127–131.
8. McChesney MB, Fujinami RS, Lampert PW, Oldstone MBA. Viruses disrupt functions of human T lymphocytes. II. Measles virus suppresses antibody production by acting on B lymphocytes. *J Exp Med* 1986;163:1331–1336.
9. Oldstone MBA, Salvato M, Tishon A, Lewicki H. Virus-lymphocyte interactions. III. Biologic parameters of a virus variant that fails to generate CTL and establishes persistent infection in immunocompetent hosts. *Virology* 1988;164:507–516.
10. Salvato M, Shimomaye E, Southern P, Oldstone MBA. Virus-lymphocyte interactions. IV. Molecular characterization of LCMV Armstrong (CTL⁺) and that of its variant, Clone 13 (CTL⁻). *Virology* 1988;164:517–522.

DISCUSSION

One discussant asked how viruses escape immunosurveillance, and what mechanisms are involved. This discussant speculated that this might be dependent upon variance in cytotoxic lymphocytes (CTLs). In this connection the discussant wondered what made CTLs attack and if this was variant-specific or effected at the level of induction of CTLs. Another discussant pointed out that the number of cells required to respond to virus is not known, perhaps being 1 in 1000 to 1 in 5000 lymphocytes. This discussant also pointed out that it is not known whether the virus is destroyed by that cell, or if the virus might be hiding in lymphocytes to escape immunosurveillance. Exactly how virus mediates immunosuppression is not known.

A participant inquired why, when infected early in life, CTLs would not recognize variants. Another participant was not sure how animals ever become (immunologically) tolerant. This participant pointed out that a vaccine wouldn't work in persistently infected animals, but that immunological reconstitution would work consistently. The vaccine then would prevent duplication of the deficit. The number of CTLs generated in response to antigen is very small compared with the amount of antigen present, thus the response may become exhausted. Virus is cleared in animals normally, but infected animals become persistently infected when the host CTLs are insufficient. Such responses may not be evident because of antigen excess.

Human Retrovirology: HTLV,
edited by William A. Blattner.
Raven Press, Ltd., New York © 1990.

The *Trans*-Regulatory Proteins of HTLV-I: Analysis of Tax and Rex

Warner C. Greene, Dean W. Ballard, *Ernst Böhnlein,
Laurence T. Rimsky, Sarah M. Hanly, Jerome H. Kim,
Michael H. Malim, and Bryan R. Cullen

Howard Hughes Medical Institute, Departments of Medicine and Microbiology-Immunology, Duke University Medical Center, Durham, North Carolina 27710

INTRODUCTION

HTLV-I, a type C human retrovirus, has been implicated as the cause of the Adult T-Cell Leukemia (ATL) and more recently linked with a progressive demyelinating syndrome termed HTLV-I–associated myelopathy (HAM) or tropical spastic paraparesis (TSP) (1–4). ATL, an often aggressive and fatal tumor of CD4[+] T-lymphocytes, occurs in areas of the world where HTLV-I infection is endemic. Clinically, this leukemia may be present with tumor infiltrates of the skin, hypercalcemia, osteolytic bone lesions, lymphadenopathy, hepatosplenomegaly, and varying degrees of immunosuppression (5). In general, patients with the acute form of ATL survive only 3–6 mo. A uniform hallmark of cell lines derived from ATL patients or established by HTLV-I infection of normal T-cells is the high-level, constitutive expression of membrane receptors for interleukin-2. Fresh leukemic cells isolated from ATL patients may also display these IL-2 receptors, although the absolute levels have proven quite variable. These findings have raised the possibility that control of the IL-2/IL-2 receptor axis is altered in ATL and that the deregulation of this growth factor/receptor system may play an important role in leukemic transformation (6–9).

Insights into the pathophysiology of ATL have emerged with the molecular analysis of HTLV-I proviral isolates. Like other retroviruses, HTLV-I contains dual long terminal repeats (LTR) as well as *gag, pol,* and *env* structural genes (10). However, unlike the other acutely transforming animal retroviruses, HTLV-I lacks a classical oncogene. In addition, HTLV-I does not appear to transform T cells by integrating at specific sites in the host genome. Together, these findings suggest that HTLV-I–induced immortalization of human T cells may involve a novel strategy. The first clue to this potential strategy emerged with the detection of the pX region of the virus positioned near the 3′ LTR (10,11). This pX region

---

* Current Address: Sandoz Research Institute, 1235 Vienna, Austria.

was found to encode at least two distinct nonstructural proteins via the translation of a doubly spliced, polycistronic mRNA (12–16). These two proteins are now referred to as Tax (formerly p40^x, *tat*-1) and Rex (formerly p27). We shall discuss salient features of the mechanism of action of each of these viral *trans*-regulatory proteins.

TAX PROTEIN OF HTLV-I: ACTIVATION OF SELECT CELLULAR GENES INVOLVED IN T-CELL GROWTH

Several groups have independently assembled data indicating that the IL-2Rα (Tac, p55) promoter is activated by the HTLV-I Tax protein in both transfected leukemic T-cell lines (Jurkat, HSB-2) and primary normal human T lymphocytes (6–9). Additionally, Inoue and colleagues (6) have reported that transient expression of the *tax* gene in Jurkat or HSB T cells leads to modest induction of the endogenous IL-2Rα gene. To analyze the persistent effects of Tax on the IL-2Rα gene, we (17) have prepared Jurkat T-cell lines stably transfected with an expression vector encoding the Tax and Rex proteins of HTLV-I. Analysis of these cells revealed the constitutive production of biologically active Tax protein as well as the induction and sustained expression of the IL-2Rα and GM-CSF cellular genes. In contrast, control cell lines containing anti-sense Tax cDNA failed to express these cellular genes. Of note, induction of the Tax cell lines with various T-cell stimuli such as OKT3, calcium ionophore, or PHA led to marked increases (100- to 1000-fold) in the expression of the IL-2 gene. These findings suggest that Tax only partially activates the IL-2 gene. In contrast, Tax appears able to fully activate the IL-2Rα gene.

Tax Activation of the IL-2Rα Gene Is Indirect, Involving the Induced Nuclear Expression of Cellular DNA Binding Proteins That Interact with a κB-like Enhancer

These matched sense and anti-sense Tax cell lines provided an excellent experimental system to investigate the biochemical basis for Tax action. Transfection of these cells with a progressively deleted series of 5' mutants of the IL-2Rα promoter revealed that sequences located upstream of base −266 were largely dispensable for Tax-induced activation (18,19). However, further deletion to base −248 virtually ablated Tax inducibility without significantly affecting basal transcription. This pattern of Tax responsiveness found in Jurkat cells was recapitulated in studies of primary human T cells. In contrast to these results with Tax, IL-2Rα promoter induction mediated by phorbol esters (PMA) or tumor necrosis factor-alpha (TNF-α) in either Jurkat T cells or primary T lymphocytes required additional 5' sequences located between bases −281 and −266 (19–21).

Radiolabeled oligonucleotide probes spanning the IL-2Rα promoter region between bases −291 and −245 were used to explore the potential binding of inducible *trans*-acting proteins in this region. In gel retardation assays, the nuclear extracts from PMA, PHA, TNF-α-induced, or Tax-expressing Jurkat T cells formed two sequence-specific, DNA-protein complexes with this −291/−245 IL-2Rα probe (18–22). These complexes were not detected with extracts from either uninduced Jurkat T cells or Jurkat cells stably transfected with the anti-sense Tax expression plasmids. Footprinting (18,24) and methylation interference studies revealed the same binding site for an inducible protein(s) in both of these retarded complexes located between bases −266 and −256 (GGGAATCTCCC).

Site-directed mutagenesis of this IL-2Rα binding site in the context of the functional −317 construct led not only to the loss of protein binding but also to the lack of inducibility by PMA, TNF-α, and Tax (21,22,26). Furthermore, insertion of 12 bp, 17 bp, or 47 bp oligonucleotides containing this IL-2Rα binding site upstream of the mitogen-unresponsive, herpes simplex virus thymidine kinase (TK) promoter proved sufficient to confer PMA, TNF-α, and Tax inducibility on this transcription unit (18–21). This IL-2Rα promoter element also exhibited enhancer-like properties as it functioned in both orientations and duplication of the binding site led to amplified stimulatory effects.

Comparison of the sequence of this IL-2Rα promoter binding site with other known DNA binding motifs revealed a striking homology with the site recognized by NF-κB. NF-κB is a DNA binding factor(s) first shown to interact with the enhancer of the kappa light chain immunoglobulin gene (25). Recent studies, however, have revealed that this inducible factor is expressed in cells other than B lymphocytes, including phorbol-ester-activated T cells. Of note, NF-κB-like binding sites have also been identified in a variety of cellular and viral enhancer elements (class I MHC genes, the β_2 microglobulin gene, HIV-1, SV40, and CMV) (25–28).

To identify and biochemically characterize the *trans*-acting proteins interacting with the IL-2Rα κB element, two experimental approaches have been employed. These include ultraviolet light–induced crosslinking of DNA-protein complexes and DNA affinity precipitation (DNAP) assays performed with radiolabeled cellular proteins and biotinylated oligonucleotide probes. In the UV crosslinking experiments, we have demonstrated the binding of at least two different proteins (50–55 kD and 80–90 kD) at the IL-2Rα κB element. In the DNA affinity precipitation assays, not only were two proteins consistent in size with these species identified (51 and 85 kD) but also proteins migrating at 65, 70, and 71 kD (22,24). Because this assay has the capability of detecting both DNA-protein and protein-protein interactions, we suspect that the 65-, 70-, and 71-kD species may be involved in protein-protein interactions with the 51- and 86-kD proteins that directly bind to nucleotides within the κB element (23,24). These latter two species correlate in size with the proteins detected in the crosslinking assays.

Recently, we have identified additional regulatory sequences in the IL-2Rα promoter that immediately flank the κB enhancer. One element is positioned immediately upstream of the κB motif (−279 to −268); this element is required for PMA- and TNF-α-mediated activation but appears largely dispensable for Tax induction (19–21,24). Protein binding at this upstream element has been detected in solution exonuclease III assays but not by gel retardation. This finding suggests a low affinity interaction (24). However, DNAP assays have demonstrated the specific binding of a 56-kD protein at this upstream site. In contrast to the κB-specific factors, this upstream factor is constitutively produced (24). In addition to the upstream element, a second regulatory sequence closely related to an Spl site is located immediately downstream of the κB element (R. Sen, et al., unpublished data). Crosslinking at this site captures three constitutively expressed DNA binding proteins, including 100-, 82-, and 76-kD species (D. W. Ballard, et al., unpublished data). It remains unknown whether the smaller proteins correspond to degraded forms of the 100-kD species, which is consistent in size with the eukaryotic Spl transcription factor. Mutations within this downstream element inhibit activation induced by PMA, TNF-α, and Tax. Together, these findings highlight the complex nature of IL-2Rα gene activation that is regulated by the specific binding of multiple inducible and constitutive factors to adjacent regions of the IL-2Rα promoter. These studies also underscore the indirect nature of Tax activation of the IL-2Rα gene involving the induction of κB-specific cellular proteins that in turn bind to the IL-2Rα promoter. Recent studies suggest that these same proteins bind to a related element in the IL-2 gene, which may contribute to the coordinate activation of this growth factor (29). Activation of IL-2 and its receptor in CD4$^+$ T cells may play a role in T-cell immortalization induced by HTLV-I; however, this process almost certainly involves multiple steps. The Tax protein may act early in this process, facilitating polyclonal proliferation of the T cells. In addition to the cellular genes, Tax also activates the HTLV-I LTR. This action of Tax appears to involve yet another set of cellular DNA binding proteins and unique 21-bp enhancer elements. These findings emphasize the capacity of this viral transactivator to interface with multiple cellular transcription factor pathways.

REX PROTEIN OF HTLV-I

The *rex* gene of HTLV-I encodes the 27-kD Rex phosphoprotein that is primarily localized in the nucleoli of expressing cells (30). Rex is produced at high levels early in the viral life cycle as the translated product of a 2.1-kb, doubly spliced, polycistronic mRNA that also encodes the Tax protein from a different reading frame (12,13,31). Though it is clear that both the *rex* and *tax* gene products are required for viral replication, the mechanism of *rex* action has remained poorly defined. Recent studies support the notion that Rex acts entirely at a posttranscriptional level (32,33). Rex is obligately required for the expression of the unspliced, 8.5-kb genomic mRNA encoding the structural gene products of *gag* and *pol,* and the singly spliced, 4.2-kb mRNA encoding the *env* gene (31,32). Simultaneous with the switch to expression of these structural gene products, *rex* decreases the level of the doubly spliced mRNA encoding the regulatory proteins Tax and Rex (31,32). Seiki *et al.* (33) have suggested that two *cis*-acting elements are required for Rex action; one region corresponds to a portion of the R region within the 3' LTR while the second element corresponds to a 5' splice donor site.

HTLV-I Rex Protein Can Functionally Replace the HIV-1 Rev Protein

Many of these properties of the HTLV-I Rex protein resemble those attributed to the Rev protein of HIV-1. Although these proteins are devoid of primary sequence homology, we have investigated the ability of Rex to substitute functionally for Rev (34). These studies utilized an assay (35) that exploits the capacity of Rev to induce the production of a truncated single exon form of the HIV-1 Tat protein (72 aa rather than 86 aa, reflecting the loss of the second coding exon). Briefly, in the presence of Rev, unspliced mRNA derived from a genomic *tat* expression vector is expressed in the cytoplasm, resulting in the induced synthesis of the 72-aa Tat protein. In the absence of Rev, the same genomic *tat* vector results in exclusive synthesis of the full-length 86-aa Tat protein. Surprisingly, Rex, like Rev, proved capable of inducing the expression of the truncated Tat protein. Furthermore, Rex rescued in trans the replication of *rev*-deficient mutants of HIV-1. The ability of the HTLV-I Rex protein to substitute for its HIV-1 counterpart, Rev, suggests a remarkable evolutionary conservation of this novel posttranscriptional regulatory pathway of eukaryotic gene expression.

Rev Cannot Replace Rex, Indicating a Nonreciprocal Pattern of Genetic Complementation by the Viral Regulatory Proteins

Previous studies of the Rev-response element (RRE) present in the genomic *tat* construct described above identified a region in the *tat* intron that coincided with a highly stable, predicted stem-loop RNA structure. Deletion of this element in the pgTAT vector (pgΔTAT) led to the loss of both Rev and Rex responsiveness (36). When the 3' LTR of HTLV-I was inserted at the 3' end of the pgΔTAT construct in lieu of rat preproinsulin polyadenylation sequences, *rex,* but not *rev,* responsiveness was reconstituted. Thus, whereas Rex can replace Rev, Rev cannot apparently replace Rex. These findings highlight a nonreciprocal pattern of genetic complementation which strengthens the possibility that these viral proteins directly interact with their respective RNA response elements.

Precise Identification of the HTLV-I Rex-Response Element

Given the similar action and functional complementarity of *rev* and *rex,* we analyzed the HTLV-I LTR for the presence of mRNA secondary structures resembling that of the *rev* response element present in the HIV-1 *env* gene (36). A thermostable stem-loop structure (279 nt) was identified and subcloned using the polymerase chain reaction. Insertion of this element within the *tat* intron of pgΔTAT resulted in Rex but not Rev responsiveness. When the RexRE was placed 3' of the polyadenylation site in the vector, Rex-responsiveness was not obtained. Similarly, the Rex response element failed to function in an anti-sense orientation, implicating its action at an RNA rather than a DNA level. Of note, the Rex protein of HTLV-II proved capable of acting through both the HTLV-I Rex and HIV-1 Rev response elements.

Rex Protein Enhances Transport of Unspliced or Partially Spliced Viral mRNAs from the Cell Nucleus to the Cytoplasm

Previous studies have not resolved whether the effects of Rex are mediated by direct suppression of mRNA splicing or by the facilitated transport of unspliced mRNA (33). Our S1 nuclease protection analyses favor the latter possibility. Analysis of viral RNA in the nucleus revealed no difference in spliced and unspliced viral mRNA levels in the presence or absence of the Rex protein. Unexpectedly high levels of unspliced mRNA were observed in the nucleus in the absence of Rex, which argues against a role for Rex as an inhibitor of splicing and suggests that the splice sites in HTLV-I, like those of other retroviruses, are inefficiently utilized. In contrast, when cytoplasmic RNAs were analyzed, virtually no unspliced viral RNA species were evident in the absence of Rex. However, high levels of unspliced mRNAs were found when Rex was present. This overall increase in unspliced RNA within the cell is compensated for by decreases in spliced transcripts. Thus the total amount of RNA is unaffected by Rex. This finding argues against models of Rex action involving changes in transcription or RNA stability. Rather, these findings suggest that Rex promotes the redistribution of unspliced mRNA from the cell nucleus to cytoplasm. Similar effects on unspliced viral mRNA transport have been described for the HIV-1 Rev protein (36). The localization of Rex within the nucleoli of expressing cells and its effect on viral mRNA transport raise still more interesting questions about its mechanism of action. We note that ribosomal RNA transport involves critical functions by proteins present in the nucleolus. It is possible that Rex allows utilization of this RNA transport pathway by the HTLV-I structural gene transcripts.

CONCLUSIONS

Tax and Rex each play an important role in the life cycle of HTLV-I and presumably in the pathogenesis of the adult T-cell leukemia. The expression of Tax transcriptionally activates replication of the virus but also induces the partial or complete expression of certain cellular genes that regulate T-cell growth (IL-2, IL-2Rα). The Rex protein facilitates the nuclear export of structural gene mRNAs acting via a Rex response element located in the 3′ LTR of the retrovirus. This element coincides with a complex RNA stem-loop structure that must be represented in the target RNAs in the correct orientation. The surprising ability of the HTLV-I Rex protein to replace the function of the HIV-1 Rev protein emphasizes the important evolutionary conservation of the pathway. The inability of Rev to replace Rex highlights the nonreciprocal nature of the interplay of these viral proteins. Future study will focus on the biochemical basis for the action of these viral trans-regulatory proteins and their role in T-cell transformation.

SUMMARY

The type I human T-cell leukemia virus (HTLV-I) encodes at least two trans-regulatory factors termed Tax and Rex. The 40 kD Tax protein both activates the HTLV-I LTR and induces the expression of select cellular genes involved in T-cell growth. Specifically, the genes encoding interleukin-2 (IL-2), the alpha subunit of the high-affinity interleukin-2 receptor (IL-2Rα), and granulocyte-macrophage colony stimulating factor (GM-CSF) are activated by Tax. Tax activation of this cellular transcription as well as the HTLV-I LTR appears to be indirect, involving various cellular DNA binding proteins. In the case of the IL-2Rα gene, Tax induces the nuclear expression of at least two host cell polypeptides that specifically interact with a κB-like enhancer present in the promoter of this gene. The 27-kD Rex protein, like Tax, is absolutely required for viral replication. In contrast to Tax, Rex acts at a posttranscriptional level, serving to activate the nucleocytoplasmic transport of the unspliced or singly spliced viral mRNAs that encode the virion structural proteins. These effects of Rex are sequence specific, involving an RNA Rex response element located in the 3′ retroviral LTR. This viral response element coincides with an energetically stable predicted RNA stem-loop structure in the viral RNA. The Rex protein not only functions in the HTLV-I system but also is capable of substituting for the apparently unrelated Rev protein of the Type 1 human immunodeficiency virus. We have recently found that this surprising genetic complementation of Rex and Rev is nonreciprocal, as the HIV-1 Rev protein is unable to act in the HTLV-I system.

ACKNOWLEDGMENTS

We thank Ms. B. Kissell and Ms. S. Goodwin for secretarial assistance in the preparation of this manuscript.

REFERENCES

1. Poiesz BJ, Ruscetti FW, Gazdar AF, Bunn PA, Minna JD, Gallo RC. Detection and isolation of type-C retrovirus particles from fresh and cultured lymphocytes of a patient with cutaneous T-cell lymphoma. *Proc Natl Acad Sci USA* 1980;77:7415.
2. Poiesz BJ, Ruscetti FW, Reitz MS, Kalyanaraman VS, Gallo RC. Isolation of a new type-C retrovirus (HTLV) in primary uncultured cells of a patient with Sezary T-cell leukemia. *Nature* 1981;254:268.
3. Yoshida M, Miyoshi IL, Hinuma Y. Isolation and characterization of retrovirus from cell lines of human T cell leukemia and its implication in disease. *Proc Natl Acad Sci USA* 1982;78:6476.
4. Gessain A, Barin F, Vernant JC, et al. Antibodies to human T-lymphotropic virus type-I in patients with tropical spastic paraparesis. *Lancet* 1985;2:407.
5. Broder S, Bunn PA, Jaffe ES, et al. T-cell lymphoproliferative syndrome associated with human T-cell leukemia/lymphoma virus. *Ann Intern Med* 1984;100:543.
6. Inoue J, Seiki M, Taniguchi T, Tsuru S, Yoshida M. Induction of interleukin-2 receptor gene expression by p40^x encoded by human T cell leukemia virus-type I. *EMBO J* 1986;5:2883.
7. Cross SL, Feinberg MB, Wold JB, Holbrook NJ, Wong-Staal F, Leonard WJ. Regulation of the human interleukin-2 receptor α promoter by the *trans*-activator gene of HTLV-I. *Cell* 1987;49:47.
8. Maruyama M, Shibuya H, Harada H, et al. Evidence for aberrant activation of the interleukin-2 autocrine loop by HTLV-I–encoded p40^x and T3/Ti complex triggering. *Cell* 1987;48:343.
9. Siekevitz M, Feinberg MB, Holbrook N, Yodoi J, Wong-Staal F, Greene WC. Activation of interleukin-2 and interleukin-2 receptor (Tac) promoter expression by the *trans*-activator (tat) gene product of human T-cell leukemia virus, type I. *Proc Natl Acad Sci USA* 1987;84:5389.
10. Seiki M, Hattori S, Hirayama Y, Yoshida M. Human adult T-cell leukemia virus: complete nucleotide sequence of the provirus genome integrated in leukemia cell DNA. *Proc Natl Acad Sci USA* 1983;80:3618.
11. Sodroski JG, Rosen CA, Haseltine WA. *Trans*-acting transcriptional activation of the long terminal repeat of human T lymphotropic viruses in infected cells. *Science* 1984;225:381.
12. Lee TH, Colligan JE, Sodroski JG, et al. Antigens encoded by the 3' terminal region of human T-cell leukemia virus: evidence for a functional gene. *Science* 1984;226:57.
13. Slamon DJ, Shimotohno K, Cline MJ, Golde DW, Chen ISY. Identification of the putative transforming protein of the human T-cell leukemia viruses HTLV-I and HTLV-II. *Science* 1984;226:61.
14. Kiyokawa T, Seiki M, Iwashita S, Imagawa K, Shimizu F, Yoshida M. p27$^{x\text{-III}}$ and p21$^{x\text{-III}}$, proteins encoded by the pX sequence of the human T-cell leukemia virus type I. *Proc Natl Acad Sci USA* 1985;82:8359.
15. Seiki M, Hikikoshi A, Taniguchi T, Yoshida M. Expression of the pX gene of HTLV-I: general splicing mechanism in the HTLV family. *Science* 1985;228:1532.
16. Wachsman W, Golde DW, Temple PA, Orr EC, Clark SC, Chen ISY. HTLV x-gene product: requirement for the env methionine initiation codon. *Science* 1985;228:1534.
17. Wano Y, Feinberg M, Hosking JB, Bogerd H, Greene WC. Stable expression of the tax gene of type I human T-cell leukemia virus in human T cells activates specific cellular genes involved in growth. *Proc Natl Acad Sci USA* 1988;85:9733.
18. Ballard DW, Böhnlein E, Lowenthal JW, Wano Y, Franza BR, Greene WC. HTLV-I tax induces cellular proteins that activate the κB element of the IL-2 receptor α gene. *Science* 1988;241:1652.
19. Lowenthal JW, Böhnlein E, Ballard DW, Greene WC. Regulation of interleukin 2 receptor α subunit (Tac or CD25 antigen) gene expression: binding of inducible nuclear proteins to discrete promoter sequences correlates with transcriptional activation. *Proc Natl Acad Sci USA* 1988;85:4468.
20. Lowenthal JW, Ballard DN, Böhnlein E, Greene WC. Tumor necrosis factor α induces κB-binding proteins that regulate interleukin 2 receptor α-chain gene expression in primary human T-lymphocytes. *Proc Natl Acad Sci USA* 1989;86:2331–2335.
21. Lowenthal JW, Ballard DW, Bogerd H, Böhnlein E, Greene WC. TNF-α activation of the IL-2 receptor α gene involves the induction of κB-specific DNA binding proteins. *J Immunol* 1989;142:3121–3128.
22. Böhnlein E, Lowenthal JW, Siekevitz M, Ballard DW, Franza BR, Greene WC. The same inducible

nuclear proteins regulate mitogen activation of both the interleukin-2 receptor-alpha gene and type 1 HIV. *Cell* 1988;53:827.
23. Böhnlein E, Siekevitz M, Lowenthal JW, et al. HTLV-I, HIV-1, and T-cell activation. In: Franza BR, Cullen BR, Wong-Staal F, eds. *The Control of Human Retrovirus Gene Expression.* Cold Spring Harbor: Cold Spring Harbor Laboratory, 1988;191.
24. Ballard DW, Lowenthal JW, Böhnlein E, Hoffman JA, Franza RB, Greene WC. (*submitted*).
25. Sen R, Baltimore D. Inducibility of κ immunoglobulin enhancer-binding protein NF-κB by a post-translational mechanism. *Cell* 1986;47:921.
26. Israel A, Kimura A, Kievan M, et al. A common positive *trans*-acting factor binds to enhancer sequences in the promoters of mouse H-2 and β2 microglobulin genes. *Proc Natl Acad Sci USA* 1987;84:2653.
27. Nabel G, Baltimore D. An inducible transcription factor activates expression of human immunodeficiency virus in T cells. *Nature* 1987;326:711.
28. Baldwin AS, Sharp PA. Two factors, NF-κB and H2TF1, interact with a single regulatory sequence in the class 1 MHC promoter. *Proc Natl Acad Sci USA* 1988;85:723.
29. Hoyas B, Ballard DW, Böhnlein E, Siekevitz M, Greene WC. Kappa B specific DNA binding proteins regulate activation of the human interleukin-2 gene. *Science* 1989;244:457–460.
30. Siomi H, Shida H, Nam SH, Nosaka T, Maki M, Hatanaka M. Sequence requirements for nucleolar localization of human T-cell leukemia virus type I pX protein, which regulates viral RNA processing. *Cell* 1988;55:197.
31. Hidaka M, Inoue J, Yoshida M, Seiki M. Post-transcriptional regulator (rex) of HTLV-I initiates expression of viral structural proteins but suppresses expression of regulatory proteins. *EMBO J* 1988;7:519.
32. Inoue J, Yoshida M, Seiki M. Transcriptional (p40^x) and post-transcriptional (p27^{x-III}) regulators are required for the expression and replication of human T-cell leukemia virus type I genes. *Proc Natl Acad Sci USA* 1987;84:3653.
33. Seiki M, Inoue J, Hidaka M, Yoshida M. Two *cis*-acting elements responsible for posttranscriptional *trans*-regulation of gene expression of human T-cell leukemia virus type I. *Proc Natl Acad Sci USA* 1988;85:7124.
34. Rimsky L, Hauber J, Dukovich M, et al. Functional replacement of the HIV-1 rev protein by the HTLV-I rex protein. *Nature* 1988;335:738.
35. Malim MH, Hauber J, Fenrick R, Cullen BR. Immunodeficiency virus *rev trans*-activator modulates the expression of the viral regulatory genes. *Nature* 1988;335:181.
36. Malim MH, Hauber J, Shu-Yun Le, Maizel JV, Cullen BR. The HIV-1 *rev trans*-activator acts through a highly structured target sequence to activate nuclear export of unspliced viral mRNA. *Nature* 1989;338:254.

DISCUSSION

A discussant inquired as to the ability to differentiate cytoplasmic instability of RNA from rex transport from the nucleus. Another discussant replied that one could not totally exclude the possibility that as soon as the viral RNA is transported from the nucleus to the cytoplasm it is immediately degraded but that one would then have to invoke a compartmentalized degradation model linked to transport. One participant commented on the model for HTLV-I carcinogenesis that was presented, stating that HTLV-I-positive cells could act as a mitogenic stimulus and then, through transactivating genes, cause cell proliferation, with a possible third event immortalizing the infected cells. It was asked if, in this context, the *tax* gene could immortalize cells. The previous discussant replied that transferring and expressing this gene to primary T cells is difficult, and thus this question is unanswered at the present time.

Human Retrovirology: HTLV,
edited by William A. Blattner.
Raven Press, Ltd., New York © 1990.

HTLV Host Cell Range and Its Receptor: Implications for Pathogenesis

A. G. Dalgleish and J. Richardson

MRC. Clinical Research Centre, Harrow, HA1 3UJ, United Kingdom

HTLV-I has a broader host cell range in vitro and probably in vivo than its name implies (1). The limitations placed on a retrovirus trying to enter and replicate in a cell include the binding penetration and fusion events (which involve a specific receptor) and the ability to integrate into the host's genome and subsequently to be expressed—features dependent on availability of appropriate cellular factors interacting with the long terminal repeat regions of the provirus.

HTLV-I and HIV are both T-cell tropic. Whereas HIV uses the CD4 molecule as its cellular receptor (2), HTLV does not. However, the nature of the HTLV-I receptor remains unclear. A feature of cells that express viral receptors on their surface is their ability to fuse with virus-infected cells to form giant, multinucleated cells known as syncytia. The ability to form such giant cells has been used as a simple screening method for infectability. Another method designed to overcome the inability of many retroviruses to be titrated by a quantitative assay is the vesicular stomatitis virus (VSV) pseudotype method first described by Zavada (3). Briefly, a pathogenic virus (VSV) is grown in cells infected with the retrovirus under study (HTLV-I). Some virus particles will be hybrid or transvestite particles (one in several thousand), consisting of VSV core particles and bearing HTLV-I envelope glycoproteins as the outer envelope (pseudotype particles). If the cells on which these VSV(HTLV-I) pseudotypes are plated express the specific receptor for HTLV-I, the pseudotype will enter, whereupon the VSV genome replicates. This can be used in a plaque assay for neutralization and to screen receptor expression on various cell types. Binding assays have also been described. Infectivity can also be assessed by immunofluorescence using a specific monoclonal antibody, i.e., p19, or infected sera. Replication can be assessed by measuring reverse transcriptase production.

Using these systems, the host cell range for HTLV-I has been examined. Briefly, a wide variety of human cells are infectable including many nonlymphoid cells. Not only can nonlymphoid cells be infected but they may be capable of producing high levels of virus, as has been demonstrated by the HOS cell line (4).

RECEPTOR INTERFERENCE

Cells chronically infected with retroviruses express viral envelope glycoproteins that mask or down-modulate receptor expression at the cell surface, rendering the cell resistant to superinfection by viruses sharing the same receptor. This phenomenon is known as receptor interference and has been used to demonstrate that HTLV-I and HTLV-II use a common receptor (1,5).

Chromosomal Assignment of HTLV-I Receptor

The plating efficiency of VSV(HTLV-I) pseudotypes on mouse cells is $\sim 1\%$ that of human cells, which has allowed the use of human-mouse somatic-cell hybrids to assign the receptor gene to a human chromosome, in a manner similar to that reported previously for other viruses. The susceptibility of these hybridomas to VSV(HTLV-I) infection has been reported by Sommerfelt and colleagues (5). Human chromosome 17 was the only common chromosome among the infectable hybridomas, although it was present in 2 of 15 cell lines which were resistant to infection. Expression of a known chromosome 17 cell antigen was demonstrated on both cell lines. The cell lines carrying only chromosome 17 were independently derived and both were infectable to both VSV(HTLV-I) and VSV(HTLV-II). Regional localization of the receptor gene to 17cen-qter is suggested from studies using hybridomas with deleted portions of chromosome 17. NGL (also known as the *C-EbB-2* or *neu* oncogene); CD7, a lymphocyte antigen; nerve growth factor receptor (NGFR); and MIC6 are all cell surface antigens coded by genes on chromosome 17. However, antibodies to NGL, MIC6, and CD7 failed to block HTLV syncytial induction (5). Transgenic mice expressing the *tax* gene of HTLV-I develop a syndrome resembling neurofibromatosis, similar to those seen in von Recklinghausen's neurofibromatosis, the coding gene of which has also been mapped to chromosome 17. This may be a chance observation or some undescribed and as yet not understood interaction between HTLV-I, its receptor, and the gene for neurofibromatosis. HTLV-I is not associated with neurofibromatosis in humans, although this association is interesting in that both neurological and proliferative diseases are recognized [tropical spastic paraparesis (TSP) and adult T cell leukemia (ATL)].

Strategies for Identification of the Receptor

To date, studies aimed at determining the HTLV-I receptor using whole virions do not appear to have been successful. A number of workers are now trying to express HTLV-I recombinant gp46 envelope in order to develop new generation ELISA assays as well as to examine its role in the development of vaccines. Pure envelope will enable the development of binding, cross-linking, and co-immunoprecipitation assays to identify receptor molecules; this may allow cloning of the genes through expression-vector strategies, not to mention the ability to screen large numbers of human-mouse hybridomas. Other approaches include 1) raising monoclonal antibodies to hybridomas which appear to express the HTLV-I receptor; 2) using an anti-idiotype approach against an anti-envelope monoclonal antibody in a similar method to that used to determine the Rheovirus receptor (6); and 3) directly cloning the gene, as recently described for other virus receptors (7).

The HIV receptor (CD4) was identified following the observation that monoclonal antibodies against CD4 inhibited the development of Syncytia in susceptible host cells (2). A similar approach has been tried with HTLV-I. A large number of monoclonal antibodies—which includes the available "CD" monoclonals obtained from the III International Leukocyte Workshop, as well as antibodies against the IL-2 and transferrin receptors against human and other species' cellular ligands—failed to inhibit syncytial formation.

HTLV-I: ITS ROLE IN THE ETIOLOGY OF NEUROLOGICAL AND MALIGNANT DISEASE IN MAN

It is not known how HTLV-I causes ATL or TSP. However, the following facts limit the potential possibilities: 1) HTLV-I does not encode any human oncogene; 2) HTLV-I does not integrate into any specific genomic site (8); 3) the *tat* (*tax*) gene can positively regulate IL-2, its receptor (IL-2R), GM-CSF, and the *c-fos* oncogene (W. Greene, this volume); 4) the activation by turning on IL-2 and IL-2R does not lead to an autocrine growth mechanism for malignancy, as both IL-2 and IL-2R are down-regulated in the malignant cells (9) (in contrast to HTLV-I infected cells); and finally, 5) ATL progresses from a polyclonal to a monoclonal proliferation with a propensity to displaying a translocation of chromosome 14, which is also common in non-HTLV-I T-cell malignancies (10).

As only ~1–5% of HTLV-I–infected people will develop ATL or TSP, it is likely that other factors are important. The perpetual activation of T helper (CD4) cells may make them more susceptible to a second "oncogenic" event, such as a chance translocation or exposure to a second oncogenic virus or carcinogens. Genetic factors such as HLA have also been reported to be associated with both susceptibility to infection and progression to disease (11), although the significance of this association remains unclear.

With regards to the development of TSP, it would appear that the same virus causes both ATL and TSP. Sequence variations in the envelope of isolates from both conditions occur, but no specific site on the envelope has been implicated as being associated with TSP or ATL. The in vivo host cell range has yet to be fully described, so it is not known whether neurological tropism is the cause of TSP (see Saida et al., this volume). PCR and in situ hybridization from post mortem material may help resolve this question. However, the immune response to HTLV-I is greater in TSP patients than asymptomatic infected patients or most patients with ATL (11–13). PCR analysis of DNA from the peripheral blood of TSP patients gives a strong signal, suggesting a large number of virus copies.

It is interesting to remember that HTLV-I can infect endothelial cells in vitro (14,15) and that these may act as a reservoir of infected cells in vivo, which can then infect CD4 cells at a later date. Interestingly, cell-free transmission of HTLV-I to endothelial but not to lymphoid cells has been described (14). These observations may be important in understanding the increasing incidence of seroconversion with age, mentioned in the chapter by Kennedy et al.

As the mechanisms whereby HTLV-I causes ATL and TSP are slowly unwound and understood, it is likely that new insights into the pathogenesis of other similar diseases will be acquired.

REFERENCES

1. Weiss RA, Clapham PA, Nagy K, Hoshino M. Envelope properties of human T-cell leukaemia viruses. *Curr Top Microbiol Immunol* 1985;115:235–245.
2. Dalgleish AG, Beverley PC, Clapham PR, Crawford DH, Greaves MR, Weiss RA. The CD4 (T4) antigen is an essential component of the receptor for the AIDS retrovirus. *Nature* 1984;312: 763–767.
3. Zavada J. Assay methods for viral pseudotypes. *Meth Virol* 1977;6:109–142.
4. Clapham P, Nagy K, Cheingsong-Popov R, Exley M, Weiss RA. Productive infection and cell free transmission of human T cell leukaemia virus in a nonlymphoid cell line. *Science* 1983;222: 1125–1127.
5. Sommerfelt MM, Williams BP, Clapham PR, Solomon E, Goodfellow PN, Weiss RA. Human T cell leukaemia viruses use a receptor determined by human chromosome 17. *Science* 1989;242: 1557–1559.
6. Noseworthy JM, Fields BN, Dichter MS, et al. Syngeneic monoclonal anti-idiotypic antibody identifies the cell surface receptor for reoviruses. *J Immunol* 1983;131:2533–2543.
7. Mendelsohn CL, Wimmer E, Racaniello VR. Cellular receptor for poliovirus: molecular cloning, nucleotide sequence and expression of a new member of the immunoglobulin superfamily. *Cell* 1989;56:855–865.
8. Seiki M, Eddy R, Shows TB, Yoshida M. Nonspecific integration of the HTLV provirus genome into adult T-cell leukaemia cells. *Nature* 1984;309:640–642.
9. Arya SK, Wong-Staal F, Gallo RC. T-cell growth factor gene: lack of expression in human T cell leukaemia lymphoma virus infected cells. *Science* 1984;223:1086–1088.
10. Sadamori N, Kusano M, Nishino K, et al. Abnormalities of chromosome 14 at band 14q11 in Japanese patients with adult T-cell leukaemia. *Cancer Genet Cytogenet* 1985;17:279–282.
11. Usuku K, Sonoda S, Osame M, et al. HLA Haplotype-linked high immune responsiveness against HTLV-I in HTLV-I associated myelopathy: comparison with adult T cell leukaemia/lymphoma. *Ann Neurol* 1988;23:S143–150.
12. Dalgleish AG, Richardson J, Matutes ES, et al. HTLV-I infection in TSP lymphocyte culture and serological response. *AIDS Res Hum Retroviruses* 1988;4(6):475–485.
13. Lolli F, Fredrickson S, Kam-Hansen S, Link H. Increased reactivity to HTLV-I in inflammatory nervous system diseases. *Ann Neurol* 1987;22:67–71.
14. Hoxie JA, Matthews DM, Clines DB. Infection of human endothelial cells by human T cell leukaemia virus type 1. *Proc Natl Acad Sci USA* 1984;81:7591–7595.
15. Ho DD, Rota JR, Hirsh MS. Infection of human endothelial cells by human T lymphotropic virus type 1. *Proc Natl Acad Sci USA* 1984;81:7588–7590.
16. Macchi BI, Popovic M, Allavera P, et al. In-vitro susceptibility of different human T cell sub-populations and resistance of large granular lymphocytes to HTLV-I infection. *Int J Cancer* 1987;40:1–6.

Human Retrovirology: HTLV,
edited by William A. Blattner.
Raven Press, Ltd., New York © 1990.

Molecular Variation of Human T-Lymphotropic Viruses and Clinical Associations

Lee Ratner

*Departments of Medicine and Molecular Microbiology,
Washington University, St. Louis, Missouri 63110*

INTRODUCTION

Members of the human T-lymphotropic virus (HTLV) family have been associated with a wide range of clinical disorders (1). The strongest associations are for HTLV-I with adult T-cell leukemia/lymphoma (ATL) and HTLV-I–associated myelopathy/tropical spastic paraparesis (HAM/TSP). HTLV-I, -II, or possibly variant viruses have also been associated with the development of B cell lymphomas in HIV-1 infected individuals, B cell chronic lymphocytic leukemia in the Caribbean, multiple sclerosis, mycosis fungoides, immunosuppression and infections by multiple opportunistic agents, bronchopneumopathy, large granular lymphocytic leukemia, immune thrombocytopenic purpura, atypical forms of hairy cell leukemia, a CD8 dermatolymphocytic infiltrative disorder, and vasculitis (see article by W. Hall, this volume). Furthermore, the majority of HTLV-infected individuals remain asymptomatic (2). Features which determine the absence or presence of clinical symptoms and the type and severity of symptoms remain to be defined. These factors may include host-specific and virus-specific factors.

Host-specific factors determining disease course may be divided into those which regulate individual steps of virus replication at a cellular level, and those which regulate the dissemination of virus or virus infected cells. Cellular regulation may occur at the step of viral entry by differences in the receptor on different cells or in different individuals, at transcriptional levels by tissue-specific differences in factors which up- or down-regulate viral RNA expression, or at any other step in the virus life cycle. Factors which may regulate dissemination of the virus include dose and site of viral entry into the body and immune regulators such as neutralizing antibodies, antibody-dependent cellular cytotoxicity, or cytotoxic T-cell activity.

Virus-specific factors which regulate qualitative or quantitative aspects of HIV-1 replication are based on sequence differences between strains. This will be the major focus of this review. First, a description of ATL cases in the United States will be presented. Familial occurrence of ATL will be highlighted and its implications for viral pathogenesis discussed. Second, molecular characterization of HTLV-I and -II strains will be provided. Third, the mechanism of generation of sequence heterogeneity of HTLVs will be discussed, and the mechanisms of restriction of sequence variation of HTLVs compared with human immunodeficiency viruses (HIV) will be emphasized. Fourth, the possible relationship of sequence variation to alterations in biological properties will be discussed.

ATL IN THE UNITED STATES

Among individuals born in the United States, there have been 39 cases of ATL reported in the literature (Table 1) (1). This includes both acute and chronic (or smoldering) forms of ATL. Chronic ATL includes HTLV-I–infected individuals in whom <10% of the nucleated blood cells are leukemic and <10% of the body surface is involved with a skin infiltrate, with or without the presence of lymphadenopathy which on biopsy is not clearly indicative of lymphoma. These individuals lack hypercalcemia and visceral organ involvement. Acute ATL is defined as a T-cell non-Hodgkin's lymphoma in a HTLV-I–infected individual with leukemic cells, hypercalcemia, or skin infiltrates.

This compilation includes seven cases described from our medical center in the last 3 y. These patients were identified from a group of 40 patients referred by local physicians for evaluation of T-cell leukemia, lymphoma, or lymphoproliferative skin disorders. Sera were screened by an HTLV-I ELISA and positive assays confirmed by virus culture, Southern blot hybridization, or polymerase chain amplification reaction (PCR) assays. For PCR assays, we have utilized primers identical or complementary to *rex* and *tax* sequences conserved between HTLV-I and -II (Fig. 1). Amplified products are confirmed as containing HTLV-I or -II products by hybridization with specific probes for each virus.

The finding of seven cases of ATL at our medical center in only 3 y suggested that either a) the prevalence of HTLV-I infection is significantly higher in the St. Louis area than the overall prevalence rate in the United States, b) the percent of ATL cases among HTLV-I infected individuals is higher in our region than other areas of the United States, c) the incidence of HTLV-I infection is increasing more rapidly in our area than in other areas of the country, and/or d) cases of ATL are more often recognized in our population than in other areas of the United States. The current prevalence of HTLV-I sero-positivity among blood donors in the St. Louis area is 0.025% (personal communication with L. Sherman), very similar to the prevalence in the United States overall. This would predict that there are ~250 HTLV-I–infected individuals in our population base of ~1 000 000 individuals. Assuming that 2%–4% of HTLV-I–infected patients develop ATL, one would predict 5–10 cases of ATL over the lifetime of the at-risk population. The number of cases of ATL seen in our area is clearly higher

TABLE 1. *Patients with ATL born in the United States*

Patient	Age	Sex	Race	Birthplace	ATL classification
CR	28	M	B	Alabama	Acute
WA	24	M	B	Georgia	Acute
PL	27	F	B	Florida	Acute
LJ	46	M	B	Washington, D.C.	Acute
OB	30	F	B	Georgia	Acute
BH	62	F	B	Texas	Acute
JN	76	M	Al	Alaska	Acute
MJ	49	M	C	Massachusetts	Chronic
1	16	F	C	Iowa	Acute
	38	M		New York	Acute
0005	50	M	B	Tennessee	Acute
0008	59	F	C	Massachusetts	Acute
0010	26	F	B	Florida	Acute
0011	58	M	C	New York	Acute
0014	63	M	B	South Carolina	Acute
0015	40	F	B	New York	Acute
0016	65	M	B	South Carolina	Acute
0017	50	F	B	South Carolina	Acute
0018	71	F	B	South Carolina	Acute
0019	51	F	B	North Carolina	Acute
0029	49	M	B	North Carolina	Acute
0032	74	M	B	South Carolina	Acute
0039	26	M	B	South Carolina	Acute
0056	58	M	A	Hawaii	Acute
0057	69	F	A	Hawaii	Acute
0061	30	M	B	Alabama	Acute
0062	55	M	B	Tennessee	Acute
0074	56	M	B	Louisiana	Acute
B11*	41	F	B	North Carolina	Chronic
B15*	38	M	B	North Carolina	Acute
B21*	38	M		North Carolina	Chronic
C5*		F		North Carolina	Chronic
1	35	M	B	Missouri	Intermediate
2	65	F	B	Tennessee	Acute
3	35	F	B	Illinois	Acute
4	31	F	B	Illinois	Chronic
5	7	F	B	Illinois	Acute
7†	43	F	B	Missouri	Acute
8†	32	M	B	Missouri	Acute

Abbreviations: M = male; F = female; B = Black; A = Asian; Al = Aleutian; C = Caucasian.
* C5 is the niece, B11 the sister, and B21 the husband of B15.
† Patients 7 and 8 were siblings.
Reprinted with permission from ref. 1.

than that predicted by these estimates. It is possible that the blood bank estimates of HTLV-I seroprevalence are significant underestimates of the seroprevalence in the overall population. Our recent data of HTLV seroprevalence rates among prisoners and urban hospital admissions are not consistent with this finding (unpublished observations). There are no data, however, to suggest an increasing incidence of HTLV in our area of the country or any other areas of the world.

a]

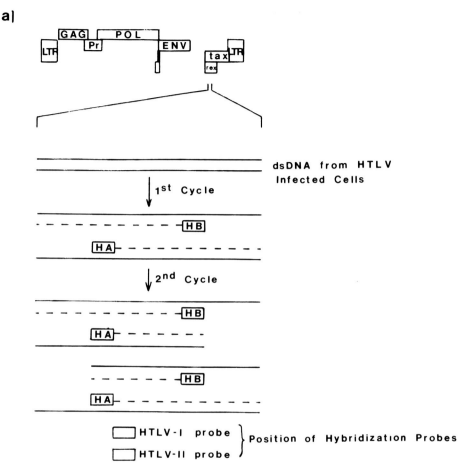

FIG. 1. Detection of HTLV-I DNA sequences by polymerase chain amplification assay using fresh tissues from patients with ATL. **A.** A schematic map of the HTLV-I genome is shown with the relative positions of each gene. The region amplified with oligonucleotides HA and HB is found within the *rex* and *tax* genes. A schematic for two cycles of amplification reaction is shown. During each cycle, DNA is denatured and annealed to the oligonucleotide primers, and a new strand of DNA is synthesized (dotted lines). The concentration of DNA sequences between the two primers increases exponentially with the number of cycles of amplification. The positions of the HTLV-I and -II hybridization probes relative to the amplified segment of DNA are indicated. **B.** DNA from peripheral blood (PBM) or bone marrow cells (BM), uninfected K562 or H9, or the HTLV-I–infected HUT 102 cell line, or cloned HTLV-I or -II DNA sequences were used for amplification. Primers are identical and complementary to HTLV-I nucleotides 7463–7486 and 7552–7572, respectively. The identical sequences are present in the HTLV-II genome. The reaction products were analyzed on a nondenaturing 5% polyacrylamide gel and ethidium-bromide stained. The gels were blotted electrophoretically onto Zeta probe membranes and hybridized with **C)** an end-labeled HTLV-I–specific probe identical to nucleotides 7487–7506, and **D)** an end-labeled HTLV-II–specific probe identical to nucleotides 7398–7417. Oligonucleotide primers, specific amplified DNA products, and products resulting from nonspecific amplification of endogenous DNA sequences are indicated. Reprinted with permission from ref. 1.

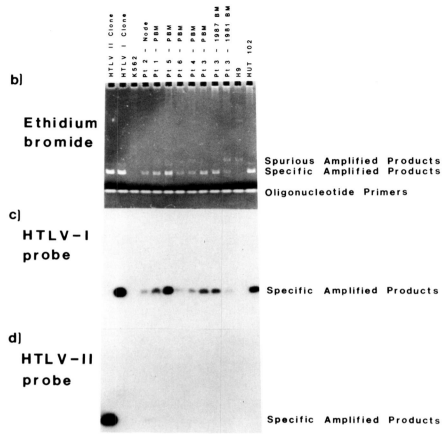

FIG. 1. *Continued.*

Among those individuals with ATL who were born in the United States, 69% were born in southeastern United States (1). Eighty percent were blacks, 11% Caucasions, 6% Asian, and 3% Aleutian. Approximately one-half were females, and 84% were diagnosed with ATL between ages 20 and 60. Only one childhood case of acute ATL has been seen.

Our data would suggest that ATL is more common in the United States than most reported studies would indicate. Serological screening of all individuals with T-lymphocyte disorders or individuals with lymphoma with lytic bone lesions, skin involvement, or hypercalcemia is likely to allow detection of most cases of ATL. Serological screening is important because the prognosis of HTLV-I–associated lymphoma is markedly worse than that of lymphoma not associated with HTLV-I. Furthermore, HTLV-I–infected individuals should be informed of the risk of viral transmission by sexual intercourse, blood transfusion, or viral transmission to the fetus by transplacental mechanisms or to the newborn by breast-feeding.

FAMILIAL ATL

An interesting and unexpected finding from our studies was the diagnosis of ATL in more than one member of the same family. The studies of Kajiyama and colleagues have demonstrated a fivefold higher seroprevalence of HTLV-I in families with an infected individual than the overall seroprevalence of the area (3). Their data demonstrate transmission primarily from infected husbands to wives and subsequent maternal-child transmission (4). There is also considerable evidence to support the requirement for acquisition of HTLV-I early in life for the development of ATL (1). This is not true, however, for HAM/TSP, which can develop within a few years after transfusion or sexual acquisition of HTLV-I (personal communication with B. Poiesz). ATL development may require a) a long latency after HTLV-I infection; b) multiple separate infections by HTLV-I, or by HTLV-I and another infectious agent; or c) infection at a time when the immune system is compromised.

In Japan, ATL is found to occur in members of the same family in <10% of cases (5–7). However, studies of patients presenting to our medical center suggested that the occurrence of ATL in more than one individual of the same family was common. Family studies of patient 3 in our series demonstrates that two of eight tested siblings were seropositive, one of whom was asymptomatic, and one of whom had chronic ATL (patient 4). The patient's husband was also seropositive but asymptomatic. Patient 7 presented to our center with acute ATL 3 yr after her brother had died of ATL (patient 8). Clinical features were similar in these two siblings in that both presented with mild leukocytosis, central nervous system involvement, hypercalcemia, and lytic bone lesions, but no skin infiltration.

The finding of familial cases of ATL in the United States, despite the rarity of this presentation in Japan, may be due at least partially to the smaller average family unit size in Japan than in the United States. These data also suggest that virus-specific or host-specific factors may predispose to the development of ATL. Further analysis for structural variants of HTLV-I among isolates from these familial cases, or for variants with quantitatively higher replicative or transforming activity, may define features important for ATL development.

HTLV-I VARIANTS

More than 100 strains of HTLV-I have now been isolated from diverse areas of the world, including Japan, the Caribbean, and the United States (8). Limited restriction enzyme mapping of many of these isolates, and detailed restriction enzyme maps of a limited number of cloned proviruses, are available (9). Four

HTLV-I proviruses are compared in Fig. 2. HTLV-I-ATK and HTLV-I-MT2 are from Japanese patients, HTLV-I-CH from a patient from the United States, and HTLV-Ib-EL from an African patient. Of 38 restriction enzymes sites mapped in HTLV-I-ATK, only 4 differences are noted in HTLV-I-CH compared with HTLV-I-ATK, including 2 additional sites and 2 fewer sites. This predicts a 1.8% nucleotide sequence divergences between these two clones.

Of 20 restriction enzymes sites mapped for HTLV-I-MT2, only 2 differences are noted compared with HTLV-I-ATK, and only 1 difference compared with HTLV-I-CH. This predicts 1.6% and 0.83% nucleotide sequence differences in HTLV-I-MT2 compared with HTLV-I-ATK and HTLV-I-CH, respectively.

Of 32 restriction enzyme sites mapped for HTLV-Ib-EL, 14 differ from HTLV-I-ATK and 12 differ from HTLV-I-CH, predicting 7.3% and 6.2% sequence divergence, respectively. Compared with the 20 sites mapped for HTLV-I-MT2, 7 differences are noted in HTLV-Ib-EL, predicting 5.8% nucleotide sequence differences.

Further analysis of HTLV-Ib-EL was undertaken because it manifested a greater degree of sequence divergence than did other HTLV-I strains. In this endeavor, 3897 nucleotides were determined, including all of the pX region situated between *env* and the 3'LTR, as well as randomly selected portions of *pol, env,* and the protease gene (Table 2) (10). This demonstrated a total of 141 nucleotide differences compared with HTLV-I-ATK, or 3.6% nucleotide differences. These differences all represented point mutations except for an 11-nucleotide deletion including the first ATG codon of the first open reading frame of pX, pX-I. Because HTLV-Ib-EL was capable of replication and transformation, this suggested that if pX-I encodes a protein product, it is dispensable for these activities. The point mutations in HTLV-Ib-EL predict 43 amino acid changes, of which 35 are nonconservative, predicting 2%–5% amino acid changes in different open reading frames compared with HTLV-I-ATK. It is interesting to note that amino acid sequence alterations occurred more than twice as frequently in the POL product than the ENV product. This is in contrast with studies of sequence heterogeneity of other retroviruses which demonstrate that *pol* is generally more conserved than *env* (11–13).

The level of sequence variation between HTLV-Ib-EL and other HTLV-I strains is overemphasized in this study, because more sequence data was obtained from regions of the genome which manifested restriction enzyme site differences than from regions such as *gag* and LTR, which manifested few differences. Thus, sequence variation of 3.5% among HTLV-I strains represents an upper limit of the level of sequence variation. This may be contrasted with sequence variation among 13 different HIV-1 proviruses which manifest up to 10% nucleotide sequence alterations with up to 27% amino acid alterations in *env* gene products of different isolates (13).

Recently, Malik and colleagues have reported the complete nucleotide sequence of a HTLV-I isolate, HTLV-I-HS35, derived from an ATL patient of Caribbean origin (14). This sequence demonstrates 2.3% variation from HTLV-I-ATK.

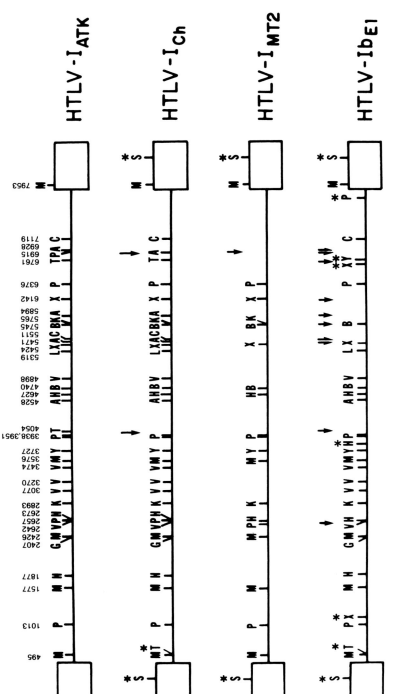

FIG. 2. Restriction enzyme maps of HTLV-I variants. Numbers indicate positions in the genome relative to the RNA initiation site. Arrows indicate sites missing in clones compared with HTLV-I-ATK, and asterisks indicate sites present in clones compared with HTLV-I-ATK. Restriction enzyme sites are a, *Aha* III; A, *Apa* I; b, *Bal* I; B, *Bam*HI; C, *Cla* I; D, *Nde* I; E, *Eco*RI; G, *Bgl* II; H, *Hind*III; K, *Kpn* I; I, *Bcl* I; L, *Sal* I; M, *Sma* I; N, *Nco* I; p, *Apa* I; P, *Pst* I; R, *Eco*RV; S, *Sac* I; T, *Sac* II; U, *Stu* I; V, *Pvu* II; X, *Xho* I; Y, *Xba* I; II, *Avr* II. Reprinted with permission from ref. 9.

TABLE 2. *Summary of sequence differences between HTLV-Ib and HTLV-I*

Region of genome	Total number nucleotides	Number of nucleotides sequenced	Number of nucleotide differences	Percentage nucleotide differences	Number of predicted amino acid differences	Percentage predicted amino acid differences	Number of predicted non-conservative amino acid differences
intervening region between *gag* and *pol*	408	408	6	1.5			
pol	2688	776	28	3.6	12	4.6	8
env	1464	905	23	2.5	6	2.0	6
intervening region between *env* and pX	190	190	21	11.1			
pX	1523	1523	64	4.5	12	3.0	10
pX-I	297	297	25	8.4	2*		2
pX-II	261	261	2	0.8	2	2.3	2
pX-III	324	324	5	1.5	5	4.6	4
pX-IV	735	735	21	2.9	8	3.3	3
LOR	1071	1071	26	2.4	10	2.7	5
Total	6268	3897	141	3.6			

* Change in pX-I open reading frame introduced by deletion.
Reprinted with permission from ref. 10.

Within the protease gene, HTLV-I-HS35 differs in 0.6% nucleotides from another Caribbean isolate, HTLV-I-1010/3, 1% from two Japanese HTLV-I isolates, HTLV-I-ATK and HTLV-I-HY4 (15), and 2.4% from the African isolate, HTLV-Ib-EL. In comparison to *pol* sequences of HTLV-I-ATK, HTLV-I-HS35 differs at 2.1% of nucleotide positions, predicting 11 amino acid changes. In *env*, HTLV-I-HS35 differs by 1.5% of nucleotides from HTLV-I-1010/3 and 2.4% of nucleotides from HTLV-I-ATK. In pX, only 1.0% nucleotide differences are noted between HTLV-I-HS35 and HTLV-I-ATK, but 2.3% sequence differences from HTLV-I-EL. An interesting observation from these sequence comparisons was the finding that HTLV-I isolates from the same geographical area were more related to one another than those from different areas.

Sequence conservation of HTLV-Is is reaffirmed by the analysis of a cloned HTLV-I provirus, H5, from a patient with HAM/TSP. Of 13 restriction enzyme sites mapped, no differences were noted in comparison to HTLV-I-ATK (16). Complete nucleotide sequences of H5 have been obtained from the protease, *env, tax,* and *rex* genes, and the LTR, and 97% homology with HTLV-I-ATK was noted (17). Of five nucleotide differences (0.7%) in the protease gene, only two amino acid alterations (0.8%) are predicted to occur. However, both of these changes are nonconservative amino acid replacements, with a leucine substitution for a proline and a tryptophan substitution for a termination codon. Fourteen nucleotide substitutions (1.0%) are noted in *env* which predict 7 amino acid substitutions (1.4%), of which 3 are nonconservative substitutions (0.6%). Four nucleotide substitutions (1.3%) are found in the pX-I open reading frame of which two predict amino acid substitutions (2.0%), which are both nonconservative replacements. In the LTR, seven nucleotide substitutions (0.9%) and five nucleotide insertions (0.7%) are noted.

These data would indicate that if distinct variants of HTLV-I cause ATL and HAM/TSP, very subtle structural alterations are responsible for markedly different clinical sequelae. Further structural studies of isolates from HAM/TSP patients may define a distinct genetic structure.

Further evidence for the conservation of HTLV viruses is the 95% nucleotide sequence similarity of simian T-cell leukemia virus (STLV) (18,19). This is to be contrasted with the 60% sequence similarity of SIV to HIV-1 and 80% sequence similarity to HIV-2 (20,21).

HTLV-II VARIANTS

Analyses of HTLV-II variants have been more limited because of the rarity of this virus. Only 11 isolates have been reported (22–25, and personal communication with Z. Salahuddin) though serological and PCR studies suggest a high prevalence in intravenous drug abusers (26–28) (see articles by Shaw, Chen, and Hall). Complete sequence data is available from only one strain (HTLV-II-MO); restriction enzyme map data is available from clone HTLV-II-JP and HTLV-II-NRA (9). Restriction enzyme map data of HTLV-II-JP fails to demonstrate differences compared with HTLV-II-MO (Fig. 3). Restriction enzyme map data of HTLV-II-NRA demonstrates three restriction enzyme site differences among eight analyzed. It would be premature to calculate nucleotide sequence divergence based on this limited data base.

The biological activity of HTLV-II in tissue culture is indistinguishable from that of HTLV-I. Both viruses immortalize T4 lymphocytes, cause multinucleated giant cells, and lead to the expression of antigens found with lymphocyte activation and novel HLA class II serologic reactivity (29,30). The overall structure of the HTLV-II genome is quite similar to that of HTLV-I (31). The most remarkable difference is the lack of a pX-I open reading frame in HTLV-II (32). However, there is only 60% nucleotide sequence conservation between HTLV-I and -II (31). Functional analyses of these viruses in a wide range of cell types in tissue culture and animal model systems may define important biological differences. Alternatively, it is possible that in fact HTLV-I and -II have identical disease induction activities. The failure to identify ATL or HAM cases associated with HTLV-II may be a result of the rarity of this virus, as well as difficulties in distinguishing HTLV-I and -II by serological means alone (33).

MECHANISM OF SEQUENCE VARIATION

Sequence variation in retroviruses is likely to be a result of multiple different mechanisms, each of which may be predominant under different circumstances. Reverse transcriptases (RT) have been shown to be more error-prone than DNA-

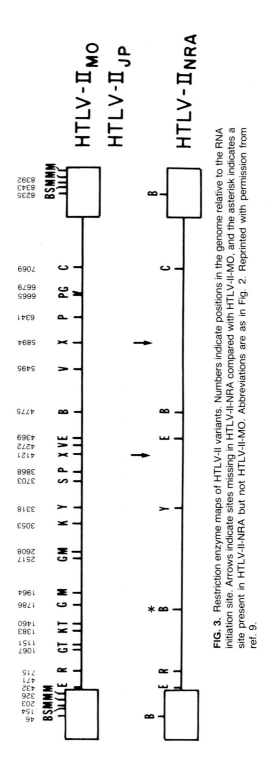

FIG. 3. Restriction enzyme maps of HTLV-II variants. Numbers indicate positions in the genome relative to the RNA initiation site. Arrows indicate sites missing in HTLV-II-NRA compared with HTLV-II-MO, and the asterisk indicates a site present in HTLV-II-NRA but not HTLV-II-MO. Abbreviations are as in Fig. 2. Reprinted with permission from ref. 9.

dependent DNA polymerases. The HIV-1 RT has been shown to have 10-fold more infidelity than Moloney murine leukemia virus (MuLV) and avian myeloblastosis virus (AMV) RTs (34,35).

Copy-choice mechanisms of RT may also lead to sequence variations as a result of alternate template copying (36). Recombination has also been shown to be important for generating viral variants in murine systems (37), and occurs with HIV-1 in vitro (38). However, no data are available to demonstrate a significant contribution of recombination to generation of sequence heterogeneity for human retroviruses in vivo.

Corrections of synthesis errors are claimed not to occur in retrovirus systems (39). However, recent studies of RNA editing mechanisms may be applicable to these virus infections (40,41).

Syntheses of altered sequences are likely to only partly account for the types of sequence variants generated in vivo. Selection for sequence variants which are not lethal for retrovirus replication is important. Furthermore, immune selection may also play an important role in sequence variation in vivo. Though immunoevasive mechanisms have been well defined for equine infectious anemia virus (42), the contributions to selection of sequence variants of human retroviruses are unclear.

Sequence variation among HTLVs is more restricted than is variation among different isolates of HIV-1 or -2. There are several possible explanations for this finding. First, HTLV proteins may be more compact and less tolerant of alterations. The HTLV envelope proteins gp46 and gp21 are smaller than the gp120 and gp41 of HIV-1 (9).

Second, the HTLV RT may be less error-prone than that of HIV-1. Assays for the rate of nucleotide substitutions, deletions, insertions, or frameshifts occurring during in vitro synthesis with the HTLV-I polymerase may be revealing.

Third, the level of unintegrated DNA species in HTLV-infected cells is markedly less than that in HIV-1–infected cells (43). Thus, recombination among HTLV variants is less likely to occur than that among HIV-1 variants.

Fourth, there may also be differences in the level of viral RNA taken up into cells infected by HTLVs than by cells infected with HIVs. Superinfection of HIV-1–infected cells may lead to massive increases in unintegrated viral DNA synthesis (44). This has not been examined in the case of HTLVs. Large numbers of viral RNA templates may promote copy-choice mechanisms of sequence alterations.

Fifth, cellular factors may play a greater role in restricting HTLV variation than in restricting HIV variation. HTLV may infect subsets of human cells that tolerate sequence variation poorly. However, the wide range of human and rabbit cells infectible by HTLVs (see article by Dalgleish in this volume), and the fact that both HIV-1 and HTLV-I can infect at least some of the same CD4$^+$ cell types argues against a major contribution at this level.

Sixth, immune regulation may restrict HTLV variation and/or promote HIV variation. Further analysis of such interactions in both systems is crucial for our understanding of the immunopathogenesis of this disease and the development of viral vaccines.

Seventh, and perhaps most important, is the fact that HTLV dissemination within an individual and transmission between individuals may differ in a fundamental way from that of HIV. HTLV in culture, and perhaps in vivo, is primarily cell-bound. RT production by HTLV-I–infected cell lines is only 1%–10% of that of HIV-1–infected cells (unpublished observations). Furthermore, HIV-1 is efficiently transmitted by cell-free blood products, whereas HTLV-I is not efficiently transmitted by this mechanism. Approximately 50%–80% of hemophiliacs are infected with HIV-1, whereas <5% are infected with HTLV-I (45). The nationwide seroprevalence of HIV-1 and HTLV-I among blood donors is 0.15% and 0.025%, respectively (46,47). Thus, HTLV-infected cultures or individuals may involve significantly fewer rounds of virus replication over a given period of time than is found with HIV-1 infection. Thus, the opportunity for generation and amplification of sequence variation may be considerably lower on this basis for HTLV than HIV.

SIGNIFICANCE OF SEQUENCE VARIATION

Sequence variation may be responsible for alterations in biological activity of HTLVs. This may include alterations of virus infectivity, replication kinetics, cytopathicity, immortalizing activity, growth factor dependence, cell tropism, immune responsiveness, or species specificity. At least some of these characteristics have been demonstrated to be altered among closely related variants of HIV-1. Surprisingly few studies have assessed altered biological properties among HTLV variants. Popovic and colleagues demonstrated differences in interleukin-2 requirements of different cultures of umbilical cord blood lymphocytes immortalized with a variety of HTLV-I strains (48). However, the variation in this phenotype may be due to virus- or cell-specific differences.

Though there are only limited sequence differences among HTLV-I strains, these may lead to significant alterations in biological activity. Only very limited (0.02%) sequence differences are noted between HIV-1 strains, with marked differences in cell-specificity of infection (our unpublished observations). Further analysis of host- and virus-specific factors will undoubtedly lead to a greater understanding of the pathogenesis of the diverse range of clinical disorders associated with these retrovirus infections.

REFERENCES

1. Ratner L, Poiesz BJ. Leukemias associated with human T-cell lymphotropic virus type I in a non-endemic region. *Medicine* 1989;67:401–422.
2. Kondo T, Kono H, Nonaka H, et al. Risk of adult T-cell leukaemia/lymphoma in HTLV-I carriers. *Lancet* 1987;2:159.

3. Kajiyama W, Kashigwagi S, Hayashi J, et al. Intrafamilial clustering of anti-ATLA positive persons. *Am J Epidemiol* 1986;124:800–806.
4. Kajiyama W, Kashiwagi S, Ikematsu H, et al. Intrafamilial transmission of adult T cell leukemia virus. *J Infect Dis* 1986;254:851–857.
5. Imamura N, Koganemaru S, Kuramoto A. T cell leukemia a few months apart in two brothers. *Lancet* 1982;2:1361–1362.
6. Ishibashi K, Iwahashi N, Nomura K, et al. The outbreak of five lymphoid malignancies in one family during seven years. *Jpn J Clin Hemat* 1985;26:374–380.
7. Taguchi H, Niya K, Kubonishi I, et al. Adult T-cell leukaemia in two siblings. Acute crisis of smouldering disease in one patient. *Cancer* 1985;56:2870–2873.
8. Gallo RC. Human T-cell leukemia-lymphoma virus and T-cell malignancies in adults. *Cancer Surv* 1984;3:113–159.
9. Ratner L, Wong-Staal F. Human T-lymphotropic viruses. *In:* O'Brien S, ed. *Genetic Maps.* Cold Spring Harbor, NY: Cold Spring Harbor Press, 1987;124–129.
10. Ratner L, Josephs SF, Starcich B, et al. Nucleotide sequence analysis of a variant of human T-cell leukemia virus (HTLV-Ib) provirus with a deletion in pX-I. *J Virol* 1985;54:781–790.
11. Alizon M, Wain-Hobson S, Montagnier L, Sonigo P. Genetic variability of the AIDS virus: nucleotide sequence analysis of two isolates from African patients. *Cell* 1986;46:63–74.
12. Chiu I-M, Callahan R, Tronick SR, et al. Major *pol* gene progenitors in the evolution of onco-viruses. *Science* 1984;223:364–370.
13. Coffin JM. Genetic variation in AIDS viruses. *Cell* 1986;46:1–4.
14. Malik KTA, Even J, Karpas A. Molecular cloning and complete nucleotide sequence of an adult T cell leukaemia virus/human T cell leukaemia virus type I (ATLV/HTLV-I) isolate of a Caribbean origin: relationship to other members of the ATLV/HTLV-I subgroup. *J Gen Virol* 1988;69:1695–1710.
15. Hiramatsu K, Nishida J, Naito A, Yoshikura H. Molecular cloning of the closed circular provirus of human T cell leukaemia virus type I: a new open reading frame in the *gag-pol* region. *J Gen Virol* 1987;68:213–218.
16. Imamura J, Tsujimoto A, Ohta Y, et al. DNA blotting analysis of human retroviruses in cerebrospinal fluid of spastic paraparesis patients: the viruses are identical to human T-cell leukemia virus type-I (HTLV-I). *Int J Cancer* 1988;42:221–224.
17. Tsujimoto A, Teruuchi T, Imamura J, et al. Nucleotide sequence analysis of a provirus derived from HTLV-I–associated myelopathy (HAM). *Mol Biol Med* 1988;5:29–42.
18. Inoue J, Watanabe T, Sato M, et al. Nucleotide sequence of the protease-coding region in an infectious DNA of simian retrovirus (STLV) of the HTLV-I family. *Virology* 1986;150:187–195.
19. Guo C, Wong-Staal F, Gallo RC. Novel viral sequences related to human T-cell leukemia virus in T-cell of a seropositive baboon. *Science* 1984;223:1195–1196.
20. Chakrabarti L, Guyader M, Alizon M, et al. Sequence of simian immunodeficiency virus from macaque and its relationship to other human and simian retroviruses. *Nature* 1987;328:543–547.
21. Fukasawa M, Miura T, Hasegawa A, et al. Sequence of simian immunodeficiency virus from African green monkey, a new member of the HIV/SIV group. *Nature* 1988;333:457–461.
22. Kalyanaraman VS, Sarngadharan MG, Robert-Guroff M, et al. A new subtype of human T-cell leukemia virus (HTLV-II) associated with a T-cell variant of hairy cell leukemia. *Science* 1982;218:571–573.
23. Kalyanaraman VS, Narayan R, Feorino P, et al. Isolation and characterization of a human T-cell leukemia virus type II from a hemophilia-A patient with pancytopenia. *EMBO J* 1985;4:1455–1460.
24. Hahn BH, Popovic M, Kalyanaraman VS, et al. Detection and characterization of an HTLV-II provirus in a patient with AIDS. *In:* Gottlieb MS, Groopman JE, eds. *Acquired Immune Deficiency Syndrome.* New York: Alan R. Liss, 1984;73–81.
25. Rosenblatt JD, Giorgi JV, Golde DW, et al. Integrated human T cell leukemia virus II genome in CD8+ T cells from a patient with "atypical" hairy cell leukemia: evidence for distinct T and B cell lymphoproliferative disorders. *Blood* 1988;71:363–369.
26. Cheingsong-Popov R, Weiss RA, Dalgleish A, et al. Prevalence of antibody to human T-lymphotropic virus type III in AIDS and AIDS-risk patients in Britain. *Lancet* 1984;2:477–480.
27. Robert-Guroff M, Weiss SH, Giron JA, et al. Prevalence of antibodies to HTLV-I, -II, and -III in intravenous drug abusers from an AIDS endemic region. *JAMA* 1986;255:3133–3137.

28. Gore I, Snyder TL, Decker WD, et al. Frequent isolation and molecular identification of human T-cell leukemia virus type-2 (HTLV-2) in three U.S. population centers. *Clin Res* 1989;37:429A.
29. Mann DL, Popovic M, Murray C, et al. Cell surface antigen expression in newborn cord blood lymphocytes infected with HTLV. *J Immunol* 1983;131:2021–2024.
30. Mann DL, Popovic M, Sarin P, et al. Cell lines producing human T-cell lymphoma virus show altered HLA expression. *Nature* 1983;305:58–60.
31. Shimotohno K, Takahashi Y, Shimizu N, et al. Complete nucleotide sequence of an infectious clone of human T-cell leukemia virus type II: an open reading frame for the protease gene. *Proc Natl Acad Sci USA* 1985;82:3101–3105.
32. Haseltine WA, Sodroski J, Patarca R, et al. Structure of 3' terminal region of type II human T lymphotropic virus: evidence for new coding region. *Science* 1984;225:419–421.
33. Saxinger C, Gallo RC. Methods in laboratory investigation: application of the indirect enzyme-linked immunosorbent assay microtest to the detection and surveillance of human T-cell leukemia-lymphoma virus. *Lab Invest* 1983;49:371–377.
34. Preston BD, Poiesz BJ, Loeb LA. Fidelity of HIV-1 reverse transcriptase. *Science* 1988;242: 1168–1171.
35. Roberts JD, Bebenek K, Kunkel TA. The accuracy of reverse transcriptase from HIV-1. *Science* 1988;242:1171–1173.
36. Meier E, Harmison GG, Keene JD, Schubert M. Sites of copy choice replication involved in generation of vesicular stomatitis virus defective-interfering particle RNAs. *J Virol* 1984;51:515–521.
37. Coffin JM. Structure, replication, and recombination of retrovirus genomes: some unifying hypotheses. *J Gen Virol* 1979;42:1–26.
38. Clavel F, Hoggan DM, Willey RL, et al. Genetic recombination of human immunodeficiency virus. *J Virol* 1989;63:1455–1459.
39. Temin HM. Is HIV unique or merely different? *J Acq Immun Def Syn* 1989;2:1–9.
40. Benne R. RNA-editing in trypanosome mitochondria. *Biochim Biophys Acta* 1989;1007:131–139.
41. Tennyson GE, Sabatos CA, Eggerman TL, Brewer HB. Characterization of single base substitutions in edited apolipoprotein B transcripts. *Nucleic Acids Res* 1989;17:691–698.
42. Salinovich O, Payne SL, Montelaro RC, et al. Rapid emergence of novel antigenic and genetic variants of equine infectious anemia virus during persistent infection. *J Virol* 1986;57:71–80.
43. Shaw GM, Hahn BH, Arya SK, et al. Molecular characterization of human T-cell leukemia (lymphotropic) virus type III in the acquired immune deficiency syndrome. *Science* 1984;226: 1165–1171.
44. Stevenson MB, Meier C, Mann AM, et al. Envelope glycoprotein of HIV induces interference and cytolysis resistance in CD4+ cells: mechanism for persistence in AIDS. *Cell* 1988;53:483–496.
45. Chorba TL, Jason JM, Ramsey RB, et al. HTLV-I antibody status in hemophilia patients treated with factor concentrates prepared from U.S. plasma sources and in hemophilia patients with AIDS. *Thromb Haemost* 1985;53:180–182.
46. Anonymous. Human immunodeficiency virus infection in the United States: a review of current knowledge. *MMWR* 1987;36(Suppl. S4–6):1–48.
47. Williams AE, Fang CT, Slamon DJ, et al. Seroprevalence and epidemiological correlates of HTLV-I infection in U.S. blood donors. *Science* 1988;240:643–646.
48. Popovic M, Lange-Wantzin G, Sarin PS, Mann D, Gallo RC. Transformation of human umbilical cord blood T cells by human T-cell leukemia/lymphoma virus. *Proc Natl Acad Sci USA* 1983;80: 5402–5406.

DISCUSSION

A discussant raised the issue of the outcome of HTLV-I infection and disease, and cited an earlier report that many of HAM/TSP patients, not considered to have leukemia, have circulating cells indistinguishable from ATL. This report also stated that smoldering ATL is diagnosed mainly by skin lesions, either abscesses or ATL cells in the skin. The discussant wanted to know what researchers in the United States should use as criteria for ATL. The discussant felt that it is very confusing if there are just circulating cells and wondered how to deal with the situation. Another discussant stated that ATL-like cells were found frequently in carriers and 50% of the HAM/TSP cases in Japan. One participant said that chronic-type ATL is similar to chronic myeloblastic leukemia, and smoldering-type ATL is similar to chronic phase disease. Another participant noted that it is important to determine the frequency of smoldering ATL and the probability that patients with these cells will develop acute ATL. This participant wanted to know what the cofactors were. One speaker asked if the infidelity of replication leads to random errors. If it is an important site, does the virus die or is there some programming that allows errors to be made in a certain area? Another speaker stated that there are not enough data to say whether there is sequence dependence for error introduction. In some studies there does appear to be sequence dependence for errors produced in vitro with the HIV reverse transcriptase with M13 or LAC-sequences. How this relates to HTLV-I sequences is unknown. A discussant agreed that moderate sequence heterogeneity might be very important biologically. The question is, how does one go about recognizing whether these minor changes are of biological significance? What systems will be used? Another discussant brought up an idea mentioned earlier, that of establishing functional clones and analyzing the properties of each of the variants. The discussant believed that variant viruses had been derived, but was not aware of any detailed studies on differences in their replicative and immortalizing activity on different cell types. An indirect approach would be to determine if there is a common structural or functional feature, for example a neurotropic isolate with a protease domain which one participant pointed out was frequently associated with neuropathic determinants.

A discussant referred to restriction-type variation in biologically interesting isolates of HTLV-I and -II, specifically the finding that there are extraordinary variations in HIV-1 and HIV-2, but when a large number of HTLV-I and HTLV-II viruses were examined, variation is the rule, not the exception. If one examines 100 consecutive isolates of HTLV-I or HTLV-II and does not find one, two, three, or four restriction-site differences between any HTLV-I or any HTLV-II isolates, then something is wrong. Although researchers find a lot of restriction-site conservation, if there isn't some variation it's usually a contaminant. Strain variation is seen at the level of restriction-site mapping, if enough restriction-enzyme cutting is done.

Human Retrovirology: HTLV,
edited by William A. Blattner.
Raven Press, Ltd., New York 1990.

Immunological and Virological Studies in HAM/TSP

Dale E. McFarlin, Ajay Gupta, David Mattson, Jonathan Harris, Jayne S. Reuben, and Steven Jacobson

Neuroimmunology Branch, National Institutes of Health, Bethesda, Maryland 20892

INTRODUCTION

After the demonstration of antibody to human T-lymphotropic virus type I (HTLV-I) in the sera and cerebrospinal fluid (CSF) of patients with tropical spastic paraparesis (TSP) from Martinique (1), other reports followed which confined and extended the observations (2,3). In addition, HTLV-I–associated myelopathy (HAM) was described (4) and more extensive studies of HTLV-I in Japan and the Caribbean were conducted (5,6). After considerable discussion and debate, it was recently concluded that TSP and HAM are the same disease related to HTLV-I or closely related agents (7). It is current practice to refer to the syndrome as HAM/TSP. In parallel, the possible association between multiple sclerosis (MS) and HTLV-I has emerged. This was based on rationale for considering a viral etiology for MS (8), clinical overlap between HAM/TSP, and reports of antibody to HTLV-I in some patients with MS (9).

Investigation of the association between HTLV-I and neurological diseases was begun in the Neuroimmunology Branch of the National Institutes of Health (NIH) ~2 yr ago. Our initial approach was to assess immune function in patients with HAM/TSP and compare the findings with those in MS. Abnormalities in cellular immune function were identified which were similar to those observed after in vitro infection of lymphocytes with HTLV-I (10). This led to successful efforts to detect virus and viral genome in the lymphocytes in the blood and CSF of these patients (11). Subsequently, research has been directed at characterizing the viruses obtained from HAM/TSP to determine if differences exist which would predispose to central nervous system (CNS) disease as has been identified in other retroviruses. This report describes findings in nine HAM/TSP patients evaluated at the NIH and additional ongoing studies in progress.

INITIAL STUDIES

Nine patients who represent the spectrum of clinical features currently associated with HAM/TSP were initially evaluated at Clinical Center of the NIH

(12). One (referred by Dr. W. Sheremata) was Haitian, two (referred by Drs. O. Morgan and P. Rodgers-Johnson) were Jamaican, five (referred by Dr. V. Zanionovic) were Colombian, and one was a Jamaican immigrant to the United States (a patient of Dr. F. Neva). All patients had antibodies to HTLV-I in serum and CSF (18). Five of nine patients had a mild pleocytosis. CSF IgG was elevated in seven patients. Oligoclonal IgG bands were present in seven patients (12). The clinical and laboratory findings are consistent with a series of patients recently evaluated in France (13).

Cytofluorometric Analysis of Peripheral Blood Lymphocytes (PBL)

HTLV-I is predominantly (but not exclusively) trophic for T lymphocytes expressing the CD4 surface antigen phenotype (14). In patients with HTLV-I–associated adult T-cell leukemia (ATL), there are profound changes in the state of T-cell activation, as demonstrated by an increase in the number of T cells that express class II HLA molecules and/or interleukin-2 receptor (IL2R) (15). Because of the association between HTLV-I and HAM/TSP, detailed studies of T cells were conducted, and, in all nine patients, evidence of activated lymphocytes was detected by flow cytometry techniques and lymphocyte proliferation.

PBL were isolated on Ficoll-Hypaque gradients (Organon Teknika Corp., Durham, NC). Light-scattering properties of these cells were analyzed on a FACS-IV (Becton-Dickinson FACS System) and plotted as forward scatter (FSC) versus side scatter (SSC). These parameters are indicators of both the size and volume of each cell analyzed. Ten thousand cells were studied for each dot display, and gates were placed to discriminate lymphocytes from "large" cells. In dual-parameter immunofluorescence studies, PBL were stained with Leu-4, Leu-2a, or Leu-3a coupled to phycoerythyrin (Becton-Dickinson, Mountain View, CA). Murine antibodies to HLA-DR (Becton-Dickinson, Mountain View, CA) and IL2R (gift from T. Waldmann) were used to stain by indirect fluorescence with sheep anti-mouse-FITC as the second antibody. In each experiment, appropriate controls were included to exclude cross-reactivity among the reagents. Fluorescence was analyzed on a FACS-IV and the percentage of positive cells in each gate was determined (12).

Light-scattering analysis of normal PBL typically shows only a few large cells; these usually include macrophages, B cells, and occasional T cells. In HAM/TSP patients, the number of such large cells was clearly abnormal (12). In order to characterize these cells, the expression of CD3, CD4, CD8, DR, and IL2R molecules on the surface of cells in each scatter gate was determined. One gate was used to assess the lymphocyte population while the other defined the larger cells. In HAM/TSP patients, the percentages of cells expressing the CD3, CD4, and CD8 molecules in the lymphocyte gate were normal, but percentages of HLA-DR$^+$ and IL2R$^+$ cells were elevated in the majority of patients (12). The occurrence of such a high proportion of IL2R$^+$ cells as well as HLA-DR$^+$ cells indicated in vivo T-cell activation (15). A significant proportion of the large cells were CD3$^+$, and, in some patients, over 60% of these cells expressed the CD8 phenotype. In addition, by dual parameter cytofluorometry all the large CD3$^+$ cells were also HLA-DR$^+$.

Among the large, activated CD3$^+$ lymphocytes, there was a high proportion of cells that expressed neither the CD4 nor CD8 T-cell phenotype. Such "double negative" T cells have been shown to express an antigen-specific T-cell receptor (TcR) consisting of a heterodimer of gamma-delta chains (16). This is in contrast to the more conventional alpha-beta TcR expressed on the majority of T cells. The presence of T cells with the gamma-delta TcR in the peripheral blood of TSP patients was confirmed with positive cytofluorometric analysis using the monoclonal antibody TA1055 (T Cell Science, Cambridge, MA) specific for the delta chain of the TcR (data not shown).

Lymphoproliferative Responses

One measure of the cellular immune status is the capacity to proliferate to an in vitro stimulus. The lymphoproliferative responses to measles, mumps, and influenza were studied in eight of the nine HAM/TSP patients and six normal, healthy, HTLV-I–seronegative controls. Cryopreserved PBL were washed in RPMI-1640 containing 10% human serum albumin and suspended at a concentration of 3×10^5 cells per well in 96-well plates (Nunclon, Roskilde, Denmark). Triplicate cultures of cells were exposed to viral antigens at a concentration of 1 plaque-forming unit/cell in a volume of 0.2 ml or treated with media alone. Media consisted of RPMI-1640 (GIBCO, Grand Island, NY) supplemented in 2% human AB serum, 1% penicillin, streptomycin, glutamine, and Hepes buffer. After 4, 5, and 6 d, the degree of lymphoproliferation was assessed by ^{3}H-thymidine incorporation. In the control group there was a vigorous proliferative response to all three viruses. In the TSP patients the most striking observation is that, in the absence of an antigenic stimulus (media alone), the mean response was significantly elevated compared with controls at all days tested. The marked spontaneous proliferation prevented detection of specific proliferation to the viral antigens and was another indication of T-cell activation. In four patients, abnormalities in the generation of cytotoxic T lymphocytes (CTL) to measles virus were observed. In two of these four patients, there were defects in the production of CTL to other viruses (12).

Detection of HTLV-I–Related Viruses

The lymphocyte abnormalities described above were believed to be consistent with infection of HTLV-I (15,17). Viral antigen was not observed in cytocentrifuge by smears of PBL examined by immunofluorescence using antibodies against HTLV-I. Consequently, it was decided to examine the PBL for proviral

DNA. In order to have sufficient material for these studies, T-cell lines were derived from the blood and CSF by stimulation with anti-CD3. As the cell lines were being expanded, the expression of HTLV-I antigen was reexamined. Gag proteins were found in a small percentage of blood cells by immunofluorescence. The percent of protein cells expressing the gag proteins increased with each passage (18). This led to the development of a systematic approach for the detection of HTLV-I antigens in lymphocytes from HAM/TSP patients (9). The expression of HTLV-I antigen varied with passage history, but thus far has been observed in 16/17 cell lines from the nine patients (18). Budding virus and infectious virus have been demonstrated in some cell lines (9).

Characterization of HTLV-I–Related Viruses

Although the viruses in the cell lines discussed above exhibit many properties consistent with HTLV-I, our group has been interested in searching for differences that might relate to neurotropism. Both biological and molecular differences have been observed, as summarized in Table 1. The cell lines derived from HAM/TSP have remained IL2 dependent and have not become transformed. Although it is possible that in the future HAM/TSP cell lines become transformed, the lack of this property contrasts to cell lines derived from ATL, which in time characteristically become transformed. The cell line derived from one individual, the Haitian patient (19), has been studied in detail. Comparison of HTLV-I envelope proteins from this HAM/TSP cell line with those of conventional HTLV-I (HUT 102) has shown differences in molecular weights (Table 1). The differences detected so far are believed to be related to variations in glycosylation, rather than nucleotide sequence changes discussed below. Another difference is that in contrast to conventional HTLV-I (HUT 102) the HAM/TSP cell lines have less effect on T-cell function. An influenza-specific $CD4^+$ CTL line, when cocultured with HUT 102, has reduced CTL function; however, no effect was observed when cocultured with a HAM/TSP cell line (18).

Restriction maps of HAM/TSP proviral DNA are consistent with HTLV-I, but differences have been observed in material from some patients (20); similar

TABLE 1. *Differences between the HTLV-I associated with HAM/TSP and conventional HTLV-I*

Observation	HTLV-I$_{TSP}$	HTLV-I$_{Hut\ 102}$
Transformation	No transformation	Immortalization
	IL2-dependent	IL2-independent
Effect on CTL function	No inhibition	Decrease CTL
Envelope protein	Smaller molecular weight	Prototype
Induction of new antigens	CD4, CD8, TcR-alpha/beta	IL2-receptor
Proviral restriction map	Differences in some patients	Prototype
Complete nucleotide sequence	Changes in Env region	Prototype

differences have recently been observed in extensive studies of a single patient (21). Elsewhere in this volume a complete proviral sequence derived from one of our patients is described (19). Although there is 98% homology with published HTLV-I sequence (22), some interesting differences that could relate to neuro-tropism have been identified (19). It is of considerable interest to determine whether these differences are unique to the clone sequenced and are a feature of HTLV-I related agents.

POLYMERASE CHAIN REACTION (PCR) FOR HTLV-I GENOME ANALYSIS

Although the tissue culture procedures outlined above have led to successful identification of HTLV-I–like viruses, these are laborious and time consuming and require tissue culture facilities. Recently, the PCR has been used to detect evidence of HTLV-I infection. This has been based on a modification of the procedure developed by Schnittman et al. (23) for the rapid detection of HIV-1 in PBL. This is illustrated in Fig. 1 and, in brief, consists of disrupting Ficoll-Hypaque purified PBL in low concentration of detergents and proteinase-K to release DNA. One-tenth of the DNA is then amplified in a DNA thermal cycler (Perkin Elmer Cetus) with the use of appropriate primers and Taq polymerase. A ^{32}P 5'-end–labeled probe is then used in a liquid hybridization protocol to visualize the amplified DNA.

In performing this procedure, it is imperative to take precautions to avoid contamination. The reaction is conducted in 0.5-ml microcentrifuge tubes, which are sterilized and stored until use. When the reaction is carried out, DNA is placed in each vial in a biological safety cabinet which has not been previously used for HTLV-I or any other human retrovirus. Although a positive control consisting of DNA from a cell line known to be infected with HTLV-I is included in every assay, this is introduced in a separate biological safety cabinet.

In preliminary experiments the sensitivity of the above procedures was assessed. In Fig. 2, the DNA from varying numbers of MT-1 cells (known to contain prototype HTLV-I at one copy per cell) was added to tubes containing the DNA from 10^6 normal PBL. The DNA was amplified and visualized by use of primers and probes to the *pol* region of HTLV-I. A significant signal was obtained with as few as 10^2 MT-1 cells. When corrected for dilution factors, these results indicate the sensitivity of this PCR assay to detect HTLV-I sequences from 1 in 10 000 cells, provided that the proviral DNA copies were the same or greater than in MT-1 cells.

Studies of an American Family

Recently, the PCR was of value in the assessment of a Caucasian American with HAM/TSP. The patient was a 42-yr-old male who joined the United States

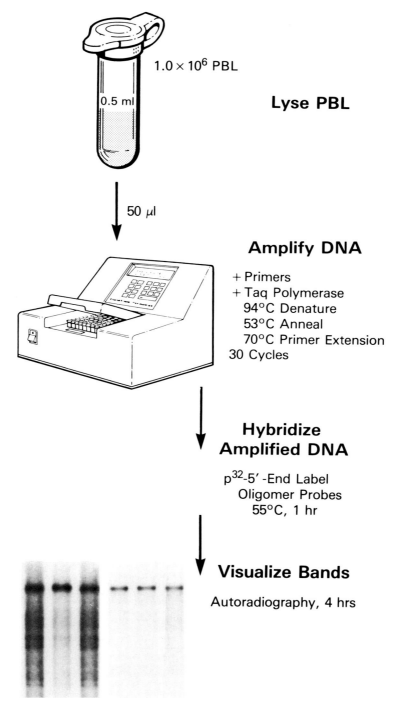

FIG. 1. Schematic representation of PCR protocol used to amplify DNA from peripheral blood lymphocytes (PBL).

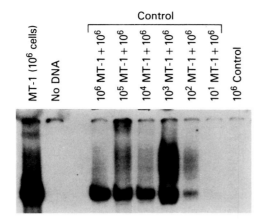

FIG. 2. Liquid hybridization of amplified DNA from varying numbers of cells of the MT-1 HTLV-I–infected cell line added to a constant number of control DNA (normal PBL). Sensitivity of this system with HTLV-I *pol* primers and probes (28) is 1 in 10,000 cells.

Navy after completion of high school and subsequently was a member of the merchant marine. He was sexually active in regions where HTLV-I is now known to be endemic. Twenty years before being evaluated at the NIH he married a Japanese woman from Kyushu. The couple's son was born ~1 yr after the marriage and was breast-fed. When evaluated, the patient gave a 4-yr history of weakness and stiffness in the lower extremities. Neurological examination showed spastic paraparesis. The patient was anergic to a battery of five skin test antigens (24). The CSF contained 10 lymphocytes but no oligoclonal bands. Both the serum and CSF contained antibodies to HTLV-I. The patient's lymphocytes underwent spontaneous proliferation when cultured in vitro (Fig. 3), and cultures of PBL expressed HTLV-I gag antigens after the first passage. Based on these findings, a diagnosis of HAM/TSP was made.

The patient's wife was clinically and neurologically normal but had antibodies to HTLV-I. Her PBL also underwent spontaneous proliferation in culture (Fig. 3) and were shown to contain HTLV-I gag antigens after one passage in tissue culture. She was also anergic. The findings in this couple have provoked several interesting questions. Does the presence of HTLV-I in the PBL relate to the anergy (25)? In this regard it is of interest that reduced skin-test responses to tuberculin have been observed in Japanese carriers of HTLV-I (26). A second question was to the origin of HTLV-I infection, and particularly, whether the infection had been transmitted to the couple's son. This 19-yr-old male was clinically normal and lacked antibody to HTLV-I. He reacted to two of the five skin test antigens, and his lymphocytes proliferated less than those of his parents but tended to be somewhat higher than controls after 7 d in culture (Fig. 3). The PCR findings were of value. Evidence of HTLV-I genome was found in both parents but not in the son (Fig. 4). Thus, no evidence of HTLV-I transmission was detected.

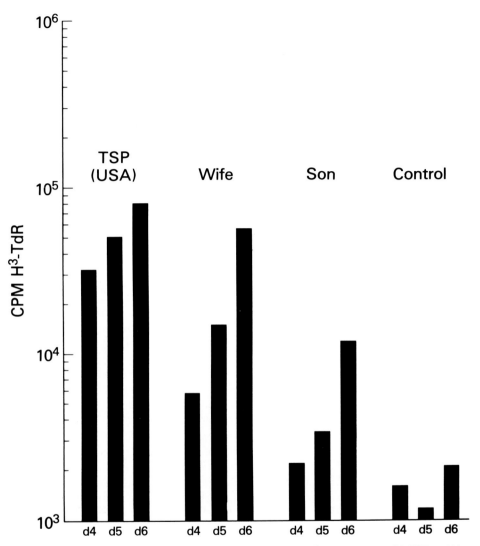

FIG. 3. Spontaneous proliferative response of a United States Caucasian man with TSP and his wife and son. The control group is a normal HTLV-I–seronegative individual with no disease. Tritiated thymidine incorporation is plotted versus 4, 5, or 6 d in culture.

The results in Fig. 4 also serve to illustrate some of the advantages and potential problems associated with the PCR assay. It was of interest to determine in what cell type HTLV-I can be detected. PBL were fractionated into T (E+)- and non-T-cell (E−) fractions by rosetting with sheep red blood cells. As seen in the wife

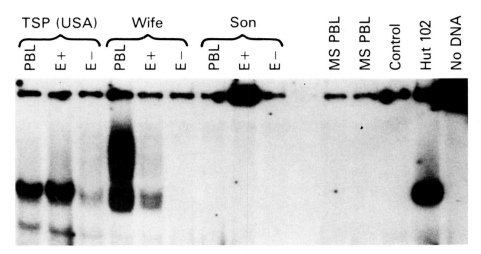

FIG. 4. Liquid hybridization of amplified DNA from a United States Caucasian man with TSP and his wife and son. DNA was amplified with HTLV-I *pol* primers from PBL as well as sheep red blood cell rosetting positive (E+) and negative (E−) cells. As negative controls, tubes with no DNA as well as control DNA from an HTLV-I–seronegative individual were amplified. The amplified DNA from the HTLV-I–positive HUT 102 cell line was used as a positive control. Amplified DNA from the PBL of two MS patients is also shown.

of the American TSP patient, an HTLV-I pol signal was only detected in the PBL and E+ T-cell population, which indicated that T cells were infected. The husband clearly generated a strong pol signal in both the PBL and E+ T-cell fraction but in addition showed a weak signal in the E− non-T-cell population which was obtained (Fig. 4). Two interpretations of these results are possible. The first is that in addition to infection of T cells (E−), cells such as macrophages and B cells are infected. The second is that a small contamination with E+ T cells in the E− non-T-cell fraction supplied enough DNA that was amplified by the PCR. Because of the profound sensitivity of the PCR, we favor the second possibility. This experiment emphasizes that caution must be exercised in interpretation of any positive result.

Studies of an Indian Patient

Another application of the PCR was in the evaluation of a patient from South India with progressive spasticity. This 46-yr-old male has an approximate 25-yr history of weakness and hyperreflexia which gradually progressed to involve both upper extremities, as well as the lower extremities; in addition, the jaw jerk was hyperreflexic. The CSF was normal and the patient did not have antibody to HTLV-I in either the blood or the CSF. The blood lymphocytes did not express increased amounts of activation molecules and did not show abnormally high spontaneous proliferation. After the seventh passage, cultures of the blood lymphocytes expressed gag proteins, and evaluation with this cell line with PCR using primers to the *pol* region showed a pattern consistent with HTLV-I. Although the spastic quadraparesis was consistent with HAM/TSP, the patient lacked serological evidence of HTLV-I infection, and neither the lymphocyte nor the CSF findings were consistent with HAM/TSP. Consequently, when a positive lymphocyte culture and PCR findings (as determined by HTLV-I specific primers and probes to the *env* region, Fig. 5) were obtained, there was concern that a laboratory error may have occurred. The patient's PBL had been cryopreserved and were used to repeat the PCR. The findings were confirmed in this original material from the patient.

The findings in this case suggest that HTLV-I genome can exist in the absence

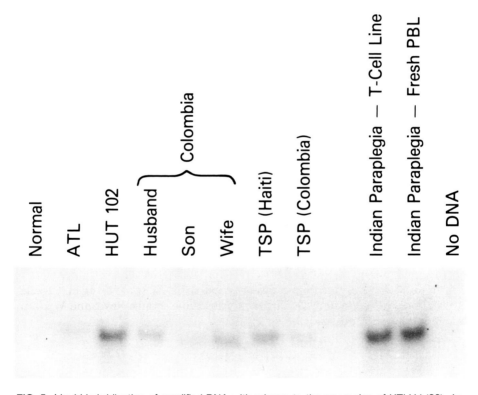

FIG. 5. Liquid hybridization of amplified DNA with primers to the *env* region of HTLV-I (28). As negative controls, tubes with no DNA as well as control DNA from an HTLV-I–seronegative, normal individual were amplified with the *env* primers. Positive controls were the amplified DNA from the HTLV-I–positive HUT 102 cell line and DNA from the PBL of a patient with ATL. The TSP patients from Haiti and Colombia as well as the Colombian family have been reported previously (11,12). Amplified DNA from both fresh PBL and an in vitro derived T-cell line from an Indian patient is shown.

of lymphocyte abnormalities and antibodies. These results have led to the speculation that such infection may have been related to the patient's neurological disorder. However, the results also emphasize that the presence of amplified DNA that reacts with an HTLV-I probe, although consistent with infection by this agent, is not absolute proof. Additional studies are needed, including sequencing of the amplified material and additional PCR studies with primers to other regions.

Studies of MS Patients

As indicated above, the possibility that MS is related to a retrovirus is attractive (9). Others have recently described the detection of HTLV-I proviral sequences by the use of PCR in some patients with MS (27,28). Our preliminary attempts using the technology described above have been, for the most part, unsuccessful. As shown in Fig. 4, DNA from the PBL of two MS patients were amplified with primers to the *pol* region of HTLV-I. No hybridizing bands were observed after liquid hybridization. These two patients were representative of 24 of 25 MS patients tested in this manner. However, in one patient, a man from West Virginia who presented with a predominantly spinal form of the disease, there was amplification of his DNA with the pol primers and subsequent hybridization with a pol probe. These results typify the patterns of data obtained by a number of laboratories. A complete sequence analysis of these hybridizing bands is currently under investigation in order to assess the relationship of these bands to known sequences of HTLV-I. In addition to the possible association between HTLV-I and MS, the involvement of other human retroviruses in this disease remains open. Clearly more extensive investigations of this important area by other laboratories will be required.

ACKNOWLEDGMENTS

The authors express their appreciation to Dr. Owen St. Clare Morgan, Pamela Rodgers-Johnson, Vladimir Zaninovic, Carlos Mora, Frank Neva, and William Sheremata for referring patients for this study.

REFERENCES

1. Gessain A, Vernant JC, Maurs L, et al. Antibodies to human T-lymphotropic virus I in patients with tropical spastic paraparesis. *Lancet* 1985;2:407–410.
2. Rodgers-Johnson P, Gajdusek DC, Morgan OS, et al. HTLV-I and HTLV-III antibodies and tropical spastic paraparesis. *Lancet* 1985;2:1247–1248.
3. Bartholomew C, Cleghorn F, Charles W, et al. HTLV-I and tropical spastic paraparesis. *Lancet* 1986;2:97–99.
4. Osame M, Usuku K, Izumo S, et al. HTLV-I–associated myelopathy, a new clinical entity. *Lancet* 1986;1:1031–1032.

5. Vernant JC, Maurs L, Gessain A. Endemic tropical spastic paraparesis associated with human T-lymphotropic virus type I: a clinical and seroepidemiological study of 25 cases. *Ann Neurol* 1987;21:123–130.
6. Osame M, Matsumoto M, Usuku K, et al. Chronic progressive myelopathy associated with elevated antibodies to human T-lymphotropic virus type I and adult T-cell leukemia like cells. *Ann Neurol* 1987;21:117–122.
7. Roman GC, Osame M. Identity of HTLV-I–associated tropical spastic paraparesis and HTLV-I–associated myelopathy. *Lancet* 1988;2:651.
8. Johnson RT. Viral aspects of multiple sclerosis. In: Vinken PJ, Bruyn GW, Klawans HL, Koetsier JC, eds. *Handbook of clinical neurology. Demyelinating diseases,* vol. 11. (47 Revised Series 3). Amsterdam, New York: Elsevier Science Publishers, 319–336.
9. Koprowski H, DeFreitas EC, Harper ME, et al. Multiple sclerosis and human T-lymphotropic retroviruses. *Nature* 1985;318:154–160.
10. Jacobson S, Zaninovic V, Mora C, et al. Immunological findings in neurological diseases associated with antibodies to HTLV-I: activated lymphocytes in tropical spastic paraparesis. *Ann Neurol* 1988;23(Suppl.):S196–S200.
11. Jacobson S, Raine CS, Mingioli ES, McFarlin DE. Isolation of an HTLV-I–like retrovirus from patients with tropical spastic paraparesis. *Nature* 1988;331:540–543.
12. Jacobson S, Gupta A, Mattson D, Mingioli ES, McFarlin DE. Immunological studies in tropical spastic paraparesis (TSP). *Ann Neurol* 1990;27:149–156.
13. Gout O, Gessain A, Bolgert F. Chronic myelopathies associated with human T-lymphocytotropic virus I: a clinical serologic and immunovirologic study of ten patients in France. *Arch Neurol* 1989;46:255–261.
14. Wong-Staal F, Gallo RC. Human T lymphocytotropic virus I. *Nature* 1985;317:395–403.
15. Greene WC, Leonard WJ, Depper JM, Nelson DL, Waldmann TA. The human interleukin-2 receptor: normal and abnormal expression in T cells and in leukemias induced by the human T-lymphocytotropic viruses. *Ann Intern Med* 1986;105:560–572.
16. Lanier LL, Serafin AT, Ruitenberg JJ, et al. The gamma T cell antigen receptor. *J Clin Immunol* 1987;7:429–440.
17. Gattolo L, Dodon MO. Direct activation of resting T-lymphocytes by human T-lymphotropic virus type I. *Nature* 1987;326:714–717.
18. McFarlin DE, Mingioli ES, Jacobson S. Isolation of HTLV-I–like agents from patients with tropical spastic paraparesis by stimulation of the CD3-T cell receptor complex. In: Roman GC, Vernant JC, Osame M, eds. *HTLV-I and the nervous system.* New York: Alan R. Liss, 1989;51: 31–37.
19. Haase A, Evangelista A, Minnigan H, et al. The issues of causation and neurotropism in neurological diseases associated with infections by retroviruses. In: Blattner WA, ed. *Human Retrovirology: HTLV.* New York: Raven Press, 1990:15–25.
20. Greenberg SJ, Jacobson S, Waldmann TA, McFarlin DE. Molecular analysis of HTLV-I proviral integration and T cell receptor beta-chain rearrangement in tropical spastic paraparesis. *J Infect Dis* 1989;159:741–745.
21. Sarin PS, Rodgers-Johnson PEB, Sun DK, et al. Comparison of a human T-cell lymphotropic virus type-I strain from cerebrospinal fluid of a Jamaican patient with tropical spastic paraparesis with a prototype human T-cell lymphotropic virus type-I. *Proc Natl Acad Sci USA* 1989;86: 2021–2025.
22. Tsujimoto A, Teruuchi TJ, Imamura J, Shimotohno K, Miyoshi I, Miwa M. Nucleotide sequence analysis of a provirus derived from HTLV-I–associated myelopathy (HAM). *Mol Biol Med* 1988;5: 29–42.
23. Schnittman SM, Psallidopoulos MC, Lane HC, et al. The reservoir for HIV in human peripheral blood is a T cell that maintains expression of CD4. *Science* (in press).
24. Kniker WT, Anderson CT, Roumiantzeff M. The multi-test system: a standardized approach to evaluation of delayed hypersensitivity and cell-mediated immunity. *Ann Allergy* 1979;43:73–79.
25. Harris J, Jacobson S, McFarlin DE. Cutaneous anergy in chronic progressive myelopathy (CPM) associated with human T-lymphotropic virus type I (HTLV-I) in a Caucasian male and his asymptomatic seropositive wife. *Ann Neurol* (in press).
26. Tachibana N, Okayama A, Ishizaki J, et al. Suppression of tuberculin skin reaction in healthy HTLV-I carriers from Japan. *Int J Cancer* 1988;42:829–831.
27. Reddy EP, Sandberg-Wollheim M, Mettus RV, et al. Amplification and molecular cloning of HTLV-I–sequences from DNA of multiple sclerosis patients. *Science* 1989;243:529–533.

28. Greenberg SJ, Ehrlich GD, Abbott MA, Hurwitz BJ, Waldmann TA, Poiesz BJ. Detection of sequences homologous to human retroviral DNA in multiple sclerosis by gene amplification. *Proc Natl Acad Sci USA* 1989;86:2878–2882.

DISCUSSION

A discussant wanted to know where and how American TSP patients acquired their infections. Another discussant replied that they got their infections in different geographic locations and probably by different routes. For example, one boy was breast-fed. The father of the child had previously given the infection to his wife.

A participant asked what happened to the HTLV-I–infected cell lines and if there was any T-cell–receptor-gene rearrangement. It was explained that there was receptor-gene rearrangement, but the mechanism was not known.

A participant asked if there were any HTLV-I–specific biological phenomena and also inquired about related phenomena affecting peripheral blood lymphocytes that might occur in acute viral infection. Another participant indicated that there were. AZT will stop spontaneous proliferation by stopping env expression. There is an associated effect on spontaneous proliferation.

There was a question whether spontaneous proliferation occurred in seronegative HTLV-I, virus-positive patients. The answer was affirmative, especially after culture. Using PCR, proviral DNA increased with culture of virus. PCR is carried out in whole PBL where no DNA had been amplified. ATL serum inhibits HTLV-I, but TSP serum does not. A discussant stated that HUT-102 is a poor HTLV-I donor and inquired whether other cell lines had been used. The answer was that with the use of HUT-102, 70%–80% of the cells became infected.

Human Retrovirology: HTLV,
edited by William A. Blattner.
Raven Press, Ltd., New York 1990.

Spontaneous Lymphocyte Proliferation Is Elevated in Asymptomatic HTLV-I–Positive Jamaicans

*Alexander Krämer, †Steven Jacobson, †Jayne S. Reuben,
*Edward L. Murphy, *Stefan Z. Wiktor, ‡Beverly Cranston,
§J. Peter Figueroa, ‡Barrie Hanchard, †Dale McFarlin,
and *William A. Blattner

*National Cancer Institute, Bethesda, Maryland; †National Institute of Neurological
and Communicative Disorders and Stroke, Bethesda, Maryland; ‡University of the
West Indies, Kingston, Jamaica; and §Jamaican Ministry of Health, Kingston, Jamaica

INTRODUCTION

The normal immune response involves a delicately balanced interplay of regulatory and effector cells. Included in this repertoire is the controlled expansion or proliferation of subsets of T and B lymphocytes as part of the immune response to foreign antigens, such as viruses. Recently, Jacobson and colleagues reported elevated spontaneous proliferation of lymphocytes in patients with HTLV-I–associated Myelopathy/Tropical Spastic Paraparesis (HAM/TSP) who were infected with human T-cell leukemia virus (HTLV-I) (1). However, this study did not address the issue of whether this phenomenon was associated with disease (HAM/TSP) or exposure (HTLV-I). The current study was undertaken to characterize spontaneous lymphocyte proliferation in HTLV-I–infected normal persons free of neurologic disease and to ascertain whether other factors (e.g., demographic, socioeconomic, etc.) influence this phenomenon.

METHODS

Subject Selection

Whole blood samples were obtained from four groups of subjects. All the Jamaican subjects were enrolled in epidemiologic studies of HTLV-I sponsored by the National Cancer Institute in collaboration with the University of the West Indies in Kingston, Jamaica, and the Jamaican Ministry of Health. These studies had been approved by the protocol review committees of both institutions, and all subjects gave informed written consent. The American control subjects were blood donors at the National Institutes of Health Blood Bank whose blood samples were designated for research purposes. The four experimental groups were

Group 1. Healthy Jamaican HTLV-I–seronegative controls. The 15 subjects in this group were randomly chosen from a nationwide serologic survey of HTLV-I antibody in Jamaica in 1985–1986. Healthy individuals were enrolled when they applied for licenses required for employment in food-handling occupations. In collaboration with the Jamaican Ministry of Health, blood samples obtained for syphilis screening were separated into aliquots for the HTLV-I study. A questionnaire concerning demographic and life-style factors and medical history and a screening physical examination were administered by trained personnel. Median age of subjects was 40; four were male; all were black.
Group 2. Healthy Jamaican HTLV-I seropositives. Fifteen healthy subjects from the same survey who tested positive for HTLV-I antibodies were randomly included in this group. Median age was 41; four were male; all were black.
Group 3. Patients with adult T-cell leukemia/lymphoma (ATL). Twelve patients with ATL were enrolled in a case-control study of all hematologic malignancies diagnosed at the University Hospital of the West Indies from 1984 through 1986 (2). All 12 had a clinicopathologic diagnosis of diffuse lymphoma, T-lymphocyte phenotype of the malignant cells, and antibodies to HTLV-I. Skin infiltration and hypercalcemia were common clinical findings in this group. Median age was 40; three were male; all were black.
Group 4. Healthy American controls. Thirteen healthy blood donors from the National Institutes of Health Blood Bank served as controls. All were seronegative for HTLV-I and HIV-1. Median age was 40; three were male; all were white.

Virologic Studies

Sera were tested for the presence of antibodies to HTLV-I with a research enzyme-linked immunosorbent assay (ELISA) that used purified, disrupted whole virus particles as antigen (DuPont Corp., Wilmington, DE). All positive samples were confirmed with a Western blot assay that used HUT102 cells as the source of antigen (Biotech Laboratories, Rockville, MD). The minimum criteria for a positive Western blot consisted of the presence of bands specific to the HTLV-I *gag* gene proteins p19 and p24 (3). Additionally, an envelope ELISA (Cambridge Bioscience, Cambridge, MA) was performed for confirmation, thus satisfying current Food and Drug Administration criteria that antibodies to two gene groups are detected in positive samples.

Immunologic Studies

Lymphoproliferative Response

Peripheral blood lymphocytes (PBL) were isolated by standard methods and viably frozen in vapor-phase liquid nitrogen until use. Cryopreserved PBL were thawed and plated in triplicate at a concentration of 3×10^5 cells per well in 96-well plates (Nunclon, Roskilde, Denmark). Media consisted of RPMI (GIBCO, Grand Island, NY) supplemented with 2% human AB serum. After 4, 5, and 6 d, wells were pulsed with 1 μCi of ^{3}H-thymidine for 4 h and harvested on a Scatron harvesting system (Skatron Inc., Sterling, VA) and counted per minute (CPM) in a liquid scintillation counter.

Statistical Methods

Spontaneous lymphoproliferation counts were log-transformed for analysis. Student's t test was used for comparing means. To evaluate risk factors for high lymphoproliferation among the 30 healthy Jamaicans, odds ratios (OR) and the appropriate 95% confidence intervals (CI) were calculated for a variety of demographic and life-style factors, including medical history. Logistic regression was carried out to determine the separate effects of HTLV-I serostatus and income on spontaneous lymphoproliferation. For these purposes, variables were dichotomized at their median.

RESULTS

In vitro spontaneous lymphocyte proliferation (in the absence of exogenous antigens) at days four and five was highly correlated with proliferation at day six ($r = 0.75$, $p = 0.0001$; and $r = 0.90$, $p = 0.0001$, respectively). The findings at day 6 are shown in Fig. 1. Mean proliferation rate ($\pm$standard error) was 62 852 ($\pm$12 569) CPM in HTLV-I–seropositive Jamaicans compared with 18 483 ($\pm$4371) CPM in HTLV-I–seronegative Jamaican controls, 12 629 ($\pm$2860) CPM in HTLV-I–seropositive ATL, and 2345 ($\pm$436) CPM in HTLV-I–seronegative American controls. Comparisons of mean responses between the various groups showed three significant differences: 1. HTLV-I–seropositive versus HTLV-I–seronegative Jamaicans ($p = 0.003$), 2. HTLV-I–seropositive Jamaicans versus HTLV-I–seropositive ATL ($p = 0.0004$), and 3. HTLV-I–seronegative Jamaicans versus American controls ($p = 0.0002$). There was no significant difference between rates of HTLV-I–seronegative Jamaican controls and HTLV-I–seropositive ATL ($p = 0.7$). There was no correlation between proliferation rate and white blood cell count or lymphocyte count for any of the four groups.

Additionally, among the 30 healthy seropositive and seronegative Jamaicans,

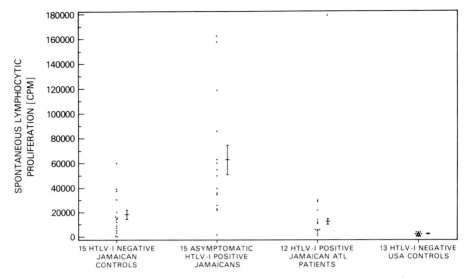

FIG. 1. Spontaneous lymphocytic proliferation rates (counts per minute [CPM] of ^{3}H-thymidine intake of lymphocytes at day 6) in the four study groups; 15 HTLV-I–seronegative Jamaicans, 15 asymptomatic HTLV-I–seropositive Jamaicans, 12 HTLV-I–seropositive Jamaican patients with ATL, and 13 HTLV-I–seronegative American controls. Vertical bars indicate mean values ± SEM. HAM/TSP patients (7) had a mean lymphocytic proliferation rate of 57 927 (±9916) which thus was similar to a rate of 62 852 (±12 569) of asymptomatic HTLV-I carriers in this study. Similar values for the controls in the two studies assure comparability of the results.

the relationship between lymphocyte proliferation and a number of demographic features (age, sex, and socioeconomic status) was explored. Of these, only low income was associated with high lymphoproliferation. However, logistic regression analysis revealed that HTLV-I seropositivity was a stronger risk factor for high lymphoproliferation (OR = 17.2, 95% CI 1.8, 164.7) than was low income (OR = 8.5, 95% CI 0.9, 81.6).

DISCUSSION

The clinical outcomes of HTLV-I infection include a type of T-lymphoproliferative malignancy (ATL) and a chronic degenerative neurologic disease (HAM/TSP). Clinical immunodeficiency associated with HTLV-I has been anecdotally reported and myriad in vitro immunologic perturbations observed (4).

Jacobson et al. have recently reported that patients with HAM/TSP have an array of immunologically abnormal cells, particularly increased numbers of activated T lymphocytes in fresh peripheral blood as measured by increased proportion of DR and Tac positive cells (1). Similarly, Yasuda et al. (5) and Usuku et al. (6) observed activated T cells in fresh isolates and cultured PBL. In addition to these perturbations in T-cell subsets, Jacobson also reported elevated spontaneous proliferation of T lymphocytes. The current study was undertaken to evaluate in vitro lymphocyte proliferation in healthy HTLV-I–positive individuals from Jamaica.

Spontaneous lymphoproliferation in asymptomatic HTLV-I–positive Jamaicans (Fig. 1) was significantly higher than among HTLV-I–negative Jamaicans. This finding was not correlated with total white blood cell or lymphocyte counts, arguing that this association was not an artifact of an overabundance of dividing cells. Rather these data suggest that elevated spontaneous lymphoproliferation may be associated with HTLV-I exposure per se. Because the rate of proliferation was indistinguishable in HTLV-I positives and HAM/TSP patients (7), a possible mechanism that may also provide insight into the pathogenesis of HAM/TSP is provided by Sonoda and colleagues (6). In their in vitro studies, exogenous HTLV-I antigens were added to resting PBL and induced proliferation to HTLV-I antigen was measured. They reported that HAM/TSP patients had a high lymphocyte proliferative response to added virus, whereas those with ATL were low responders. Family members who shared certain HLA-haplotypes with HAM/TSP or ATL patients showed parallel responses, i.e., those with the HAM/TSP HLA-haplotype were high responders, whereas those with the ATL HLA-haplotype were low responders. Perhaps the presence of virus infection in our study subjects results in viral antigen expression in vitro which stimulates cells to spontaneous proliferation. Whatever the mechanism, such a perturbation could lead to altered cell growth as well as immunologic dysregulation, with obvious implications for HTLV-I disease pathogenesis.

The low spontaneous lymphocytic proliferation in HTLV-I–positive ATL likely reflects an overabundance of HTLV-I–positive tumor cells that have lost their capacity for proliferation. The frequent finding of opportunistic infections in ATL supports the concept that ATL patients have an altered immune response.

Recent reports have suggested that persons co-infected with HTLV-I and HIV-1 may progress to AIDS at an accelerated rate (8–10). HTLV-I–induced spontaneous proliferation of blood lymphocytes may provide a mechanism to explain this. Specifically, for HIV-1 to kill a T lymphocyte, current evidence suggests that that cell must undergo cell division, which promotes viral gene expression, an overproduction of viral particles, and cell death. HTLV-I may act as a T-4–cell mitogen that stimulates cell division, both accelerating cell death as well as creating activated T cells, which are suitable targets for HIV-1 infection.

Of the two HTLV-I–uninfected groups, HTLV-I–negative Jamaicans had significantly higher lymphoproliferation compared with American controls. The reason for this finding is unclear. Because all Jamaicans were black and all American controls were white, genetic factors may account for this difference (11). However, it is also possible that spontaneous proliferation is influenced by the level of acute and chronic exposure to a variety of different antigens and pathogens. Multiple infections are more common in the tropical country of Jamaica, which has a significantly lower per capita income and standard of living than the United States. Because low income may be a surrogate marker for antigenic stimulation by various pathogens, it is interesting that in this study low income was an additional risk factor for high lymphoproliferation.

We conclude that 1. HTLV-I infection is the main source for increased spontaneous lymphoproliferation, a functional abnormality of T lymphocytes occurring in the absence of exogenous immune stimulators; and 2. environmental factors, such as other antigens (parasites, viruses, dietary components, and others) may modify this effect. Additional prospective data are needed to further clarify the significance of spontaneous lymphocytic proliferation in the natural history of HTLV-I infection and disease.

SUMMARY

In vitro elevated spontaneous lymphocytic proliferation (in the absence of exogenous antigens), first reported in patients with HTLV-I–associated myelopathy/tropical spastic paraparesis (HAM/TSP), was measured in asymptomatic HTLV-I–seropositive and –negative Jamaicans. Mean spontaneous lymphocytic proliferation rate was 62 852 [±12 569 SEM] CPM in 15 seropositives compared with 18 483 [±4371] CPM in 15 seronegatives ($p = 0.003$). Among 12 HTLV-I–seropositive Jamaican patients with ATL, the counts were 12 629 (±2860) CPM, thus significantly lower than asymptomatic Jamaican HTLV-I seropositives ($p = 0.0004$), but similar to seronegative Jamaican controls ($p = 0.7$). Thirteen HTLV-I–seronegative American controls had a mean proliferation value of 2345 (±436) CPM, significantly lower compared with Jamaican seronegative controls ($p = 0.0002$). Elevated spontaneous lymphocytic proliferation in asymptomatic Jamaican seropositives appears to be due to HTLV-I infection per se, whereas the absence in seropositive ATL may reflect overproduction of functionally deficient tumor cells. The higher rate in seronegative Jamaicans compared with American controls may reflect racial or environmental exposure differences, because markers of lower social class in Jamaica were moderately associated with increased proliferation.

REFERENCES

1. Jacobson S, Zaninovic V, Mora C, et al. Immunologic findings in neurological diseases associated with antibodies to HTLV-I: activated lymphocytes in tropical spastic paraparesis. *Ann Neurol* 1988;23(suppl):S196–S200.
2. Gibbs WN, Lofters WS, Campbell M, et al. Non-Hodgkin lymphoma in Jamaica and its relation to adult T-cell leukemia-lymphoma. *Ann Intern Med* 1987;106:361–368.
3. Agius G, Biggar RJ, Alexander SS, et al. Human T-lymphotropic virus type I antibody patterns: evidence of difference by age and risk group. *J Infect Dis* 1988;158:1235–1244.
4. Popovic M, Flomenberg N, Volkman DJ, et al. Alterations of T-cell functions by infection with HTLV-I or HTLV-II. *Science* 1984;226:459–462.
5. Yasuda K, Yoshitatsu S, Yokoyama MM, Tanaka K, Hara A. Healthy HTLV-I carriers in Japan: the haematological and immunological characteristics. *Br J Haematol* 1986;64:195–203.

6. Usuku K, Sonoda S, Osame M, et al. HLA haplotype–linked high immune responsiveness against HTLV-I in HTLV-I–associated myelopathy: comparison with adult T-cell leukemia/lymphoma. *Ann Neurol* 1988;23(suppl):S143–S150.
7. Jacobson S, Gupta A, Mattson D, Mingioli E, McFarlin DE. Immunological studies in tropical spastic paraparesis (TSP). *Ann Neurol* 1989 (*in press*).
8. Bartholomew C, Blattner W, Cleghorn F. Progression to AIDS in homosexual men co-infected with HIV and HTLV-I in Trinidad. *Lancet* 1987;2:1469.
9. Hattori T, Koito A, Takatsuki K, et al. Frequent infection with human T-cell lymphotropic virus type I in patients with AIDS but not in carriers of human immunodeficiency virus type 1. *J AIDS* 1989;2:272–276.
10. Weiss SH, French J, Holland B, et al. HTLV-I/II co-infection is significantly associated with risk for progression to AIDS among HIV+ intravenous drug abusers. *Fifth International Conference on AIDS,* Montreal 1989, Abstract Th.A.O.23, p.75.
11. Tollerud DJ, Clark JW, Morris Brown L, et al. The influence of age, race, and gender on peripheral blood mononuclear-cell subsets in healthy nonsmokers. *J Clin Immunol* 1989;9:214–222.

Human Retrovirology: HTLV,
edited by William A. Blattner.
Raven Press, Ltd., New York 1990.

Altered Cellular Gene Expression in Human Retroviral-Associated Leukemogenesis

*Steven J. Greenberg, *||Craig L. Tendler, †Angela Manns,
‡Courtenay F. Bartholomew, §Barrie Hanchard,
†William A. Blattner, and *Thomas A. Waldmann

Metabolism Branch, and †Environmental Epidemiology Branch, National Cancer Institute, National Institutes of Health, Bethesda, Maryland; ‡University of West Indies, Port of Spain, Trinidad; and §University of West Indies, Kingston, Jamaica; ||St. Jude Fellow of the Pediatric Scientist Training Program

INTRODUCTION

The first pathogenic human retrovirus discovered, human T-cell leukemia virus type I (HTLV-I) (1), was isolated from a patient with a variant form of cutaneous T-cell lymphoma (2). Subsequently it was recognized that HTLV-I–associated disease expression is highly pleiotropic. Although the majority of infections are asymptomatic, HTLV-I is thought to cause the hematologic malignancy adult T-cell leukemia (ATL) (3), the progressive neurologic condition tropical spastic paraparesis (TSP) (4,5) or HTLV-I–associated myelopathy (HAM) (6), and is associated with an immunosuppressed state reflected in vitro and in vivo (7–14). Therefore, HTLV-I may be oncogenic, neuropathic, or immunosuppressive in its host; this degree of diversity at the primary clinical level is amplified by secondary cellular physiologic derangements which often accompany HTLV-I–associated disease.

The mechanisms by which HTLV-I infection contributes to malignant transformation and subversion of the cellular biochemistry of the host are unclear, but are thought to be related to the activation of cellular gene expression by a viral transactivating function. However, attempts to demonstrate viral regulatory RNA transcripts in vivo have been unsuccessful. More recently, advances in technology have enabled one to detect minute quantities of viral message as well as cellular mRNAs. This article will examine the current data pertaining to altered cell expression in HTLV-I retroviral infection. The clinical and physiologic consequences of HTLV-I infection are introduced first, followed by a review of the molecular biological evidence supporting a role for altered cell expression by the unique retroviral transactivator. A modification of the technique of gene amplification by the polymerase chain reaction targeting splice excluded transcription (SET-PCR) products was developed. This highly sensitive technique was employed to demonstrate the high level of constitutively expressed interleukin-2 receptor alpha (IL-2Rα) gene in lymph node as well as circulating mononuclear cells in ATL. We also report its successful application for the detection of HTLV-I–transactivator-associated message in select ATL tissues ex vivo. The demonstration of the retroviral regulatory transcript supports a role for transactivation in altered cellular gene expression in viral leukomogenesis.

CLINICAL MANIFESTATIONS OF HTLV-I INFECTION

The clinical features of ATL are that of an adult onset, subacute or chronic leukemia that invariably terminates in a rapidly progressive malignant course, frequently associated with lymphadenopathy and hepatosplenomegaly, disseminated skin lesions, and a proclivity toward invasion of the central nervous system, lungs, and gastrointestinal tract (13,15). In most cases infection with HTLV-I can be affirmed by the presence of circulating antibodies specific for HTLV-I antigens and by demonstration of HTLV-I proviral DNA integration into host genome on Southern hybridization (3). Laboratory diagnosis of ATL is confirmed by i) a lymphocytosis accompanied by circulating abnormal mature lymphocytes which tend to display multilobulated nuclei, dense clumped nuclear chromatin, and inconspicuous nucleoli (13,15); ii) a predominant helper T-cell phenotype expressing IL-2Rα chain (CD2$^+$ CD3$^+$ CD4$^+$ CD8$^-$ CD25$^+$) (14,16,17); iii) elevated serum, soluble IL-2R levels (18); and iv) T-cell clonal expansion defined using an analysis of beta chain gene rearrangement (19). In addition to leukemia, paraneoplastic states are frequently associated with ATL, reflecting altered regulation of cellular genes, and include hypercalcemia, unexplained rashes, and infiltrative skin lesions; hepatic dysfunction (13,15); and encephalopathy in the more malignant terminal stages of illness.

Disease onset in HAM/TSP is usually insidious, although in some cases sudden events of illness reminiscent of a subacute transverse myelitis occur (20). Spasticity is usually profound and may overshadow weakness. Motor strength is compromised more in the lower than upper extremities. Initially the distribution of weakness may be asymmetric, slowly progressing to affect both limbs. Bladder and bowel dysfunctions are frequently encountered. Dysesthesias and numbness are experienced by a majority of patients, but abnormal sensory findings are minimal or absent, being limited to a decrease of vibratory or pin-prick perception distally in the feet, and may reflect subclinical involvement of spinal sensory pathways revealed by somatosensory evoked potentials (21,22). On examination there is a paucity of cranial nerve findings, although prolonged latencies on testing visual- and brain stem auditory-evoked potentials are detected in a subset of patients (23,24). In one study, mild to moderate electroencephalographic abnormalities were found in a majority of patients screened (25), but cognitive deficits are not a recognized feature of HAM/TSP. Electrophysiologic testing may disclose an associated radiculopathy or mild peripheral sensorimotor poly-neuropathy (21–23,26), and brain magnetic resonance imaging (MRI) studies have occasionally disclosed foci of increased signal intensity in periventricular white matter and brain stem (22,24). However, the preponderant clinical presentation in HAM/TSP strongly localizes in the neuraxis to the spinal cord.

Cellular and humoral immune responses are markedly impaired in patients with ATL, but varying degrees of immunosuppression have also been observed among HAM/TSP patients (27) and occasionally in otherwise healthy HTLV-I–seropositive carriers (28). Abnormalities in ATL include disordered helper and suppressor T-cell function, mitogen responsiveness, killer cell induction, and B-cell immunoglobulin synthesis (7,8,29). HTLV-I–infected ATL cells and supernatants of ATL lines suppress the lectin-induced immunoglobulin synthesis of cocultured mononuclear cells (9–11). Delayed-type hypersensitivity by skin testing routinely reveals anergy in both ATL and HAM/TSP patients (12,27). Dysregulation of the immune response in ATL manifests clinically by the frequent occurrence of opportunistic infection, and includes *Pneumocystis carinii* and cytomegalovirus pneumonias, cryptococcal meningitis, fungal and bacterial sepsis, and parasitic gastroenteritis (12–14). Indeed, individuals without malignancy or neurologic disease who are infected with HTLV-I may be more prone to infection with various pathogenic organisms (28).

Multiple clinical conditions, including hematologic malignancy and associated paraneoplastic syndromes, neurologic spastic paraparesis, and immunosuppression, in association with the same or similar viral agent, raise many provocative questions concerning viral infectivity and host-immune responsiveness (Table 1). The capacity of HTLV-I elements to transactivate viral and cellular genes has led to speculation that this function contributes to the panoply of disease states associated with HTLV-I infection.

LEUKEMOGENESIS AND CELLULAR GENE ACTIVATION

Seroepidemiologic and molecular genetic data firmly establish an association between HTLV-I and ATL. ATL is endemic in regions of Japan, the Caribbean basin, and central Africa; the geographic distribution for HTLV-I seroprevalence is concordant with that for ATL (30–32). With rare exceptions (33), proviral genome and antibodies against HTLV-I are detected in patients with ATL (34); and, in all cases, HTLV-I is detected in neoplastic cells but not in the majority of nonleukemic cells of patients with ATL (3,35). Furthermore, analysis of proviral genome integration into host cellular DNA reflects a pattern of clonal expansion of the leukemic cells among all ATL cases (3). In vitro cocultivation of primary lymphocytes with x-irradiated HTLV-I–infected cells has been used to transform recipient T cells across species barriers (36–39). The universal finding of HTLV-I among ATL patients, the presence of HTLV-I in neoplastic cells, and the monoclonal nature of the leukemic/lymphoma cells with regard to the

TABLE 1. *Putative induction of cellular biomolecules by HTLV-I tax*

Cultured HTLV-I infected cells

Macrophage migration inhibitory factor*
Leukocyte migration inhibitory factor*
Leukocyte migration enhancing factor*
Colony stimulating factor*
Growth and maturation activity*
Differentiation inducing*
Macrophage activating factor
Fibroblast activating factor
Platelet derived growth factor
Interleukin-1
Interleukin-2 (T-cell growth factor)
Interleukin-3
Interleukin-6 (B-cell growth factor)
Interleukin-2 receptor alpha
Interferon-gamma (gamma-IFN)
Lymphotoxin (TNF-beta)
Tumor necrosis factor (TNF-alpha, cachectin)
Granulocyte/macrophage-colony stimulating factor

HTLV-I infection or tax gene transfection

Interleukin-2
Interleukin-3
Interleukin-4
Interleukin-2 receptor alpha
Lymphotoxin
Tumor necrosis factor
Granulocyte/macrophage-colony stimulating factor
Vimentin

A compilation of cellular products identified as constitutively expressed in supernatants of cultured HTLV-I–infected cell lines (18,43–50) or from transformed cells infected by coculture or transfected with the *tax* cDNA (51–58). Those with asterisks were less well characterized and may represent factors subsequently identified as IL-3, IL-6, or GM-CSF.

provirus integration site tend to imply that T-cell infection with HTLV-I is an initial event that contributes some essential element leading to the development of malignant transformation.

Retroviral-induced malignancy may occur via a number of general mechanisms (40). Rapid cell transformation quickly leading to malignancy may occur through a series of events initiated by products encoded by viral oncogenes. Alternatively, site-specific integration into the host genome may lead to aberrant *cis*-activation of an adjacent cellular protooncogene by the proviral LTR. However, there is no evidence for the existence of a human cellular homologue to any of the nonstructural HTLV-I genes (35,41), and a viral promoter site-specific insertion mechanism is precluded by virtue of the fact that a common integration site of HTLV-I provirus in primary leukemic lymphocytes does not exist (42). These conclusions support the notion that a transacting viral function may be operative in regulating viral expression and in the leukemogenesis of HTLV-I–associated ATL.

A transactivating mechanism is also suspected to activate the expression of certain cellular genes, especially those associated with T-cell activation and proliferation. HTLV-I–transformed T-cell lines have been shown to constitutively produce various biologically active molecules, including lymphokines, interleukins, colony stimulating factors, and the IL-2Rα chain subunit (18,43–50). Infection with HTLV-I by cocultivation or transfection with the HTLV-I transactivator *tax* (see below) results in the activation of certain cellular gene regulatory elements, including IL-2 (T-cell growth factor) and IL-2Rα (51–54), interleukin-3 (55), interleukin-4 (56), granulocyte-macrophage-colony stimulating factor (55), tumor necrosis factor (56), lymphotoxin (56,57), and vimentin (58). The expansion of activated, antigen-reactive T-cell populations is associated with the sequential induction and coordinate expression of a battery of lymphokines, cytokines, and the IL-2Rα. The altered regulation of such cellular genes by viral transactivation may have profound consequences and account for the subverted cellular physiology associated with HTLV-I infection (Table 2).

TRANSACTIVATION OF VIRAL AND CELLULAR GENES

Regulation of HTLV-I production is mediated by the nonstructural gene products Tax (previously Tat, x-lor, p40^x)—a transcriptional activator (59,60) and essential for efficient transcription of the viral genome—and Rex (previously p27^{x-III})—a posttranscriptional regulator (61,62) that up-regulates *gag* and *env* protein expression but suppresses expression of regulatory proteins. The derivation of these retroviral regulatory mRNAs focuses attention on the unique pX region within the HTLV-I genome. Three viral mRNAs, 8.5, 4.2, and 2.1-kb, are required for viral replication (42). The 8.5-kb mRNA is the primary transcript and is used for expression of *gag* and *pol* proteins and also as the progeny viral genome. The two subgenomic mRNAs are formed by various splicing mechanisms: the 4.2-kb mRNA is formed by a single-step splicing event and encodes for the *env* protein; the 2.1-kb mRNA is formed by a double splicing event and encodes for the pX regulatory proteins Tax and Rex. Similar pX genes and corresponding proteins present in HTLV-II (63,64), bovine leukemia virus (65,66), and simian T-cell leukemia virus type I (67,68) attest to the highly conserved nature of these sequences among retroviruses and suggests a fundamental role in viral expression.

From comparative sequencing studies of the retroviral long terminal repeats (LTR), limited regions of homology between HTLV-I and HTLV-II were identified which were thought to contain sequences essential for the regulation of proviral expression in *cis* (69). An arrangement of three imperfect 21-nucleotide repeats in the U3 region of the LTR conformed to an enhancer-like element

TABLE 2. *Altered cell function in HTLV-I–associated disease*

ATL	
T-cell proliferation	(IL2-Rα, IL-2 ?)
Hypercalcemia	(TNF, LT, Gamma-IFN ?)
Dermal lesions	
Pruritus, urticaria	(IL-3 ?)
Infiltrative leukemia	(EGF ?)
KS	(Angiogenic factors ?)
Hepatic dysfunction	(TNF ?)
Encephalopathy	(IL-1, IL-2, TNF ?)
Immunosuppression	
Leukemogenesis	
TSP	
Spontaneous proliferation and T-cell activation	(IL-2Rα, IL-2 ?)
Modulation of MHC molecules in CNS	(Gamma-INF ?)
Seropositive carriers	
Occasional immunosuppression	
Spontaneous proliferation	(IL-2Rα, IL-2 ?)

Various subverted physiologic states associated with three different HTLV-I–infected populations. In parentheses, some cellular products putatively induced by HTLV-I *tax* transactivation are offered as potential mediators.

and deletion of these sequences from the LTR-abolished *tax*-induced retroviral gene expression (70–72).

The Tax protein also acts in *trans* to induce the expression of host cellular genes, especially those that play crucial roles in cell proliferation and differentiation, such as IL-2 and IL-2Rα (51,52,73). In contrast to ATL, T lymphocytes in HAM/TSP appear to be polyclonal with regard to HTLV-I proviral integration and T-cell receptor arrangement (74,75) but demonstrate a significantly elevated level of spontaneous proliferation in vitro (27,76,77). Some degree of augmented spontaneous proliferation is also noted in the HTLV-I carrier state (76), and, although leukocyte counts and differentials are normal, the percentage of CD2$^+$ and CD25$^+$ (Tac-positive, IL-2Rα positive) cells is elevated as well (78). It has been suggested that there may be a phase of autocrine growth in the early stages of HTLV-I infection, where *tax* gene activity induces the production of IL-2 and its receptor, leading to polyclonal T-cell proliferation (18). It is not known whether the spontaneous proliferation in HAM/TSP and the HTLV-I carrier state is an in vitro consequence of T-cell transactivation that is virally driven, or due to a physiologic response of viral antigen immune recognition and subsequent activation and expansion of cytotoxic T cells. A second, currently undefined event (or series of events) seems to be required for leukemogenesis, rendering the ATL lymphocytes autonomous and independent of IL-2 stimulation.

Although transactivation of host genes by the Tax element is a very attractive possibility, previous attempts to identify Tax protein or *tax* message in freshly obtained circulating cells have been unsuccessful. We wished to reexamine this issue looking at other body sites (e.g., lymph nodes versus circulating mononuclear cells) for *tax* expression. Furthermore we wished to use a method, SET-PCR, that is more sensitive than the Northern blot analysis method that was used previously.

AMPLIFICATION OF SET-PCR

The technique of gene amplification by PCR amplifies a sequence of interest by repeating n times a three-step cycle which results in a theoretical 2^n increase in target (79). The DNA sequence, if present, is amplified enzymatically by a thermostable DNA polymerase under the direction of two synthetic oligonucleotide primers. The primers are constructed complementary to opposite strands of the DNA and flank a target sequence usually 100–1000 base pairs in length. Through repetitive cycles of denaturation, annealing, and polymerization, DNA is synthesized which is identical to the region included by the primers. The amplified DNA generated in this manner is positively identified by hybridization to oligonucleotide probes which lie between and are not complementary to the primers (80). PCR is useful in identification of mutational and transformational genetic events, in rapid direct sequencing analysis, and in well identification of low-copy-number genetic sequences, particularly infectious agents such as human retroviruses present at very low multiplicities of infection. This technology provides direct analysis at the molecular genetic level, supersedes many of the constraints imposed by serological data, and is vastly superior in sensitivity to analysis by routine DNA or RNA blot hybridization methods (81).

PCR can be applied to detect unique messenger RNA (mRNA) species (82) by initially subjecting the mRNA of interest to primer-specific reverse transcription. The resultant complementary DNA (cDNA) sequence serves as template in the first polymerization step in PCR to produce a double-stranded DNA molecule. Both DNA strands comprise the primer-specific target sequence that is exponentially amplified through subsequent polymerization cycles. In this way efficient cDNA amplification would proceed simultaneously with that of any genomic or proviral DNA that contaminates the preparation.

To distinguish between genomic or proviral DNA–derived amplified product and mRNA-derived amplified product, we have devised a strategy that constrains the PCR to amplify only transcriptionally spliced genetic sequences characteristic of mRNA. This approach takes advantage of the fact that the PCR reaction is only effective over short distances (usually <2000 bp) and that elements that are far apart in genomic DNA may be brought close together in spliced messenger RNA. With appropriate primers and probes mRNA, but not genomic DNA, will be amplified and detected. This is accomplished by constructing primers homologous to exons, in the genomic sense, or to nonadjacent promoter and open reading frame regions, in the proviral sense, which flank intron or non-transcribed stretches of genetic material. In this manner, the intervening distance between primer pairs is large, i.e., on the order of a few thousand to tens of thousands of bases. By further limiting primer extension time, the ability to fully polymerize complementary DNA sequences by PCR is exceeded and exponential amplification of genomic or proviral DNA is prevented. However, subsequent to transcriptional splicing events, the primer-related regions are juxtaposed and the rearranged product provides a template that permits efficient gene amplification (Fig. 1). Therefore, amplification of SET-PCR is selective for cDNA. Further confirmation of the nature of the amplified sequence is obtained from sizing analysis and by the use of internal probes which recognize exon or intron regions, but not both.

Analysis by SET-PCR is very sensitive and permits detection of as few as 2000 molecules of specific mRNA in 1 μg of total RNA. This level of detection is also quantitative and allows the discrimination between constitutive production and low-level induction of cellular and viral messages, and we have studied the expression of various cytokines and lymphokines by this method.

DETECTION OF CELLULAR AND VIRAL mRNA BY SET-PCR

The induction and expression of IL-2Rα are pivotal events in the antigen reactive T-lymphocyte proliferative response. However, circulating leukemic cells in ATL represent mature, terminally differentiated T cells, and the acquisition of IL-2Rα probably occurs as an early event soon after mitosis in the lymph node. This is supported by the observation that freshly prepared ATL lymph nodes are strongly IL-2Rα positive on immunohistochemical staining (E. Jaffe, personal communication) and suggests that the ATL lymph node would be a rich source of IL-2Rα mRNA. Whether IL-2Rα transcription at this early stage of T-cell maturation is phasic or whether IL-2Rα mRNA production is dysregulated such that it remains constitutively expressed is not known. We applied the technique of SET-PCR to define the level of in vivo cellular gene expression of IL-2Rα by comparison of specific mRNA produced in ATL lymph node and PBL with that of normal tissue. IL-2Rα mRNA was strongly expressed and at comparable levels in ATL lymph node and circulating mononuclear cells, approaching the level of transcription in the HTLV-I–infected T-cell line HUT 102 (Fig. 2). By contrast, the level of IL-2Rα transcription in normal lymph node and circulating mononuclear cells was nominal.

We next attempted to detect the presence of retroviral pX message by using SET-PCR. As previously discussed, the HTLV-I *tax* gene product is capable of transactivating heterologous eukaryotic promoters in addition to the HTLV-I LTR. However, its role in the induction and persistent expression of various cellular genes, including IL-2Rα, is unclear because, although an attractive candidate as noted above, previous attempts to demonstrate Tax protein (83–85) or *tax* message (83) in freshly prepared ATL cells have been unsuccessful. mRNA was extracted directly from freshly prepared circulating ATL mononuclear cells, pX primer-specific reversed transcribed, and amplified for pX cDNA by SET-PCR. In none of eight ATL cases was pX message detected in fresh leukemic cells using SET-PCR.

The peripheral leukemic lymphocytes in ATL represent terminally differentiated T cells, and the expression of pX may segregate instead to tissues of high cell turnover, as in the lymph nodes. As a regulatory message, the half-life of pX mRNA may be short and therefore lost by the time the leukemic cells enter the circulation. To define whether pX was expressed in the malignant lymph nodes of ATL patients, freshly biopsied nodes obtained from patients with ATL were snap frozen in liquid nitrogen. mRNA was extracted from nodal tissue and subjected to gene amplification as above. In two of six cases, sequences hybridized with the HTLV-I probe were generated of proper molecular weight which are representative of the doubly spliced pX message (Fig. 3). This demonstration of HTLV-I pX message present in lymph node tissue from some individuals with ATL provides evidence for the existence of a functional transactivator, although perhaps tissue specific, during the maintenance phase of leukemia/lymphoma in these patients.

DISCUSSION

A fundamental issue in defining the molecular mechanism by which HTLV-I causes transformation and alters cellular gene expression focuses on the role the pX encoded regulatory proteins play at the time of induction or during maintenance of the leukemogenic state. The high level of constitutive expression of IL-2Rα in lymph node and circulating peripheral blood mononuclear cells in ATL might suggest an autocrine method of perpetuating T-cell proliferation. However, it has been reported that IL-2 mRNA was not detected in fresh leukemic cells of most patients, and only extremely low levels of transcripts were present in a minority of HTLV-I–positive cell lines (14). If this is so, then a simple autostimulation model seems unlikely for the majority of patients in the advanced stage of their disease. We are currently screening freshly obtained ATL cells for mRNA IL-2 expression by SET-PCR and shall address whether IL-2 mRNA production is tissue specific. The IL-2R is constitutively expressed in ATL cells. Theoretically this could reflect transactivation of the IL-2Rα gene by HTLV-I–encoded Tax. Transactivation of IL-2Rα expression early in the course of HTLV-I infection may result in cycles of T-cell activation and proliferation which would tend to prime these cells to a second event which ultimately results in leukemogenesis. A corollary of this hypothesis suggests that Tax may be driving lymphocytes, in the case of HAM/TSP or the infected carrier populations, toward polyclonal activation which may be reflected in vitro by the spontaneous proliferative responses others have observed.

However, the possibility that *tax* expression persists during the maintenance

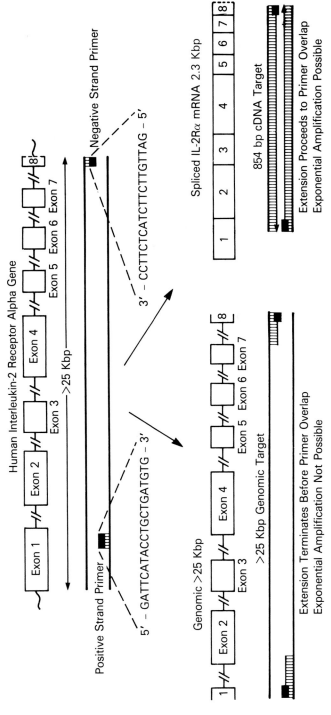

FIG. 1. Strategy for gene amplification of splice-excluded transcription by polymerase chain reaction (SET-PCR). **Above:** Application of SET-PCR for identification of a specific cellular mRNA, in this case human IL-2Rα. Twenty base oligomers were synthesized to complimentary regions spanning a 3' region within exon 1 and 5' region within exon 8 (positive and negative primers, respectively). **Below:** Application of SET-PCR for identification of the retroviral pX mRNA. Twenty base oligomers were synthesized to complimentary regions spanning a sequence within the first spliced segment consisting of 118 nucleotides derived from the R regions of the LTR and a 5' sequence within the third spliced segment consisting of 191 nucleotides 3' in the *pol* region and includes the initiation codon ATG for the *env* gene.

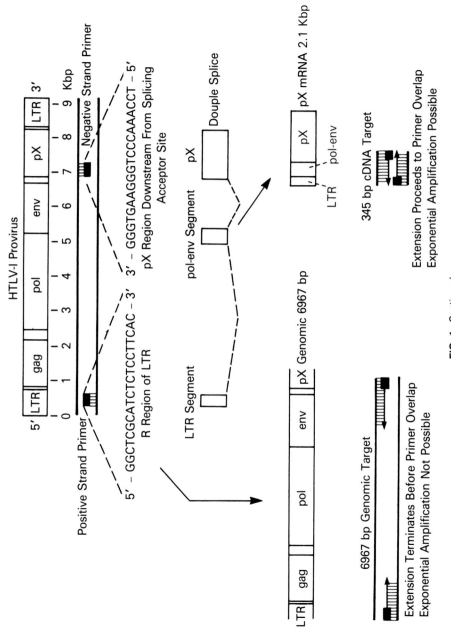

FIG. 1. Continued.

97

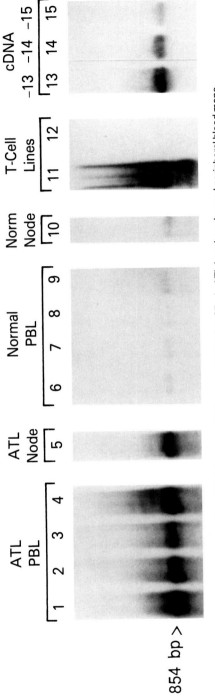

FIG. 2. Demonstration of high-level constitutive expression of IL-2Rα in ATL lymph nodes and peripheral blood mononuclear cells. Blot hybridization of SET-PCR amplified cDNA using exon 1/exon 8 specific primers to IL-2Rα. Probe was generated from a sequence within exon 2. The intense 854-bp signal corresponds to the expected IL-2Rα mRNA-targeted sequence and was observed in all ATL circulating leukemia cells and lymph nodes tested (lanes 1–5). Very weak signals were generated from normal circulating, unstimulated mononuclear cells and normal lymph node specimens (lanes 6–10). Strong signal generated by HTLV-I–infected T-cell line, HUT 102 (lane 11), was comparable with that observed for ATL specimens. Control, non-HTLV-I–infected T-cell line, Jurkat (lane 12), did not generate a signal. Quantitative serial dilutions of 10^{-13}, 10^{-14}, and 10^{-15} g of cloned IL-2Rα cDNA correspond to a sensitivity of detection of 2000 molecules of specific polyadenylated RNA.

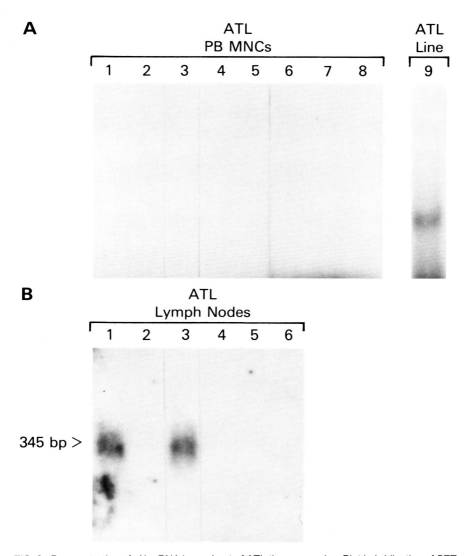

FIG. 3. Demonstration of pX mRNA in a subset of ATL tissue samples. Blot hybridization of SET-PCR–amplified cDNA using LTR/pX specific primers to HTLV-I. Probe was generated from a sequence within the *pol* region comprising the second pX mRNA segment. **A.** 345-bp signal corresponding to the pX-targeted mRNA sequence was generated from a cultured ATL T-cell line (lane 9), but not from mRNA extracted from ATL cells (lanes 1–8) or normal PBMNCs (not shown). **B.** Two of six lymph node specimens obtained from patients with ATL displayed a band pattern at 345 bp (lanes 1 and 3) and demonstrate the presence of pX mRNA in some ATL tissues ex vivo.

state of leukemia has broader implications for the altered cellular physiology that accompanies ATL. In addition to a role in the induction of IL-2Rα expression, pX expression may lead directly to the induction of various cellular bioactive molecules, resulting in an associated paraneoplastic state. For example, biopsy specimens of lytic bone lesions from ATL patients exhibiting hypercalcemia routinely fail to show involvement by tumor and display instead microcystic resorption and increased osteoclastic activity. These findings suggest that a humoral factor, a product of altered cellular gene expression associated with the leukemogenic state, is responsible for the increased osteoclast activation, bone resorption, and hypercalcemia.

Our ability to demonstrate pX message, albeit in only a minority of ATL lymph nodes examined, raises the possibility that within certain microenvironments other than the blood, e.g., within the lymph node or other sites, the expression of the pX-encoded Tax protein may play a vital role in altered cellular gene expression in human retroviral–associated leukemogenesis.

REFERENCES

1. Poiesz BJ, Ruscetti FW, Reitz MS, Kalyanaraman VS, Gallo RC. Isolation of a new type C retrovirus (HTLV) in primary uncultured cells of a patient with Sezary T-cell leukemia. *Nature* 1981;294:268–271.
2. Poiesz BJ, Ruscetti FW, Gazdar AF, Bunn PA, Minna JD, Gallo RC. Detection and isolation of type C retrovirus particles from fresh and cultured lymphocytes of a patient with cutaneous T cell lymphoma. *Proc Natl Acad Sci USA* 1980;77:7415–7419.
3. Yoshida M, Seiki M, Yamaguchi K, Takatsuki K. Monoclonal integration of HTLV in all primary tumors of adult T-cell leukemia suggests causative role of HTLV in the disease. *Proc Natl Acad Sci USA* 1984;81:2534–2537.
4. Roman GC, Schoenberg BS, Madden DL, et al. Human T-lymphotropic virus type I antibodies in serum of patients with tropical spastic paraparesis in the Seychelles Islands. *Arch Neurol* 1987;44:605–607.
5. Vernant JC, Maurs L, Gessain A, et al. Endemic tropical spastic paraparesis associated with human T-lymphotropic virus type I: a clinical and seroepidemiological study of 25 cases. *Ann Neurol* 1987;21:123–130.
6. Osame M, Matsumoto M, Usuku K, et al. Chronic progressive myelopathy associated with increased antibodies and human T-lymphotropic virus type I and adult T-cell leukemic cells. *Ann Neurol* 1987;21:117–122.
7. Morimoto C, Matsuyama T, Oshige C, et al. Functional and phenotypic studies of Japanese adult T-cell leukemia cells. *J Clin Invest* 1985;75:836–843.
8. Uchiyama T, Sagawa K, Takatsuki K, Uchino H. Effect of adult T-cell leukemia cells on pokeweed mitogen-induced normal B-cell differentiation. *Clin Immunol Immunopathol* 1978;10:23–24.
9. Tagawa S, Sawada M, Tokumine Y, et al. Severe immunosuppressive acidic protein in adult T-cell leukemia (ATL). *Scand J Haematol* 1985;34:360–369.
10. Shirakawa F, Tanaka Y, Oda S, et al. Immunosuppressive factors from adult T-cell leukemia cells. *Cancer Res* 1986;46:4450–4462.
11. Waldmann TA, Greene WC, Sarin PS, et al. Functional and phenotypic comparison of human T cell leukemia/lymphoma virus positive adult T cell leukemia with human T cell leukemia/lymphoma virus negative Sezary leukemia, and their distinction using anti-Tac: monoclonal antibody identifying the human receptor for T cell growth factor. *J Clin Invest* 1984;73:1711–1718.
12. Greenberg SJ, Davey MP, Zierdt WS, Waldmann TA. Isospora belli enteric infection in patients with HTLV-I associated adult T-cell leukemia. *Am J Med* 1988;85:435–438.
13. Bunn PA, Schechter GP, Jaffe E, et al. Clinical course of retrovirus-associated adult T-cell lymphoma in the United States. *N Engl J Med* 1983;309:257–264.
14. Broder S, Bunn PA, Jaffe ES, et al. T-cell lymphoproliferative syndrome associated with human T-cell leukemia/lymphoma virus. *Ann Intern Med* 1984;100:543–557.

15. Uchiyama T, Yodoi J, Sagawa K, Takasuki K, Uchino H. Adult T-cell leukemia: clinical and hematological features of 16 cases. *Blood* 1977;50:481–492.
16. Uchiyama T, Hori T, Tsudo M, et al. Interleukin-2 receptor (Tac antigen) expressed on adult T cell leukemia cells. *J Clin Invest* 1985;76:446–453.
17. Depper JM, Leonard WJ, Kronke M, Waldmann TA, Greene WC. Augmented T-cell growth factor receptor expression in HTLV-I–infected human leukemic T-cells. *J Immunol* 1984;133: 1691–1695.
18. Greene WC, Leonard WJ, Depper JM, Nelson DL, Waldmann TA. The interleukin-2 receptor: Normal and abnormal expression in T cells and leukemias induced by the human T-lymphotropic retroviruses. *Ann Intern Med* 1986;105:560–572.
19. Waldmann TA, Davis MM, Bongiovanni KF, Korsmeyer SJ. Rearrangement of genes for the antigen receptor on T cells as markers of lineage and clonality in human lymphoid neoplasms. *N Engl J Med* 1985;313:776–783.
20. Roman GC, Spencer PS, Schoenberg BS, et al. Tropical spastic paraparesis in the Saychelles Islands: a clinical case-control neuroepidemiologic study. *Neurology* 1987;37:1323–1328.
21. Arimura K, Rosales R, Osame M, Igata A. Clinical electrophysiologic studies of HTLV-I–associated myelopathy. *Arch Neurol* 1987;44:609–612.
22. Ludolph AC, Hugon J, Roman GC, Spenser PS, Schoenberg BS. A clinical neurophysiologic study of tropical spastic paraparesis. *Muscle Nerve* 1988;2:393–397.
23. Bhagavati S, Ehrlich G, Kula RW, et al. Detection of human T-cell lymphoma/leukemia virus type I DNA and antigen in spinal fluid and blood of patients with chronic progressive myelopathy. *N Engl J Med* 1988;318:1141–1147.
24. Newton M, Cruickshank K, Miller D, et al. Antibody to human T-lymphotropic virus type I in West-Indian-born UK residents with spastic paraparesis. *Lancet* 1987;1:415–416.
25. Yonenaga Y, Arimura K, Suehara M, Arimura Y, Osame M. Electroencephalographic abnormalities in human T-cell lymphotropic virus I–associated myelopathy. *Arch Neurol* 1989;46: 513–516.
26. Said G, Goulon-Goeau C, Lacroix C, Feve A, Descamps H, Fouchard M. Inflammatory lesions of peripheral nerve in a patient with human T-lymphotropic virus type I–associated myelopathy. *Ann Neurol* 1988;24:275–277.
27. Jacobson S, Zaninovic V, Mora C, et al. Immunological findings in neurological diseases associated with antibodies to HTLV-I: activated lymphocytes in tropical spastic paraparesis. *Ann Neurol* 1988;23(suppl):S196–S200.
28. Essex ME, McLane MF, Tachibana N, Francis DP, Lee T-H. Seroepidemiology of human T-cell leukemia virus in relation to immunosuppression and the acquired immunodeficiency syndrome. In: Gallo RC, Essex M, Gross L, eds. *Human T-cell leukemia lymphoma virus.* Cold Spring Harbor, New York: Cold Spring Harbor Laboratory, 1984:355–362.
29. Shaw GM, Broder S, Essex M, Gallo RC. Human T-cell leukemia virus: its discovery and role in leukemogenesis and immunosuppression. *Adv Intern Med* 1984;30:1–27.
30. Tajima K, Tominaga S, Suchi T, et al. Epidemiological analysis of the distribution of antibody to adult T-cell leukemia virus. *Gann* 1982;73:893–901.
31. Blattner WA, Kalyanaraman VS, Robert-Guroff M, et al. The human type C retrovirus, HTLV, in blacks from the Caribbean region, and relationship to adult T cell leukemia/lymphoma. *Int J Cancer* 1982;30:257–264.
32. Hunsmann G, Schneider J, Schmitt J, Yamamoto N. Detection of serum antibodies to adult T-cell leukemia virus in non-human primates and in people from Africa. *Int J Cancer* 1983;32: 329–332.
33. Shimoyama M, Kagami Y, Shimotohno K, et al. Adult T-cell leukemia/lymphoma not associated with human T-cell leukemia virus type I. *Proc Natl Acad Sci USA* 1986;83:4524–4528.
34. Yoshida M, Miyoshi I, Hinuma Y. Isolation and characterization of retrovirus from cell lines of human adult T cell leukemia and its implication in the disease. *Proc Natl Acad Sci USA* 1982;79: 2031–2035.
35. Wong-Staal F, Hahn H, Manzari V, et al. A survey of human leukemia for sequences of a human retrovirus. *Nature* 1983;302:626–628.
36. Miyoshi I, Kubonishi I, Yoshimoto S, et al. Type C virus particles in a cord T cell line derived by cocultivating normal human cord leukocytes and human leukemic T cells. *Nature* 1981;294: 770–771.
37. Yamamoto N, Okada M, Koyanagi Y, Kannagi M, Hinuma Y. Transformation of human leu-

kocytes by cocultivation with an adult T cell leukemia virus producer cell line. *Science* 1982;217: 737–739.

38. Miyoshi I, Tauchi H, Fujishita M, et al. Transformation of monkey lymphocytes with adult T-cell leukemia virus. *Lancet* 1982;1:1016.
39. Tateno M, Kondo N, Itoh T, Chubachi T, Togashi T, Yoshiki T. Rat lymphoid lines with human T-cell leukemia virus production. I. Biological and serological characterization. *J Exp Med* 1984;159:1105–1116.
40. Bishop JM. Viral oncogenes. *Cell* 1985;42:23–38.
41. Seiki M, Hattori S, Hirayama Y, Yoshida M. Human adult T cell leukemia virus: complete nucleotide sequence of the provirus genome integrated in leukemia cell DNA. *Proc Natl Acad Sci USA* 1983;80:3618–3622.
42. Seiki M, Eddy R, Show TB, Yoshida M. Nonspecific integration of the HTLV-I provirus genome into adult T-cell leukemia cells. *Nature* 1984;309:640–642.
43. Gootenberg JE, Ruscetti FW, Gallo RC. A biochemical variant of human T cell growth factor produced by a cutaneous T cell lymphoma cell line. *J Immunol* 1982;129:1499–1505.
44. Le J, Pronsky W, Henriksen D, Vilcek J. Synthesis of alpha and gamma interferons by a human cutaneous lymphoma with helper T-cell phenotype. *Cell Immunol* 1982;72:157–165.
45. Salahuddin SZ, Markham PD, Lindner SG, et al. Lymphokine production by cultured human T cells transformed by human T-cell leukemia-lymphoma virus-I. *Science* 1984;223:703–706.
46. Chan JY, Slamon DJ, Nimer SD, Golde DW, Gasson J. Regulation of expression of human granulocyte/macrophage colony-stimulating factor. *Proc Natl Acad Sci USA* 1986;83:8669–8673.
47. Sugamura K, Matsuyama M, Fujii M, Kanagi M, Hinuma Y. Establishment of human cell lines constitutively producing immune interferon: transformation of normal T cells by a human retrovirus. *J Immunol* 1983;131:1611–1612.
48. Noma T, Mizuta T, Rosen A, Hirano T, Kishimoto T, Honjo T. Enhancement of the interleukin 2 receptor expression on T cells by multiple B-lymphotropic lymphokines. *Immunol Lett* 1987;15: 249–253.
49. Azuma C, Tanabe T, Konishi M, et al. Cloning of cDNA for human T-cell replacing factor (interleukin-5) and comparison with the murine homologue. *Nucleic Acids Res* 1986;14:9149–9158.
50. Shimizu K, Hirano T, Ishibashi K, et al. Immortalization of BGDF (BCGF II)- and BCGF-producing T cells by human T cell leukemia virus (HTLV) and characterization of human BGDF (BCGF II). *J Immunol* 1985;134:1728–1733.
51. Inoue J, Seiki M, Taniguchi T, Tsuau S, Yoshida M. Induction of interleukin-2 receptor gene expression by p40^x encoded by human T-cell leukemia virus type I. *EMBO J* 1986;5:2883–2888.
52. Maruyama M, Shibuya H, Harada H, et al. Evidence for aberrant activation of the interleukin-2 autocrine loop by HTLV-I encoded p40^x and T3/Ti complex triggering. *Cell* 1987;48:343–350.
53. Siekevitz M, Feinberg MB, Holbrook N, et al. Activation of interleukin-2 and interleukin-2 receptor (Tac) promoter expression by the trans-activator (tat) gene product of human T-cell leukemia virus, type I. *Proc Natl Acad Sci USA* 1987;84:5389–5393.
54. Cross SL, Feinberg MB, Wolf JB, et al. Regulation of the human interleukin-2 receptor alpha-chain promoter: activation of a nonfunctional promoter by the transactivator gene of HTLV-I. *Cell* 1987;48:343–350.
55. Miyatake S, Seiki M, Malefijit RD, et al. Activation of T cell-derived lymphokine genes in T cells and fibroblasts: effects of human T cell leukemia virus type I p40^x protein and bovine papilloma virus encoded E2 protein. *Nucleic Acids Res* 1987;16:6547–6567.
56. Tschachler E, Robert-Guroff M, Gallo RC, Reitz MS. Human T-lymphotropic virus I-infected T cells constitutively express lymphotoxin in vivo. *Blood* 1989;73:194–201.
57. Paul NL, Ruddle NH. HTLV-I regulation of lymphotoxin production. (Abstract) *J Cell Biochem* (Suppl) 1989;13B:258.
58. Duc Dodon M, Lilienbaum A, Paulin D, Gazzolo L. Effect of human T cell leukemia virus, type I tax protein on the activation of the human vimentin gene. (Abstract) *J Cell Biochem* (Suppl) 1989;13B:268.
59. Fujisawa J-I, Seiki M, Kiyokawa T, Yoshida M. Functional activation of the long terminal repeat of human T-cell leukemia virus type I by a trans-activating factor. *Biochemistry* 1985;82:2277–2281.

60. Sodroski JG, Rosen CA, Haseltine WA. Trans-acting transcriptional activation of the long terminal repeat of human T lymphotropic viruses in infected cells. *Science* 1984;225:381–385.
61. Inoue J-I, Yoshida M, Seiki M. Transcriptional (p40^x) and post-transcriptional (p27^{x-III}) regulators are required for the expression and replication of human T-cell leukemia virus type I genes. *Proc Natl Acad Sci USA* 1987;84:3653–3657.
62. Hidaka M, Inoue J-I, Yoshida M, Seiki M. Post-transcriptional regulator (rex) of HTLV-I initiates expression of viral structural proteins but suppresses expression of regulatory proteins. *EMBO J* 1988;7:519–523.
63. Shimotohno K, Wachsman W, Takahashi Y, et al. Nucleotide sequence of the 3' region of an infectious human T-cell leukemia virus type II genome. *Proc Natl Acad Sci USA* 1984;81:6657–6661.
64. Haseltine WA, Sodroski J, Patarca R, Briggs D, Perkins D, Wong-Staal F. Structure of 3' terminal region of type II human T lymphotropic virus: evidence for new coding region. *Science* 1984;225:419–421.
65. Rice NR, Stephens RM, Covez D, et al. The nucleotide sequence of the env and post-env region of bovine leukemia virus. *Virology* 1984;138:82–93.
66. Sagata N, Yasunaga T, Tsuzuku-Kawamura J, Ohnishi K, Ogawa Y, Ikawa Y. Complete nucleotide sequence of the genome of bovine leukemia virus: its evolutionary relationship to other retroviruses. *Proc Natl Acad Sci USA* 1985;82:677–681.
67. Watanabe T, Seiki M, Tsujimoto H, Miyoshi I, Hayami M, Yoshida M. Sequence homology of the simian retrovirus genome with human T-cell leukemia virus type I. *Virology* 1985;144:59–65.
68. Guo HG, Wong-Staal F, Gallo RC. Novel viral sequences related to human T-cell leukemia virus in T-cells of a seropositive baboon. *Science* 1984;223:1195–1197.
69. Shimotohno K, Golde DW, Miwa M, et al. Nucleotide sequence analysis of the long terminal repeat of human T-cell leukemia virus type II. *Proc Natl Acad Sci USA* 1984;81:1079–1083.
70. Ohtani K, Nakamura M, Saito S, et al. Identification of two distinct elements in the long terminal repeat of HTLV-I responsible for maximum gene expression. *EMBO J* 1987;6:389–395.
71. Rosen CA, Sodroski JG, Haseltine WA. Location of cis-acting regulatory sequences in the human T-cell leukemia virus type I long terminal repeat. *Proc Natl Acad Sci USA* 1985;82:6502–6506.
72. Brady J, Jeang KT, Duvall J, Khoury G. Identification of p40^x-responsive regulatory sequences within the human T-cell leukemia virus type I long terminal repeat. *J Virol* 1987;61:2175–2181.
73. Greene WC, Leonard WJ, Wano Y, et al. Transactivator gene of HTLV-II induces IL-2 receptor and IL-2 cellular gene expression. *Science* 1986;232:877–880.
74. Greenberg SJ, Jacobson S, Waldmann TA, McFarlin DE. Molecular analysis of HTLV-I proviral integration and T cell receptor arrangement indicates that T cells in tropical spastic paraparesis are polyclonal. *J Infect Dis* 1989;159:741–744.
75. Yoshida M, Osame M, Usuku K, Matsumoto M, Igata A. Viruses detected in HTLV-I–associated myelopathy and adult T-cell leukemia are identical on DNA blotting. *Lancet* 1987;1:1085–1086.
76. Itoyama Y, Minato S, Kira J, et al. Spontaneous proliferation of peripheral blood lymphocytes increased in patients with HTLV-I associated myelopathy. *Neurology* 1988;38:1302–1307.
77. Kitajima I, Osame M, Izumo S, Igata A. Immunological studies of HTLV-I associated myelopathy. *Autoimmunity* 1988;1:125–131.
78. Yasuda K, Sei Y, Yokoyama MM, Tanaka K, Hara A. Healthy HTLV-I carriers in Japan: the haematological and immunological characteristics. *Br J Haematol* 1986;64:195–203.
79. Mullis KB, Faloona FA. Specific synthesis of DNA in vitro via a polymerase-catalyzed chain reaction. *Methods Enzymol* 1987;155:335–350.
80. Abbott MA, Poiesz BJ, Byrne BC, Kwok S, Sninsky JJ, Ehrlich GD. Enzymatic gene amplification: qualitative and quantitative methods for detecting proviral DNA amplified in vitro. *J Infect Dis* 1988;158:1158–1169.
81. Kwok S, Ehrlich G, Poiesz BJ, Kalish R, Sninsky JJ. Enzymatic amplification of HTLV-I viral sequences from peripheral blood mononuclear cells and infected tissues. *Blood* 1988;72:1117–1123.
82. Newman PJ, Gorski J, White GC II, Gidwitz S, Cretney CJ, Aster RH. Enzymatic amplification of platelet-specific messenger RNA using the polymerase chain reaction. *J Clin Invest* 1988;82:739–743.
83. Clarke MF, Trainor CD, Mann DL, Gallo RC, Reitz MS. Methylation of human T-cell leukemia

virus proviral DNA and viral RNA expression in short- and long-term cultures of infected cells. *Virology* 1984;135:97–104.

84. Kitamura T, Takano M, Hoshino H, et al. Methylation pattern of human T-cell leukemia virus in vivo and in vitro: pX and LTR regions are hypomethylated in vivo. *Int J Cancer* 1985;35: 629–635.

85. Kyokawa T, Seiki M, Iwashita S, Imagawa K, Shimizu F, Yoshida M. p27^{x-III} and p21^{x-III}, proteins encoded by the pX sequence of human T-cell leukemia virus type I. *Proc Natl Acad Sci USA* 1985;82:8359–8363.

DISCUSSION

One speaker felt that the absence of RNA, according to the very sensitive PCR methods, was more remarkable than the finding of mRNA in two of six lymph node samples. Another speaker thought the failure was due to technical problems rather than the absence of viral replication. This speaker was asked to expand on the observation of lymphotoxin detection in lymph nodes by PCR and peripheral blood cells from patients and was further asked if lymphotoxin was also found in these same tissues from normal individuals. The speaker answered that such message in normal cells as part of a pool of constitutive mRNA cytokines had been found, but there was a difference in levels between normal and leukemic tissues. One participant asked if the tumor cells from the patient with Kaposi's sarcoma produced the growth factor identified in one laboratory. The speaker stated that this was unknown.

Human Retrovirology: HTLV,
edited by William A. Blattner.
Raven Press, Ltd., New York 1990.

Constitutive Expression of Lymphotoxin (Tumor Necrosis Factor β) in HTLV-I–Infected Cell Lines

Erwin Tschachler, Robert C. Gallo, and Marvin S. Reitz, Jr.

*Laboratory of Tumor Cell Biology, Division of Cancer Etiology,
National Cancer Institute, Bethesda, Maryland 20892*

INTRODUCTION

Human T-cell leukemia virus type I (HTLV-I) (1) is the etiologic agent of adult T-cell leukemia (ATL), a frequently rapidly fatal malignancy of adults, which is prevalent in various localized regions of the world. Infection in vitro with HTLV-I results in the establishment of immortalized T-cell lines, and similar infected cell lines can be established from the blood of ATL patients (2–4). These cell lines have many characteristics in common with leukemic peripheral blood cells from ATL, including morphology and surface markers.

ATL is frequently preceded by or associated with hypercalcemia and lytic bone lesions (5–7), which—particularly in more aggressive cases—are difficult to control and can be a factor in the mortality of the disease. Serum parathyroid and 1,25-dihydroxy vitamin D levels, which are often high in other conditions involving bone destruction, have been reported to be normal (8,9). Secretion of a soluble factor that is able to induce bone resorption has been reported (5,10) from HTLV-I–infected ATL cells, and may be of significance in the bone lesions and hypercalcemia of ATL patients.

Tumor necrosis factor (TNFα) and lymphotoxin (TNFβ) are related but distinct cytokines that mediate a range of similar biologic effects. These include cytotoxicity for some tumor cells (11–13) and a negative effect on expression of some viruses (14,15). TNFβ is produced only by activated T and B lymphocytes (16). TNFα is thought to be produced primarily by macrophages (17), although under appropriate conditions peripheral blood T cells have also been shown to produce it (18). TNFα and β have also been reported to be able to activate osteoclasts in vitro (19,20), and to potentiate the activity of other osteoclast-activating factors (21). The bone destruction seen in multiple myeloma has been linked to TNFβ produced by leukemic cells. We therefore thought it of interest to analyze TNFα and -β expression in HTLV-I–infected T cells.

105

EXPRESSION OF TNFα AND β mRNA IN
HTLV-I–INFECTED CELL LINES

As shown in Fig. 1a, RNA from cell lines established from normal cord blood T cells by infection with HTLV-I was positive in nine of nine cases for the expression of TNFα mRNA (22). RNA from the T-cell line PEER was negative, as was RNA from the T-cell line MOLT 3, either uninfected or infected with the HTLV-IIIB strain of HIV-1. In all cases, the signal was at least as strong as that from the positive control, HL60 cells stimulated with phorbol myristate acetate. Similarly, all nine tested cell lines were also positive for the expression of TNFβ mRNA (Fig. 1b), although in one case (ECl-55) the signal was rather weak. MOLT 3 cells, whether uninfected or infected with HIV-1, were negative, whereas the PEER cells were very slightly positive. Similar results were also obtained from HTLV-I–infected T-cell lines established directly from ATL patients (Fig. 2). Five of five tested were positive for both TNFα and β, whereas T cells from the Jurkat cell line were negative for both.

Functional T-cell clones established from normal adult peripheral blood can be infected in vitro with HTLV-I (23), leading eventually to loss of immune functions. We wished to compare expression of TNFα and -β in identical cells which differed only by infection with HTLV-I. We therefore compared expression of mRNA for both cytokines in one such clone, TM-11 (24), and its infected counterpart, TM-11/HTLV-I. TNFα expression, shown in Fig. 3, was actually lower in the HTLV-I–infected TM-11 cells. This is most likely due to the presence of antigen-presenting accessory cells, which must be added to the uninfected TM-11 cultures in combination with tetanus toxin and IL-2 to maintain their growth. (The infected TM-11 cells grow without the need for antigenic stimulation.) In contrast, TNFβ mRNA expression was evident only in the HTLV-I

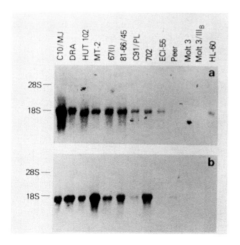

FIG. 1. Detection of TNFα and -β mRNA in HTLV-I–infected T-cell lines. RNA from the indicated cell lines was analyzed by Northern blotting and hybridized with oligomeric probes for TNFα (Panel a) or TNFβ (Panel b). The first nine samples are HTLV-I–infected T-cell lines, PEER and MOLT 3 are uninfected T-cell lines, MOLT 3/IIIB is infected with the HTLV-IIIB strain of HIV-1, and HL-60 is an uninfected monocyte-granulocyte cell line. The positions of 18S and 28S ribosomal RNA markers are indicated.

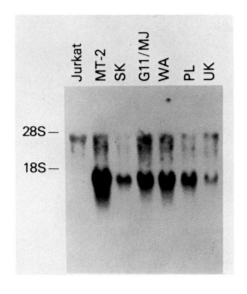

FIG. 2. Detection of TNFβ mRNA in T-cell lines derived from ATL patients. RNA from the indicated cell lines was analyzed by Northern blotting for the presence of TNFβ mRNA. Jurkat is an uninfected T cell line. All other cell lines are derived from ATL patients and infected with HTLV-I. The position of 18S and 28S ribosomal RNA markers is shown.

cells, indicating that a T cell not expressing TNFβ is induced to do so after infection with HTLV-I.

TNFβ mRNA EXPRESSION IS NOT ASSOCIATED WITH GENE REARRANGEMENT

To test whether or not the high apparently constitutive expression of TNFβ mRNA in HTLV-I–infected cells could be the result of a disturbance of the gene locus—as would be the case if, for example, proviral integration took place near or within the gene—we analyzed DNA from four infected cell lines for evidence of gene rearrangement. After digestion with several different restriction endonucleases, in no case was there any evidence of rearrangement of the TNFβ gene. This is consistent with previous data that HTLV-I integration is seemingly at random and even occurs on different chromosomes in different cases of ATL (25).

TNFβ mRNA EXPRESSION MAY BE ACTIVATED BY THE HTLV-I *TAX* GENE

The cell line 81-66/45 is infected with HTLV-I and expresses the *tax* gene product, but does not express any of the viral structural proteins (26). Expression of both TNFα and -β mRNA is high in 81-66/45 cells, suggesting that expression is a consequence of transactivation mediated by *tax*. The *tax* gene product also has been shown to activate transcription of other cellular genes, including the

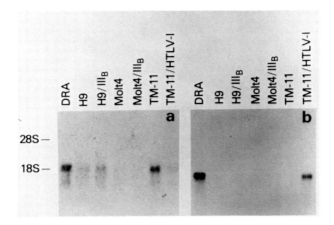

FIG. 3. Expression of TNFα and β mRNA in a T-cell clone before and after infection with HTLV-I. RNA from the indicated cell lines was analyzed by Northern blotting and hybridized to an oligomeric probe for TNFα (Panel a) or TNFβ (Panel b). H9 and MOLT 4 are uninfected T-cell lines; the corresponding infected cell lines are infected with the HTLV-IIIB strain of HIV-1. DRA is an infected T-cell line established by in vitro infection with HTLV-I. TM-11 is a functional clonal helper T-cell line specific for tetanus toxoid (24). TM-11/HTLV-I is its HTLV-I–infected counterpart.

IL-2 receptor (27), and *tax*-mediated induction of cellular genes is one possible mechanism by which HTLV-I is able to transform T cells.

TNFβ IS SECRETED BY HTLV-I–INFECTED T CELLS

Media from HTLV-I–infected cell lines were assayed for TNF activity by their ability to kill mouse L929 cells. Killing was readily observed with serial dilutions of culture fluid (22). In contrast, culture fluid from H9, Jurkat, MOLT 3, and PEER cells had little effect, even when used undiluted. Rabbit anti-TNFβ antiserum completely inhibited this activity in media from two of four tested infected lines (81-66/45 and EC1-55) and reduced the activity to <50% with the other two lines (MT-2 and C10/MJ). A mouse monoclonal anti-TNFα antibody had no effect when added alone; when added in combination with the TNFβ antiserum, however, a virtually complete elimination of L929 cell killing was noted with the C10/MJ and MT-2 culture fluids. This suggests that these two cell lines are secreting both TNFα and -β.

TNFβ-related protein secretion by HTLV-I cell lines was analyzed by radioimmunoprecipitation of media from metabolically labeled cells using the rabbit anti-TNFβ antiserum (22). As shown in Fig. 4, TNFβ-related proteins were detected with all three tested cell lines. Interestingly, the sizes differed. Cell line 81-66/45 (lane 5) produced a 25-kd protein, whereas the one from MT-2 was somewhat larger (lane 3). MT-2 cells produced two smaller proteins of ~22

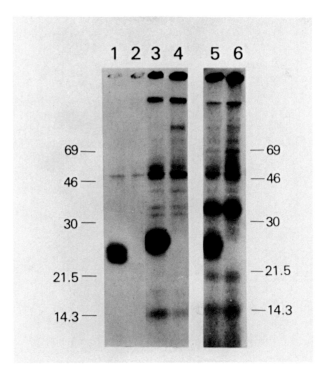

FIG. 4. Detection of TNFβ in the supernatant of HTLV-I–infected T-cell lines by immunoprecipitation. Cells were grown in the presence of methionine-depleted media supplemented with [^{35}S]-methionine (100 μCi/ml) for 18 h. Five hundred microliters of media was reacted with preimmune rabbit IgG (lanes 2, 4, and 6) or rabbit antihuman TNFβ (lanes 1, 3, and 5), then precipitated with protein A-Sepharose and analyzed on a 12% PAGE-SDS gel. Media were from the HTLV-I–infected T-cell lines MT-2 (lanes 1 and 2), C10/MJ (lanes 3 and 4), and 81-66 (lanes 5 and 6). The positions of protein molecular weight markers, whose size is given in kilodaltons, are indicated.

and 24 kd. The variation in the observed size of these proteins may be due to some post-translational modification, such as glycosylation, because there does not appear to be any great difference in the size of the mRNA in these cells.

LACK OF TNFβ EXPRESSION BY UNCULTURED PERIPHERAL BLOOD CELLS FROM ATL PATIENTS

We tested peripheral blood cells from three ATL patients for the presence of mRNA for TNFβ. In contrast to HTLV-I–infected cell lines, all three were negative for TNFβ mRNA. In addition, they were negative for viral RNA. Consistent with earlier results (28), when the cells from one such patient were put into culture, viral RNA began to be expressed. At 48 h after culture, both HTLV-I RNA and TNFβ mRNA were readily detected by Northern blotting (not shown).

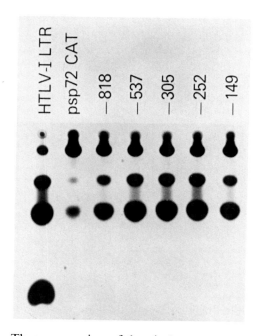

FIG. 5. Activation of TNFβ promoter in HTLV-I–infected T cells. MT-2 cells were transfected with plasmids containing the chloramphenicol acetyl transferase (*CAT*) gene in the absence of a RNA polymerase promoter (psp72 *CAT*) or downstream of the HTLV-I promoter (HTLV-I LTR) or the region 5′ of the initiation of transcription of the TNFβ gene. The 5′ TNFβ region was deleted to different extents as indicated above. The regions included contained sequences from the indicated position, given relative to initiation start site, to the start site itself. Thus, −149 contained the upstream 149 nucleotides and −818 the upstream 818 nucleotides.

Thus, expression of the viral genome appears to be necessary for expression of TNFβ in ATL cells.

The significance of the inability to detect TNFβ mRNA in uncultured ATL cells is not clear. It could be present at biologically significant levels which are below our detection limits. This is particularly true in view of our use of low numbers of cryopreserved cells, which do not give as high a quality of RNA as do fresh cells, and in view of the heterogeneity of peripheral blood cells from ATL patients who did not have high white blood cell counts. Alternatively, because TNFβ mRNA could be detected when viral mRNA became detectable, after a short time in culture, virus expression may be a prerequisite for TNFβ expression. This points to a paradox in the involvement of HTLV-I in ATL, which is that virus expression in circulating leukemic cells is low to negative (28) until cells are grown in culture, suggesting that HTLV-I is not required for the maintenance of the disease. Antibodies to the virus, however, remain persistently high, suggesting that virus replication in ATL remains active at some as-of-yet-undetermined site(s). It is here that one might expect to also find expression of TNFβ.

To help determine whether or not the expression of TNFβ is mediated at the level of transcription in HTLV-I–infected cells, we constructed a plasmid containing the region 5′ from the coding sequences for the human TNFβ gene (generously provided by Dr. Thomas Speiess, Harvard University, Cambridge, Massachusetts) placed upstream from a reporter gene (*CAT*). CAT activity was evident after transfection of HTLV-I, but not uninfected or HIV-1–infected T-cell lines, as shown in Fig. 5. This strongly suggests that TNFβ expression is indeed mediated at the level of RNA synthesis. No CAT activity was observed after transfection into Jurkat cells constitutively expressing a transfected *tax* gene (provided by Dr. Warner Greene, Duke University, Durham, North Carolina), and these cells also do not express TNFβ. Neither they nor untransfected Jurkat cells can be induced to express TNFβ. This indicates that *tax* alone is not sufficient for TNFβ induction and that some cellular factor(s) are also required. Deletion mutagenesis of the lymphotoxin gene 5' of the coding sequences showed that the promoter-enhancer region contains both positive and negative regulatory regions (not shown), in that some deletions resulted in an increased expression of CAT activity. It thus appears that *tax* expression alone is not sufficient to activate the HTLV-I promoter. If the CAT assay is an accurate reflection of the induction of expression, a combination of *tax* and cellular factors present in those types of cells which can normally be induced to express TNFβ may interact in such a way as to activate to TNFβ promoter.

CONCLUSIONS

We have looked at a series of T-cell lines for expression of lymphotoxin and tumor necrosis factor (TNFβ and -α). These cell lines include uninfected cell lines, cell lines infected by HIV-1, and cell lines established either by in vitro HTLV-I infection or directly from ATL patients. All cell lines infected with HTLV-I constitutively expressed high levels of TNFβ RNA, protein, and activity. Many also expressed TNFα. In contrast, none of the other cell lines, whether uninfected or infected by HIV-1, expressed detectable levels of either lymphokine. A functional helper-T-cell clone expressed no detectable lymphotoxin before infection with HTLV-I; after infection, high levels of expression were noted. This indicates that infection of a particular cell type by HTLV-I can lead directly to TNFβ expression.

One cell line which expressed the transactivator gene, *tax*, but did not express any structural virus proteins also expressed high levels of lymphotoxin, suggesting that transactivation of the lymphotoxin promoter is the mechanism of HTLV-I induction of lymphotoxin expression. Indeed, we found that transfection of a plasmid containing the promoter and neighboring regions of the human TNFβ gene into cells infected with HTLV-I was transcriptionally activated. This activation, however, is not simply due to *tax*, because transfection into Jurkat cells stably expressing an introduced *tax* gene did not result in significant CAT activity. Moreover, deletion of some of the sequences upstream from the promoter increases the observed CAT activity. The data thus seem to show that both positive and negative regulatory regions exist for the TNFβ promoter, and that expression of *tax* alone is not sufficient for mRNA expression. If *tax* indeed is involved in the induction of TNFβ expression in HTLV-I–infected cells, then it must only act indirectly or in concert with cellular factors.

The question of the relationship of the observed expression of TNFβ to the profound hypercalcemia and lytic bone lesions frequently observed in ATL is still not answered, although the current data lend support to the idea that it may play an important role. The lack of detection of TNFβ mRNA in frozen peripheral blood cells from ATL is contrary to this idea, but may reflect a sensitivity problem. The same cells were also negative for viral RNA until after a short time in culture, at which time TNFβ mRNA also became detectable. In view of the persistent nature of the antibody response to the virus in ATL, it seems likely that HTLV-I is expressed at some currently unrecognized site. If so, the same site would also be likely to serve as a source for TNFβ, and its identification would be of considerable interest.

REFERENCES

1. Poiesz BJ, Ruscetti FW, Gazdar AF, Bunn PA, Minna JP, Gallo RC. Detection and isolation of type-C retrovirus particles from fresh and cultured lymphocytes of a patient with cutaneous T-cell lymphoma. *Proc Natl Acad Sci USA* 1980;77:7415–7419.
2. Markham P, Salahuddin Z, Kalyanaraman VS, Popovic M, Sarin PS, Gallo RC. Infection and transformation of fresh human umbilical cord blood cells by multiple sources of human T-cell leukemia/lymphoma virus (HTLV). *Int J Cancer* 1983;31:413–420.
3. Miyoshi I, Kubonishi I, Yoshimoto S, et al. Type C virus particles in a cord T-cell line derived by co-cultivating normal human cord blood leukocytes and human leukemic T-cells. *Nature* 1981;294:770–771.
4. Popovic M, Sarin PS, Robert-Guroff M, et al. Isolation and transmission of human retrovirus (human T-cell leukemia virus). *Science* 1983;219:856–859.
5. Bunn PA, Schechter GP, Jaffe E, et al. Clinical course of retrovirus-associated adult T-cell lymphoma in the United States. *N Engl J Med* 1983;309:257–264.
6. Grossman B, Schechter G, Horton JE, Pierce L, Jaffe E, Wahl L. Hypercalcemia associated with T-cell lymphoma-leukemia. *Am J Pathol* 1981;75:149–155.
7. Kuefler PR, Bunn PA. Adult T cell leukaemia/lymphoma. *Clin Haematol* 1986;15:695–727.
8. Broder S, Bunn PA, Jaffe ES, et al. NIH conference. T-cell lymphoproliferative syndrome associated with human T-cell leukemia/lymphoma virus. *Ann Intern Med* 1984;100:543–557.
9. Dodd RC, Winkler CF, Williams ME, Bunn PA, Gray TK. 1,25-dihydroxyvitamin D levels in hypercalcemia patients with adult T-cell lymphoma. *J Clin Endocrinol Metab* 1986;146:1971–1978.
10. Fujihira T, Eto S, Sato K, et al. Evidence of bone resorption-stimulating factor in adult T-cell leukemia. *Jpn J Clin Oncol* 1985;15:385–391.
11. Haranaka K, Satomi N, Sakurai A. Antitumor activity of murine tumor necrosis factor (TNF) against transplanted murine tumors and heterotransplanted human tumors in nude mice. *Int J Cancer* 1984;34:263–267.
12. Old LJ. Tumor necrosis factor (TNF). *Science* 1985;230:630–632.
13. Wang AM, Creasy AA, Ladner MB, et al. Molecular cloning of the complementary DNA for human tumor necrosis factor. *Science* 1985;228:149–154.
14. Mestan J, Digel W, Mittnacht S, et al. Antiviral effects of recombinant tumour necrosis factor in vitro. *Nature* 1986;323:816–819.
15. Wong GHW, Goeddel DV. Tumour necrosis factor alpha and beta inhibit virus replication and synergize with interferons. *Nature* 1987;323:819–822.
16. Ruddle NH, Waksman BW. Cytotoxicity mediated by soluble antigen and lymphocytes in delayed hypersensitivity. III. Analysis of mechanism. *J Exp Med* 1968;128:1267–1279.
17. Carswell EA, Old LJ, Kassel RL, Green S, Fiore S, Williamson B. An endotoxin-induced serum factor that causes necrosis of tumors. *Proc Natl Acad Sci USA* 1975;72:3666–3670.
18. Cuturi MC, Murphy M, Costa-Giomi MP, Weinmann R, Perussia B, Trinchieri G. Independent

regulation of tumor necrosis factor and lymphotoxin production by human peripheral blood lymphocytes. *J Exp Med* 1987;165:1581–1594.

19. Bertolini DR, Nedwin GE, Bringman TS, Smith D, Mundy GR. Stimulation of bone resorption and inhibition of bone formation in vitro by human tumor necrosis factor. *Nature* 1983;319: 516–518.

20. Garrett IR, Durie BGM, Nedwin GE, et al. Production of lymphotoxin, a bone-resorbing cytokine, by cultured human myeloma cells. *N Engl J Med* 1987;317:526–532.

21. Stashenko P, Dewhirst FE, Peros WJ, Kent RL, Ago JM. Synergistic interactions between interleukin 1, tumor necrosis factor, and lymphotoxin in bone resorption. *J Immunol* 1987;138: 1464–1468.

22. Tschachler E, Robert-Guroff M, Gallo RC, Reitz MS. Human T-lymphotropic virus I–infected cells constitutively express lymphotoxin. *Blood* 1989;73:194–201.

23. Mitsuya H, Guo H-G, Cossman J, Megson M, Reitz MS, Broder S. Functional properties of antigen-specific T cells infected by human T-cell leukemia-lymphoma virus (HTLV-I). *Science* 1984;225:1484–1486.

24. Matsushita S, Mitsuya H, Reitz MS, Broder S. Pharmacological inhibition of in vitro infectivity of human T lymphotropic virus type I. *J Clin Invest* 1987;80:394–400.

25. Seiki M, Eddy R, Shows TB, Yoshida M. Nonspecific integration of the HTLV provirus genome into adult T-cell leukemia cells. *Nature* 1984;309:640–642.

26. Sodroski JG, Goh WC, Rosen CA, et al. Characterization of human T-cell leukemia virus–transformed non-producer cell lines. *J Virol* 1985;55:831–839.

27. Siekevitz M, Feinberg MB, Holbrook N, Wong-Staal F, Greene WC. Activation of interleukin 2 and interleukin 2 receptor (Tac) promotor expression by the trans-activator (tat) gene product of human T-cell leukemia virus, type I. *Proc Natl Acad Sci USA* 1987;84:5389–5393.

28. Clarke MF, Trainor CD, Mann DL, Gallo RC, Reitz MS. Methylation of human T-cell leukemia virus proviral DNA and viral RNA expression in short and long term culture of infected cells. *Virology* 1984;135:97–102.

DISCUSSION

One speaker referred to a report of release of an osteoclastic factor by HTLV-I–infected cells and asked whether or not that had been investigated in the system studied. One of the participants replied that this was being investigated but that parathyroid hormone levels have been reported to be normal in some patients with ATL. Another participant inquired as to the correlation of hypercalcemia with increased TNFβ levels in patients with ATL. The previous discussant stated that TNFβ activity was difficult to measure because of levels of stability. It was noted that in an investigation of TNFβ expression in peripheral blood cells from ATL patients, mRNA levels were not detectable. However, TNFβ message was found after short-term culture with the concomitant expression of HTLV-I viral RNA.

Human Retrovirology: HTLV,
edited by William A. Blattner.
Raven Press, Ltd., New York © 1990.

Concomitant Infections with Human T-Cell Leukemia Viruses (HTLVs) and Human Immunodeficiency Virus (HIV): Identification of HTLV-II Infection in Intravenous Drug Abusers (IVDAs)

William W. Hall, Mark H. Kaplan, *S. Zaki Salahuddin,
Naoki Oyaizu, *Curado Gurgo, Maria Coronesi,
†Kazuo Nagashima and †Robert C. Gallo

*Division of Infectious Disease, North Shore University Hospital,
Manhasset, New York 11030; *Laboratory of Tumor Cell Biology, National Cancer
Institute, National Institutes of Health, Bethesda, Maryland; †Department of Pathology,
University of Hokkaido, Sapporo 060, Japan*

INTRODUCTION

The human T-cell leukemia viruses Type I (HTLV-I) and Type II (HTLV-II) are members of a family of oncogenic retroviruses having similar structural and biological properties, a tropism for T lymphocytes, and an association with lymphoproliferative disease (1–3). HTLV-I infection is endemic in parts of southern Japan, the Caribbean, South America, southeastern United States, and central Africa. In these areas the virus has been shown to be the causative agent of adult T-cell leukemia (ATL)—a mature T cell malignancy (4,5)—and a chronic myelopathy known both as tropical spastic paraparesis (TSP) and HTLV-I–associated myelopathy (HAM) (6,7). HTLV-II infection, in contrast, is considered rare, and the literature contains only four reports on the isolation of this virus (8–13). Infection is not known to be endemic in any geographical location, and its role in human disease remains poorly defined. Although the virus has been isolated from two patients with atypical T-cell variants of hairy cell leukemia (HCL) (8–11), its role in the pathogenesis of this disease remains to be established. Whereas these two patients were seropositive for HTLV-II, a recent study has shown that the majority of patients with HCL are not (14). Subsequent evaluation of one of the two patients showed the coexistence of a CD8$^+$ lymphoproliferative disorder with viral integration in the abnormal CD8$^+$ cell population but not in the circulating hairy cells. The relationship, if any, of this to the development of HCL remains unknown; however, it demonstrated for the first time that HTLV-II infection can produce lymphoproliferative disorder.

Serological studies have suggested that HTLV-I/HTLV-II infection is not limited to individuals with rare lymphoproliferative or neurological disorders or to areas which up to now have been considered to be endemic. This is based on the observations that certain individuals at risk of or with human immunodeficiency virus (HIV) infection, notably intravenous drug abusers (IVDAs), are also seropositive for HTLV. Seroprevalence rates have been found to range from 3% in the United Kingdom (15) to 24% in certain parts of the U.S. (16,17). In addition to these seroepidemiological studies, there have been a number of reports from various parts of the world (18–20) describing individual IVDAs with concomitant HIV and HTLV-I infection.

Dual infection has also been reported to occur in other groups at risk for HIV infection. In Trinidad, where HTLV-I is endemic, a study of 100 bisexual and/ or homosexual males showed that 15% were seropositive for HTLV-I, compared with 2.4% of the general population (21). Forty percent of the cohort were HIV positive and 6% had evidence of dual infection. In contrast, in nonendemic areas, infection with HTLV among homosexuals is relatively rare, with seroprevalence rates of <0.05% reported in a group of HIV-infected homosexual males in California (17). Similarly, heterosexual spread appears to occur only infrequently. Although elevated seroprevalence rates have been described in female prostitutes (22), the significance of these findings is unclear because of the probable but unknown amount of drug abuse in this population.

The major limitation of most of the seroepidemiologic studies reported to date has been the inability to differentiate between infection by HTLV-I and HTLV-II by serologic means because of their antigenic cross-reactivity. Earlier, two studies on HTLV-seropositive IVDAs using competition immunoassays suggested that HTLV-II may be more common (15,16). More recently a study of New Orleans drug abusers utilized the polymerase chain reaction (PCR) has suggested a high prevalence of HTLV-II in that population (34). We have recently initiated a systematic study, which included all known risk groups for this infection, to determine the seroprevalence of HTLV in a large number of HIV-infected patients from the New York City area. In this report we describe the initial results obtained on 489 HIV-positive sera and show that, in agreement with previous studies, IVDAs have much higher seroprevalence rates than any other risk group. In addition, we describe the results of preliminary studies using virus isolation and molecular analysis to specifically identify the HTLVs involved.

SEROLOGICAL STUDIES

Initial screening was carried out using a commercially available HTLV-I enzyme-linked immunosorbent assay (ELISA, Cellular Products); repeatedly reactive samples were confirmed by radioimmunoprecipitation assay (RIPA) using

TABLE 1. Serological evidence for concomitant infection with HTLV-I and/or HTLV-II in HIV-positive patients

Risk group	Sex:	White		Hispanic		Black		Total
		M	F	M	F	M	F	
Homo/bisexual male		0/130	—	0/13	—	3/25	—	3/168
IVDA		7/87	4/53	1/11	2/6	8/32	4/20	26/209
Transfusion		0/18	0/8	—	—	—	0/1	0/27
Heterosexual (spouse HIV+)		0/2	0/30	—	1/4	0/1	1/48	2/85

radiolabeled ([^{35}S] methionine) HTLV-I (NS-RC) infected cells. Samples demonstrating immunoreactivity to the *gag* p24 and the *env* gp 61/68 and/or gp46 were considered positive, as has been recently recommended by the Centers for Disease Control Public Health Service Working Group (23). Sera demonstrating other immunoreactivities were designated "indeterminate" and for the purpose of this study were considered negative. Table 1 shows the results obtained on 489 HIV positive sera. In the group of homosexual or bisexual males only 3/168 (<2%) were seropositive. Within this population, however, 0/130 whites showed evidence of infection, and the three patients with positive serologies were black (3/18). In contrast, those individuals with a history of IVDA showed much higher HTLV seroprevalence rate, with 25/181 (12.5%) being seropositive. It was also clear that blacks, both male and female, had a higher seropositivity (12/52; 23%) than whites (11/140; 7%). IVDAs of Hispanic origin also showed an intermediate to high seroprevalence rate of HTLV infection, with 3/17 (17.5%) being seropositive—although the population so far tested is small.

The study also suggests that heterosexual transmission of HTLV occurs in this population. Two females (one black, one Hispanic) who were sexual partners of IVDAs but were otherwise risk-free showed dual seropositivity for both HIV and HTLV. Thus, although the rate of heterosexual transmission of HTLV in this population is low (2/85; 2.3%), in the case of partners of IVDAs, this certainly occurs. Of 27 cases of HIV infection resulting from blood transfusion, none was seropositive for HTLV. Table 2 shows the relationship of seropositivity to age

TABLE 2. Relationship of HTLV-I and/or HTLV-II seropositivity to age in HIV-positive IVDAs

Age	White		Hispanic		Black		Total	
	M	F	M	F	M	F	Number	Percentage
20–29	2/23	0/22	0/1	0/1	—	1/4	3/51	6
30–39	1/59	3/30	0/6	2/5	4/22	1/7	11/129	7.6
40–49	5/13	1/3	1/2	—	2/8	1/1	10/27	41
>49	—	—	—	—	2/2	—	2/2	100
Total	8/95	4/55	1/9	2/6	8/32	3/12	26/209	

in the IVDA population; it is clear that seropositivity increases with age in all IVDAs, both white and black. A similar increase in seropositivity with age is seen in areas endemic for HTLV-I infection. Unfortunately, because of unreliable histories obtained from our patient population, we were unable to determine whether the increase in seropositivity with age could be correlated to total years of drug abuse.

VIRUS ISOLATION

Because of their antigenic cross-reactivity, serological methods cannot consistently differentiate infection by HTLV-I from HTLV-II. This requires virus isolation and molecular analyses. A major difficulty in isolating HTLV-I and/or HTLV-II from peripheral blood mononuclear cells (PBMC), which are also infected with HIV, is that the latter is cytopathic and cultures are difficult to maintain because of continuous cell death. To overcome this we have attempted to cocultivate the initial PBMC cultures with cell lines that do not have CD4$^+$ receptors, the rationale being that HIV will not be able to infect these cells, whereas they may support the replication of HTLV-I and/or II. In studying a number of cell lines we observed that one, BJAB, a continuous EBV-negative B-cell line (24), developed syncytia in a number of cultures initially established from seropositive IVDAs 2–10 days after cocultivation. With continued cocultivation with fresh BJAB cells, syncytia persisted, and a number such cultures are now established as continuous cell lines.

To determine whether the cells were infected with HTLV-I and/or HTLV-II, culture supernatants were assayed for reverse transcriptase (RT) activity. Culture supernatants were clarified and concentrated, and all four were found to have high levels of Mg^{2+}-dependent RT activity. Electron microscopic examination of the cocultures demonstrated the presence of typical C type virus particles (Fig. 1). These results demonstrated that the BJAB cells were infected and had in effect "captured" either HTLV-I and/or HTLV-II from the PBMC, which were also initially infected with HIV.

CHARACTERIZATION OF VIRUS ISOLATES

Four isolates (referred to as NS-AM, NS-DP, NS-DE, and NS-WD) have been characterized by restriction enzyme analysis and Southern hybridization, employing probes specific for HTLV-I and HTLV-II. DNA was isolated from the four lines and from two control cell lines known to be infected with HTLV-I (Hut 102 and NS-RC). Hybridization analyses were performed using 3' and 5' probes, which together encompassed the entire HTLV-II (Mo) provirus genome and a probe for the entire HTLV-I (PMT-2) genome. With the use of stringent washing conditions, all four isolates were shown to be HTLV-II. With the HTLV-II 3' probe a 3.5-kb fragment defined by the known two *Bam*HI restriction sites

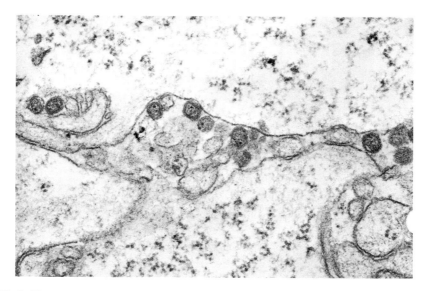

FIG. 1. Electron microscopic examination of cocultures of PBMC from a HIV- and HTLV-seropositive IVDA with BJAB cells. PBMC from dually seropositive individuals were prepared from heparinized blood samples by centrifugation on Ficoll-Hypaque density gradients. Cells were washed in RPMI 1640 medium containing 20% heat-inactivated FCS and 0.001 M L-glutamine and resuspended in the same containing phytohemagglutinin (PHA) 0.005 mg/ml at a final concentration of 10^6 cells/ml. Cultures are maintained at 37°C for 3–4 d, at which time the medium containing PHA was removed and replaced with medium containing 10% interleukin-2 (IL-2) (Cellular Products). Cells were maintained at 37° and fresh IL-2 added every 3 d. Cell lysis probably secondary to the cytopathic effects of HIV is generally observed within 1–2 wk. Surviving cells were, at this time, cocultivated with an equal number of BJAB cells and maintained in RPMI with 10% heat inactivated FCS. Cytopathic effects as noted above were usually observed within 4–6 wk following cocultivation. For electron microscopy, cells were collected by centrifugation, washed three times in ice-cold PBS, and fixed in PBS containing 2.5% glutaraldehyde. Thin sections are prepared, stained with uranyl acetate, and examined in a Zeiss electron microscope.

in the 3′ half of the genome could be identified in all four isolates (Fig. 2A). Hybridization with the HTLV-II 5′ probe showed the expected 4.7 kb fragment defined by the two *Bam*HI sites previously described for HTLV-II (Mo) in three of the isolates. However, hybridization of one isolate (NS-DP) showed an additional *Bam*HI site (Fig. 2B). No signal was observed with *Bam*HI digests of the control HTLV-I DNAs or in the four HTLV-II isolates when the HTLV-I probe was employed (data not shown).

Restriction-enzyme mapping was used to determine the relatedness of two of our isolates to the two prototype HTLV-II isolates (Mo and NRA) from patients with HCL Fig. 3. It could be seen that one isolate (NS-AM) had our identical map to that previously reported for Mo. The NS-DP isolate, however, showed differences from both Mo and NRA. As noted above, the provirus had an additional *Bam*HI site in the 5′ end in a similar location reported for NRA. Fur-

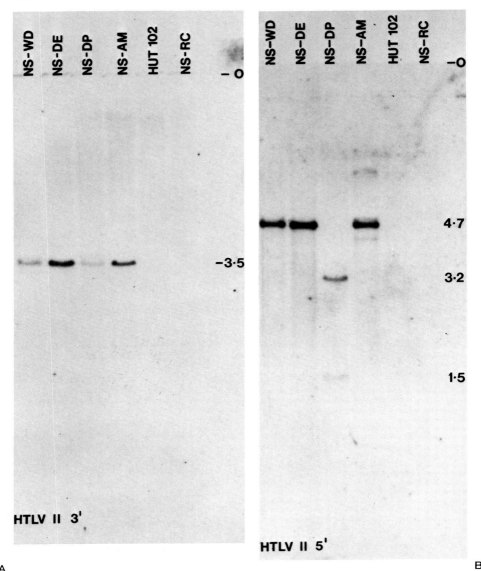

FIG. 2. Southern hybridization analysis of DNA from PBMC-BJAB cocultures established from four IVDAs (NS-AM, NS-DP, NS-DE, and NS-WD). DNA from HTLV-I–infected cells (Hut 102 and NS-RC) were used as controls. Analysis was performed with 3′ and 5′ probes, each defined by two *Bam*HI sites, and which together encompassed the entire HTLV-II (Mo) provirus genome. Cellular DNAs were isolated by phenol-chloroform extraction, digested with *Bam*HI and electrophoresed on 0.8% agarose gels. DNA was transferred to Gene-Screen Plus (Dupont) membranes by capillary transfer. Membranes were incubated in prehybridization buffer (50% formamide, 10% Dextran, 3× SSC 1% SDS, 0.2 mg/ml Salmon Sperm DNA, 0.5% Milk Powder) for 4 h at 42°C. Probes labeled with γ-^{32}P [CTP] using the random primer method were added, and hybridization continued overnight at 42°C. Following hybridization, membranes were washed twice with 2× SSC containing 0.1% SDS for 15 min and then with 0.1× SSC containing 0.1% SDS at room temperature for 20 min and at 60°C for 45 min. Membranes were exposed to X-Omat AR film with intensifying screens at −70°C.

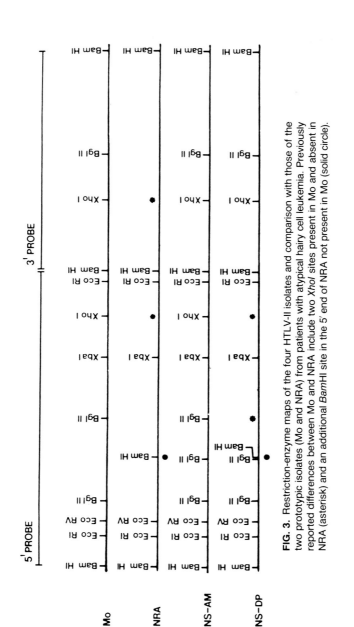

FIG. 3. Restriction-enzyme maps of the four HTLV-II isolates and comparison with those of the two prototypic isolates (Mo and NRA) from patients with atypical hairy cell leukemia. Previously reported differences between Mo and NRA include two *Xho*I sites present in Mo and absent in NRA (asterisk) and an additional *Bam*HI site in the 5' end of NRA not present in Mo (solid circle).

thermore, like NRA, it did not have a Xhol site in the 5' end which is present in Mo. However, NS-DP differed from NRA in that it did have a Xhol site in the 3' end which has been described in Mo. NS-DP also differed from NS-AM in that it had two *Bgl*II sites as opposed to three in the 5' end of the genome. The differences observed in the physical maps of Mo, NRA, and NS-DP suggest that genetic heterogeneity may exist among the HTLV-II group of viruses, which is in contrast to HTLV-I, where only minor differences among numerous isolates have been described. Characterization of a large number of isolates currently in progress should determine the significance of these preliminary findings.

PCR CHARACTERIZATION

Because virus isolation is often difficult and time consuming, more rapid and sensitive techniques need to be developed for virus identification. DNA amplification using the PCR offers much promise in this regard, and recently there have been a number of reports using PCR to detect HTLV-I in both fresh and cultured cells from patients with ATL and HAM/TSP (25–29) and in tissue of one patient with lymphoma (29). Furthermore, by amplifying relatively non-homologous regions of the *pol* gene, it was reported that infection with HTLV-I could be distinguished from HTLV-II in DNA extracted from continuous cell lines (28,29). Using the four new isolates of HTLV-II described above and the same primer/probe combinations, we confirmed these findings. DNA isolated from the four BJAB-HTLV-II cocultures and the two control HTLV-I cell lines was amplified through 30 cycles with primers specific for both viruses, and amplified DNA was detected with end-labeled oligonucleotide probes with the use of dot-blot hybridization. Figure 4 shows that only the DNA amplified from the HTLV-II–infected cells gave a positive signal with the HTLV-II primer/probe combination. No signal was observed in DNA amplified from the HTLV-I cell lines. Similarly, only the HTLV-I primer/probe combination gave a positive result with HTLV-I DNA samples and not with HTLV-II. Using similar amplification conditions we analyzed DNA isolated from fresh PBMC of two IVDAs, one of whom was seropositive for HIV and HTLV, the other being seropositive for HIV alone (Fig. 4, a and b). The former showed the presence of HTLV-II sequences, whereas the latter showed no evidence of HTLV-I or HTLV-II infection. These results demonstrate the capability of PCR amplification to detect HTLV sequences in PBMC DNA, and its application to a larger number of patients should allow the determination of the prevalence of HTLV-I and HTLV-II infection in our study population.

DISCUSSION

In the seroprevalence study described, it is clear that, of the various risk groups for HIV infection, concomitant infection with HTLV-I and/or HTLV-II occurs

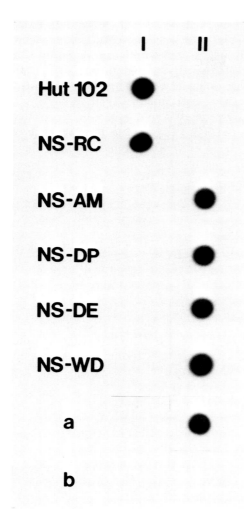

FIG. 4. PCR amplification of DNA from HTLV-I– and HTLV-II–infected cells (previously described in Fig. 2) and from DNA isolated from fresh PBMC of two IVDAs, one of whom was dually seropositive for HIV and HTLV-I/HTLV-II (a) and the other of whom was seropositive for HIV only (b). Primers specific for the amplification of HTLV-I (I) and HTLV-II (II) were as described by Kwok and co-workers (27). Amplification of DNA (1 μg) was performed in a total volume of 100 μl in a reaction mixture containing 225 μM each of dATP, cCTP, dGTP, and dTTP; 125 pmole of each primer; 50 mM KCl; 2.5 mM MgCl$_2$; 10 mM Tris-HCl (pH 8.3); and 6 units of *Thermus Aquaticus (TaQ) Polymerase* (Cetus). Solutions were covered with mineral oil to prevent condensation and 30 cycles of denaturation for 2 min at 94°C, primer annealing for 2 min at 52°C, and chain elongation for 3 min at 72° were carried out in a DNA thermal cycler (Perkin-Elmer Cetus). Primers for the specific amplification of HTLV-I and HTLV-II were as described by others and are as follows: HTLV-I, SK54 (bp 3365–3384) and SK55 (bp 3465–3483); HTLV-II, SK58 (bp 4198–4127) and SK59 (bp 4281–4300). The probes used for detection were SK56 (bp 3426–3460; HTLV-I) and SK60 (bp 4237–4276; HTLV-II). Aliquots (10 μl) of amplified products were denatured with NaOH, neutralized, and spotted onto Nytran membranes (Schleicher and Shuell) which had been presoaked in 6% SSPE. Membranes were exposed to UV light for 10 min, dried by heating in a vacuum oven for 1 h at 80°C, and prehybridized in 4× SSPE, 5% Denhardt's, 25% formamide, and 0.5% SDS for 1 h at 42°C. Hybridization was continued overnight at 42°C in the same solution containing the corresponding oligonucleotide probe end-labeled with γ-^{32}P[ATP]. Following hybridization, membranes are washed twice at room temperature with 2× SSPE containing 0.1% SDS for 5 min, twice with 0.2× SSPE containing 0.1% SDS for 15 min, and finally once with 0.2× SSPE containing 0.1% SDS at 56°C for 15 min. Membranes were exposed to XAR-2 Kodak film with intensifying screens at −70°C overnight.

primarily in IVDAs. Within this population blacks and—to a lesser extent—Hispanics demonstrated much higher seroprevalence rates than whites. These findings are consistent with an earlier study on a smaller number (56) of IVDAs from the New York City area, where 27% of blacks were dually seropositive for HIV and HTLV-I and/or HTLV-II compared with 0% of whites. Our finding of 7% dual seropositivity in white IVDAs is unlikely to represent recent transmission of HTLV into this subgroup but is more likely due to the larger number of patients included in the present study. The overall seroprevalence rate of 12% in IVDAs is much less than the 24% recently reported in a large study of Californian IVDAs (17). However, these are difficult to compare because the age, race, and HIV status of these patients was not documented.

Our study also shows that seropositivity increases with age in IVDAs, which is similar to what has been reported for HTLV-I infection in endemic areas of the Caribbean and Japan. At present the significance of this is unknown and because of incomplete and/or inconsistent histories obtained from our patients, it is unclear if this increase with age can be attributed to an increase in the total years of drug abuse.

The present study also showed an approximate 2% seroprevalence rate among HIV-seropositive, homosexual males. This is somewhat higher than previous reports from the United States (17), where seropositivity was noted to be rare (0.05%), but lower than that reported from Trinidad, an endemic area for HTLV-I infection where dual infection of HIV and HTLV-I was present in 4 of 100 individuals (21). It is interesting to note that, as was observed as in the IVDA population, all three seropositive patients were black.

The present study also provides evidence of heterosexual HTLV transmission with two females, both of whom were without risk factors for HIV infection, being seropositive for both HTLV and HIV. In endemic areas, sexual transmission of HTLV-I seems to be relatively inefficient and occurs primarily from male to female. In both cases the sexual partners were IVDAs who were seropositive for HIV and HTLV, suggesting that within this population, heterosexual transmission may be significant. Although our studies on sexual (both homo- and hetero-) transmission involved a relatively small number of patients, a 2% seropositivity rate would appear to be significant. Although the results of large-scale systematic studies on the seroprevalence of HTLV infection in the United States are not yet available, a study of ~40 000 randomly chosen blood donors from eight U.S. cities showed the very low seropositivity rate of 0.025% (30).

Dual infection with HIV and HTLV is not surprising in view of the fact they can be transmitted in similar ways, and in IVDAs this is almost certainly through blood contamination of shared needles and syringes. Studies from endemic areas in Japan have shown that blood transfusion is a major mode of transmission of HTLV-I, with seroconversion rates of 63% in recipients of cellular-containing blood products (31,32). In contrast, transfusion of plasma—which certainly leads to transmission of HIV—does not result in HTLV-I infection. The relative inefficiency of HTLV-I transmission compared with HIV is related to the fact that HTLV-I is highly cell-associated whereas HIV is cytopathic, producing high levels of cell-free virus. The relative inefficiency of HTLV-I transmission is probably responsible for the lower seroprevalence rates observed in HIV-seropositive homosexuals compared with IVDAs.

Recent evidence has suggested that IVDAs are becoming the group at greatest risk for acquiring HIV infection, and it has been estimated that 60% of IVDAs in New York City are HIV seropositive (33). The present study—demonstrating a high seroprevalence rate for HTLV in the same population—suggests that, despite relatively inefficient transmission, a second epidemic of HTLV infection could eventually occur in this population.

The major drawback of the present and most of the previously published seroprevalence studies is that the serological methods available do not adequately or consistently distinguish infection by HTLV-I from HTLV-II because of their antigenic cross-reactivity. This requires virus isolation and/or molecular analysis. As part of our preliminary studies, we have developed a novel method using cocultivation to selectively "capture" and isolate the HTLVs from peripheral blood lymphocytes, which were also infected with HIV. In preliminary studies using this method we successfully isolated HTLV-II from four seropositive IVDAs. Using the four new isolates, we confirmed previous reports that PCR amplification of relatively nonhomologous regions of the *pol* gene allowed the differentiation of HTLV-I from HTLV-II. We now intend to use this to directly identify the viruses in peripheral blood and have already shown that in one HIV-seropositive IVDA HTLV-II could be detected. The detection, using a combination of virus isolation with provirus mapping and PCR amplification methods of HTLV-II in five IVDAs, suggests that infection by this virus may be more common than previously thought. This is supported by independent reports presented at this symposium by Shaw, Chen, Lee, and their respective co-workers on the isolation and detection of HTLV-II in a large number of HIV-negative IVDAs from several large United States urban areas. Taken together, the findings indicate that HTLV-II infection is clearly not limited to individuals with rare lymphoproliferative disorders and may in fact be the prominent HTLV infection in IVDAs and perhaps other seropositive individuals in the United States. Further studies using PCR amplification on a large number of HTLV-seropositive patients, independent of their HIV seropositivity status, should rapidly determine if this is indeed the case. The findings also suggest that previously reported studies describing concomitant infection with HIV and HTLV-I in IVDAs on the basis of serology alone may have to be reevaluated.

PCR studies will also allow a more accurate assessment of those individuals with "indeterminate" serology. At the present time, it is unclear if sera immunoreactive to other virus proteins—for example, the *gag* p19 or *gag*-derived p28, but not p24, and *env* gp products which were used as confirmatory "positives" in this study—reflect actual virus infection. A correlation of such serological and molecular studies should provide a more accurate assessment of HTLV infection and may show that exclusion of certain immunoreactivities has led to an underestimation of the prevalence of HTLV infection.

The identification of HTLV-II infection in this population should allow, with prospective clinical and immunological studies, a better understanding of the role of this virus in human disease. Furthermore, with the concomitant transmission of HIV and HTLV (perhaps HTLV-II) in IVDAs, this may allow an appreciation of unique clinical entities occurring in this population.

SUMMARY

Sera from 489 HIV-infected patients were analyzed for antibody to HTLV-I and/or HTLV-II. IVDAs have much higher seroprevalence rates than any other AIDS risk group. Black and Hispanic IVDAs have appreciably higher seropositivity rates than whites and seropositivity increases with age. Homosexual and heterosexual transmission of HTLV-I and/or HTLV-II also occurs but is significantly less than that seen in IVDAs. Using virus isolation with provirus mapping and PCR amplification, we identified HTLV-II in five HTLV-seropositive IVDAs. These findings suggest that HTLV-II infection may be more common than previously thought and is not limited to individuals with rare lymphoproliferative disorders. Prospective clinical and immunological studies in this patient population should allow a better understanding of the role of HTLV-II in human disease.

ACKNOWLEDGMENTS

This work was supported by the Jane and Dayton Brown and Dayton Brown Jr. Virology Laboratory. We thank Carole Cristiano and Rosanne Rybacki for preparation of the manuscript.

REFERENCES

1. Wong-Staal F, Gallo RC. *Blood* 1985;65:253–263.
2. Wong-Staal F, Gallo RC. *Nature* 1985;317:395–403.
3. Rosenblatt JD, Chen ISY, Wachsman W. *Semin Hematol* 1988;25:230–246.
4. Poiesz BJ, Ruscetti FW, Gazdar AF, Bunn PA, Minna JD, Gallo RC. *Proc Natl Acad Sci USA* 1980;77:7415–7419.
5. Miyoshi I, Kubonishi I, Yoshimoto S, et al. *Nature* 1981;294:770–771.
6. Gessain A, Vernant JC, Maurs L, et al. *Lancet* 1985;2:407–409.
7. Osame M, Usuku K, Izumo S, et al. *Lancet* 1986;1:1031–1032.
8. Kalyanaraman VS, Sarngadharan MG, Robert-Guroff M, et al. *Science* 1982;218:571–573.
9. Chen ISY, McLaughlin J, Gasson JC, Clark SC, Golde DW. *Nature* 1983;305:502–505.
10. Gelman EP, Franchini G, Manzari V, Wong-Staal F, Gallo RC. *Proc Natl Acad Sci USA* 1984;81:993–997.
11. Rosenblatt JD, Golde DW, Wachsman W, et al. *N Engl J Med* 1986;315:372–377.
12. Hahn BH, Popovic M, Kalyanaraman VS, et al. In: Gottlieb MS, Groopman JE, eds. *Acquired Immunodeficiency Syndrome.* New York: Alan R. Liss, 1984;73–81.
13. Kalyanaraman VS, Narayanan R, Feorino P, et al. *EMBO J* 1985;4:1455–1460.
14. Rosenblatt JD, Giorgi JV, Golde DW, et al. *Blood* 1988;71:363–369.
15. Tedder RS, Shanson DC, Jeffries DJ, et al. *Lancet* 1984;2:125–128.
16. Robert-Guroff M, Weiss SH, Giron JA, et al. *J Am Med Assoc* 1986;255:3133–3137.
17. Gallo D, Hoffman MN, Lossen CK, et al. *J Clin Microbiol* 1988;26:1487–1491.
18. Gradilone A, Zani M, Barillari G, et al. *Lancet* 1986;2:753–754.

19. Getchell JP, Health JL, Hicks DR, Sporborg C, Mann JM, McCormick JB. *J Infect Dis* 1987;155: 612–616.
20. von derHelm K, von derHelm D, Deinhardt F. *J Infect Dis* 1988;157:205–207.
21. Bartholomew C, Saxinger CW, Clark JW, et al. *J Am Med Assoc* 1987;257:2604–2608.
22. Khabbaz RF, Darrow WW, Lairmore M, et al. *Fourth International Conference on AIDS.* Book 1, 270 Stockholm, June 12–16, 1988.
23. Public Health Service Working Group. *MMWR* 1988;37:736–747.
24. Klein G, Lindahl T, Jondal M, et al. *Proc Natl Acad Sci USA* 1974;71:3283–3286.
25. Bhagvati S, Ehrlich B, Kula RW, et al. *N Engl J Med* 1988;318:1141–1147.
26. Abbott MA, Poiesz BJ, Byrne BC, et al. *J Infect Dis* 1988;158:1158–1169.
27. Kwok S, Kellogg D, Ehrlich G, et al. *J Infect Disease* 1988;158:1193–1197.
28. Kwok S, Ehrlich G, Poiesz B, et al. *Blood* 1988;72:1117–1123.
29. Duggan DB, Ehrlich GD, Dowey CP, et al. *Blood* 1988;71:1027–1032.
30. William AE, Fang CT, Slamon DJ, et al. *Science* 1988;240:643–646.
31. Brown LS, Battjes R, Primm BJ, Foster K, Chu A. *Proceedings of the Fourth International Conference on AIDS.* Book 2, 8543. Stockholm, June 12–16 1988.
32. Okochi K, Sato H, Hinuma Y. *Vox Sang* 1984;46:245–253.
33. Kajiyama W, Kashiwaga S, Ikematsu H, Hayashi J, Nomura H, Okochi K. *J Infect Dis* 1986;154: 851–857.
34. Lee H, Swanson P, Shorty VS, Zack JA, Rosenblatt JD, Chen ISY. *Science* 1989;244:471–475.

DISCUSSION

One discussant pointed out that when various at-risk populations were screened serologically for HTLV-I and HTLV-II in London, HTLV-II was found exclusively in the drug abuser population. Another discussant expressed hope that there would be enough genetic variability to allow use of molecular tools to examine the epidemiology of some of these viruses. This discussant pointed out that molecular methods are being used to address discrepancies identified by the serology during well-conceived epidemiological studies. A speaker noted that by using these methods, investigators in the United Kingdom or Italy might be able to determine if the HTLV virus present in their countries was either imported from the United States through servicemen or was due to endogenous virus. Questions also are raised by data about efficacies of transmission of HTLV-II and HTLV-I, which are unknown and should be examined. A discussant reminded the group that patients who were coinfected with HIV-1 and HTLV-II did better clinically than patients who were coinfected with HIV-1 and HTLV-I. Another discussant replied that patients who are either coinfected with HTLV-I and HIV-1 or coinfected with HTLV-II and HIV have been identified. This discussant agreed that patients infected with HTLV-I and HIV do very poorly. It appears that HTLV-II is somehow protective in HIV disease, in terms of patient survival, the type of infections contracted, and clinical sequelae. One participant noted that there is no infectious molecular clone of HTLV-I and questioned its difficulty and if there had been progress in obtaining one. Another participant answered that recently full-length HTLVs had been cloned into plasmids.

Human Retrovirology: HTLV,
edited by William A. Blattner.
Raven Press, Ltd., New York © 1990.

ATL and HAM/TSP in African, French West Indian, and French Guianese Patients Studied in France: Serological and Immunological Aspects and Isolation of HTLV-I Strains

* **Antoine Gessain, *Fortuna Saal, †Olivier Gout,
‡Christiane Caudie, §Jean Michel Miclea, ‖Mabel Cruz,
#Guy de Thé, **François Sigaux, and *Jorge Peries

*UPR 0043 CNRS "Rétrovirus et Rétrotransposons des Vertébrés", Hôpital Saint-Louis, 75010 Paris; †Clinique de Neurologie et de Neuropsychologie, Hôpital de la Salpétrière, 75013 Paris; ‡Laboratoire d'Immuno-Biologie, Hôpital P. Whertheimer, 69, Lyon Cedex; §Département d'Hématologie, Hôpital Saint-Louis, 75010 Paris; France; ‖Karolinska Institutet, Department of Neurology, Huddinge University Hospital, Stockholm, Sweden; #Laboratoire d'Epidémiologie et d'Immuno-Virologie des Tumeurs, Faculté de Médecine Alexis Carrel, 69372 Lyon; and **Laboratoire d'Hématologie Moléculaire et Laboratoire Central d'Hématologie, Hôpital Saint-Louis, 75010 Paris

INTRODUCTION

Human T-cell leukemia/lymphoma virus type I (HTLV-I), a type C retrovirus discovered in 1980 (1), is currently recognized as the etiological agent of the adult T-cell leukemia (ATL) (2,3). This lymphoproliferative malignancy, occurring in clusters in high HTLV-I endemic areas, is characterized by a clonal expansion of $CD4^+$ lymphocytes associated with a monoclonal integration of HTLV-I in the tumoral cells. Furthermore, since 1985 (4,5), this retrovirus has also been linked with a chronic progressive myelopathy referred to as tropical spastic paraparesis (TSP) in tropical areas (6–8) and HTLV-I–associated myelopathy (HAM) in Japan (9,10).

In this chapter, we will summarize some of our ongoing collaborative studies concerning the serological, immunological, and viral features of ATL and HAM/TSP patients seen in France.

ATL IN FRANCE

Epidemiological and Clinical Features

Most of the patients with ATL seen in France originated either from French West Indies (11,12), French Guiana (13), or African countries (14). Comparable data have been reported in the United Kingdom (15), where nearly all the patients with ATL are immigrants from West Indies, mostly from Jamaica. Nevertheless, rare cases of ATL associated with HTLV-I have been reported in white Italian patients without known risk factors for HTLV-I infection (16), suggesting the possible presence of microfoci of HTLV-I infection in European countries.

In the last 2 yr, we have studied in St. Louis Hospital (Paris, France) eight consecutive cases of ATL. As seen in Table 1, there were five women and three men, and the mean age at onset was 40 yr. Four of the patients were natives of French West Indies (Martinique and Guadeloupe), but three of these lived in metropolitan France for ≥ 15 yr. The four remaining patients originated from African countries, three from Ivory Coast and one from Mauritania. From an epidemiological viewpoint, the presence of these four African patients suggests that—despite the fact that only 10 typical ATL have been described up to now in African patients (14,17,18)—ATL cases must be much more frequent in Africa than reported. Worthwhile to note here is that four other ATL cases in African patients have been recently seen in Paris hospitals, one from Gabon, one from Algeria, and two from Ivory Coast (B Rio and M Tulliez, personal communications).

On a clinical point of view, five patients had an acute ATL type, with a high white blood cell count, peripheral lymph-node enlargement, hepatosplenomegaly, and hypercalcemia at the time of admission or during clinical course of the illness, but none of them had skin lesions. Their evolution was poor, and three of them died within the year following the diagnosis.

The three other patients had a smoldering ATL type, with specific cutaneous lesions as the predominant and premonitory clinical feature, associated with a low percentage of cells with abnormal nuclei in their peripheral blood for a long period.

We report here details of three of these ATL patients presenting a worthy case.

Case 1. This young woman from Ivory Coast had a typical ATL of acute type, with 50 000/mm^3 of abnormal CD4$^+$ lymphocytes with convoluted nuclei, a monoclonal integration of HTLV-I proviral DNA in the tumoral cells with two viruses (one being defective), and with T-cell receptor β and γ gene rearrangement. Various opportunistic infections were present in this young woman at the onset and during the course of her illness, and she was found to be HIV-2 seropositive. When we tried to isolate these viruses in cultured cells, we did not succeed either in isolating HIV-2 or in establishing an HTLV-I–infected T-cell line (despite transient positivity in HTLV-I–specific immunofluorescence and the presence of type C retroviral particles in electronic microscopy), and

TABLE 1. Clinical features and immunovirological findings of 8 patients with adult T-cell leukemia studied in Saint Louis Hospital, Paris (1987–1989)

Case	Sex and age	Geographical origin	ATL type	HTLV-I monoclonal integration	Viral detection short-term culture		HTLV-I-producing T-cell line	Other characteristics
					IF	EM		
1	F/19	Ivory Coast	Acute	Leukemic cells	+	+	–	Concomitant infection by HIV-2 and HHV-6
2	F/45	Ivory Coast	Acute	Leukemic cells	+	+	+	*Pneumocystis carinii*
3	M/33	Martinique	Acute	Leukemic cells lymph node	+	+	+	Initial strongyloides stercoralis
4	M/53	Mauritania	Smold.	Cutaneous tumors	nd	nd	nd	Hodgkin's disease 5 yr before
5	F/45	Martinique	Smold.	Cutaneous tumors	–	–	–	
6	F/43	Martinique	Acute	Leukemic cells	+	+	+	Initial strongyloides stercoralis and *Pneumocystis carinii*
7	M/52	Guadeloupe	Acute	Leukemic cells	+	+	+	Initial hypercalcemia
8	F/35	Ivory Coast	Smold.	Cutaneous tumors	+	+	+	

IF: Indirect immunofluorescence; EM: electron microscopy; HIV-2: human immunodeficiency virus Type 2; HHV-6: Human herpes virus type 6; nd: not done.

the cells died in a few days with appearance of giant cells. A human herpes virus 6 (HHV-6) strain was then isolated and characterized from these cells by Luc Montagnier's group in Pasteur Institute (19). It is interesting to note that recently other HHV-6 strains have been isolated from cultured cells from HIV-2 or HTLV-I–infected African patients (20).

Despite treatment with a polychemotherapy, AZT and DCF, the patient rapidly died, with a high tumoral load and hypercalcemia. In the family members, four subjects were HTLV-I seropositive; none were infected by HIV-2, but three were HHV-6 seropositive (14).

Case 4. In 1982, a mixed-cellularity Hodgkin's disease, stage II Aa, was diagnosed in this Mauritanian man of 33 yr, and a treatment with a combined chemotherapy and irradiation led to a complete remission for >5 yr. In 1987, he developed multiple tumors of the face and the trunk. The histology revealed an important epidermotropism with an infiltration by CD4$^+$ cells. At that time he was found seropositive for HTLV-I. By Southern blot analysis (Fig. 1), we demonstrated a clonal T-cell population with TCRγ gene rearrangement. Furthermore, a monoclonal integration of two HTLV-I proviruses was found in the skin biopsy tumoral DNA. Peripheral blood mononuclear cells were not available for molecular biology studies but numerous cells resembling Sézary cells were present on his peripheral blood film at that time. The patient returned to Mauritania, where he died of opportunistic infections.

Case 5. This case illustrates the need of molecular biology tools to prove the role of HTLV-I in smoldering ATL cases with a sometimes-difficult clinical diagnosis at the onset of the disease. This 45-yr-old woman from Martinique visited St. Louis Hospital in February 1988, complaining of multiple plaques and nodules over her whole body. She was HTLV-I seropositive and had very few cells with abnormal nuclei on her peripheral blood film and a normal CD4/ CD8 ratio with no detectable activated T cells (Tac and DR expression was normal). The skin biopsy of a nodule showed an epidermotropism of CD4$^+$ cells with typical "Pautrier" microabscess. Using Southern blot analysis, we were able to demonstrate the monoclonality of the tumoral cells in the skin biopsy by showing TCR β and γ gene rearrangements. Furthermore, the monoclonal integration of one HTLV-I provirus was clearly shown using various restriction enzymes and two HTLV-I probes (*env* and LTR)—only in the skin lesions' DNA, not in the peripheral blood mononuclear cells' (PBMCs') DNA. This technique can be useful for analyzing the HTLV-I specificity and viral origin of the cutaneous lesions in smoldering ATL and can also differentiate this ATL type from a real mycosis fungoides or Sézary syndrome or from other cutaneous T-cell lymphomas, even those occurring in an HTLV-I–seropositive carrier.

Immunological and Viral Studies

In the eight cases, the diagnosis of ATL was highly suspected on the basis of the clinical features, the origin of the patients, the HTLV-I–positive serology

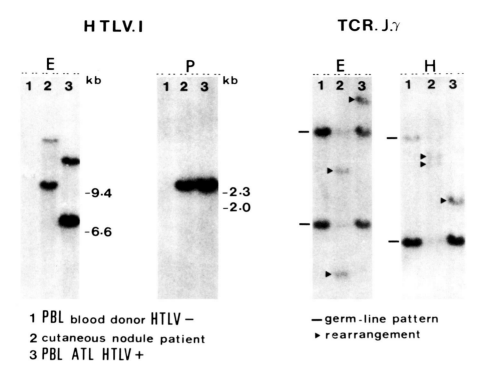

HTLV. I   **TCR. J.**γ

1 PBL blood donor HTLV −   —germ-line pattern

2 cutaneous nodule patient   ▶ rearrangement

3 PBL ATL HTLV +

FIG. 1. Southern blot analysis. High-molecular-weight DNA was extracted from 1, peripheral blood lymphocytes of an HTLV-I–seronegative blood donor; 2, cutaneous nodule (case 4); and 3, Lymphoid cells of an ATL (case 1). DNA samples were digested with E, *Eco*RI; P, *Pst*I, H, *Hind*III restriction endonucleases. Restriction DNA fragments were separated by electrophoresis, transferred to nylon membrane, and hybridized with appropriate probes as previously described (21). HTLV-I: the probe used is the PATK 06 subcloned *Bam*HI-*Bam*HI fragment of the PAT K01 clone. TCR J γ: this T-cell antigen receptor (TCR) probe is a subcloned *Eco*RI-*Hind*III fragment of the MH 60 J γ clone. Left: study of HTLV-I proviral integration. HTLV-I provirus are vizualised in both digests from patient's cutaneous nodule DNA. As there is no known *Eco*RI site in HTLV-I provirus, the observation of two bands denotes clonal integration of two viral genomes. Right: study of TCR rearrangement.—, germ line fragments; ▶, rearranged fragments. A clonal rearrangement pattern is observed in patient's cutaneous nodule and ATL DNA samples.

and the presence of a CD4$^+$ lymphoproliferation in blood, node, or skin, associated with typical ATL cells in the peripheral blood. Nevertheless, in all the cases, the diagnosis of ATL was definitively confirmed by the demonstration of a clonal integration of one or two HTLV-I proviruses in the tumoral cells.

In seven cases (all except case 4), PBMCs were put in culture with 10% of recombinant Interleukin 2 (IL-2) after initial stimulation by phytohemagglutinin P for 3 d. Indirect immunofluorescence (IF) on cultured, acetone-fixed cells, using either HTLV-I polyclonal sera (TSP and ATL sera) or monoclonal HTLV-I anti-p19 and anti-p24 antibodies, was performed at different periods of culture, as previously described (21). Furthermore, electron microscopy studies were regularly done. In all the cultures except case 5, viral detection was successful. Thus, after some days of culture (from 3 to 15), HTLV-I–specific antigens were detected in the cultured cells by IF and type-C retroviral particles were seen in the extra-cellular space by electron microscopy. Long-term IL-2–dependent T-cell lines (CD4$^+$, CD25$^+$, DR$^+$) producing HTLV-I were established in cases 2, 3, 6, 7, and 8.

HTLV-I SEROLOGICAL FEATURES AND CEREBROSPINAL FLUID (CSF) ANALYSIS IN HAM/TSP

HTLV-I Serological Aspects of HAM/TSP

During the last 4 yr, we had the opportunity to perform detection and analysis of HTLV-I antibodies in 70 cases of HAM/TSP, most originating from HTLV-I–endemic areas. Similar serovirological patterns were found in all the patients (22) despite their different geographical origin. The enzyme-linked immunosorbent assay (ELISA, Dupont) titers were on the average 50 to 100 times higher in serum (1:100 to 1:10 240) than in CSF (1:2 to 1:320). When tested by particle agglutination (PA) (Fujirebio), the titers ranged from 1:16 348 to 1:31 072 in sera and from 1:312 to 1:4096 in CSF. By IF using HUT 102 cell line as HTLV-I antigen, the titers ranged from 1:400 to 1:3200 in serum and from 1:8 to 1:32 in CSF. By Western blot (WB), reactivities against *gag*-encoded proteins p19, p24, and their precursor pr53 were always detected in serum and CSF. Detection of *env*-encoded glycoprotein gp46, gp61-62, although found in most of the sera and CSF, was easier using radioimmunoprecipitation assay (RIPA) (22). Other proteins, p26, p28, p32, p36, and p40, were also generally present in serum of HAM/TSP patients depending on the tested sera. A main point is that both sera and CSF from HAM/TSP patients exhibited antibodies to HTLV-I–encoded polypeptides in a pattern qualitatively indistinguishable from West Indian or African patients with ATL. The only difference was a quantitative one and the intensity of the WB reaction was always high, with a clear resolution, when we used TSP sera; whereas with ATL sera, the WB reaction was sometimes very faint and difficult to analyze, the ELISA, PA, and IF titers from ATL patients (sera) being much lower than those from TSP. Furthermore, no HTLV-I anti-bodies were detected at a significant level in our experience in the few tested CSF samples from ATL by ELISA, PA, or IF assays.

Intratechal IgG Synthesis and CSF IgG Oligoclonal Bands

We studied the sera and CSF samples of 42 HAM/TSP patients using tech-niques previously described (22). The origin of the patients and the results of the CSF immunological study are summarized in Table 2. The IgG index was elevated in 36 of the cases tested and in the normal range in 6 patients. The

TABLE 2. *CSF analysis of 42 patients with HAM/TSP*

Geographical origin	No. cases tested	HTLV-I antibodies, ELISA and WB, Serum/CSF	Elevated IgG index no. cases	Elevated IBBB IgG synthesis no. cases	Presence of IgG oligoclonal no. cases
Martinique	20	+/+	15	13	14
Guadeloupe	7	+/+	7	7	7
French Guiana	5	+/+	4	4	3
Haiti	1	+/+	1	1	1
Dominica	1	+/+	1	1	1
Ivory Coast	3	+/+	3	3	3
Senegal	1	+/+	1	1	1
Zaire	1	+/+	1	1	1
Central African Rep.	1	+/+	1	1	1
France	2	+/+	2	2	2

The HTLV-I antibody index was calculated for 25 cases and was found elevated (>2) in 23 cases. For the IgG index and the IBBB IgG synthesis, upper normal reference value in our laboratory was, respectively, 0.70 and 3.5 mg/day.

total intra blood-brain barrier (IBBB) IgG synthesis was elevated in 34 of the cases. The ratio of CSF albumin to serum albumin, taken as an indicator of the integrity of the blood-brain barrier, was within the normal range in 38 of 42 patients, borderline for 1, and abnormal in 3. IgG oligoclonal bands, defined as discrete bands in the gamma region of the CSF electrophoretic pattern that were not present in the serum pattern, were observed by agarose gel electrophoresis in 33 patients and confirmed in all these by immunofixation. When the light chains of IgG were identified, only kappa type was found in all 12 tested patients. The characterization of the IgG oligoclonal bands of 22 of these HAM/TSP was performed in the laboratory of Prof. H. Link by Dr. M. Cruz and Dr. S. Kam-Hansen, as reported elsewhere (23). In 19 patients, IgG oligoclonal bands consisting of HTLV-I–specific antibodies were observed mostly in CSF but also in some sera (Fig. 2). These findings, also reported by others (24), are in favor of the presence of HTLV-I active infection within the central nervous system of HAM/TSP patients, supporting a causal relationship between HTLV-I and this type of chronic neuromyelopathy.

HAM/TSP IN FRANCE

Epidemiological and Clinical Features

As reported in this volume by G. The et al., to appreciate the magnitude of this disease in an a priori nonendemic area, we have conducted a large seroepidemiological study in Pitié-Salpétrière Hospital in Paris during 3 yr follow-up (1986–1988). During this period, 14 patients were diagnosed as HAM/TSP on the basis of clinical and biological criteria (4,5,25). As for ATL patients, the majority of

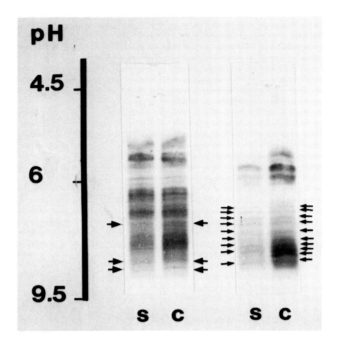

FIG. 2. IgG oligoclonal bands analysis in an African patient with HAM/TSP. Left: oligoclonal IgG pattern of serum (s) and cerebrospinal fluid (c). Right: oligoclonal anti-HTLV-I IgG pattern from the corresponding serum and CSF.

them originated from French West Indies, French Guiana, or Africa (Sénégal, Zaire, or Central African Republic). Two patients only (cases 7 and 8) were born in—and had always lived in—metropolitan France (26).

ATL-like Cells and Cell Surface Phenotypes

In 1986 Osame et al. (9,10) first detected the presence of few lymphoid cells with nuclear convolutions, comparable with the ATL abnormal lymphoid cells, on peripheral blood smears and CSF samples of HAM patients. Subsequent reports from Japan confirmed this peculiar feature in most HAM patients. We have studied the cell morphology by light microscopy on May Grünwald Giemsa stained peripheral blood cells of 10 HAM/TSP patients (27). In all cases, lymphoid cells with an abnormally shaped nucleus were detected in 5% to 15% of the total lymphoid cells. Furthermore, typical ATL-like cells with hyperconvoluted nuclei were less frequently (2%) seen in all but one case. This is in line with the recent detection of such cells in West Indian–born, UK-resident TSP patients studied in London (28).

The phenotypic analysis of PBMC of 10 HAM/TSP patients was performed using a panel of monoclonal antibodies. The major findings were an elevated

CD4/CD8 ratio associated with a high percentage of DR+ cells (15% to 38%) which largely exceeds those of B and monocytes in all the patients. This indicates that DR-expressing T cells were present in all the patients' peripheral blood. Similar immunological features have recently been found in Japanese HAM patients (29).

HTLV-I Proviral Integration in Fresh PBMC

To study the status of HTLV-I proviral integration in uncultured PBMC of HAM/TSP patients, we have extracted DNA from 10 of the 14 patients followed in Pitié-Salpétrière Hospital and performed on these DNA a Southern blot analysis using different restriction endonuclease enzymes and HTLV-I probes (27). In all the Pst1 digests, a clear band was observed using an *env* probe. This proviral integration was polyclonal because no band was detected in the *Eco*RI digests using the same probe. Furthermore, by dilutions experiments, we were able to estimate that this polyclonal integration was present in 3% to 15% of the HAM/TSP PBMC, irrespective of the patients' geographical origin, duration of illness, and HTLV-I antibody titers. These results are comparable with those reported by Yoshida et al. (30), who found such an HTLV-I proviral polyclonal integration in uncultured PBMC DNA of 8 of 9 Japanese patients with HAM and by Yamaguchi et al. (this volume). This high–HTLV-I proviral DNA load with polyclonal integration of HTLV-I in the PBMC, which seems a constant feature of HAM/TSP patients, might play an important, but as yet unclear, role in the pathogenesis of this illness. Thus the cytological, immunological, and virologic Japanese studies and ours are consistent with the recent proposition that TSP/HTLV-I and HAM represent the same clinicovirological entity.

Establishment of HTLV-I–Producing T-Cell Lines

In all except two patients, long-term T-cell lines were established from the PBMC or CSF cells of patients with HAM/TSP. Most of these IL-2–dependent cell lines exhibited a pattern characteristic of CD4$^+$-activated T cells, with a high expression of CD2, CD3, and CD4 antigens associated with a strong density of TAC (CD25) and DR molecules. CD8 antigen was found expressed in one cell line even after several months of culture. These immunological phenotypes remain stable on successive analysis.

Although HTLV-I antigens were never detected in uncultured PBMC of HAM/TSP, they were expressed in a few cells (0.5% to 5%) after short-term culture. In all the T-cell lines, the majority of the cells were HTLV-I positive by IF after 4 mo of culture using either a HTLV-I polyclonal serum or HTLV-I anti-p19 monoclonal antibody. Electron microscopy performed at regular intervals of culture revealed numerous type-C retroviral particles. They were present mostly in extracellular spaces. In long-term cultures, clusters of viral particles close to the cell surface and typical buddings can be found.

Characterization of Viral Isolates

Although HTLV-I–like retrovirus strains have been isolated from cultured PBMC or CSF cells from patients with TSP (21,31–33) or HAM (34,35), the basic mechanisms involved in the pathogenesis of these chronic neuromyelopathies are unknown. To show whether or not the viruses isolated from HAM/TSP or ATL patients exhibit the same immunological and molecular characteristics, we have compared different HTLV-I viral strains obtained in our laboratory.

TSP and ATL cell lines as well as HTLV-I–infected (HUT 102 and MT2) or –uninfected control cells (CEM), were studied by WB using HTLV-I positive polyclonal sera (TSP and ATL). In all the studied cell lines, bands corresponding to p53, p24, and p19 core polypeptides were observed as well as a possible *env*-encoded, 46-kd polypeptide. WB and RIPA of purified viruses revealed the same gp46, p24, p19, and pr53 gag proteins, similar to those detected in HUT 102 and MT2 cell lines (21). By labeling with C14 manose, and using a RIPA technique, the analysis of one TSP cell line versus HUT 102 and MT2 isolates revealed minor differences in the molecular weight (21) of the env-coded precursor. Comparable minor differences between TSP and ATL isolates could also be observed on cell lysates with the use of WB (36).

DNA analysis of two long-term cell lines established from PBMC of HAM/TSP patients, using various restriction enzymes, shows, as expected, a clonal integration pattern of a proviral DNA closely related to the HTLV-I leukemogenic isolate (21). Furthermore, an evolution from a polyclonal state to a monoclonal HTLV-I integration was observed in the two cell lines studied, with appearance of a clonal rearrangement pattern for T-cell receptor β and γ genes. All these data, also reported by others (31–35), demonstrate that a type-C retrovirus, closely related on an immunological and molecular basis to leukemogenic HTLV-I strains, can be isolated from culture of PBMC or CSF cells from HAM/TSP patients, irrespective of their geographical origin and the duration of their illness.

It is not known whether or not the slight differences observed either in molecular weight of env precursor proteins or in restriction enzyme pattern analysis are involved in the genesis of HAM/TSP or ATL. This might only reflect viral strain differences depending on geographical or individual HTLV-I variability, as observed for other human retroviruses. Studies involving viral isolates from different clinical conditions (ATL, HAM/TSP, healthy carriers) of various geographical origins are necessary for a better understanding of the pathogenesis of these retrovirus-induced diseases.

ACKNOWLEDGMENTS

We are indebted to Professor O. Lyon Caen for his continuous interest in and support of this work. We thank also Dr. M. T. Daniel for performing excellent cytological analysis of HAM/TSP and ATL cases and Dr. J. Lasneret for the electron microscopy study. We are also grateful to F. Stenger and Y. Poirot for serological and technical assistance. We thank Dr. I. Moulonguet, Dr. A. Diallot, Dr. T. De Revel, Prof. G. Auzanneau, Dr. O. Pattey, and Prof. P. Morel for having provided us with clinical and biological specimens from some ATL patients. This work was supported in parts by grants from The Centre National de La Recherche Scientifique, Association pour la Recherche sur le Cancer (Contract no. 6692), Programme National de La Recherche sur le SIDA et les Retro Virus Humains (960166), and the Fondation Contre la Leucémie.

REFERENCES

1. Poiesz BJ, Ruscetti FW, Gazdar AF, Bunn PA, Mina JD, Gallo RC. Detection and isolation of type C retrovirus particles from fresh and cultured lymphocytes of a patient with cutaneous T-cell lymphoma. *Proc Natl Acad Sci USA* 1980;77:7415–7419.
2. Takatsuki K, Uchiayama T, Sagawa K, Yodoi J. Adult T-cell leukemia in Japan. In: Seno S, Takaku F, Irino S, eds. *Topics in Hematology.* Amsterdam: Excerpta Medica 1977;73–77.
3. Yamamoto N, Hinuma Y. Viral aetiology of adult T-cell leukaemia. *J Gen Virol* 1985;66:1641–1660.
4. Gessain A, Barin F, Vernant JC, et al. Antibodies to human T lymphotropic virus type I in patients with tropical spastic paraparesis. *Lancet* 1985;2:407–410.
5. Vernant JC, Gessain A, Gout O, et al. Paraparesies spastiques tropicales en Martinique. Etude clinique. Haute prévalence d'anticorps anti-HTLV-I. *Press Med* 1986;15(9):419–422.
6. Rodgers-Johnson P, Gajdusek DC, Morgan OSC, Zaninovic V, Sarin PS, Graham DS. HTLV-I and HTLV-III antibodies and tropical spastic paraparesis. *Lancet* 1985;ii:1247–1248.
7. Gessain A, Francis H, Sonan T, et al. HTLV-I and tropical spastic paraparesis in Africa. *Lancet* 1986;2:698.
8. Roman GC. The neuroepidemiology of tropical spastic paraparesis. *Ann Neurol* 1988;23(Suppl): S113–S120.
9. Osame M, Usuku K, Izumo S, et al. HTLV-I associated myelopathy, a new clinical entity. *Lancet* 1986;i:1031–1032.
10. Osame M, Matsumoto M, Usuku K, et al. Chronic progressive myelopathy associated with elevated antibodies to HTLV-I and adult T-cell leukemia-like cells. *Ann Neurol* 1987;21:117–122.
11. Gessain A, Jouannelle A, Escarmant P, Calender A, Schaffar Deshayes L, De-Thé G. HTLV antibodies in patients with non-Hodgkin lymphomas in Martinique. *Lancet* 1984;1:1183–1184.
12. Gessain A, Plumelle Y, Sanhadji K, et al. Leucémie/lymphome T de l'adulte associé au virus HTLV-I en Martinique: a propos de deux cas. *Nouv Rev Fr Hematol* 1986;28:107–113.
13. Dombret H, Leblond V, Raphael M, Merle-Beral H, Frances C, Binet JL. Lymphomes T leucémiques de l'adulte avec sérologie HTLV-I positive. A propos d'un cas. *Nouv Rev Fr Hematol* 1985;27:84.
14. Baurmann H, Miclea JM, Ferchal F, et al. Adult T-cell leukemia associated with HTLV-I and

simultaneous infection by human immunodeficiency virus type 2 and human herpesvirus 6 in an African woman: a clinical, virologic and familial serologic study. *Am J Med* 1988;85:853–857.

15. Greaves MF, Verbi W, Tilley R, et al. Human T-cell leukaemia virus (HTLV) in the United Kingdom. *Int J Cancer* 1984;33:795–806.
16. Manzari V, Gradilone A, Barillari G, et al. HTLV-I is endemic in southern Italy: detection of the first infectious cluster in a white population. *Int J Cancer* 1985;36:557–559.
17. Fleming AF, Maharajan R, Abraham M, et al. Antibodies to HTLV-I in Nigerian blood donors, their relatives and patients with leukemias, lymphomas and other diseases. *Int J Cancer* 1986;38:809–813.
18. Williams CKO, Alabi GO, Junaid TA, et al. Human T-cell leukaemia virus associated lymphoproliferative disease: report of two cases in Nigeria. *Br Med J* 1984;288:1495–1496.
19. Agut H, Guétard D, Collandre H, et al. Concomitant infection by human herpesvirus 6, HTLV-I and HIV-2. *Lancet* 1988;1:712.
20. Becker WB, Engelbrecht S, Becker MLB, Piek C, Robson BA, Wood L, Jacobs P. New T-lymphotropic human herpesvirus. *Lancet* 1989;1:41.
21. Gessain A, Saal F, Morozov V, et al. Characterization of HTLV-I isolates and T lymphoid cell lines derived from French West Indian patients with tropical spastic paraparesis. *Int J Cancer* 1989;43:327–333.
22. Gessain A, Caudie C, Gout O, et al. Intrathecal synthesis of antibodies to HTLV-I and presence of IgG oligoclonal bands in the cerebro-spinal fluid of patients with endemic tropical spastic paraparesis. *J Inf Dis* 1988;157:1226–1234.
23. Link H, Cruz M, Gessain A, Gout O, De-Thé G, Kam-Hansen S. Chronic progressive myelopathy associated with HTLV-I: oligoclonal IgG and anti-HTLV-I IgG antibodies in cerebrospinal fluid and serum. *Neurology* 1989;39:1566–1571.
24. Ceroni M, Piccardo P, Rodgers-Johnson P, et al. Intrathecal synthesis of IgG antibodies to HTLV-I supports an etiological role for HTLV-I in tropical spastic paraparesis. *Ann Neurol* 1988;23(Suppl.):S188–S191.
25. Vernant JC, Maurs L, Gessain A, et al. Endemic tropical spastic paraparesis associated with human T-cell leukemia virus type I. A clinical and sero-epidemiological study of 25 cases. *Ann Neurol* 1987;21:123–130.
26. Gout O, Gessain A, Bolgert F, et al. Chronic myelopathies associated with HTLV-I. A clinical, serologic and immunovirologic study of 10 patients seen in France. *Arch Neurol* 1989;46:255–260.
27. Gessain A, Saal F, Gout O, et al. High HTLV-I proviral DNA load with polyclonal integration in peripheral blood mononuclear cells of French West Indian, Guianese and African patients with tropical spastic paraparesis. *Blood* (in press).
28. Dalgleish A, Richardson J, Matutes E, et al. HTLV-I infection in tropical spastic paraparesis: lymphocyte culture and serological response. *AIDS Res Hum Retroviruses* 1988;4:475–485.
29. Itoyam Y, Minato S, Kira J-i, et al. Altered subsets of peripheral blood lymphocytes in patients with HTLV-I associated myelopathy (HAM). *Neurology* 1988;38:816–820.
30. Yoshida M, Osame M, Usuku K, Matsumoto M, Igata A. Viruses detected in HTLV-I associated myelopathy and adult T-cell leukemia are identical on DNA blotting. *Lancet* 1987;1:1085–1086.
31. Jacobson S, Raine CS, Mingioli ES, McFarlin DE. Isolation of an HTLV-I-like retrovirus from patient with tropical spastic paraparesis. *Nature* 1988;331:540–543.
32. Sarin P, Rodgers-Johnson P, Sun DK, et al. Comparison of a human T-cell lymphotropic virus type I strain from cerebrospinal fluid of a Jamaican patient with tropical spastic paraparesis with a prototype human T-cell lymphotropic virus type I. *Proc Natl Acad Sci USA* 1989;86:2021–2025.
33. Reddy PE, Mettus RV, DeFreitas E, Wroblewska Z, Cisco M, Koprowkski H. Molecular cloning of human T-cell lymphotropic virus type I-like proviral genome from the peripheral lymphocyte DNA of a patient with chronic neurologic disorders. *Proc Natl Acad Sci USA* 1988;85:3599–3603.
34. Hirose S, Uemura Y, Fujishita M, et al. Isolation of HTLV-I from cerebrospinal fluid of a patient with myelopathy. *Lancet* 1986;2:397–398.
35. Nishimura M, Akiguchi I, Takigawa M, Fukita M, Kameyama M, Maeda M. Human T cell lines established from the cerebrospinal fluid of patients with human T lymphotropic virus type I–associated myelopathy (HAM). *J Neuroimmunol* 1988;17:229–236.

36. Gessain A, Saal F, Giron ML. Cell surface phenotype and HTLV-I antigen expression in 12 T cell lines derived from peripheral blood and cerebro-spinal fluid of West Indian, Guianese and African patients with tropical spastic paraparesis. *J Gen Virol* (in press).

DISCUSSION

One speaker was asked about earlier statements that the molecular weight of the envelope product from a HAM isolate was larger than that from Hut-102 and larger than that from most of the other ATL cases that the speaker had seen. This speaker had also mentioned that GP-66 from HAM isolates were consistently larger than those expressed in Hut-102 cells. The speaker replied that by radioimmune precipitation there is a slight difference between the molecular weight of the envelope glycoprotein, and that's all. A discussant asked if the isolates obtained from ATL patients exhibit a consistently lower molecular weight envelope product than those from HAM isolates. The speaker replied that 12 HAM isolates were examined, and in nearly all cases they were higher, and from three ATL lines, two were lower (GP60, 65), and one was unchanged (GP66). The questioner asked if they all are grown and compared on the same cell line or different cell lines and was told different cell lines. A participant pointed out that in the case of HIV isolates from Africa there is considerable sequence diversity, and that one African HTLV-I isolate that has been characterized also exhibits more sequence diversity than either the American or Japanese isolates. This participant wanted to know if the speaker had any detailed restriction maps of the isolates to quantitate the amount of sequence diversity. The speaker replied that there were no data on the amount of sequence diversity but that restriction mapping was being done. Another participant asked if there were any data that would suggest that the isolates coming from the TSP patients might have a higher totipotence potential to reflect its replicative function and wondered if it is easier to isolate. The speaker did not know, but had tried virus isolations on some T-cell leukemia cases and 15 TSP cases. In all cases of TSP except two, the speaker was able to isolate the virus. The speaker stated that it has been proven more difficult to obtain isolates from asymptomatic HTLV-I carriers.

Human Retrovirology: HTLV,
edited by William A. Blattner.
Raven Press, Ltd., New York © 1990.

Human T-Lymphotropic Virus V: HTLV-V

Vittorio Manzari, *Enrico Collalti, *Ida Silvestri,
*Andrea Modesti, *Angela Santoni, and *Luigi Frati

*Department of Experimental Medicine and Biochemical Sciences, II University, 00173
Rome; and Department of Experimental Medicine, I University, 00161 Rome, Italy*

Human retroviruses have been eluding search for many years. Only at the end of the 1970s was the first human retrovirus isolated and characterized. This retrovirus, associated with adult T-cell leukemia (ATL), was named HTLV (1,2) and later HTLV-I after the discovery of a new member of the family, HTLV-II (3,4). Two further viruses were added to the family (HTLV-III and -IV), but they were renamed human immunodeficiency virus (HIV) because they belong to a different family (5,6).

Hints of the presence of other retroviruses related to HTLVs were indicated in different laboratories, but no producing cell lines or direct evidence of viral infection were obtained (7,8). Presence of virus in a subset of T-cell cutaneous lymphomas was indicated both by serological cross-reactivities in serological tests for HTLV-I infection detection and by cross-hybridization of neoplastic genomic DNA with HTLV-I probes (9–11).

Definite proof of the presence of a retrovirus was obtained by lymphocyte culture obtained from the blood of a patient affected by cutaneous lymphoma/leukemia (GB): cells were put in culture in RPMI 1640 combined with 20% fetal calf serum (FCS), and an immortalized cell line appeared after ~90 d in culture. After >6 mo, the line began to produce a retrovirus identified by presence of reverse transcriptase associated with particles with a density of 1.16–1.19 g/ml and by electron microscopy. Cell DNA contained the same sequences hybridizing with HTLV-I probe that are present in fresh cells; the protein pattern of the purified virus indicated a major band (possibly major core protein) slightly bigger than HTLV-I or HIV-1 p24 (12).

Cells were characterized by fluorescence-activated cell sorter analysis and appeared to be phenotypically completely different from original neoplastic cells. Whereas neoplastic cells were T lymphocytes, CD4 positive, and TAC negative, cultured cells showed a set of B-cell markers and produced antibodies. It was apparent that the cell line developed was derived from a minor population of B cells present in the patient's blood. It is not yet possible to explain why T cells died out and B cells were immortalized, but two hypotheses can be drawn: one, that culture conditions were inadequate for neoplastic T cells, whereas B cells doubly infected by the retrovirus and by Epstein-Barr virus (EBV) could grow out; two, that the virus is cytopathic for T cells when expressed in vitro, and infected neoplastic cells die out in culture.

Since then, sequences homologous to HTLV-I have been identified in neoplastic T cells in more ATL patients and were demonstrated exogenous in at least two cases of lymphomas without peripheral blood involvement, because DNA extracted from cutaneous lesions hybridized to HTLV-I probes (13), whereas DNA extracted from normal PBL obtained from the same patient did not. Sera from these patients gave low positivity in enzyme-linked immunosorbent assay (ELISA) both against HTLV-I and HIV-1 antigens, and in Western blot, p24 was faintly evidenced in both cases; immunocompetition, however, gave completely negative results.

The virus was considered distantly related to HTLV-I by molecular hybridization, but the pathology was clearly distinct from ATL: although all cases but one showed cutaneous involvement and the original diagnosis was *Mycosis fungoides,* they showed a less aggressive clinical course, and all were TAC negative. In eight cases cutaneous involvement was followed by a leukemic phase $\sim$ 2 yr later and the patients died in 8 mo after onset (6 to 12). On the basis of biological and clinical data, the virus was provisionally named HTLV-V.

To test infectivity, cord blood lymphocytes were utilized by cocultivation with lethally irradiated infected GB cells at a 5:1 ratio. A cell line derived from cord blood was obtained, but even in this case the line was B, and both HTLV-V and EBV DNAs were present in the genome; virus production was evidenced by reverse transcriptase (R.T.) and electron microscopy, but R.T. positivity was limited to a short period, whereas few viral buddings could be evidenced for a longer period by electron microscopy.

The main problem to date has been to get a cell line producing continuously adequate amounts of virus and to get a non-B cell line useful to make immunofluorescence for a first serological screening.

Although many leukemic cell lines have been tested for infection, it has been possible to infect only a human promyelocytic cell line (HL 60), which appears to be a good indicator of infection both after coculture with GB cells or exposure to concentrated supernatant from the cells.

Briefly, in the first case, 2×10^6 GB cells were irradiated (3000 rads, 7 min); after 24 h the cells were washed in serum-free medium and added to HL 60 (GB:HL 60 ratio, 1:5) in RPMI 1640, 20% FCS. In the second case, 10^7 HL 60 cells in exponential growth were washed with serum-free medium and treated with 2 μl/ml polybrene 30 min at 37°C, washed and adsorbed to GB cell supernatant concentrated from 360 ml for 1.5 h at 37°C.

Syncytia and giant cells well distinguishable from the original line appeared in both cases $\sim$ 72 h after infection. Virus production, however, was low because R.T. level was constantly, for at least 5 mo, double that of background in cocultivated cells and background in cells infected by free virus; by electron microscopy, budding of virus from infected cells could be evidenced.

We can conclude that HTLV-V is an infectious exogenous retrovirus, but optimal conditions for in vitro growth have not yet been defined. HL-60 represents a good recipient cell line and can become an easy marker of infection. Identification of different cell lines susceptible to infection is underway to better characterize the virus and to permit a serological analysis of the diffusion of infection in population.

ACKNOWLEDGMENTS

We thank R. C. Gallo for discussion and G. Cappelletti for secretarial assistance. This work was partially supported by a grant from Progetto Finalizzato Oncologia Consiglio Nazionale delle Ricerche.

REFERENCES

1. Poiesz BJ, Ruscett FW, Mier GW, et al. *Proc Natl Acad Sci USA* 1980;77:7415.
2. Takatsuki K, Uchiyama J, Sagawa K, Yodoi J. In: Seno S, Takaku F, Irino S, eds. *Topics in hematology.* Amsterdam: Excerpta Medica, 1977;73.
3. Blattner WA, et al. *J Infect Dis* 1983;147:406.
4. Gelmann EP, Franchini G, Manzari V, Wong Staal F, Gallo RC. *Proc Natl Acad Sci USA* 1984;81:993.
5. Barré Sinoussi F, et al. *Science* 1983;220:868.
6. Popovic M, Sarngadharan MG, Read E, Gallo RC. *Science* 1984;224:497.
7. Saxinger WC, et al. In: Gallo RC, Essex ME, eds. *Human T-cell leukemia virus.* Cold Spring Harbor, NY: Cold Spring Harbor Laboratory, 1984;323.
8. Saxinger WC, Lange Wantzin G, Thomsen K, Hoh M, Gallo RC. *Scand J Haematol* 1985;34:455.
9. Manzari V, Fovzio VM, Martinotti S. *Int J Cancer* 1984;34:891.
10. Pandolfi F, DeRoss G, Laurie F. *Lancet* 1985;2:633.
11. Manzari V, Gradilone A. *Int J Cancer* 1985;36:430.
12. Manzari V, Gismonoli A, Barilleri. *Science* 1987;238:1581.
13. Manzari V, Wong Staol F, Franchini G. *Proc Natl Acad Sci USA* 1983;80:1574.

DISCUSSION

A discussant brought up the point that in identification of a new retrovirus, high stringency hybridization conditions are required. There are many artifacts seen with low stringency hybridization conditions that have been pointed out in the literature. The discussant asked if there were any data from either cloned provirus from HTLV-V or PCR products from portions of the HTLV-V genome. A speaker replied that an attempt to clone some small sequences that would have given high stringency hybridization had been made, but to date a complete clone had not been obtained. There are some small pieces that exhibit high stringency for DNA and RNA but no clones. Another discussant stated that photos of the HL-60 cells cocultivated were very interesting. This discussant wanted to know what would be seen if sera from these mycosis fungoides patients were taken and compared with a variety of other sera that control against HL-60 reactivity using indirect immunofluorescence. The speaker replied that sera from these patients not only yielded immunofluorescence with the infected HL-60 cells, but also a high background with the uninfected HL-60 cells. With sera from these patients, specific immunofluorescence is consistently observed. The speaker was asked if it was possible to metabolically radiolabel infected HL-60 cells and then use the patient serum to precipitate viral specific protein from the cell lysate. The speaker responded that it had not been tried.

Human Retrovirology: HTLV,
edited by William A. Blattner.
Raven Press, Ltd., New York © 1990.

Analysis of T Cells and Long-Term T-Cell Lines from Patients with Sézary Syndrome

*J. Todd Abrams, *†Stuart Lessin, *†William Ju,
‡Eric Vonderheid, †Alain Rook, §Peter Nowell,
‖Marshall Kadin, and *Elaine DeFreitas

*The Wistar Institute, Philadelphia, Pennsylvania 19104; †Department of
Dermatology, University of Pennsylvania, Philadelphia, Pennsylvania 19104;
‡Department of Medicine, Division of Dermatology, Hahnemann University,
Philadelphia, Pennsylvania 19102; §Department of Pathology,
University of Pennsylvania, Philadelphia, Pennsylvania 19104; and
‖Beth Israel Hospital, Boston, Massachusetts 02215

INTRODUCTION

Cutaneous T-cell lymphomas (CTCL) are lymphoproliferative disorders in which the malignant cells appear to be epidermotropic. Sézary syndrome, a variant of CTCL, is an intermediate-grade malignant lymphoma defined by the clinical triad of exfoliative erthyroderma, generalized lymphadenopathy, and circulating neoplastic cells characterized by atypical cerebriform nuclei. With a single exception (1), attempts to establish T-cell lines derived from peripheral blood of non-HTLV-I–infected patients with Sézary syndrome have met with failure (2–7). Although not well understood, Sézary T-cell lines cannot be generated using experimental conditions that are sufficient for the propagation of CD4-positive cell lines from blood of healthy donors (4). This defect may be due to an inability of Sézary cells to be activated by conventional cell mitogens and/or a requirement for a growth factor distinct from interleukin 2 (IL-2) (6).

To confirm studies of others, we have examined the responsiveness of Sézary patient peripheral blood mononuclear cells (PBMC) to a variety of mitogenic stimuli. Furthermore, using a combination of a factor produced by the PBMC of Sézary syndrome patients and recombinant (r) IL-2, we have established IL-2–responsive, continuously growing T-cell lines, some of which contain cells with morphologic and genetic characteristics of neoplastic Sézary cells. We describe here the characteristics of the cell lines generated from Sézary patients and discuss our observations concerning the clonality of this disease.

RESPONSIVENESS OF SÉZARY PBMC TO MITOGENS

It has been reported, albeit with little detail, that PBMC from Sézary patients do not respond to mitogens as well as PBMC isolated from healthy donors (2–7). To investigate whether Sézary patients are capable of responding to mitogenic signals, and to extend studies to other mitogens, the responses of Sézary PBMC and PBMC from random healthy donors to mitogenic stimuli were compared. Various concentrations of phytohemagglutinin (PHA), concanavalin A (Con A), anti-CD3, and phorbol myristic acetate (PMA) and Ionomycin with and without exogenous rIL-2 were mixed with 10^5 PBMC/well in 96-well flat-bottom plates. PBMC were recovered from frozen vials from patients with Sézary syndrome and from healthy donors. Cultures were maintained for 3 d, with tritiated thymidine added during the final 4 h of culture.

In these preliminary experiments, we found that for certain mitogens, lymphocytes from Sézary patients did not incorporate as much tritiated thymidine per cell as those from healthy donors. Responses to Con A and anti-T3 exhibited only 9% (±8%) and 8.3 (±8%) of the control response, respectively. The response to PHA was less severely and more variably reduced, giving 30% (±33%) of the control response, whereas the response to PMA + Ionomycin was 63% (±85%) of the control response, suggesting little or no reduction of the Sézary PBMC response for this stimulus. In general the Sézary PBMC response to PMA + Ionomycin is better than that observed for other mitogenic stimuli but still impaired compared with controls.

Spontaneous tritiated thymidine incorporation by Sézary PBMC is on the average 10% (±7%) of that observed with healthy PBMC. Therefore, if the data are expressed as a ratio of mitogen stimulation divided by background (stimulatory index, SI), many Sézary patients exhibit values approaching healthy controls. For example, average SI for four Sézary patients in response to PHA and anti-T3 was 9.1 and 7.1, respectively, whereas two normal donors, whose thymidine incorporation was much greater, only produced an SI of 9.3 and 10.9, respectively, to the same stimuli. Although Sézary syndrome patients do indeed have impaired responses to several T cell–directed mitogenic signals, not all patients can be characterized as "unresponsive." Tumor burden alone cannot explain this impairment because differences in responsiveness between patients did not correlate with total white blood cell count (WBC) or percentage of atypical cells in the PBMC.

Production and Effect of Conditioned Media from Sézary PBMC

Besides initiating proliferation of T cells, conventional T-cell mitogens also induce normal T cells to secrete a variety of growth factors. To determine whether Sézary PBMC can produce T-cell activating factors, we examined the ability of conditioned media (CM) from Con A–treated PBMC from Sézary patient 1 (SZ-1) to stimulate T-cell proliferation. After culturing SZ-1 PBMC at 4×10^6/ml with 2 μg/ml Con A for 48 h, the supernatant was collected. Alpha-methyl mannoside (1 mg/ml) was added to inhibit Con A activity.

CM from mitogen-activated Sézary PBMC (SAF-CM) was added to PBMC from a Sézary patient. Results show (Table 1A) that CM induces IL-2 receptors (R) on these cells in that T cells show increased Tac expression of IL-2R, (light chain) from 12% to 43% positive. The induced IL-2R appear to be functional because these cells show enhanced proliferation compared with that found in the cells cultured in the presence of rIL-2R alone.

INITIATION OF CELL LINES FROM PATIENTS WITH SÉZARY SYNDROME

Because we found Sézary patient PBMC responsive to SAF-CM, we attempted to use SAF-CM to establish long-term lines from individuals suffering from this disease. PBMC from Sézary patients were recovered from heparinized blood by Ficoll-Hypaque density centrifugation. Twenty million cells were placed in media containing 15 U/ml rIL-2 and 10% CM from Con A–treated Sézary PBMC. Cultures were fed two to three times per week with media containing rIL-2, and split at a 1:2 ratio when cells began to grow. When a particular culture's growth

TABLE 1. *Induction of functional IL-2R on Sézary cells*

Treatment	Tac (%)[a]	3H-TdR incorporation (cpm)[b]
A. Induction of Tac and proliferation on PBMC from a patient with Sézary syndrome		
None	6.0	321
rIL-2[c] alone	4.4	1357
SAF[d] alone	29.8	410
SAF + IL-2	42.3	40 706
B. Reinduction of Tac expression and proliferation on a long-term T-cell line established from a Sézary patient		
None	4.8	985
IL-2 alone	7.0	6895
SAF[e] alone	17.8	9068
SAF + IL-2	ND	49 319

[a] Tac measured by immunofluorescence of a cytofluorograf 3 d after initiating the culture.

[b] Proliferation measured by addition of 1 uCi 3H-thymidine 16 h before termination of the 3-d culture.

[c] 25 units recombinant IL-2 added to the culture. 1 unit is 50% maximal proliferation of CTLL cells.

[d] Partially purified SAF recovered from gel filtration column, used at 10% concentration.

[e] Whole CM containing SAF used at 0.2% final concentration.

slowed, as determined by tritiated thymidine incorporation and by viable cell counts, the cells were restimulated. Restimulation consisted of addition of either CM alone or both CM and irradiated random allogeneic PBMC. Cell lines typically required restimulation every 6–8 wk.

RESPONSE OF LONG-TERM SÉZARY T-CELL LINES TO SAF-CM

SAF-CM was added to a long-term culture that was derived from the blood of a patient with Sézary syndrome. Results show that CM induces functional IL-2R on these cells, in that the T cells showed increased Tac expression and would proliferate when provided rIL-2. This experiment confirms the observation made with Sézary PBMC that rIL-2 cannot replace SAF-CM in either induction of Tac or proliferation of these cells. However, it should be noted that rIL-2 is required for proliferation to occur, suggesting that neither Sézary PBMC nor this long-term line is capable of producing IL-2 (Table 1B).

Phenotypic Analysis

Cell lines established from patients with Sézary syndrome were examined for cell surface markers with several mouse anti-human monoclonal antibodies which were detected with a fluorescein isothiocyanate (FITC)-conjugated goat anti-mouse IgG and analyzed on an Othro cytofluorograf cell analyzer. The results (Table 2) show that 23 of 24 of the cell lines examined expressed both CD3 and CD2. A majority expressed predominantly CD4, with some of these individuals expressing both CD4 and T8 (SZ-6 and 7). Only seven individuals expressed predominantly CD8. Sézary cells in the peripheral blood typically are CD3 and CD4 positive and CD8 negative (8), although CD8-positive leukemic cells have been reported (3). Interestingly, SZ-13, a cell line with low expression of these T-cell markers, was found to contain an abnormal clone by cytogenetic analysis and also a clone by Southern blot analysis (see below).

Fourteen of these 24 lines contained >50% cells expressing BE2, a monoclonal antibody–defined marker whose reactivity was originally associated with Sézary cells (9). However, we found that BE2 expression was variable with time in culture on any one cell line and that T-cell lines from healthy donors also expressed BE2. Thus, BE2 expression appears to be related to the state of cell activation rather than to disease status. Leu 8 and Leu 9 expression was variable and did not correlate with the presence of the malignant clone.

Cytogenetic Analysis

Peripheral blood specimens were cultured for 72–96 h with a mitogen combination consisting of PHA, tetradecanol-O-phorbol-13-acetate (TPA), and IL-

TABLE 2. *T-cell– and tumor-associated antigen expression on long-term line established from patients with Sézary syndrome*

	T3	T4	T8	T11	DR	BE2	LEU8	LEU9	B1	TAC
SZ-1	98	98	33	ND[b]	ND	97	ND	84	0	98
SZ-2	99	95	9	99	ND	96	3	3	ND	86
SZ-4[a]	97	96	3	98	95	97	32	38	5	97
SZ-5	99	19	96	64	70	45	ND	ND	12	ND
SZ-6	99	97	26	99	51	74	ND	97	18	96
SZ-7	99	99	33	99	66	62	ND	98	8	91
SZ-8	99	77	21	72	68	42	ND	ND	6	32
SZ-9	99	98	9	92	57	77	ND	ND	ND	37
SZ-10	99	9	69	99	71	82	ND	82	13	46
SZ-13[a]	31	34	5	45	90	78	38	9	5	ND
SZ-14	98	91	50	ND	ND	96	ND	95	12	ND
SZ-15	98	94	4	ND	ND	92	ND	86	7	78
SZ-17	99	7	78	74	71	94	ND	69	5	79
SZ-20	58	16	38	77	57	41	2	37	7	6
SZ-25	84	19	48	98	71	16	3	41	2	2
SZ-26	100	92	17	100	85	20	57	85	1	ND
SZ-27	97	68	16	98	71	40	47	74	3	15
SZ-28[a]	88	79	4	99	69	99	75	43	5	55
SZ-29[a]	99	77	27	100	92	23	66	91	0	ND
SZ-30	99	99	5	100	16	65	ND	64	1	15
SZ-31	99	11	33	100	35	35	ND	85	1	22
SZ-32[a]	99	98	3	93	58	64	ND	75	3	61
SZ-34	52	43	18	77	31	8	19	38	1	5
SZ-35	85	35	44	93	11	10	6	57	2	2
N[c]-1	99	68	46	100	86	28	47	94	3	ND
N[c]-2	98	33	31	95	68	66	16	88	3	ND

[a] Cell line contained a malignant clone as determined by karyotypic and/or genotypic analysis at the time phenotype was performed.
[b] ND, not done.
[c] Cell line established from healthy donor.

2 as previously described (10,11). Standard trypsin-Giemsa–banded chromosome preparations were made (11). At least 30 counts and 4 karyotype analyses were done on each sample to characterize the chromosomally abnormal clone. Cell lines were examined after at least 60 d in culture and were similarly processed without additional mitogenic stimulation.

Table 3 shows that in 5 of the 11 cases in which both blood and the cell line were examined, the same or related karyotype was found. Nine of the 13 blood samples examined showed cytogenetic abnormalities, with cell lines from 3 of these individuals containing identical or related abnormalities, although one of the latter subsequently reverted to normal in vitro (SZ-6). From one individual who showed no abnormality in the blood, a cell line was established with an abnormality (SZ-13). Note that no other cell line developed cytogenetic abnormalities, including cell lines established from healthy donors ($N = 3$) or from the blood of other Sézary syndrome patients that did not contain a chromosomally abnormal clone.

TABLE 3. *Comparison of karyotypes from blood and cell line in patients with Sézary syndrome*

Patient	Karyotype blood	Karyotype[b] line
SZ-2	Hypotetraploid,1p−,abn2,10p−,10p+	Discordant (normal)
SZ-4	46,XX,1p−,3q−,4q−,+?4q−,abn5, −10,16q+,iso[17q],−18,21q+	Hypotetraploid, same abnormalities Concordant
SZ-5	46,X,−Y,abn1,t(8;17),t(6;11),del 10q−	Discordant (normal)
SZ-6	46,X,−Y,1p−,6q−,6p+q−, +8,−9, 9p−,−10,+10,11q−,15q+,−15, 17p−,+3mar	Pseudodiploid, same clone[a]
SZ-7	46,XX	Concordant
SZ-8	49,XY,2p−,17p−,19p+,?+18,+2mar	Discordant (normal)
SZ-9	46,XX	Concordant
SZ-10	46,XY	Concordant
SZ-12	45,XY,−1,−1,−2,−3C,−D,−17, 11q−,12p−,+7mar	Discordant (normal)
SZ-13	46,XY	Discordant Hypotetraploid,1p−,2q+, 6q−,8p−,10q−, 11p+,iso17q
SZ-17	46,XY,6q−,inv6,+8	Discordant (normal)
SZ-18	47,XX,+7,t(1;8),6q−	Pending
SZ-20	ND	46,XX
SZ-22	Hypotetraploid,1p−,3q+,	Discordant (normal)
SZ-28	Hypotetraploid,2p−,17p+,11p+qabn	Concordant Hypotetraploid same clone

[a] Subsequently, abnormal clone lost from culture.
[b] Concordant or discordant, with respect to karyotype observed in blood.

These data demonstrate that although all the individuals studied have Sézary syndrome with circulating atypical lymphocytes, cytogenetic abnormalities are either not detected or are absent from the malignant cells in 30% of these individuals.

T-Cell Receptor Gene Rearrangement: Measure of Clonality for Sézary T Cells

Southern blot analysis was performed on the blood and cell line of 19 individuals. DNA was extracted and digested with the appropriate enzyme, run on an agarose gel, then transferred to nitrocellulose and probed with a radiolabeled cDNA probe coding for the constant region of the beta chain of the human T-cell receptor gene (TCR) (12,13,14). The DNA was digested with either *Eco*RI to demonstrate $C\beta1$ gene rearrangements, *Hin*dIII to detect $C\beta2$ rearrangements, or *Bam*HI to detect both $C\beta1$ and $C\beta2$ rearrangements. In comparing the TCR autoradiographs of the cell lines to that of the PBMC, we have identified six patterns. The six cell lines described below exemplify these patterns. A summary of all the findings is shown in Table 4.

TABLE 4. Profile of PBMC from Sézary syndrome patients and summary of TCR gene rearrangement analysis of blood and corresponding cell line

	White blood count (per cmm)	Lymph (%)	Lymph (per cmm)	Atypical lymph (%)	TCR blood	TCR line
SZ-1	26 000	73	18 980	NA	C[a]	C (eq)[b]
SZ-2	8100	73	6913	62	ND[c]	C
SZ-4	12 600	38	4788	16	C	C (same)[d]
SZ-5	11 500	37	4255	10	GL[e]	GL
SZ-6	12 300	43	5289	6	C	C (diff)[f]
SZ-7	11 900	16	1904	30	C	C (diff)
SZ-8	10 500	48	5040	50	C	C (diff)
SZ-10	15 200	27	4104	30	C	PC[g]
SZ-13	7300	37	2701	44	PC	C
SZ-17	10 200	42	4284	63	C	C (diff)
SZ-18	18 800	43	8087	59	GL	GL
SZ-20	9000	32	2880	37	GL	GL
SZ-22	5500	6	330	45	C	C (eq)
SZ-23	8200	17	1148	5	ND	ND
SZ-25	11 100	77	8547	75	PC	C
SZ-26	5400	11	594	4	PC	ND
SZ-27	10 100	18	1818	5	GL	GL
SZ-28	34 400	40	13 760	85	C	C (same)
SZ-29	24 600	80	19 680	80	C	C (same)
SZ-30	19 000	61	11 590	17	GL	GL
SZ-31	9800	18	1764	0	ND	ND
SZ-32	11 300	39	4400	70	C	C (same)
SZ-33	38 000	78	29 640	62	C	ND
SZ-34	7600	33	2508	2	GL	C
SZ-35	28 000	27	7560	14	C	ND
SZ-36	4600	20	920	2	ND	ND
Nor. (n = 3)	10 000	30–40	4000	0	PC	PC

[a] Clonal.
[b] Equivocal result showing same rearrangement in one allele but not the other.
[c] Not done.
[d] Same rearrangement in blood and cell line.
[e] Germline.
[f] Different rearrangement in blood and cell line.
[g] Polyclonal.

Identical clones in blood and cell line. SZ-4 showed a 10-kb rearranged band in *Bam*HI-digested DNA from both cell line and blood and an 8-kb fragment in *Hin*dIII-digested DNA from both blood and cell line (Fig. 1). These data indicate that both the blood and cell line contain an identical clone utilizing the $C\beta2$ gene. Four individuals showed this pattern.

Different clones in blood and cell line. SZ-7 blood and cell line contained a rearrangement that appeared identical when examined in *Bam*HI digests, but not when examined with *Eco*RI- and *Hin*dIII-digested DNA. The blood contained a clone with the $C\beta1$ gene rearranged, whereas the cell line did not. In addition, the blood contained a $C\beta2$ gene rearrangement, whereas the cell line

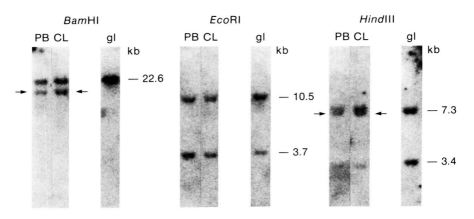

FIG. 1. DNA extracted from SZ-4 cell line and the patient's PBMC were subjected to Southern blot analysis using a probe for the C-beta region of the TCR. *Bam*HI-digested DNA show the germline band at 22.6 kb and a 10-kb rearrangement in both the blood and cell line. The blood and cell line provide germline configurations in the *Eco*RI-digested DNA. *Hind*III digestion produced a 7-kb fragment in the DNA from both cell line and blood.

contained two different $C\beta2$ gene rearrangements. Thus, the blood and cell lines contain clones with different TCR rearrangements (data not shown). This pattern was observed with four patients.

No clone in blood, clone in line. SZ-13 cell line contained a clone utilizing the $C\beta2$ gene, whereas the blood appeared not to contain any predominant rearrangements (data not shown). This pattern was found in three patients.

No clone in blood or cell line. SZ-20 had no detectable predominant rearrangement in either the blood or cell line. This was observed in spite of the fact that this individual has high Sézary-like atypical cell counts (data not shown). Five individuals were observed to have this pattern.

Equivocal pattern. SZ-22 showed a complex pattern. The blood was found to contain a clone with both $C\beta1$ and $C\beta2$ alleles rearranged, assumably one productively and one nonproductively. The cell line contained a $C\beta1$ rearrangement with an *Eco*RI fragment identical in size with that observed in the blood; however, it did not contain a $C\beta2$ rearrangement (data not shown). This suggests that either the cell line deleted the $C\beta2$ gene or it represents a different clone. In this case, we found a cytogenetically abnormal clone in the blood not found in the cell line, indicating the cell line's $C\beta1$ rearrangement represents a different clone than that observed in the blood. An equivocal pattern was observed with two patients.

Clone in blood, no clone in line. SZ-10 contained a clone in the blood, whereas the cell line contained no clonal rearrangement (data not shown). This individual is the one example of this pattern.

Polyclonal in blood, polyclonal in line. All three PBMC and cell lines from healthy donors showed this pattern.

Southern blot analysis of TCR gene rearrangements (15), using a probe for the Cβ gene, indicates that the blood of 13 of 22 patients contains a predominant clonal rearrangement. In five of the nine patients with no detectable clone, the percent of atypical cells should have been sufficient for the detection of a dominant rearrangement in these samples (>5%). This observation suggests that abnormal morphology may overestimate the number of neoplastic cells in the blood and may not provide the best means for enumerating malignant cells.

Cell lines from 11 of the 13 individuals containing clones in their blood were examined for TCR gene rearrangements. Four of the cell lines appeared to contain identical rearrangements to those found in their respective PBMC (SZ-4, SZ-28, SZ-29, and SZ-32); two contained identical rearrangements with one allele but not the other (SZ-1 and SZ-22); four contained clones with a different pattern from that observed in the blood (SZ-6, SZ-7, SZ-8, and SZ-17); and one line showed no rearrangement (SZ-10). The equivocal pattern observed in SZ-1 and SZ-22 demonstrates the importance of examining TCR rearrangements with all three enzymes. In equivocal situations, other techniques such as cytogenetic analysis or identification of the V-region utilized by the clone might distinguish these possibilities.

We examined cell lines from eight of the nine individuals in which no clone was detected in the blood. Five of the cell lines showed the germline or polyclonal pattern found in the blood (SZ-5, SZ-18, SZ-20, SZ-27, and SZ-30), whereas three of the cell lines contained a detectable clone (SZ-13, SZ-25, and SZ-34). In the case of SZ-13, the T-cell clone detected in the cell line was found to contain cytogenetic abnormalities, whereas no clone was found in the patient's PBMC on five separate occasions (twice for TCR rearrangement and three times by cytogenetic analysis).

CHARACTERISTICS OF A CELL LINE CONTAINING THE MALIGNANT CLONE

SZ-4 is a T-cell line established from the PBMC of a patient with Sézary syndrome. The growth characteristics of this cell line over a 4-mo period after initial stimulation with CM are depicted in Fig. 2. The cell line has a doubling time of ~30 h, which appears to be maintained throughout the period shown.

Phenotypic analysis of SZ-4 cell line, shown in Table 2, indicates that the cell line has a mature helper T-cell expressing CD2, CD3, and CD4 but not the cytotoxic/suppressor cell marker CD8. This phenotype correlates well with the cells in the blood of patient SZ-4, in which 90% of the cells reacted with anti-CD3, 81% with anti-CD4, and 9% with anti-CD8.

The cell line also expresses a variety of activation markers, including HLA-DR, transferrin receptors, CD5 (T10), and CD25 (Tac, light chain of the IL-2 receptor). In addition, these cells were reactive with BE2. The cell line had partial reactivity with Leu 8 (38%), CD7 (Leu 9, 3A1) (38%), lymphocyte function–

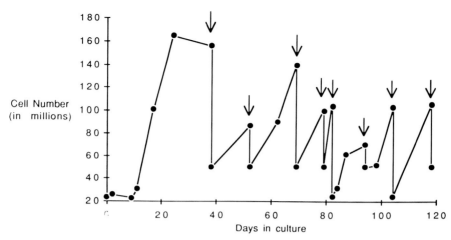

FIG. 2. The growth pattern of a clonal concordant T-cell line established from Sézary PBMC. The growth pattern of SZ-4 cell line over a 120-d period is depicted. Twenty-five million SZ-4 cells were placed in culture with 15 U/ml rIL-2 and 0.2% CM until day 3, when the cells were cultured with rIL-2 only. The arrows represent removal of cells from the culture.

associated antigen 1 (LFA-1) (60%), and anti-beta TCR antibody (41%) but was found negative for CD16 (NK marker), CD20 (B-cell marker), and TCR delta chain.

The entry of normal T cells from G1 to S phase of the cell cycle is dependent on the interaction with either endogenous or exogenous IL-2 along with other signals (16–21). Because T-cell dependence on IL-2 for growth is a hallmark of normal but not necessarily of tumor cells (22), we wished to determine whether SZ-4 was dependent on IL-2 for its continuous proliferation. To investigate the IL-2 requirements of this cell line, we performed serial dilutions of rIL-2. Figure 3A demonstrates that the line is responsive to IL-2 but continues to synthesize DNA in the absence of exogenous rIL-2. In contrast, CTLL cells, known to be strictly dependent on IL-2, show a marked decrease in DNA synthesis when rIL-2 is withdrawn.

To confirm the IL-2 independence of this line and to eliminate the possibility that the line is producing IL-2 sufficient to stimulate the cell division observed in the no IL-2 group, we cultured the line for 2 d without IL-2 and then with varying amounts of hyperimmune goat anti-human IL-2 in media without rIL-2. The data in Fig. 3B show that the proliferation of SZ-4 is not affected by the antiserum, suggesting that the line is proliferating in an IL-2–independent manner. It is possible that the antiserum is not capable of completely inhibiting autocrine IL-2 stimulation. Nonetheless, SZ-4 does not behave like a normal T-cell line in that it has both constitutive IL-2R expression and long periods of sustained proliferation in rIL-2 without restimulation. Thus, this cell line may qualify as "transformed." Because of the IL-

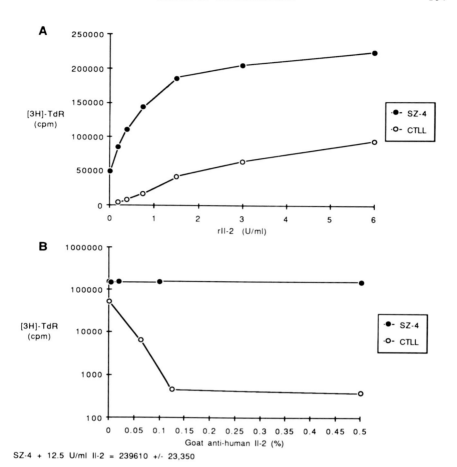

SZ-4 + 12.5 U/ml Il-2 = 239610 +/- 23,350

FIG. 3. A. Response of SZ-4 T-cell line to IL-2. SZ-4 cells (closed circle) were cultured in media and without rIL-2 for 48 h before initiation of the experiment. Cells were then cultured at 5 × 10⁴ cells/dilutions of rIL-2 in triplicate. IL-2–dependent CTLL cells (open circles) were removed from an exponentially growing culture and seeded as described for SZ-4. All groups were cultured for 40 h with 2uCi/well ³H-TdR added for the last 4 h. Data represent the mean of triplicate cultures. **B.** Effect of anti-rIL-2 antisera on proliferation of SZ-4 cells (closed circles) and CTLL cells (open circles). SZ-4 cells were deprived of IL-2 for 48 h, then cultured without rIL-2 for an additional 48 h in dilutions of goat anti-rIL-2 antisera. CTLL cells were given 15U/ml rIL-2 and various dilutions of anti-rIL-2 antisera as indicated. The conditions of culture were identical to those described in Fig. 2A. Response of SZ-4 cell line when provided 15U/ml rIL-2 was 239 610 cpm.

2 independence of SZ-4 and because HTLV-I infection has been detected in some patients with CTCL (23), we tested serum from this individual for reactivity against HTLV-I proteins by Western blot (24). Reactivity against the HTLV-I *gag* protein (p24) was observed, but little or no reactivity against the *env* or other mature proteins was found. The cell line was also tested for expression of the *gag* proteins p19 and p24 of HTLV-I and -II and *env* protein

gp46 of HTLV-I using the alkaline phosphatase anti-alkaline phosphatase (APAAP) system (24). The cell line did not contain cells positive for any of these proteins (data not shown). In addition, culture supernatants were found devoid of reverse transcriptase activity (data not shown). Thus, the apparent transformed phenotype of this line appears not to be the result of HTLV-I or -II infection.

DISCUSSION AND CONCLUSIONS

Response of Sézary PBMC to Mitogenic Signals and Establishment of Cell Lines

We have confirmed that Sézary PBMC have impaired mitogen responses and have extended these studies to include mitogenic anti-T3 monoclonal antibody, PMA, and calcium ionophore stimulation. We found that the response to anti-T3 is markedly impaired, whereas the response to PMA and Ionomycin is not as strongly diminished.

We have found that Sézary cells appear to produce a potentially novel factor we call Sézary activating factor (SAF), which can induce functional IL-2R on these cells. The paradox of producing such a substance and the general difficulty in establishing T-cell lines from the same individuals remains to be explained. However, it is not known how much of the product is made in vivo, or if sufficient levels of IL-2 are available in situ to allow proliferation. Nonetheless, in 25% of the cases where a clonal marker was detected in blood, long-term T-cell lines were established containing the same clone. Most of these lines express normal helper T-cell markers CD2, CD3, CD4 and activation antigens HLA-Dr, transferrin receptors, BE2, and Tac. To our knowledge these are the first non-HTLV–transformed Sézary T-cell lines to express these antigens and be derived from the neoplastic clone. Success in establishing these lines may relate to the SAF-containing CM we used to initiate the culture, and/ or to the aggressive nature of these tumor clones. Although we and others have failed to establish Sézary tumor cell lines in culture through the addition of mitogens (2–7), SAF may activate Sézary cells through a different pathway from mitogens.

SZ-4 cell line, of clonal origin as determined by karyotype and TCR genotype, demonstrates two features that may be characteristic of malignant Sézary T cells: first, expression of CD3, CD4, and CD11; and second, this cell line does not require the addition of exogenous rIL-2 to proliferate, although not infected with HTLV-I or -II. The apparent contradiction that Sézary T cells may be transformed and yet difficult to establish in culture is still unanswered. It is conceivable that some of the characteristics of SZ-4 may not be typical of malignant cells from all Sézary syndrome patients.

Implications of Southern Blot Analysis of TCR Gene Rearrangement

Although Sézary syndrome is reported to be a clonal malignancy (11,25), Southern blot analysis of TCR beta chain gene rearrangements on patients in which the percent of atypical lymphocytes is known indicates that only 13 of 22 individuals with diagnosed Sézary syndrome contain a detectable clone. Lack of clonality because of low numbers of atypical lymphocytes might explain only four of these cases. In five of the nine patients without detectable clonal rearrangements, well over 5% (lower level of detectability in this assay) of the total DNA was theoretically derived from the atypical lymphocyte population. In addition, individuals without a TCR beta clone were also found devoid of gamma/delta rearrangements. Thus, no TCR clonal rearrangement was observed in these individuals. Furthermore, in 4 of the 12 individuals found to contain a detectable clone, the percent of the total DNA represented by the atypical lymphocytes was <5%. These observations suggest that in almost one-half of the cases of Sézary syndrome examined here, Sézary cells are not clonally derived. Furthermore, the number of circulating atypical lymphocytes does not appear to correlate with the number of malignant clonally derived T cells.

Of the 13 individuals with Sézary syndrome containing a detectable clone in their PBMC by Southern blot analysis, we have established four long-term T-cell lines containing clonal TCR gene rearrangements identical to those found in the PBMC. Therefore, SAF-CM may contain an agent which allows the propagation of malignant long-term lines from ∼30% of patients with Sézary syndrome. Furthermore, a clone was observed in three of the cultures established from individuals with no detectable clone in their PBMC (SZ-13, SZ-25, and SZ-34), with one of these lines containing cytogenetic abnormalities (SZ-13). However, without a clonal marker in the patient, or a Sézary-specific cell-surface marker, it is not possible to determine the relationship between the T cells in culture and the neoplastic cells in these patients. We conclude that, although cell lines containing malignant clones can be established from patients with Sézary syndrome, lines established from individuals without a detectable clone in their PBMC—despite the high number of atypical cells (e.g., SZ-25 75%; SZ-13 44%)—cannot be presently evaluated for their relationship to the putative malignant cells in the patient.

Conclusions

The long-term Sézary T-cell lines described here may prove useful in understanding the nature of this disease through examination of their responsiveness to, and production of, cytokines. Although the chromosomal abnormalities in patients with Sézary syndrome appear to be mostly random (11), analysis of the specific abnormalities found in these clones may provide clues to their growth advantage in culture and possible IL-2 independence. In addition, determination of the specific V-region family used by the TCR of these clones compared with those used by other epidermatotropic T cells from individuals suffering from other dermatologic diseases may provide clues as to whether a subtype of T cell, or a specific antigen, is involved in this disease.

ACKNOWLEDGMENTS

We wish to thank Dr. G. Trinchieri and Dr. P. Wettstein of The Wistar Institute, Dr. T. Palker and Dr. B. Haynes of Duke University, and Dr. C. Berger of Columbia University for providing antibodies. Expert technical assistance was provided by Reneé Henry, Dawn Demangone, Diane Wittaker, and Lisa Moreau. This work was funded by CA 60335 (ED), ARO 1645 (SL), and CA 15822 (PN) from The National Institutes of Health; a grant from the Eleanor Naylor Dana Foundation (ED); and a Society for Investigative Dermatology Research Fellowship from the Dermatology Foundation (SL).

REFERENCES

1. Kaltoft K, Bisballe S, Fogh Rasmussen H, Thestrup-Pedersen K, Thomsen K, Sterry W. A continuous T cell line from a patient with Sézary syndrome. *Arch Dermatol Res* 1987;279:293–298.
2. Foa R, Catovsky D, Incarbone E, et al. Chronic T cell leukemias. III. T-colonies, PHA response and correlation with membrane phenotype. *Leuk Res* 1982;6:809–814.
3. Kaltoft K, Thestrup-Pedersen K, Jenson JR, Bisballe S, Zachariae H. Establishment of T and B cell lines from patients with mycosis fungoides. *Br J Dermatol* 1984;3:303–308.
4. Gazdar AF, Carney DN, Bunn PA, et al. Mitogen requirements for the in vitro propagation of cutaneous T-cell lymphomas. *Blood* 1980;55:409–417.
5. Gazdar AF, Carney DN, Russel EK, Schechter GP, Bunn PA Jr. In vitro growth of cutaneous T-cell lymphomas. *Cancer Treat Rep* 1979;63:587–590.
6. Golstein MM, Farnarier-Seidel C, Daubney P, Kaplanski S. An OKT4+ T-cell population in Sézary syndrome: attempts to elucidate its lack of proliferative capacity and its suppressive effect. *Scand J Immunol* 1986;23:53–64.
7. Jones CM, Prince CA, Langford MP, Hester JP. Identification of a human monocyte cytotoxicity-inducing factor from T cell hybridomas produced from Sézary cells. *J Immunol* 1986;137:571–577.
8. Broder S, Edelson RL, Lutzner MA, et al. The Sézary syndrome: a malignant proliferation of helper T cells. *J Clin Invest* 1976;58:1297–1306.
9. Berger CL, Morrison S, Chu A, et al. Diagnosis of cutaneous T-cell lymphoma by use of monoclonal antibodies reactive with tumor-associated antigens. *J Clin Invest* 1982;70:1205–1215.
10. Nowell PC, Finan JB, Vonderheid EC. Clonal characteristics of cutaneous T cell lymphomas: cytogenetic evidence from blood, lymph nodes, and skin. *J Invest Dermatol* 1982;78:69–75.
11. Nowell PC, Vonderheid EC, Besa E, Hoxie JA, Moreau L, Finan JB. The most common chromosome change in 86 chronic B cell or T cell tumors: a 14q:32 translocation. *Cancer Genet Cytogenet* 1986;19:219–227.
12. Southern EM. Detection of specific sequences among DNA fragments separated by gel electrophoresis. *J Mol Biol* 1975;98:503–517.
13. Isobe M, Erikson J, Emanuel BS, Nowell PC, Croce CM. Location of gene for B subunit of human T cell receptor at band 7q35, a region prone to rearrangements in T cells. *Science* 1985;228:580–582.
14. Rigby PWJ, Dieckmann M, Rhodes C, Berg P. Labeling deoxyribonucleic acid to high specific activity in vitro by nick translation with DNA polymerase I. *J Mol Biol* 1977;113:237–251.

15. Waldmann TA, Davis MM, Bongiovanni KF. Rearrangements of genes for the antigen receptor on T-cells as markers of lineage and clonality in human lymphoid neoplasia. *N Engl J Med* 1985;313:776–783.
16. Mizel SB, Mizel D. Purification to apparent homogeneity of murine interleukin 1. *J Immunol* 1981;126:834–837.
17. Meuer SC, Buschenfelde KHM. T cell receptor triggering induces responsiveness to interleukin 1 and interleukin 2 but does not lead to T cell proliferation. *J Immunol* 1986;136:4106–4112.
18. Nisbet-Brown E, Lec JWW, Cheung RK, Gelfand EW. Antigen-specific and -non specific mitogen signals in the activation of human T cell clones. *J Immunol* 1987;138:3713–3719.
19. Oudrhiri N, Faret JP, Gourdin MF, et al. Mechanism of accessory cell requirement in inducing IL-2 responsiveness by human T4 lymphocytes that generate colonies under PHA stimulation. *J Immunol* 1985;135:1813–1818.
20. Schwab R, Crow MK, Russo C, Weksler ME. Requirements for T cell activation by OKT3 monoclonal antibody: role of modulation of T3 molecules and interleukin. *J Immunol* 1985;135: 1714–1718.
21. Neckers LM, Crossman J. Transferrin receptor induction in mitogen-stimulated human T lymphocytes is required for DNA synthesis and cell division and is regulated by interleukin 2. *Proc Natl Acad Sci USA* 1983;80:3493–3498.
22. Arya SK, Wong-Staal F, Gallo RC. T-cell growth factor gene: lack of expression on human T-cell leukemia lymphoma virus–infected cells. *Science* 1984;223:1086–1087.
23. Poiesz BJ, Ruscetti FW, Gazdar AF, Bunn PA, Minna JD, Gallo RC. Detection and isolation of type C retrovirus particles from fresh and cultured lymphocytes of a patient with cutaneous T-cell lymphoma. *Proc Natl Acad Sci USA* 1980;77:7415–7419.
24. DeFreitas E, Wrobleska Z, Maul G, et al. HTLV-I infection of cerebrospinal fluid T cells from patients with chronic neurologic disease. *AIDS Hum Retroviruses* 1987;3:19–31.
25. Berger CL, Eisenberg A, Soper L, et al. Dual genotype in cutaneous T cell lymphoma: immunoglobulin gene rearrangement in clonal T cell malignancy. *J Invest Dermatol* 1988;90:73–77.

Human Retrovirology: HTLV,
edited by William A. Blattner.
Raven Press, Ltd., New York © 1990.

Pathogenesis of Adult T-Cell Leukemia from Clinical Pathologic Features

Kazunari Yamaguchi, Tetsuyuki Kiyokawa, Genjiro Futami, Toshinori Ishii, and Kiyoshi Takatsuki

Blood Transfusion Service and The Second Department of Internal Medicine, Kumamoto University Medical School, Honjo 1-1-1, Kumamoto 860, Japan

Adult T-cell leukemia (ATL), a disease entity first described by Takatsuki et al. (1), is endemic in southwestern Japan and the Caribbean Islands. Since the human T-lymphotropic virus type I (HTLV-I) (2,3)—a retrovirus discovered by Gallo et al.—was found to be linked to the etiology of ATL, many investigators have been interested in ATL and HTLV-I. The diagnosis of ATL is made from the characteristic clinical findings, the detection of serum antibodies to HTLV-I, and, when necessary, the confirmation of monoclonal integration of HTLV-I proviral DNA in cellular DNA of ATL cells (4,5).

CLINICAL FEATURES OF ATL

Eight percent of the population over 40 yr old in Kyushu and Okinawa tested positive for the HTLV-I antibody. We studied 187 patients with ATL in Kyushu: 113 males and 74 females, whose age at the onset ranged from 27 to 82 yr, with a median age of 56 yr.

The predominant physical findings were peripheral lymph node enlargement (72%), hepatomegaly (47%), splenomegaly (26%), and skin lesions (53%). Hypercalcemia (28%) was frequently associated with ATL. White blood cell counts ranged from normal to 500 000. Leukemic cells resembled Sézary cells, having indented or lobulated nuclei. The surface phenotype of ATL cells characterized by monoclonal antibodies was $T3^+$, $T4^+$, $T8^-$ and Tac^+.

Anemia and thrombocytopenia were rare. The survival time in acute and lymphoma type ATL ranged from 2 wk to more than 1 yr. The causes of death were pulmonary complications, including *Pneumocystis carinii* pneumonia; hypercalcemia; cryptococcus meningitis; disseminated herpes zoster; and disseminated intravascular coagulopathies.

CLASSIFICATION OF ATL

ATL patients are classified into four subtypes according to the clinical picture: acute, chronic, smoldering (6), and lymphoma type. The acute type is the so-called prototypic ATL, which exhibits increased ATL cells, skin lesions, systemic lymphadenopathy, and hepatosplenomegaly. Most of these are resistant to combination chemotherapy using, for example, vincristin, cyclophosphamide, prednisolone, adriamycin, and sometimes methotrexate. In general, a poor prognosis is indicated by the elevation of serum lactic dehydrogenase (LDH), calcium, and bilirubin, as well as by high white blood cell counts. Chronic-type ATL patients suffer from increased white blood cell counts, cough, and skin disease. In a few of these patients, slight lymphadenopathy and hepatosplenomegaly are observed, and elevation in serum LDH is also noted, but this is not associated with hypercalcemia or hyperbilirubinemia. Smoldering ATL is characterized by the presence of a few ATL cells in the peripheral blood for a long period. Patients frequently have skin lesions as premonitory symptoms. The serum LDH values are within normal range and not associated with hypercalcemia. Lymphadenopathy, hepatosplenomegaly, and bone marrow infiltration are very slight. Smoldering- and chronic-type ATL often progress to ATL after a long duration. Lymphoma-type ATL is characterized by prominent lymphadenopathy. This type has been diagnosed as nonleukemic malignant lymphoma.

FATAL COMPLICATIONS IN ATL

Pulmonary complication is one of several fatal complications in ATL (9). Among 29 patients with ATL, pulmonary complications were seen in 26, leukemic pulmonary infiltration in 13, bleeding in 1, interstitial pneumonitis in 1, and pulmonary infection in 13. In 10 of 13 patients with pulmonary infiltration it occurred in the early stage, and 6 of them were diagnosed as having "chronic lung disease" before the diagnosis of ATL. ATL cell infiltration to the lung and fibrosis were confirmed by transbronchial lung biopsy (TBLB). TBLB is useful for lung diagnosis of infiltration, or infection.

Another serious complication is hypercalcemia (10). The incidence of hypercalcemia in ATL patients is 28% at the time of admission and over 50% during clinical course. To clarify the mechanism of hypercalcemia, 18 autopsied patients were reviewed clinicopathologically. Eight of nine patients with terminal phase hypercalcemia had osteoclastic bone resorptions. None of the normocalcemic patients had bone resorption. It was concluded that hypercalcemia in ATL was associated with osteoclastic bone resorption. The mechanism of osteoclastic activation in ATL is not clear. Osteoclast activating factors like substances and PTH–related protein (PTH-RP) have been suggested as candidates.

Opportunistic infection is also a serious complication.

MORPHOLOGY OF ATL CELLS

Numerous abnormal lymphocytes, which vary considerably in size (mean diameter 15 μm) and cytoplasmic basophility, are seen in acute-type ATL. Cells from 30% of this type possess small vacuoles but not azurophilic granules. Most of the cells characteristically exhibit lobular division of their nuclei; most are multifoliate. Cells with a nuclear configuration are known as flower cells.

Cells from chronic-type ATL are relatively uniform in size and nuclear configuration. Cells are smaller than those from acute- or smoldering-type ATL (mean diameter 13 μm). Cells rarely possess small vacuoles and do not possess azurophilic granules. Cells in this type also exhibit lobular division of their nuclei. However, most of the lobulated nuclei are bi- or trifoliate. Many cells exhibit deep indentation rather than lobulation. The nuclear chromatin is in coarse strands and deeply stained. The nucleocytoplasmic ratio is larger than that in the normal lymphocytes.

The proportion of abnormal cells in the peripheral blood varies from 0.5% to 3% in smoldering-type ATL. Cells from this type are relatively large (mean diameter 15 μm). These cells do not have cytoplasmic granules or vacuoles. The lobulated nuclei are bi- or trifoliate. The nucleocytoplasmic ratio is large. The nuclear chromatin is in coarse strands and deeply stained.

FAMILY STUDY

The pedigree of a family having two ATL patients examined for HTLV-I antibody has been reported (11). Twenty-six family members and relatives of two ATL patients were available for examination of serum HTLV-I antibody: they were all healthy and hematologically normal. This family included two siblings who developed ATL; of 26 healthy persons, 7 were found to have HTLV-I antibody. Eleven children and their mothers were negative, but one daughter of a patient had HTLV-I antibody. The married couples in this study were classified as follows: husband(+)/wife(+), two couples; husband(−)/wife(+), three couples; husband(+)/wife(−), not found (the wives of ATL patients were all positive). This family study suggested two main routes of transmission of HTLV-I: one vertical from parents to children, and the other horizontal among spouses, especially from husband to wife. The HTLV-I antibody-positive family can be considered a high-risk group for ATL.

Another family study reported that three sisters, ranging in age from 56 to 59 years old, developed lymphoma type ATL during a 19-mo period (12). The patients were born in the Amakusa area of Kumamoto Prefecture, an area where the incidence of ATL is high. As their lives and adult environments were clearly different, HTLV-I infection may have occurred in their childhood. This finding suggests that the disease developed after a long latent period following the first viral infection. Their elder brother also had HTLV-I antibody. However, he is currently well and is a healthy carrier. Further studies are needed to elucidate the role of genetic and environmental factors in the etiology of familial ATL.

COMPARISON OF JAPANESE AND CARIBBEAN ATL PATIENTS

We had an opportunity to research Caribbean ATL in the United Kingdom in 1986, when we compared clinical, hematological, immunological, cytogenetic, and viral features of Japanese ATL patients with those of Caribbean ATL patients. Lymphadenopathy, skin lesion, poor prognosis, subtype of ATL, frequency of hypercalcemia, opportunistic infection, cell morphology, and surface phenotype were the same in the Japanese and Caribbean patients. The only difference was age of onset: the average for the Caribbean and African ATL was 43 (range 19–62), whereas the Japanese patients were older (average 56, range 27–82). If HTLV-I is transmitted in early childhood, from mother to child via breast milk, why should ATL appear earlier in Caribbean and African patients than in Japanese? We suggest some cofactors for HTLV-I–induced leukemogenesis. Strongyloides stercoralis is one of the candidates for cofactor. We studied the relationship between strongyloidiasis and HTLV-I in Okinawa, an area where both conditions are endemic (14). Thirty-six patients with strongyloidiasis were seropositive for HTLV-I and suffered from several related clinical complications. Fourteen of these patients (39%) were shown to have monoclonal integration of HTLV-I proviral DNA in their blood lymphocytes, a condition designated as smoldering ATL.

Filariasis, malaria, other parasites, viruses, and bacterias are also possible candidates for acceleration of HTLV-I infected cells. An analogous situation is the relation in African Burkitt's lymphoma between Epstein-Barr virus and malaria, and between that virus and the dietary intake of *Euphorbiaceae* plants that contain phorbol esters.

NATURAL HISTORY OF HTLV-I INFECTION

Infection with HTLV-I is a direct cause of ATL. Furthermore, infection with this virus can indirectly cause many other diseases via the induction of immunodeficiency (Fig. 1), such as chronic lung diseases, opportunistic lung infections, cancer of other organs (15), monoclonal gammopathy (16), chronic renal failure (17), strongyloidiasis (14), nonspecific dermatomycosis, nonspecific lymph node swelling, and HTLV-I–associated myelopathy (HAM/TSP) (18). Figure 2 shows the natural course of HTLV-I infection. The clinical stage gradually progresses from carrier to intermediate, smoldering, chronic, and acute-type ATL. Mode of HTLV-I proviral DNA integration also changes from not detectable to polyclonal, monoclonal, and monoclonal malignant transformation. Recently, we proposed the concept of an intermediate state (19) between the healthy carrier and smoldering ATL. Patients in this state correspond to a clinically more pro-

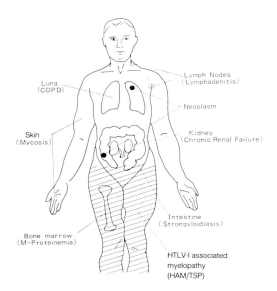

Lung
(COPD)

Lymph Nodes
(Lymphadenitis)

Neoplasm

Skin
(Mycosis)

Kidney
(Chronic Renal Failure)

Intestine
(Strongyloidiasis)

Bone marrow
(M-Proteinemia)

HTLV-I associated
myelopathy
(HAM/TSP)

FIG. 1. HTLV-I and secondary immunodeficiency: clinical features of HTLV-I infection.

gressive group than carriers but lack the features of smoldering ATL. Clinical manifestations, the evidence for immunodeficiency, percentage of IL-2 receptor-positive cells, and the density of T3 antigen suggest that the patients in this state of HTLV-I infection appear to be intermediate between carriers and smoldering ATL. Intermediate-state individuals have polyclonal integration of HTLV-I proviral DNA. On the other hand, in most cases with HAM/TSP, provirus genomes are detected as polyclonal in the peripheral blood. The fact that proviruses are detected in most HAM/TSP patients indicates that the viruses are widespread among lymphocytes.

Using an enzyme-linked immunosorbent assay (ELISA), we measured the level of soluble IL-2 receptor in the serum of 48 asymptomatic HTLV-I carriers, 11 patients with HAM/TSP, and 39 patients with ATL (4 smoldering, 10 chronic, 9 lymphoma, and 16 acute type). The highest levels of sIL-2 receptor were observed in patients with acute- and lymphoma-type ATL as opposed to those with chronic and smoldering ATL. The levels of sIL-2 receptor in patients with HAM were between those of smoldering ATL and healthy carriers. Serum sIL-2 receptor levels of healthy carriers were also elevated compared with normal controls. These observations suggest that the measurement of sIL-2 receptor levels in patients with ATL can be useful as a noninvasive measure of tumor burden, because it represents a phenotypic feature of the malignant cells in this disease, and should aid in the understanding of the natural history of HTLV-I infection leading to the development of ATL. The natural history of HTLV-I infection leading to ATL is multistage leukemogenesis.

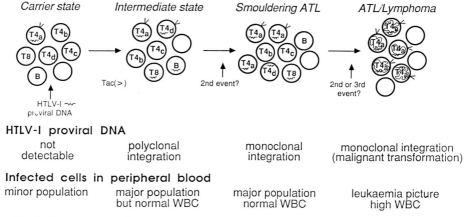

FIG. 2. Natural course of HTLV-I infection leading to adult T-cell leukemia (Multistage leukemogenesis).

PREVENTION AND TREATMENT OF ATL

To prevent infection with HTLV-I, all samples of donated blood collected at nationwide blood centers were subjected to HTLV-I antibody testing by gelatin particle agglutination beginning in November 1986. A similar test should also be carried out for blood collected within hospitals.

To prevent vertical transmission, HTLV-I antibody positive women are now instructed to refrain from breast-feeding in some local communities within the framework of a pilot study. This recommendation will be made on a nationwide level. Transmission between spouses is transmission between persons of the same generation: hence there is hardly any risk of ATL onset even when this form of infection occurs. Therefore, to prevent HTLV-I infection in the next generation, HTLV-I–antibody-positive mothers should refrain from breast-feeding. Although experiments have disclosed that vaccination against HTLV-I is possible, there seems to be no necessity for vaccination.

The results of ATL treatment in the past have been unfavorable. In general, patients with acute- and lymphoma-type ATL should be treated with combination chemotherapy directed toward achieving a cure. However, those patients showing hypercalcemia, high LDH levels, and an abnormal increase in white blood cells have a 50% survival time of less than 6 mo (20). Even if aggressive combination chemotherapy such as CHOP, VEPA, or COMLA is given, the prognosis does not improve. Patients often die of severe respiratory infection or hypercalcemia. On the other hand, regardless of treatment, chronic-type and smoldering-type ATL have a longer course. Aggressive chemotherapy may induce severe respiratory infection. Therefore, an independent treatment protocol for chronic- and smoldering-type ATL, different from that for acute- and lymphoma-type ATL, must be established. We are now starting three new treatment protocols.

Deoxycoformycin (DCF) was successfully used as a single agent to treat a patient with acute-type ATL, resulting in an apparent long-lasting remission. DCF, a nucleoside analogue produced by *Streptomyces antibioticus* or *Aspergillus nidulans,* is a potent tight-binding inhibitor of the enzyme adenosine deaminase (ADA). The therapeutic selectivity of DCF for lymphoid neoplasms was inferred from the preferential lymphoid impairment in congenital ADA deficiency. Seven patients with ATL refractory to conventional chemotherapy were treated with 5 mg/m^2 of DCF. Three patients showed a good response, and four were resistant to DCF. Two patients with ATL receiving DCF had a continuous remission without further therapy (21). Phase II of the clinical study of DCF in Japan was started in 1988.

The second protocol is extracorporeal photochemotherapy. Because the combination of administration of a photoactive drug [8-methoxypsoralen (8-MOP)] and exposure to a long-wave ultraviolet-light system resulted in complete response to psoriasis, this psoralen plus ultraviolet A (PUVA) therapy has been performed in the area of dermatology. Edelson et al. at Yale University developed extracorporeal photochemotherapy, photopheresis, and applied it to cutaneous T-cell lymphoma (CTCL) (22). We have started to apply it to four patients, two with chronic- and two with smoldering-type ATL. Photopheresis is performed for 2 successive days every 4 weeks, repeated after 5 months. Although the observation period is 9–15 wk, the skin lesion of one patient is improved and cell marker analysis is improved in three patients. Decrease of CD4/CD8 ratio and normalization of serum soluble IL-2 receptor level were observed after photopheresis treatment.

The third protocol is a diphtheria toxin–related interleukin 2 fusion protein (IL-2 toxin) therapy. The inhibitory effect of IL-2 toxin on protein synthesis in ATL cells was examined in vitro. Williams et al. have recently described the genetic construction of chimeric toxin, in which the diphtheria toxin receptor-binding domain has been genetically replaced with IL-2 sequences (23). IL-2 toxin has been shown to bind to the high-affinity form of the IL-2 receptor, to be internalized by receptor-mediated endocytosis, and to inhibit protein synthesis.

The effect of IL-2 toxin was examined on ATL cells freshly isolated from 12 patients with different clinical stages of disease. IL-2 toxin inhibited protein synthesis in ATL cells in six patients of acute-type ATL from 20% to 57% compared with the untreated control cultures. It is of interest to note that lymph-node T cells from acute- and lymphoma-type ATL patients were highly sensitive to IL-2 toxin. Peripheral blood T cells from chronic- and smoldering-type ATL patients were more resistant to the action of IL-2 toxin, and protein synthesis was inhibited by only 1%–13%. In addition, we examined the effect of IL-2 toxin on T cells from normal volunteers, lymph-node T cells from a patient with reactive lymphadenopathy following viral infection, and cells from a patient with non-Hodgkin's lymphoma.

REFERENCES

1. Takatsuki K, Yamaguchi K, Kawano F, et al. Clinical diversity in adult T-cell leukemia-lymphoma. *Cancer Res* 1985;(Suppl)45:4644–4645.
2. Poiesz BJ, Ruscetti FW, Fazdar AF, et al. Detection and isolation of type C retrovirus particles from fresh and cultured lymphocytes of a patient with cutaneous T-cell lymphoma. *Proc Natl Acad Sci USA* 1980;77:7415–7419.
3. Hinuma Y, Nagata K, Hanaoka M, et al. Antigen in an adult T-cell leukemia cell line and detection of antibodies to the antigen in human sera. *Proc Natl Acad Sci USA* 1981;78:6476–6480.
4. Yoshida M, Miyoshi I, Hinuma Y. Isolation and characterization of retrovirus from cell lines of human adult T-cell leukemia and its implication in the disease. *Proc Natl Acad Sci USA* 1982;79:2031–2035.
5. Yamaguchi K, Seiki M, Yoshida M, et al. The detection of human T-cell leukemia virus proviral DNA and its application for classification and diagnosis of T-cell malignancy. *Blood* 1984;63:1235–1240.
6. Yamaguchi K, Nishimura H, Kohrogi H, et al. A proposal for smoldering T-cell leukemia: a clinicopathologic study of 5 cases. *Blood* 1983;62:758–766.
7. Yamaguchi K, Yoshioka R, Kiyokawa T, et al. Lymphoma type adult T-cell leukemia—a clinicopathologic study of HTLV related T-cell type malignant lymphoma. *Hematol Oncol* 1986;4:59–65.
8. Kawano F, Yamaguchi K, Nishimura H, et al. Variations in the clinical courses of adult T-cell leukemia. *Cancer* 1985;55:851–856.
9. Yoshioka R, Yamaguchi K, Yoshinaga T, et al. Pulmonary complications in patients with adult T-cell leukemia. *Cancer* 1985;55:2491–2494.
10. Kiyokawa T, Yamaguchi K, Takeya M, et al. Hypercalcemia and osteoclast proliferation in adult T-cell leukemia. *Cancer* 1987;59:1187–1191.
11. Miyamoto Y, Yamaguchi K, Nishimura H, et al. Familial adult T-cell leukemia. *Cancer* 1984;55:181–185.
12. Yamaguchi K, Lee SY, Shimizu T, et al. Concurrence of lymphoma type adult T-cell leukemia in three sisters. *Cancer* 1985;56:1688–1690.
13. Yamaguchi K, Matutes E, Catovsky D, et al. Strongyloides stercoralis as candidate co-factor for HTLV-I induced leukaemogenesis. *Lancet* 1987;2:94.
14. Nakada K, Yamaguchi K, Furugen S, et al. Monoclonal integration of HTLV-I proviral DNA in patients with strongyloidiasis. *Int J Cancer* 1987;40:145–148.
15. Asou N, Kumagai T, Uekihara S, et al. HTLV-I seroprevalence in patients with malignancy. *Cancer* 1986;58:903–907.
16. Matsuzaki H, Yamaguchi K, Kagimoto T, et al. Monoclonal gammopathies in T-cell leukemia. *Cancer* 1985;56:1380–1383.
17. Lee SY, Matsushita K, Yamaguchi K, et al. Human T-cell leukemia virus type I infection in hemodialysis patients. *Cancer* 1987;60:1474–1478.
18. Osame M, Usuku K, Izumo S. HTLV-I associated myelopathy; a new clinical entity. *Lancet* 1986;1:1031–1032.
19. Yamaguchi K, Kiyokawa T, Nakada K, et al. Polyclonal integration of HTLV-I proviral DNA in lymphocytes from HTLV-I seropositive individuals: an intermediate state between the healthy carrier state and smoldering ATL. *Br J Haematol* 1988;68:169–174.
20. Shimoyama M, Yamaguchi K, Takatsuki K, et al. Major prognostic factors of adult patients with advanced T-cell lymphoma/leukemia. *J Clin Oncol* 1988;6:1088–1097.
21. Yamaguchi K, Takatsuki K, Dearden C, et al. Chemotherapy with deoxycoformycin in mature T-cell malignancies. In: Kimura K, et al. *Cancer chemotherapy.* Tokyo: Excerpta Medica, 1988;216–220.
22. Edelson R, Berger C, Gasparro F, et al. Treatment of cutaneous T-cell lymphoma by extracorporeal photochemotherapy. *N Engl J Med* 1987;316:297–303.
23. Williams DP, Parker K, Bacha P, et al. Diphtheria toxin receptor binding domain substitution with interleukin 2: genetic construction and properties of a diphtheria toxin-related interleukin 2 fusion protein. *Protein Eng* 1987;1:493–498.
24. Kiyokawa T, Shirono K, Hattori T, et al. Cytotoxicity of interleukin-2 toxin toward lymphocytes from patients with adult T-cell leukemia. *Cancer Res* 1989; in press.

DISCUSSION

A discussant asked whether the skin lesions persisted in chronic and smoldering ATL. Another discussant replied that spontaneous remissions occur and that the lesions may disappear with therapy. One speaker raised the following questions: Has HTLV-II been identified in any patients with ATL or other malignancies in Japan, and are there patients in Japan with an ATL-like picture where HTLV-I is not present in the tumor cells? In the discussion that followed, it was stated that there was no information about the incidence of HTLV-II infection in the Japanese population and that about 10% of patients with ATL-like diseases were HTLV-I negative by serology and by molecular analysis. In further comments about HTLV-II, it was stated that the virus had not been detected in any of the tumors that had been screened. The question of transplacental transmission as a means of infecting offspring was raised. A discussant stated that the primary mode was via mother's milk. One participant questioned the coincidence of tropical spastic paraparesis and ATL in the same patient. Another discussant replied that in 350 patients with ATL of the chronic or smoldering type, 1 or 2 had neurologic diseases.

Human Retrovirology: HTLV,
edited by William A. Blattner.
Raven Press, Ltd., New York 1990.

Adult T-Cell Leukemia/Lymphoma (ATL) in Jamaica

B. Hanchard, W. N. Gibbs, W. Lofters, M. Campbell,
E. Williams, N. Williams, *E. Jaffe, B. Cranston,
L. D. Panchoosingh, L. LaGrenade, R. Wilks,
*E. Murphy, *W. Blattner, and *A. Manns

*Pathology Department, University of the West Indies, Jamaica; and *National Cancer
Institute, National Institutes of Health, Bethesda, Maryland*

INTRODUCTION

That Jamaica was likely to be a source of HTLV-I–associated lymphoreticular malignancies was first suggested by Catovsky et al. in 1982 when they documented Japanese-like adult T-cell leukemia/lymphoma (ATL) in a group of 16 West Indian migrants in Great Britain, 4 of whom were Jamaican (1).

The first reports from Jamaica were published in 1983 when a combined University of the West Indies/National Institutes of Health (UWI/NIH) group, headed by Prof. W. Gibbs and Dr. W. Blattner, respectively, reported that 11 of 16 non-Hodgkin's lymphoma (NHL) patients attending the Haematology Service at the University Hospital of the West Indies were seropositive for HTLV-I antibodies. These patients had both the clinical features and the poor prognosis of ATL (2).

Since then Jamaica has been shown to be an endemic area for HTLV-I, based on widespread seropositivity in a number of cohort groups throughout the island. Viral antibody is present in 3%–7% of the population, as revealed from various studies (3–5).

It is not surprising, therefore, that we have been able to document increasing numbers of patients with ATL in Jamaica. In 1987, the combined UWI/NIH study group reported the clinicopathological features of 27 patients with ATL of 95 consecutively diagnosed patients with NHL between February 1982 and August 1985 (6). ATL was shown to have a similar natural history to that seen elsewhere, with an acute fatal form with a mean survival of 14 wk and a chronic, smoldering type with a mean survival of 81 wk. The major difference in the disease was that systemic opportunistic infections were rarely seen in our patients, contrasting with the American and Japanese experience, where systemic opportunistic infections often occur (7,8).

TABLE 1. *Parish of birth of ATL patients (n = 82)*

Parish	No. of patients	Parish	No. of patients
Kingston & St. Andrew	17	St. Thomas	2
Portland	4	St. Mary	8
St. Ann	4	Trelawny	2
St. James	2	Hanover	6
Westmoreland	2	St. Elizabeth	2
Manchester	3	Clarendon	7
St. Catherine	6	Unknown	17

CLINICOPATHOLOGICAL STUDIES

Diagnosis of ATL

The diagnosis of ATL was made in HTLV-I seropositive patients who had morphologic evidence of a post-thymic lymphoid malignancy with two of the following four clinical features: hypercalcemia; histologically proven skin infiltration; leukemia and bone marrow infiltration, providing that the pattern of infiltration and the morphology of the infiltrating cells was typical of ATL. Included was a small group that was regarded as "consistent" with ATL where only one of the above was present.

Demographic Features of ATL

There were 40 males and 42 females (M:F ratio = 1:1), ranging in age from 17 to 75 yr (mean 43 yr, median 39.5 yr). With respect to the place of birth, there was roughly equal distribution throughout the rural parishes in the island. Kingston and St. Andrew, a combined predominantly urban parish, containing roughly one-quarter of the island's population, showed the largest number of cases (Table 1).

TABLE 2. *Clinical signs in ATL*

Clinical sign	Percent
Lymphadenopathy	87
Hypercalcemia	65
Leukemia	60
Bone marrow involvement	63
Hepatosplenomegaly	37
Skin involvement	35
Lytic bone lesions	15
Lung infiltration	10

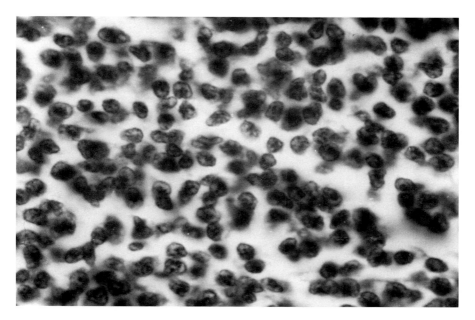

FIG. 1A. Medium-sized cell lymphoma. Lymph node (H&E, ×800).

Clinical Features of ATL

Table 2 lists the clinical signs which comprise the clinical spectrum of the disease.

Lymphadenopathy. Lymphadenopathy was present in 71 (87%) of patients at presentation. Of these, generalized lymphadenopathy was present in 61 (75%). In the remainder, there was localized involvement of less than three lymph node groups, cervical and/or axillary adenopathy being most frequent.

Histologic classification was based on the recommendations of Rappaport (9), the National Cancer Institute (NCI) Working Formulation (10), and the Lymphoma Study Group in Japan (11). We found that the classification most suited for the morphologic appraisal was that of the Lymphoma Study Group in Japan. Neither the Rappaport nor the Working Formulation of the NCI recognize the category medium-sized cell lymphoma, which comprises a well-defined, distinct group in our population of ATLs. On the other hand, we found only two cases of small-cell lymphoma in our survey.

All the lymphomas were pleomorphic and polymorphous, and lymph node involvement was leukemic and nondestructive. The neoplastic cell that was most often present was a medium-sized, indented, convoluted lymphocyte, which appeared to be the tissue counterpart of the polylobated cell in the blood of leukemic patients. A predominance of these cells was typical of the medium-

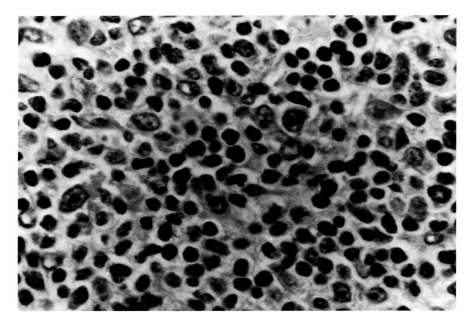

FIG. 1B. Mixed-cell lymphoma. Lymph node (H&E, ×800).

sized cell lymphoma (Fig. 1A). Other lymphomas showed an admixture of larger transformed cells. These larger transformed cells, when seen in a frequency of about eight per high-power field, typified the mixed lymphoma (Fig. 1B), and, when they comprised the majority, typified the large-cell lymphomas (Fig. 1C). The pleomorphic lymphoma contained, in addition to the above elements, larger cells with atypical, pleomorphic cells with prominent nucleoli, often resembling Reed-Sternberg cells (Fig. 1D).

The distribution of subtypes in the various classifications is shown in Table 3. There were 19 cases in which classification was not possible because of insufficient or inadequate material or because the diagnosis was made on liver or skin biopsies.

Hepatosplenomegaly. This was seen in 30 (37%) patients at presentation. There were 18 cases with hepatomegaly without splenomegaly, and 2 patients had splenomegaly alone. In the majority of cases, involvement of the liver was associated with alterations of tests of liver function. In liver biopsies, infiltration of the liver was either portal, sinusoidal, or a mixture of both. Portal tract involvement was more often present when there was no associated leukemia, but, in the presence of leukemia, infiltration was mainly sinusoidal.

Bone Marrow Involvement. Fifty-two patients (63%) had bone marrow involvement at the time of presentation. Involvement of bone marrow, like lymph

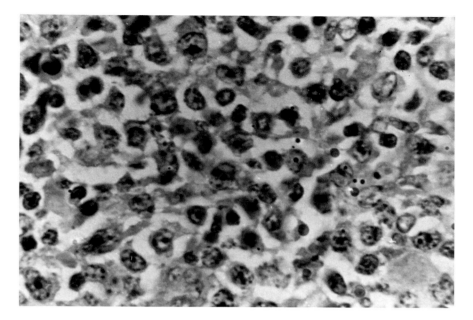

FIG. 1C. Large-cell lymphoma. Lymph node (H&E, ×800).

nodes, was diffuse, with the neoplastic cells showing the same polymorphism. Notably, the infiltrates were not paratrabecular but medullary, displacing marrow elements.

Lytic Bone Lesions. Twelve patients (15%) had radiographic evidence of lytic bone lesions. These were seen in the long bones, skull, ribs, and patella. Involvement of a single site was seen in eight patients, the remainder showing involvement of multiple sites. Bone biopsies in seven patients confirmed infiltration by neoplastic lymphocytes.

Infections. Intercurrent opportunistic infections involving single organs were seen in seven patients. In three they involved the lung (two patients with viral pneumonia and one with fungal pneumonia), and in the remaining four they involved the skin (three had crusted "Norwegian" scabes and one had chronic tinea corporis). However, the majority of infections were bacterial, involving either the lungs or urinary tract, and these were often associated with septicemia.

Hypercalcemia. This was present in 53 (65%) patients at the time of diagnosis. The mean level was 3.64 mmol/L with a range from 2.80 to 4.87 mmol/L. In one patient there were no accompanying clinical signs, but in the remainder there was polyuria, polydipsia, and central nervous system symptoms ranging from mental confusion to coma.

Skin Involvement. Twenty-nine (35%) of the ATL patients had skin involvement, and this was the only presenting feature in one. Lesions were papulonodu-

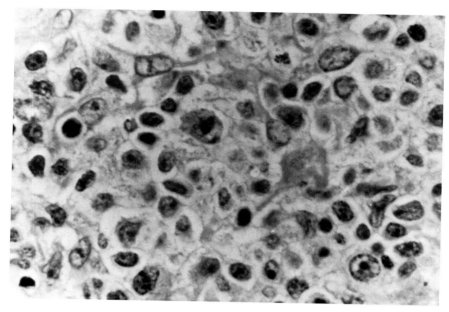

FIG. 1D. Pleomorphic lymphoma. Lymph node (H&E, ×800).

lar, ulcerative or plaque-like. Biopsies of the skin showed either superficial or deep dermal infiltration of neoplastic cells. As in lymph nodes, the polymorphous nature of the infiltrate was evident, and, in many, there was evidence of epidermotropism with Pautrier microabscess formation (Fig. 1E).

Leukemia. Leukemia was present in 49 (60%) of the patients at initial diagnosis but developed in most during the course of the disease. Total counts ranged from 2.2×10^9/L to 365.0×10^9/L with a mean of 60.5×10^9/L. Leukemic cells showed the typical polylobated appearance and were relatively uniform in

TABLE 3. *Histological classification of ATL*

Rappaport	Working formulation	Lymphoma Study Group (Japan)	No.	%
Well-Dif. Lymphcyt.	Small cell	Small cell	2	3
Poorly Dif. Lymphcyt.	Unclassifiable	Medium sized cell	7	11
Mixed cell	Mixed small & large	Mixed cell	10	16
Histiocytic/Undiff.	Large cell	Large cell	33	52
Histiocytic/MCT	Large cell immunoblast.	Pleomorphic	11	18

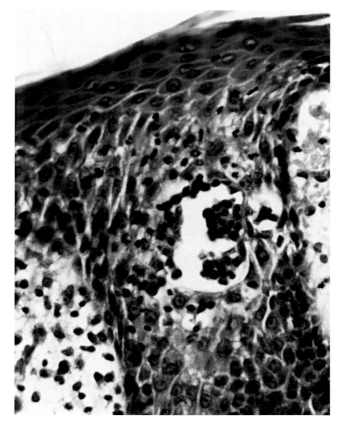

FIG. 1E. Skin showing Pautrier's abscess (H&E, ×800).

size. In some cases, there was a variable admixture of somewhat larger, more ovoid transformed lymphocytes showing little or no nuclear indentations.

Pulmonary Infiltration. Eight (10%) of our patients had symptoms of lung disease. Shortness of breath and a chronic nonproductive cough were the most frequent complaints, and there was radiographic evidence of pulmonary infiltration often accompanied by pleural effusion. This was seen particularly in patients with widespread disease and in association with many of the other typical features of ATL. Lung biopsy in two patients showed diffuse infiltration by malignant cells.

Stage at Presentation. The majority (93%) of our patients presented with advanced disease (Stage III and Stage IV). Five (6.1%) presented in Stage II, and only one patient presented in Stage I.

Immunological Marker Studies

Immunofluorescence and/or immunoperoxidase marker studies were performed on peripheral blood lymphocytes in forty-five leukemic patients. The neoplastic lymphocytes were shown to be peripheral T cells with markers for the T4 subtype. Immunoperoxidase markers applied to frozen sections of lymph nodes in these forty-five patients and an additional nine non-leukemic patients confirmed T-cell lymphoma.

HTLV-I Seropositivity in ATL

All 82 cases included in the study were seropositive for HTLV-I antibodies, and there has been 100% retrieval of proviral HTLV-I sequences in DNA from neoplastic lymphocytes in 27 cases tested so far (12). Excluded from the study, however, were a small group who, although showing some of the clinical features of ATL, were seronegative for HTLV-I antibodies. It is expected, however, that some of these will have proviral sequences in their lymphocytes by further analysis. These patients (HTLV-I seronegative, proviral DNA positive) may form a group parallel to the seropositive ATLs and are currently being followed separately.

Causes of Death

The main categories into which the causes of death could be classified were (a) tumor progression (29.5%); (b) infection with or without septicemia (27.3%); and (c) hypercalcemia (9.1%). Many patients had combinations of these three groups (27.3%). Other patients died as a result of opportunistic infections (4.5%) and therapy complications (2.3%).

Survival

A review of the status of our ATL patients as of December 1988 shows that of the total of 82 patients, 84% are dead, 8% are still alive, and 8% are of unknown status, these being lost to follow-up. Survival rates for ATL patients show a median survival of 16 wk.

CONCLUSIONS

ATL in Jamaica shows a similar clinicopathologic profile to that seen in other countries (13–16). Males and females are affected equally (M:F ratio = 1:1), and there is a wide range in the ages of those affected (17–75 yr). Lymphadenopathy, hepatosplenomegaly, skin infiltration, bone marrow infiltration, lytic bone lesions, and hypercalcemia remain the common clinical signs. Lung involvement, although occurring in only 10% of our patients, is nevertheless an important clinical phenomenon, and symptoms of lung disease may be the initial presenting complaint. Nevertheless, most patients with lung involvement are found to have widespread and advanced disease.

We found that the classification that was most useful in the histological appraisal of the lymphomas was that of the Japanese Lymphoma Study Group, as it recognizes the medium-sized cell lymphoma which is a distinct subtype unclassifiable by the Rappaport and the NCI Working Formulation. Nevertheless, most of our lymphomas were large-cell type (52%), as distinct from the Japanese experience where the pleomorphic and medium-sized cell types are preponderant (42% and 36%, respectively) (11). However, the histologic type does not appear to influence the prognosis.

We have elected to define ATL as a post-thymic lymphoid malignancy in patients who are seropositive for HTLV-I antibodies with specific clinicopathologic features. This excludes a small number of seronegative patients who also have post-thymic lymphoid malignancies and similar clinicopathologic features. These are being followed in a parallel study in the expectation that some of these will show HTLV-I proviral sequences in their neoplastic lymphocytes.

The prognosis remains poor. The median life expectancy is 16 wk despite therapy, but there is a small group that shows a protracted course typical of chronic ATL. Death is mainly due to intercurrent infection, hypercalcemia, tumor progression, or a combination of these. Unlike the experience in Japan or the United States, however, systemic opportunistic infections are rare.

ACKNOWLEDGMENTS

Supported by NCI/NIH grant, Contract No.: NO1-CP-31006. Thanks to the staff of the HTLV Project, Pathology Department, UWI, Jamaica; the staff of the Pathology Department, UWI, Jamaica; the staff of Research Triangle Institute, Bethesda, Maryland; and colleagues at the National Institutes of Health.

REFERENCES

1. Catovsky D, Greaves MF, Rose M, et al. Adult T-cell lymphoma/leukemia in blacks from the West Indies. *Lancet* 1982;1:639–643.
2. Blattner W, Gibbs W, Saxinger C, et al. Human T-cell leukemia/lymphoma virus–associated lymphoreticular neoplasia in Jamaica. *Lancet* 1983;2:61–64.
3. Murphy E, Figueroa JP, Gibbs WN, et al. Retroviral epidemiology in Jamaica, West Indies: introduction of HIV into an HTLV-I endemic island. *Proceedings of the 3rd International Conference on AIDS, Washington, DC, USA,* (abstract TP48) 1987;70.
4. Murphy E, Figueroa JP, Gibbs WN, et al. HIV and HTLV-I infection among homosexual men in Kingston, Jamaica. *J AIDS* 1988;1:143–149.
5. Manns A, Murphy E, Wilks R, et al. Prospective cohort study of transfusion-mediated HTLV-I transmission in Jamaica. *Proceedings of the IVth International Conference on AIDS, Stockholm, Sweden,* (abstract 5568) 1988;255.

6. Gibbs W, Lofters W, Campbell M, et al. Non-Hodgkin lymphoma in Jamaica and its relationship to adult T-cell leukemia/lymphoma. *Ann Intern Med* 1987;106:361–368.
7. Blayney DW, Jaffe ES, Blattner WA, et al. The human T-cell leukemia/lymphoma virus associated with American adult T-cell leukemia/lymphoma. *Blood* 1983;62:401–405.
8. Sato E, Hasui K, Tokunga M. Autopsy findings of adult T-cell lymphoma/leukemia. In: Hanaoka M, Takasuki K, Shimoyama M eds. *Adult T-Cell Leukemia and Related Diseases.* (GANN monograph on cancer research, no. 28). London: Plenum Press, 1982;51–64.
9. Rappaport H. Tumours of the haematopoietic system. In: *Atlas of Tumor Pathology.* Section 3, fascicle 8. Washington, DC: Armed Forces Institute of Pathology, 1966;97–98.
10. The non-Hodgkin's lymphoma pathologic classification project. National Cancer Institute sponsored study of classification of non-Hodgkin's lymphomas: summary and description of a working formulation for clinical usage. *Cancer* 1982;49:2112–2135.
11. Hanaoka M, Sasaki M, Matsumoto H, et al. Adult T-cell leukemia: histological classification and characteristics. *Acta Pathol Jpn* 1979;29:723–738.
12. Clark JW, Gurgo C, Francini G, et al. Molecular epidemiology of HTLV-I–associated non-Hodgkin's lymphomas in Jamaica. *Cancer* 1988;61:1477–1482.
13. Tajima K, Tomimaga S, Suchi T. Clinico-epidemiological analysis of adult T-cell leukemia. *Gann Monogr Cancer Res* 1982;28:197–210.
14. Takatsuki K, Yamaguchi K, Kawano F, Hattori T, Nishimura H, Tsuda H, Sanada I. Clinical aspects of adult T-cell leukemia/lymphoma (ATL). In: Miwa M, Sugano H, Sugimura T, Weiss R, eds. *Retroviruses in Human Leukemia/Lymphoma.* Tokyo: Japan Science Soc. Press, 1985;51–57.
15. Jaffe E, Blattner W, Blayney D, et al. The pathologic spectrum of adult T-cell leukemia/lymphoma in the United States. *Am J Surg Path* 1984;8:4:263–275.
16. Matutes E, Brito-Babapulle V, Catovsky D. Clinical, immunological, ultrastructural and cytogenetic Studies in black patients with adult T-cell leukemia/lymphoma. In: Miwa M, Sugano H, Sugimura T, Weiss R, eds. *Retroviruses in Human Leukemia/Lymphoma.* Tokyo: Japan Science Soc. Press, 1985;59–70.

DISCUSSION

One discussant referred to a few patients from the southwest Pacific coast in the lowlands of Colombia with HTLV-I+ serology and T-cell NHL. In a recent study, 2 patients (1 black and 1 mestizo) out of 41 NHL-tested (including 8 with T-NHL) were shown to be serologically HTLV-I+ by ELISA and Western blot. Also, 1 out of 35 acute-leukemia patients was found to be HTLV-I+, but it was assumed that in this patient HTLV-I was acquired by blood transfusion. It was concluded that a closer survey of lymphoid malignancies must be undertaken in this area of Colombia.

One participant referred to data from the United States Registry of ATL, established with the aim of improving disease definition and identification of new foci and possible risk factors. This participant reported details of two cases: 1) a patient from Alaska with an immunoblastic T-NHL and low levels of HTLV-I antibody; 2) a 34-year-old female, sister of an ATL patient, found to have antibodies to HTLV-I. She presented with a skin rash that faded with corticosteroid therapy, had 1% "flower" (ATL-like) cells in the blood, which after culture expressed the HTLV-I antigen. This case brought up the question as to whether she was just an HTLV-I carrier or whether her status corresponded to that of smoldering ATL. This participant questioned this latter clinical form of ATL and pointed out that a more precise definition of HTLV-I carrier and smoldering ATL should be established. Another discussant argued that cytogenetic and molecular analysis can demonstrate the presence of an abnormal clone in smoldering ATL. A monoclonal pattern of proviral DNA integration is found only in smoldering ATL but not in the carriers. These two parameters should be used to distinguish between these two disease states.

One speaker reported the association of HTLV-I and infestation by Strongyloides in Jamaica. This speaker's data showed a high incidence of antibodies to HTLV-I in 27 patients with Strongyloides (44%), whereas 13 patients with other gastrointestinal disorders (control group) were HTLV-I–antibody-negative. Patients with HTLV-I+ often failed to respond to treatment for strongyloidiasis. The speaker commented that the relationship between Strongyloides and HTLV-I is still uncertain, because in his series of cases there was a major age difference between the control group (median 38 years) and patients with Strongyloides (median 50 years).

Another speaker described seven HTLV-I+ and three HTLV-II+ cases in New York. Among the HTLV-I cases, three patients (from Haiti, Nigeria, and Jamaica) had atypical NHL, two patients (from Chile and Angola) presented with typical features of tropical spastic paraparesis, and the remaining two suffered from skin lesions. One of the three HTLV-II+ patients was referred to as hairy cell leukemia. The speaker also reported a number of double infections: nine cases of HTLV-II+, HIV+ in intravenous drug abusers; and two black patients with double infection for HTLV-I and HIV, one with chronic $CD8^+$ T-cell lymphocytosis and the other with disfunction of the lower extremities— spastic gait—but without evidence of features characteristic of tropical spastic paraparesis.

A number of participants referred to single cases with double infections (HTLV, HIV), which appear to be more common in intravenous drug abusers.

Human Retrovirology: HTLV,
edited by William A. Blattner.
Raven Press, Ltd., New York 1990.

Adult T-Cell Leukemia in Trinidad and Tobago

*Farley R. Cleghorn, †W. Charles, ‡W. Blattner,
and §C. Bartholomew

*Caribbean Epidemiology Centre, Port of Spain, Trinidad; †Department of
Haematology, General Hospital, Port of Spain, Trinidad; ‡National Cancer Institute,
Bethesda, Maryland; and §Department of Medicine, University of the West Indies,
General Hospital, Port of Spain, Trinidad

INTRODUCTION

Trinidad and Tobago are the two southernmost islands of the Caribbean with a combined area of 1980 sq mi and a population of 1.25 million people. Trinidad is by far the larger of the two islands (area 1864 sq mi) and lies at latitude $10\frac{1}{2}°$ north and longitude $61\frac{1}{2}°$ west. The population of Trinidad is quite cosmopolitan and is composed of people of African descent (41%), Indian descent (41%), Caucasians (1%), Chinese (1%), and those of mixed ethnic background and others (16%). Tobago, on the other hand, is populated primarily by people of African descent.

People of African descent were brought to Trinidad and Tobago as slaves starting in the late 17th century. The route for many was not direct, but via other islands in the West Indies, where sugar plantations were the mainstay of the economy. People of Indian descent came to Trinidad beginning in 1845 as indentured laborers on the sugar estates after the emancipation of slaves in 1834.

Although the determination of cell surface phenotypes was rarely available in Trinidad until recently, it has observed that the course and prognosis of non-Hodgkin's lymphomas in Trinidad were substantially worse than that described in the medical literature from developed countries and similar to that described from places like Jamaica (1).

After the description of the clinical entity adult T-cell leukemia/lymphoma (ATL) in Japan in 1977 (2) and the subsequent isolation of the first known human retrovirus, HTLV-I (3), in 1983 Catovsky in London described similar cases of ATL in West Indian immigrants (4). These cases were shown to be identical clinically and pathologically to cases of ATL in Japan. One of the cases recorded in this paper, a 21-yr-old black female, represents the first report of a Trinidadian with ATL. Subsequently, another case report of ATL in a Trini-

TABLE 1. *HTLV-I seropositivity in Trinidad and Tobago*

| Sex | Ratio of seropositivity to number tested, by age range | | Total |
	<40 yr	≥40 yr	
Trinidad*			
Male	6/415 (1.5)	11/222 (4.9)	18/637 (2.8)
Female	9/301 (3.0)	7/87 (8.1)	16/388 (4.1)
Total	15/716 (2.1)	18/309 (5.8)	33/1025 (3.2)
Tobago†			
Male	0/15 (0)	5/39 (12.8)	5/54 (9.2)
Female	0/36 (0)	12/59 (22)	12/95 (13.6)
Total	0/51 (0)	17/98 (17.3)	17/149 (11.4)

The prevalence of HTLV-I in Tobagonians of African descent was significantly higher than in Trinidadians of African descent in the age group > 40 ($p = 0.0001$). There was no difference in those < 40. Values in parentheses are percentages.
   * Within Trinidad prevalence of HTLV-I was not significantly higher in women than in men (age adjusted OR = 1.9, 95% CI = 0.6, 5.9).
   † Within Tobago prevalence of HTLV-I was not significantly higher in women than in men (age adjusted OR = 1.9, 95% CI = 0.6, 5.9).

dadian was described from the Massachussetts General Hospital in 1984 (5), a 29-yr-old black female who had migrated to the USA 4 yr earlier.

In 1985 Bartholomew et al. described the first 12 cases of ATL in Trinidad (6). There were eight females and four males, all of African descent, whose ages ranged between 22 and 84 yr, with a mean of 49. All 12 cases had HTLV-I antibodies in the peripheral blood. In addition the clinical features of these initial cases were identical to those described from Japan and Jamaica.

This firmly established Trinidad as an endemic area for HTLV-I and led in 1985 to a collaborative contract between the University of the West Indies in Trinidad and the Section of Viral Epidemiology of the National Cancer Institute, Bethesda, Maryland to study HTLV-I and its clinical outcome in Trinidad and Tobago.

HTLV-I IN TRINIDAD AND TOBAGO

Utilizing a serosurvey conducted in 1982 for hepatitis markers where collected sera were stored at the Caribbean Epidemiology Centre at $-20°C$, HTLV-I antibodies were detected in 55/1729 (3.2%) of samples tested from all races (7). All of these samples were from individuals 20 yr of age and over. In Trinidad, 36/1025 (3.5%) of people of African or mixed African descent were seropositive, whereas only 1 of 448 (0.2%) Indo-Trinidadians was HTLV-I positive. This 50-yr-old man had traveled extensively and had had sexual contact with prostitutes in Japan as well as the Caribbean islands. In both Trinidad and Tobago, subjects more than 40 yr of age were significantly more likely to be seropositive than those less than 40. Women had a higher rate of seropositivity than men, particularly after age 40 (the median age), but this did not reach statistical significance in either island. However, in the rural coastal village in Tobago (where the study was conducted) 18/149 (12%) were HTLV-I antibody seropositive, a threefold higher seroprevalence than that in Trinidad (Table 1).

As elsewhere (8,9), the epidemiology of HTLV-I in these islands displayed a preponderance of females and a rise in the seroprevalence curve with age. In addition, there was an almost exclusive restriction of seropositivity to people of African descent. The possibility of an area of high seroprevalence in Tobago remains to be explored with a more detailed study of that island.

ATL IN TRINIDAD AND TOBAGO

On October 1, 1985, a study of blood and lymph disease was initiated in Trinidad and Tobago to investigate the role of HTLV-I in the etiology of all hematological malignancies in the two islands. This was a hospital-based, case-control study covering all newly diagnosed cases in individuals 15 yr and older. Age- and sex-matched controls were enrolled from the medical wards at the two general hospitals in the country.

Up to June 30, 1989, a period of 45 mo, a total of 206 eligible cases have been enrolled. Initially, these represented all types of hematological malignancies, but after February 1, 1988, only cases of lymphoid malignancies were enrolled. Overall, approximately two-thirds (66%) of all cases are classifiable as "lymphoid," the remaining cases being myeloid in origin or Hodgkin's disease. Non-Hodgkin's (including primary cutaneous) lymphoma was the single largest category of lymphoid malignancy and accounted for 40% of the total enrollment. During the period of the study, 102 such cases were consecutively studied (Table 2).

The diagnosis of ATL was made on clinical and pathological findings, the prototypic case being an aggressive non-Hodgkin's lymphoma of mature T-cell lineage with a high incidence of skin and visceral involvement and hypercalcemia (10). Forty-eight cases satisfied the clinicopathological definition of ATL. This represents 47% of all non-Hodgkin's lymphomas enrolled in the study. Not all these patients have had complete immunophenotyping of their tumors at the time of writing, however, and this is in progress at present. Ninety-four percent

TABLE 2. *A prospective study of hematologic malignances*

	Trinidad and Tobago, October 1, 1985–June 30, 1989
Non-Hodgkin's lymphoma	102
Clinicopathological ATL	48 (47)
HTLV-I positive	45/48 (94)

Values in parentheses are percentages.

TABLE 3. *Clinical features of ATL in Trinidad*

Lymphadenopathy	56/64 (87.5)
Hepatomegaly	33/64 (51.6)
Splenomegaly	21/64 (32.8)
Skin involvement	25/64 (39.0)
Hypercalcemia	29/64 (45.3)

N = 64; cases enrolled through June 30, 1989. Values in parentheses are percentages.

of cases of ATL were seropositive for HTLV-I. It is probable that the cases of ATL seronegative for virus are either non-antibody producers or carry variants of the virus that are not detectable by current methods.

When we combine all cases of ATL diagnosed in Trinidad thus far (pre- and poststudy), the clinical features are similar to those described elsewhere (Table 3). Lymphadenopathy (88%) was the most common presenting symptom, followed by skin involvement (39%). Hepatomegaly was found in 49%, splenomegaly in 32%, and hypercalcemia in 42%. Three patients had atypical peripheral and central nerve involvement, including bilateral facial palsy.

The data generated by this study have given rise to some interesting and unique epidemiological findings. The distribution of ATL in Trinidad and Tobago reflects the ethnic reservoir of HTLV-I in these islands, all cases being of African or mixed-African descent, supporting the hypothesis that HTLV-I came to the West Indies via the slave trade.

As in other studies (11,12), the incidence of ATL in Trinidad and Tobago appears to be much lower than that suggested by HTLV-I seroprevalence. Approximately 12–15 cases of ATL are diagnosed each year in Trinidad. Using figures obtained from the 1980 Population and Census Report (13) and the crude prevalence rate of HTLV-I antibodies in people of African descent, the incidence rate of ATL is ∼4 per 100 000 per year in people of African and mixed-African descent and 100 per 100 000 HTLV-I carriers per year. Table 4 shows the comparison with the island of Kyushu in southern Japan, reflecting a somewhat higher occurrence of ATL in Trinidad.

Crude annual incidence rates can be calculated for males and females by age

TABLE 4. *Comparison of HTLV-I carriers in Trinidad and Kyushu*

	Trinidad	Kyushu
Population at risk	321 288	9.7 million
HTLV-I seroprevalence	3.5%	5.8%
HTLV-I carriers	11 245	561 000
ATL cases per 100 000 population	4	5.7
Incidence of ATL per year in carriers	1:886 (0.1)	1:1700 (0.06)

Data from adults >20 yr old. Kyushu data reported in Kondo et al. (14). Values in parentheses are percentages.

TABLE 5. *Crude annual incidence rates for ATL in Trinidad*

Age group	Population	Rate of HTLV-I	ATL cases/yr per 100 000 population
		Males	
20–29	53 995	1.4	2.0
30–39	34 629	1.7	0.78
40–49	23 104	4.5	6.9
50–59	18 236	4.8	8.7
60+	21 216	7.2	7.5
Total	151 169	2.7	4.0
		Females	
20–29	52 583	2.2	1.5
30–39	33 721	4.6	2.4
40–49	23 081	14.6	0
50–59	17 734	18.2	4.5
60+	15 691	—	15.3
Total	142 810	5.9	3.6

Data from adults > 20 yr.

group as shown in Table 5. Although there is the suggestion that males have a higher incidence of ATL (14), data presented from Jamaica suggest that childhood infection with HTLV-I as a prerequisite for leukemogenesis is a better explanation of the sex difference (12). In comparison, the excess of female cases of tropical spastic paraparesis, as shown in Table 6, may be explained by relatively more efficient male-to-female transmission of HTLV-I in adult life and the shorter latent period for this disease.

There appear to be differences in the incidence of ATL between Trinidad and Tobago. Given the high seroprevalence of HTLV-I in a single town in Tobago (12%), and that not a single case of ATL has occurred in a resident Tobagonian during the tenure of this study, these differences suggest the existence of cofactors in Trinidad which may not obtain in Tobago. Further studies are in progress to ascertain whether this is so.

TABLE 6. *Comparison of demographic features of HTLV-I–positive subjects in Trinidad and Tobago*

	Carriers	ATL	TSP
Median age	49	46	50.5
Age range	23–88	15–80	17–78
Male/female	22/32	36/28	10/21
Race		All African or mixed African descent	
HTLV-I	54/54	60/64 (94)	26/31 (84)

Values in parentheses are percentages.

CONCLUSION

HTLV-I contributes significantly to the occurrence of lymphoreticular malignancy in Trinidad. The clinical features of ATL in Trinidad appear similar to those described elsewhere. Continued prospective study of lymphoid neoplasia in these islands will help to determine the disease load as a result of HTLV-I endemicity, as well as methods of interrupting HTLV-I transmission so as to eradicate this virus.

REFERENCES

1. Charles W. Port of Spain General Hospital, 1982, unpublished observations.
2. Takatsuki K, Uchiyama T, Sagawa K, Yodoi J. Adult T-cell leukemia in Japan. In: Seno S, Takaku F, Irino S, eds. *Topics in Haematology.* Amsterdam: Excerpta Medica, 1977;73–77.
3. Poiez BJ, Ruscetti FW, Gazder AF, Bunn PA, Minna PA, Gallo RC. Detection and isolation of type C retrovirus particles from fresh and cultured lymphocytes of a patient with T-cell lymphoma. *Proc Natl Acad Sci USA* 1980;77:7415–7419.
4. Catovsky D, Greaves MF, Rose M, et al. Adult T-cell lymphoma-leukemia in blacks from the West Indies. *Lancet* 1982;1:639–643.
5. Scully RE, Mark EJ, McNeely BU, eds. Case Records of the Massachusetts General Hospital. Weekly Clinicopathological Exercises. *N Engl J Med* 1984;14:906–916.
6. Bartholomew C, Charles W, Saxinger C, et al. Racial and other characteristics of human T-cell leukemia/lymphoma (HTLV-I) and AIDS (HTLV-III) in Trinidad. *Br Med J* 1985;290:1243–1246.
7. Bartholomew C, Charles W, Gallo RC, Blattner WA. The ethnic distribution of HTLV-I and HTLV-III associated diseases in Trinidad, West Indies. In: Clumeck N, Thiry L, Burny A, eds. *International Symposium on African AIDS,* Brussels, 1985.
8. Clark J, Saxinger C, Gibbs WN, et al. Seroepidemiologic studies of human T-cell leukemia/lymphoma virus type I in Jamaica. *Int J Cancer* 1985;36:37–41.
9. Tajima K, Tominaga K, Suchi T, et al. Epidemiological analysis of the distribution of antibody to adult T-cell leukemia virus associated antigen: possible horizontal transmission of adult T-cell leukemia virus. *Gann* 1982;73:893.
10. Kawano F, Yamaguchi K, Nishimura H, Tsuda H, Takatsuki K. Variation in the clinical course of adult T-cell leukemia. *Cancer* 1985;55:851–856.
11. Tajima K, Kuroishi T. Estimation of the rate of incidence of ATL among ATLV (HTLV-I) carriers in Kyushu, Japan. *Jpn J Clin Oncol* 1985;15(2):423–430.
12. Murphy EL, Hanchard B, Figueroa JP, Gibbs WN, Goedert JJ, Blattner WA. Modelling the risk of adult T-cell leukemia/lymphoma in persons infected with human T-lymphotropic virus type I. *Int J Cancer* 1989;43:250–253.
13. *Population and Census Report, 1980.* Central Statistical Office of the Office of the Prime Minister, Government Printery, Port of Spain, Trinidad, 1981.
14. Kondo T, Kono H, Nonaka H, et al. Risk of adult T-cell leukemia/lymphoma in HTLV-I carriers. *Lancet* 1987;2:259.

Human Retrovirology: HTLV,
edited by William A. Blattner.
Raven Press, Ltd., New York © 1990.

Review of WHO Kagoshima Meeting and Diagnostic Guidelines for HAM/TSP

Mitsuhiro Osame

World Health Organization Collaborating Centre for Retroviral Infections Associated with Neurological Diseases, The Third Department of Internal Medicine, Faculty of Medicine, Kagoshima University, Kagoshima, Japan

INTRODUCTION

Involvement of the nervous system by the human T-lymphotropic virus type I (HTLV-I) was independently demonstrated in the tropics and Japan in two chronic neurologic disorders, tropical spastic paraparesis (TSP) and HTLV-I–associated myelopathy (HAM) (1–8). In this regard, the World Health Organization (WHO) designated the Third Department of Internal Medicine, Faculty of Medicine, Kagoshima University, Kagoshima, Japan, as a WHO Collaborating Centre for Human Retroviral Infections Associated with Neurological Disorders, with Dr. Mitsuhiro Osame as Head of the Collaborating Centre, on October 13, 1988. The meeting of the Scientific Group on HTLV-I Infections and its Associated Diseases was then held, convened by the Regional Office for the Western Pacific of the WHO. This paper reviews this meeting and introduces the outline of the report of the meeting, including the diagnostic guidelines of HAM/TSP (9).

MEETING OF THE SCIENTIFIC GROUP ON HTLV-I INFECTIONS AND ASSOCIATED DISEASES

The meeting was held in Kagoshima, Japan, from December 10–15, 1988.

Dr. S. T. Han, special representative of the Director-General, World Health Organization, gave the introductory remarks. Dr. Hiroshi Nakajima, Director-General, opened the meeting.

Dr. Mitsuhiro Osame was appointed Chairman; Dr. Pamela Rodgers-Johnson, Vice Chairperson; and Dr. Guy de The, Dr. Gustavo Roman, and Dr. Yuzo Iwasaki, Rapporteurs.

The list of participants is shown in Table 1, and the group picture is shown in Fig. 1.

TABLE 1. *List of temporary advisers, members, observers, and secretariat*

Temporary advisers	Members	Observers	Secretariat
Dr. C. G. Gajdusek	Dr. D. Babona	Dr. H. Shoji	Dr. H. Suzuki
United States	Papau New Guinea	Japan	Philippines
Dr. A. Igata	Dr. G. de Thé	Dr. I. Maruyama	Dr. T. Umenai
Japan	France	Japan	Philippines
	Dr. A. Diwan	Dr. N. Ohba	
	United States	Japan	
	Dr. Y. Iwasaki	Dr. S. Akizuki	
	Japan	Japan	
	Dr. P. Rodgers-Johnson	Dr. E. Sato	
	United States/Jamaica	Japan	
	Dr. K. S. Mani	Dr. Y. Itoyama	
	India	Japan	
	Dr. I. Miyoshi	Dr. M. Mori	
	Japan	Japan	
	Dr. O. Morgan	Dr. T. Tabira	
	Jamaica	Japan	
	Dr. S. Nagataki	Dr. K. Arimura	
	Japan	Japan	
	Dr. K. Okochi	Dr. A. Manns	
	Japan	United States	
	Dr. M. Osame	Dr. T. Miyamoto	
	Japan	Japan	
	Dr. G. C. Roman	Dr. N. Mueller	
	United States	United States	
	Dr. T. Saida	Dr. N. Tachibana	
	Japan	Japan	
	Dr. S. Sonoda		
	Japan		
	Dr. K. Sugamura		
	Japan		
	Dr. K. Tajima		
	Japan		
	Dr. K. Takatsuki		
	Japan		
	Dr. K. Tashiro		
	Japan		
	Dr. J-C. Vernant		
	Martinique		
	Dr. M. Yoshida		
	Japan		
	Dr. V. Zaninovic		
	Colombia		

RECOMMENDATIONS

The following recommendations were proposed by the Group:

1. As HAM and TSP are clinically and pathologically identical diseases occurring in different geographical locations, both temperate and tropical, the name "HAM/TSP" should be used for the time being.

FIG. 1. Scientific group on HTLV-I infections and its associated diseases in Kagoshima, Japan, from December 10–15, 1988.

2. The Group recommended immediate dissemination of the information on HAM/TSP among physicians throughout the world in view of the high prevalence already found in Japan, the Caribbean, and South America and the worldwide sporadic occurrence of the disease.
3. The Group recommended continuous epidemiological surveillance of HTLV-I infection and its associated disorders.
4. The Group recognized the occurrence of a wide spectrum of the HTLV-I–associated diseases and recommended the search for and elucidation of cofactors that precipitate and/or modify the disease process.
5. The Group recommended the development of rapid, sensitive, and more standardized procedures for laboratory diagnosis of the infection by HTLV-I and further delineation of virus serotypes.
6. The Group recognized the four modes of HTLV-I transmission: breast-feeding, blood transfusion, sexual intercourse, and intravenous drug abuse; the exercise of appropriate measures to control the transmission of the virus was urged.
7. The Group encouraged the further development of therapeutic regimens for HAM/TSP.

DIAGNOSTIC GUIDELINES FOR HAM/TSP

The diagnostic guidelines for HAM have been established and used in Japan (10). Based on the recognition described in recommendation 1 above, the group discussed the guidelines for HAM/TSP. The spectrum of clinical features (such as pulmonary involvement, Sjogren's syndrome, uveitis, arthropathy, ichthyosis, neuropathy, encephalopathy, and polymyositis) that has been recognized to be associated with HAM/TSP (11–15) is wider than previously thought. Cases mimicking multiple sclerosis or amyotrophic lateral sclerosis have also been reported (11).

The following diagnostic guidelines for HAM/TSP were agreed on by the Group:

I. *Clinical criteria*

The florid clinical picture of chronic spastic paraparesis is not always seen when the patient first presents. A single symptom or physical sign may be the only evidence of early HAM/TSP.

A. *Age and sex incidence*

Mostly sporadic and adult, but sometimes familial; occasionally seen in childhood; females predominant.

B. *Onset*

This is usually insidious but may be sudden.

C. *Main neurological manifestations*

1. Chronic spastic paraparesis, which usually progresses slowly, sometimes remains static after initial progression.
2. Weakness of the lower limbs, more marked proximally.

3. Bladder disturbance usually an early feature; constipation usually occurs later; impotence or decreased libido is common.
4. Sensory symptoms such as tingling, pins and needles, burning, etc. are more prominent than objective physical signs.
5. Low lumbar pain with radiation to the legs is common.
6. Vibration sense is frequently impaired; proprioception is less often affected.
7. Hyperreflexia of the lower limbs, often with clonus and Babinski's sign.
8. Hyperreflexia of upper limbs; positive Hoffmann's and Tromner signs frequent; weakness may be absent.
9. Exaggerated jaw jerk in some patients.

D. *Less frequent neurological findings*
   Cerebellar signs, optic atrophy, deafness, nystagmus, other cranial nerve deficits, hand tremor, absent or depressed ankle jerk.
   Convulsions, cognitive impairment, dementia, or impaired consciousness are rare.

E. *Other neurological manifestations that may be associated with HAM/TSP*
   Muscular atrophy, fasciculations (rare), polymyositis, peripheral neuropathy, polyradiculopathy, cranial neuropathy, meningitis, encephalopathy.

F. *Systemic nonneurological manifestations which may be associated with HAM/TSP*
   Pulmonary alveolitis, uveitis, Sjogren's syndrome, arthropathy, vasculitis, ichthyosis, cryoglobulinemia, monoclonal gammopathy, adult T-cell leukemia/lymphoma.

II. *Laboratory diagnosis*

A. Presence of HTLV-I antibodies or antigens in blood and cerebrospinal fluid (CSF).
B. CSF may show mild lymphocyte pleocytosis.
C. Lobulated lymphocyte may be present in blood and/or CSF.
D. Mild to moderate increase of protein may be present in CSF.
E. Viral isolation when possible from blood and/or CSF.

COMMENTS

The cases which satisfy the clinical criteria but not the laboratory diagnosis should be treated as problem cases; further follow-up studies will be needed.

ACKNOWLEDGMENTS

We would like to thank all the attendees of the meeting who made the meeting so fruitful. We are especially grateful for Drs. T. Umenai and H. Suzuki, of the

WHO Regional Office for the Western Pacific, who led the meeting with such success. Our thanks also to Corazon R. Omega and Michiko Sameshima for their technical assistance.

REFERENCES

1. Gessain A, Barim F, Vernant JC, et al. Antibodies to human T-lymphotropic virus type-I in patient with tropical spastic paraparesis. *Lancet* 1985;2:407–410.
2. Rodgers-Johnson P, Gajdusek DC, Morgan O, et al. HTLV-I and HTLV-III antibodies and tropical spastic paraparesis. *Lancet* 1985;2:1247–1248.
3. Bartholomew C, Cleghorn F, Charles W, et al. HTLV-I and tropical spastic paraparesis. *Lancet* 1986;2:99–100.
4. Zaninovic V. Spastic paraparesis: a possible sexually transmitted viral myeloneuropathy. *Lancet* 1986;2:697–698.
5. Vernant JC, Maurs L, Gessain A, et al. Endemic tropical spastic paraparesis associated with human T-lymphotropic virus type I: a clinical and seroepidemiological study of 25 cases. *Ann Neurol* 1987;21:124–130.
6. Osame M, Usuku K, Izumo S, et al. HTLV-I associated myelopathy. A new clinical entity. *Lancet* 1986;1:1031–1032.
7. Osame M, Matsumoto M, Usuku K, et al. Chronic progressive myelopathy associated with elevated antibodies to HTLV-I and adult T-cell leukemia like cells. *Ann Neurol* 1987;21:117–122.
8. Roman GC, Osame M. Identity of HTLV-I–associated tropical spastic paraparesis and HTLV-I–associated myelopathy. *Lancet* 1988;1:651.
9. Report of World Health Organization Scientific Group on HTLV-I Infections and Associated Diseases (Kagoshima, Japan, 10–15 December, 1988) Manila, March 1989.
10. Osame M, Igata A, Matsumoto M. HTLV-I–associated myelopathy (HAM) revisited. In: Roman G, Vernant JC, Osame M, eds. *HTLV-I and the nervous system.* New York: Alan R. Liss, 1989;213–223.
11. Roman GC, Vernant JC, Osame M, ed. *HTLV-I and the nervous system.* New York: Alan R. Liss, 1989.
12. Sugimoto M, Nakashima H, Watanabe S, et al. T-lymphocyte alveolitis in HTLV-I–associated myelopathy. *Lancet* 1987;2:1220.
13. Vernant JC, Buisson G, Magdeleine J, et al. T-lymphocyte alveolitis, tropical spastic paraparesis, and Sjogren syndrome. *Lancet* 1988;1:177.
14. Kitajima I, Maruyama I, Maruyama Y, et al. Polyarthritis in HTLV-I associated myelopathy (HAM). *Arthritis Rheum* 1989;32:1342–1344.
15. Nishioka K, Maruyama I, Sato K, et al. Chronic inflammatory arthropathy associated with HTLV-I. *Lancet* 1989;1:441.

DISCUSSION

A discussant mentioned a project at the Scripps Institute in which two brains were studied: one HAM and one TSP brain. The study showed TAX of HTLV-I in tissue taken from various regions of the brain. It was suggested that this observation constitutes evidence that virus is present in the brain. It was done by extraction of nucleic acids from the tissue, not amplified by PCR. *In situ* hydrization was not successful.

Another discussant inquired about the best way to identify cells in the CSF under light microscopy. It was explained that the best way to identify such cells was with Giemsa staining of cytospun specimens.

A speaker said that there is a lot of indirect evidence that transfusion is associated with virus transmission and risk of disease. Two alternative hypotheses for explaining risk of disease transmission were offered: 1) that the virus is in the bag of blood; 2) that in latent infection perhaps associated with an immunological catastrophe, a co-factor may play a role in activating the virus. It may take five years for seroconversion to occur. The speaker asked whether in Kagoshima, Japan, there has been an instance where a seronegative person has become positive after being transfused with seronegative blood. No responses were forthcoming.

One participant wanted to know if there was any possible connection between HAM and subacute myeloopticoneuropathy (SMON), which was described in Japan in the 1960s. Despite clinical similarities between the two disorders, there was not thought to be any relationship between them. A minor percentage of cases of HAM were probably diagnosed as SMON. However, the epidemic of SMON occurred earlier, and an association with a specific drug was well established by epidemiological survey. Following withdrawal of the offending drug, SMON was no longer seen in Japan.

Human Retrovirology: HTLV,
edited by William A. Blattner.
Raven Press, Ltd., New York © 1990.

Tropical Spastic Paraparesis Clinical Features

Owen Morgan

Department of Medicine, University of the West Indies,
Mona, Kingston 7, Jamaica, West Indies

INTRODUCTION

In 1956 Cruickshank described a chronic neurological disorder of uncertain etiology (1), Jamaican neuropathy, which he had studied at the University Hospital of the West Indies. He defined the elements of the syndrome and its two clinical groups, the spastic and the ataxic forms.

Jamaican neuropathy is but one of the family of tropical myeloneuropathies which itself embraces two overlapping syndromes, tropical spastic paraparesis (TSP) and tropical ataxic neuropathy (TAN). TSP occurs in either epidemic or endemic form and remains a common problem in Jamaica. TAN, on the other hand, is largely endemic and has declined in prevalence.

Recent studies have shown that the human T-cell lymphotropic virus (HTLV-I) is the causative agent of endemic TSP (2), and a similar disorder HAM/TSP has also been described in Japan (3), where HTLV-I infection is endemic.

As awareness of the disease has grown so also has interest in its clinical features. Clinicians today, especially those familiar with the disease, have the responsibility to evaluate its clinical features accurately and to construct the full spectrum of its clinical forms.

This paper reviews the clinical features of the disease in patients studied at the Neurological Clinic of the University of the West Indies (UHWI).

GEOGRAPHY

Jamaica is a small but mountainous island located in the northwestern sector of the Caribbean archipelago near Cuba and Haiti. It covers an area of 4400 sq mi and lies between latitudes 17° and 48° north, and longitudes 76° and 78° west. The island experiences high uniform temperatures consistent with its tropical location, average temperatures of 80°F and 40°F being recorded consistently at sea level and in the mountains, respectively.

The island is divided into 14 parishes and has a population of 2.3 million people, 50% of whom reside in the greater Kingston area. The society is multiracial and consists primarily of blacks of African descent (90%), East Indians, Chinese, Syrians, and Caucasians.

STUDY POPULATION

The study population consisted of 145 Jamaican TSP patients who were seen between March 1984 and December 1988. The majority of patients affected were black and from the lower socioeconomic groups; few were of mixed race. They originated from the southeastern parishes of St. Catherine, Kingston and St. Andrew, and St. Thomas. There is, however, no reason to believe that pockets of disease exist within the island; rather, the apparent clustering is the result of the more easy access to specialist services in these areas.

Cases included in the study corresponded to patients with TSP defined on the following criteria: (1) spastic paraplegia/paraparesis in either sex; (2) absence of a history of neurological disease in the family; (3) a minimum of three of the following complaints at onset: bladder dysfunction, impotence in males, low back pain, weakness in the legs, or dysasthesiae in the feet; (4) absence of Argyll Robertson pupils; and (5) impaired vibration or position sense in the feet.

The HTLV-I antibody status of the patients was not used as a criterion.

Only five of our patients had ever been transfused with blood.

Age and Sex Incidence

The age of onset of symptoms ranged from 14 to 78 yr with the majority (62, 43%) presenting in the 40–49 age group. Only three patients were under the age of 20 yr. There was a female preponderance of cases (108 females and 37 males) in a ratio of 3:1, contrasting with an almost equal sex incidence reported in earlier series.

Disease duration varied from 3 mo to 40 yr.

Presentation

The onset was gradual in most cases but was occasionally abrupt. The nature of the onset dictated the course which the illness would pursue, a very slow onset heralding a slowly advancing disease, and a quick beginning, rapidly advancing disease.

Accompanying and sometimes preceding the major symptoms were certain sensory symptoms which, from the confusion they caused, deserve emphasis.

Lumbar backache was present in 70% of patients and either remained static or radiated into the legs. Complaints of burning, pins and needles, cramps, numbness, or tingling occurred in 40% of patients. These persisted for several months or for the duration of the illness but declined in severity with time.

Impairment of sphincteric control was present in 75% of cases. Bladder symptoms comprising nocturia frequency, urgency, and incontinence of micturition preceded the appearance of constipation. Males invariably complained of impotence early in the disease; women remained fertile.

More often disability due to weakness or stiffness in the legs (80%) and a resulting gait disorder (60%) resulted in "dragging of the legs." One leg was usually involved before the other by as much as 6 mo, but invariably both were eventually affected. Leg weakness seldom appeared as an isolated phenomenon, but occurred in association with sensory and/or bladder symptoms.

The upper limbs were to a large extent symptom free. Less frequent complaints were dimness of vision (18%) and impaired hearing (5%).

The pattern of progression of symptoms was variable. Weakness and stiffness of the legs gradually worsened, reaching a plateau in 6 mo. It was unusual for further deterioration to continue after this time but in a few patients, it continued for many years.

In 10% of cases, a remitting-relapsing course was pursued with patients experiencing phases of virtually complete recovery. (The resulting disability was also variable: 60% of our patients were able to walk unaided, 20% could walk with support, and 20% were bedridden.)

Neurological Examination

By definition all were patients with spastic paraplegia/paraparesis. Spasticity of the lower limbs was present in 75% of patients. Gait was typically spastic; some patients were unable to walk unaided, whereas others required varying degrees of support. The knee and ankle tendon reflexes were increased in 84% of patients and the plantar responses extensor in 95%. The abdominal reflexes were lost in 50% of cases. Weakness, as expected, was of pyramidal distribution and was often Gd 3-4 Medical Research Council (MRC).

Cruickshank regarded peroneal muscle wasting as a feature of the disease, but we encountered this sign in 8% of our cases. Distal muscle wasting and fasciculation were conspicuously absent from either the upper or lower limb musculature. Hyperreflexia with a positive Hoffmann's sign (82%) was the major abnormality in the upper limb.

Intention tremor was present in 7% of cases.

Objective abnormalities of sensation were present in the lower limbs and to a minor degree in the upper limbs. Cutaneous sensory loss was not common, but careful examination often revealed areas of skin in the distal extremities where sensation was impaired. Occasionally the loss was severe enough to suggest a spinal cord tumor. The sense of position and passive movement was seldom seriously affected, but impairment or loss of vibration sense was found in 41%.

Less frequently encountered but important physical signs were cerebellar ataxia (7%) and lower motor neuron facial weakness. Ocular movements were intact except for the presence of nystagmus in 20% of cases. Ptosis was an uncommon finding, present in two patients. The pupils were normal. Frank optic atrophy was present in only 15% of patients. Sensorineural deafness present in 10% of cases was bilateral and severe. Mental symptoms were never present.

DIAGNOSIS

It is important to realize that the disease is not confined to black patients of lower socioeconomic status from tropical countries.

TSP is found among native Japanese and middle-class Caucasians from such temperate climates as Peru, France, and Italy, in patients who have never traveled outside these countries.

In a tropical environment where the majority of cases reside, Treponemal syndromes, nutritional disorders, vitamin B_{12} deficiency, and arachnoiditis are important causes of progressive paraplegia.

Spinal syphilis has many features which are common to the spastic type of Jamaican neuropathy but can readily be differentiated by appropriate serological tests. Our patients were neither undernourished nor B_{12} deficient.

Spinal arachnoiditis, a very common cause of spinal cord compression in Jamaica, resembles TSP, but spinothalamic signs are prominent and contrast myelography is abnormal.

Multiple sclerosis is rare in the tropics but its more chronic spinal form bears many similarities to TSP which itself may have an acute presentation, prominent cerebellar signs, and may pursue an intermittent course.

Investigations

Forty percent of our patients had polylobulated lymphocytes on the peripheral blood smear. The cerebrospinal fluid (CSF) changes were unremarkable except for a moderate elevation of lymphocytes and protein in some patients. Examination of the spinal column by X-rays of the spinal cord by myelography was normal. Magnetic resonance imaging (MRI) demonstrated periventricular lesions in the few patients in whom it was performed.

Immunological studies have shown that in 91 TSP patients, IgG antibodies were present in high concentration in 91% of sera and 84% of CSFs by ELISA and Western blot. The *gag*-encoded p19 and p24 proteins were always identified, whereas detection of the env-coded gh46 and gh60 glycoproteins and p42 varied from specimen to specimen.

Virus has been isolated from the CSF and peripheral blood of some patients and its presence observed by electron microscopy and supported by detection of reverse transcriptase in tissue fluid.

Oligoclonal bands of IgG antibodies to HTLV-I were found in CSF in the presence of an intact blood-brain barrier, thus supporting a role for HTLV-I or an antigenically related virus.

Course and Prognosis

The course of the disease varied from a few to several years. It progressed rapidly in some before stabilizing; in others it pursued an indolent course over many years before arresting. Corticosteroids favorably influenced some features of the disease in some patients, especially paraplegia and bladder symptoms, but were required in high doses (40 mg/d) for improvement to be maintained.

SUMMARY

TSP is an easily recognizable disease in its usual form with its classical clinical features. Diagnosis may be difficult, however, when presentation is atypical. Several immunological abnormalities exist in the disease, which is thought to be caused by HTLV-I. Effective therapeutic regimens and preventive measures are required if the impact of the disease is to be reduced.

REFERENCES

1. Cruickshank EK. A neurological syndrome of uncertain origin. *West Indian Med J* 1956;5:147–158.
2. Rodgers-Johnson P, Morgan OStC, Mora C, et al. The role of HTLV-I in tropical spastic paraparesis in Jamaica. *Ann Neurol* 1988;23(suppl):S121–S126.
3. Osame M, Usuku K, Izumo S, et al. HTLV-1 associated myelopathy, a new clinical entity. *Lancet* 1986;1:1031–1032.

DISCUSSION

In the experience of one discussant, cerebellar symptoms were present in a percentage of the Colombian patients with TSP. Nystagmus, too, was seen in a number of patients. This discussant stressed that in very early TSP, alcohol often brought out signs of weakness. The history and observation that alcohol produced gait difficulty and weakness was of value diagnostically. Another participant challenged these observations and stated that in dealing with TSP patients in London, cerebellar symptoms were distinctly absent. This participant thought that such findings must be unusual.

A speaker asked whether blood transfusion data were available and was told that only four patients had received transfusion. Another speaker asked if a case control study had been carried out and if the prevalence rate for lymphoma and TSP had been established.

One discussant commented that tropical ataxic neuropathy (TAN) had been described earlier, but many fewer cases had occurred in recent years. Only two cases of TAN had been seen in the last ten years.

A participant remarked that Jamaican patients with TSP were seen in Miami, Florida, and they too had a slowly progressive spastic paraparesis as their major manifestation. However, the character of illness in a number of Haitians, also seen in Miami, was different. In each instance TSP was much more rapidly progressive in the Haitians, leading to functional paraplegia in a couple of years. Of six Haitian patients with TSP, five experienced onset in their midtwenties. All Haitian patients were Western blot positive. MRI studies were abnormal in half the patients, revealing increased T2 signal in both brain and cervical spinal cord. The proportion of abnormal MRI studies in Haitians was equal to that seen in other TSP patients from the Caribbean.

Human Retrovirology: HTLV,
edited by William A. Blattner.
Raven Press, Ltd., New York 1990.

Tropical Spastic Paraparesis and HTLV-I–Associated Myelopathy—Clinical and Laboratory Diagnosis

Pamela E. B. Rodgers-Johnson, Steven Ono,
Clarence J. Gibbs Jr., and D. Carleton Gajdusek

Laboratory of Central Nervous System Studies, National Institute of Neurological Disorders and Stroke, National Institutes of Health, Bethesda, Maryland 20892

Although much has been written about the tropical myeloneuropathies—epidemic and endemic tropical spastic paraparesis (TSP) and tropical ataxic neuropathy (TAN), these disorders remained orphans in the field of neurology. They were not recognized by the majority of neurologists in the developed world; this may have been in part because of their undetermined etiology and their classification as neurological disorders that occurred in tropical regions among persons of low income and of mainly African descent (1–3). Interest was stimulated in TSP in 1985 when IgG antibodies to HTLV-I were found in the serum of 59% of Martinique patients with myelopathy (4) and shortly after antibodies to HTLV-I were found in cerebrospinal fluid (CSF) and sera of TSP patients from Jamaica and Colombia (5,6). A subsequent report by Osame et al. described patients similar to TSP in the nontropical HTLV-I–endemic region of southern Japan and they named the disease HTLV-I–associated myelopathy (HAM) as only HTLV-I–positive patients were included (7). It was soon realized that the overall clinical picture of HAM and HTLV-I–positive TSP were the same and the name HAM/TSP was introduced (8). Before the link with HTLV-I, the diagnosis of TSP was based on the clinical picture and absence of a specific etiological factor by laboratory or other tests; today it is based on the clinical picture as well as the detection of HTLV-I antibody or antigen in CSF and serum. However, in several countries there still exists a group of patients with classical TSP who are HTLV-I–negative by routine serological methods (6,9–12). Because these patients may be anergic, more specific tests for HTLV-I antigen, such as gene amplification by the polymerase chain reaction (PCR), could be most useful (13).

NOTE ON EPIDEMIOLOGY

The neuroepidemiology of HAM/TSP (14) has been facilitated by extensive epidemiological surveys that identified HTLV-I–endemic regions and patients with adult T-cell leukemia (ATL) (15). HAM/TSP is now known to occur in Caucasians; the first such patient was a male who had traveled to endemic regions and who had mycosis fungoides, an HTLV-I–associated T-cell lymphoma (16). Since then it has been diagnosed in Caucasians in nonendemic regions in temperate climates and has been reported in varying numbers from more than 40 countries (Fig. 1); it is therefore evident that country, climate, and race are not barriers to the disease (17).

The known modes of HTLV-I transmission are by sexual intercourse (more easily effected from male to female), from mother to child in breast milk (18), via blood transfusion (19), and by the sharing of needles by parenteral drug abusers. Preventive measures can therefore be effectively used. Because 26% of Japanese HAM/TSP patients had previous blood transfusions it was mandated that all blood donors be screened (19); similarly, restriction or cessation of breast-feeding was introduced to prevent mother-to-child transmission of HTLV-I infection (20).

CLINICAL FEATURES OF HAM/TSP

There is a long latent period from the time of infection to disease onset (14,21), but this is shorter when transmission is from infected mothers' milk (18) and blood transfusion (19). There is a 3:1 female preponderance (9) with age of onset most frequent in 35- to 49-yr-old age group, although it can occur in children (22) and is sometimes familial (23). The predominant neurological findings are lower lumbar pain, spastic paraparesis of the lower extremities with ankle clonus and positive Babinski, variable impairment of superficial and deep sensation, and interference of bladder and bowel function. Hyperreflexia of the upper limbs occurs in most patients, a positive jaw jerk is often present, and ataxia and tremor of the hands occur less frequently. HAM/TSP is not a disease which rapidly progresses to death. Patients may live for 30–40 yr, and those who die early succumb to urinary infections or pulmonary emboli.

Some of the less frequently seen physical signs reported in individual series were not initially classified as part of the HAM/TSP syndrome until several neurologists familiar with the disorder met at the World Health Organization (WHO) meeting on "Human Retroviral Infections associated with Neurological Disorders" held at Kagoshima University, Kagoshima, Japan in December 1988, and agreement was reached on a spectrum of clinical signs (report printed by the Regional Office for the Western Pacific of the WHO, Manila, Philippines, March 1989). These less frequently seen clinical signs may show regional variations in incidence; included in this group are visual disturbance, optic atrophy, deafness, nystagmus, a variety of cranial nerve deficits including pseudobulbar palsy, and absent or depressed ankle jerks. Convulsions, cognitive impairment, dementia, and impaired consciousness are rare. Other neurological manifestations which may be associated with HAM/TSP are muscular atrophy (sometimes with

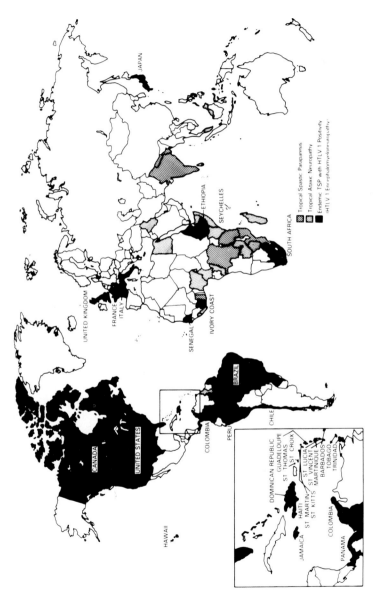

FIG. 1. Map of the world showing the areas from which the various types of tropical myelopathies have been reported. Inset shows the Caribbean basin and islands with HTLV-I–positive tropical spastic paraparesis.

fasciculations), polymyositis, peripheral neuropathy, polyradiculopathy, cranial neuropathy, meningitis, and encephalopathy. Systemic manifestations which have been associated with HAM/TSP are ATL, pulmonary alveolitis, uveitis, Sjögren's syndrome, arthropathy, vasculitis, icthyosis, cryoglobulinemia, and monoclonal gammopathy.

There is no guaranteed successful treatment available for HAM/TSP patients. Physiotherapy combined with muscle relaxants and symptomatic treatment of bladder and bowel symptoms remains the mainstay, although improvement may be temporary. Corticosteroids have been the most useful drug to date: improvement in mobility has been reported in Japanese (27) and Colombian patients (V. Zaninovic, personal communication), and improvement of bladder disturbance with little or no improvement in motor disability has been reported in Jamaican patients (O. Morgan, personal communication). A few patients have had plasmapheresis, and small trials with zidovudine (azidothymidine, AZT) are in progress.

INVESTIGATIONS

No specific lesions are found on roentgenographic examination of the spine, and myelography has been normal. Magnetic reasonance imaging shows small periventricular lesions and spinal atrophy in some patients (21). Electrophysiological studies show normal nerve conduction in the majority of patients (24), although a few patients have delayed nerve conduction.

Nonspecific CSF changes of mild lymphocytosis with moderate increase of protein are sometimes present, and lobulated lymphocytes may be found in peripheral blood smears and in CSF. There is intrathecal synthesis of IgG, and HTLV-I–specific IgG oligoclonal bands are found in CSF. Screening for HTLV-I infection is usually done by an enzyme-linked immunosorbent assay (ELISA), gel particle agglutination test, and indirect immunofluoresence; confirmatory tests are Western immunoblot and radioimmunoprecipitation assays. However, none of these tests distinguishes HTLV-I and -II, and "HTLV-I confirmed positive tests" could be due to crossreactivity with HTLV-II. Specific testing to distinguish these retroviruses is therefore indicated in infected patient populations (14).

In culture media, HAM/TSP mononuclear cells show striking spontaneous proliferation, which does not occur with ATL cells. Several isolates have been grown from serum and CSF samples in a number of countries; antigen has been identified by immunofluorescence assay and type-C retroviral particles visualized by electron microscopy; and a number of cell lines have been established. To detect any differences between the virus that causes HAM/TSP and that which causes ATL, Nakamura et al. compared virus isolates from ATL and HAM/TSP patients. They established a T-cell line from CSF of a HAM/TSP patient and were unable to distinguish any differences to the HTLV-I genome of ATL patients (25). However, Sarin et al. reported that a typical type C retrovirus isolated from the CSF of a Jamaican patient was related to, but not identical with, prototype HTLV-I from ATL patients by restriction mapping (26). The most recent diagnostic technique is the polymerase chain reaction (PCR), which should prove valuable for the detection of proviral and viral DNA sequences and—because the technique is applicable to formalin-fixed tissue—will facilitate the examination of autopsy material obtained from TSP patients who were never tested for HTLV-I infection.

HISTOPATHOLOGY

On light microscopy there is a chronic inflammatory reaction with perivascular cuffing with lymphocytes and demyelination. These changes are more marked in the spinal cord than in the brain (28) and can also be found in the posterior horns and in peripheral nerve roots (29). Electron microscopic studies of cord tissue have shown the presence of HTLV-I–like viral particles (30).

COFACTORS IN DEVELOPMENT OF HAM/TSP

Many persons who live in endemic regions and have high HTLV-I antibody titers are free of disease; this suggests that HTLV-I infection may not be the sole determinant for disease development. Susceptibility to HTLV-I infection is increased by alteration in host immune response resulting from previous T-cell activation, which can be caused by several agents—including other infections, blood transfusions, toxins, dietary factors, and genetic predisposition. Of prime interest is the possibility that infection with another retrovirus could trigger or enhance the disease process. Both HTLV-I and HTLV-II have been identified in a spastic paraplegic patient from Africa (31), and both viruses are known to exist in drug abusers and homosexuals who may or may not be HIV positive. No single cofactor is likely to be the precipitating agent; strongyloidiasis has been associated with HTLV-I infection in Japan, Jamaica, and Martinique, and treponemal infection could be a cofactor in Jamaica, Colombia, and the Seychelles. Of increasing importance is another spirochete, *Borrelia burgdorferi*, which causes Lyme disease, and this infectious process could trigger or mimic HAM/TSP. The neurological syndromes of Lyme disease are important in the differential diagnosis of HAM/TSP, especially as Lyme disease is treatable. Serum samples of HAM/TSP patients from three countries showed a 37–45% positivity for antibody to *Borrelia burgdorferi* and 15% positive in CSF samples from one country tested. In support of a genetic predisposition to disease is the finding of a high percentage of seropositive healthy donors among close family members of ATL and HAM/TSP patients; it was also found that HLA typing of two ethnic groups in southern Kyushu showed that they were endowed with different susceptibilities to HAM and ATL (32). Nutritional and toxic factors cannot be implicated at this time, although it has been suggested that improved diet lessens the severity of neurological lesions (33).

COMMENTS

It has been suggested that the neurologic manifestations of TSP are due to an HTLV-I–mediated autoimmune mechanism, but it is still uncertain whether there is direct invasion of the central nervous system by the virus. HTLV-I has not yet been identified in central nervous system tissue by in situ hybridization studies, but, by the use of the PCR, it may be possible to detect DNA in brain or spinal cord. Although HTLV-I is thought to be the causative agent of HAM/TSP, it has antigenic cross reactivity with HTLV-II, and standard tests are unhelpful in differentiation. HTLV-II and HTLV-I DNA can be differentiated by the PCR; this will be invaluable in determining how frequently both infections exist in patients and carriers. Studies are in progress to compare the molecular structure of virus isolates from carriers with those from ATL and HAM/TSP patients in several countries. Should these isolates be the same in carriers and patients, it would indicate that there must be a cofactor which triggers the disease mechanism. The new data from these studies will make a significant contribution to the understanding of the pathogenesis.

REFERENCES

1. Cruickshank EK. A neuropathic syndrome of uncertain origin. *West Indian Med J* 1956;5:147–158.
2. Rodgers PEB. The clinical features and aetiology of the neuropathic syndrome in Jamaica. *West Indian Med J* 1965;14:36–47.
3. Roman GC, Spencer PS, Schoenberg BS. Tropical myeloneuropathies: the hidden endemias. *Neurology* 1985;35:1158–1170.
4. Gessain A, Barin F, Vernant JC, et al. Antibodies to human T-lymphotropic virus type-I in patients with tropical spastic paraparesis. *Lancet* 1985;2:40–41.
5. Rodgers-Johnson P, Gajdusek DC, Morgan OStC, Zaninovic V, Sarin PS, Graham DS. HTLV-I and HTLV-III antibodies and tropical spastic paraparesis. *Lancet* 1985;2:1247–1248.
6. Zaninovic V, Arango C, Biojo R, et al. Tropical spastic paraparesis in Colombia. *Ann Neurol* 1988;23(suppl):127–132.
7. Osame M, Usuku K, Izumo S, et al. HTLV-I associated myelopathy, a new clinical entity. *Lancet* 1986;1:1031–1032.
8. Roman GC, Osame M. Identity of HTLV-I associated tropical spastic paraparesis and HTLV-I associated myelopathy. *Lancet* 1988;1:651.
9. Rodgers-Johnson PE, Garruto RM, Gajdusek DC. Tropical myeloneuropathies—a new aetiology. *Trends Neurosci* 1988;2:526–532.
10. Vernant JC, Maurs L, Gout O, et al. HTLV-I associated tropical spastic paraparesis in Martinique: a reappraisal. *Ann Neurol* 1988;23(suppl):133–135.
11. Roman G, Spencer P, Schoenberg BS, et al. Tropical spastic paraparesis in the Seychelles Islands. A clinical and case-control neuroepidemiologic study. *Neurology* 1987;37:1323–1328.
12. Itoyama Y, Minato S, Goto I. HTLV-I associated myelopathy (HAM) and seronegative spastic spinal paraparesis. In: Roman GC, Vernant JC, Osame M, eds. *HTLV-I and the nervous system.* New York: Alan R. Liss, 1989;209–212.
13. Bhagavati MD, Ehrlich G, Kula RW, et al. Detection of human T-cell lymphoma/leukemia

virus type I DNA and antigen in spinal fluid and blood of patients with chronic progressive myelopathy. *N Engl J Med* 1988;318:1141–1147.

14. Roman G. The neuroepidemiology of tropical spastic paraparesis. *Ann Neurol* 1988;23(suppl): S113–S120.
15. Blattner WA. Retroviruses. In: Evans AS, ed. *Viral infections of humans: epidemiology and control.* New York: Plenum Publishers, 1989 (*in press*).
16. Lee JW, Fox EP, Rodgers-Johnson P, et al. A case report implicating HTLV-I as a cause of three diseases. *Ann Intern Med* 1988;110:239–241.
17. Rodgers-Johnson PEB. Tropical myeloneuropathies in Jamaica (Jamaican neuropathy). In: Roman GC, Vernant JC, Osame M, eds. *HTLV-I and the nervous system.* New York: Alan R. Liss, 1989:123–138.
18. Osame M, Igata I, Usuku K, Rosales R, Matsumoto M. Mother-to-child transmission in HTLV-I associated myelopathy. *Lancet* 1987;1:106.
19. Osame M, Igata A, Matsumoto M, Izumo S, Kubota H. Transfusion and HTLV-I myelopathy in Japan. In: Roman GC, Vernant JC, Osame M, eds. *HTLV-I and the nervous system.* New York: Alan R. Liss, 1989:547–549.
20. Hino S, Doi H. Mechanisms of HTLV-I transmission. In: Roman GC, Vernant JC, Osame M, eds. *HTLV-I and the nervous system.* New York: Alan R. Liss, 1989:495–501.
21. Newton M, Cruickshank K, Miller D, et al. Antibody to human T-lymphotropic virus type-I in West-Indian-born UK residents with spastic paraparesis. *Lancet* 1987;1:415–416.
22. McKhann G, Gibbs CJ, Mora C, et al. Isolation and characterization of HTLV-I from symptomatic family members with tropical spastic paraparesis (HTLV-I encephalomyeloneuropathy). *J Infect Dis* 1989;160:371–379.
23. Shoji H, Kuwasaki N, Natori H, Kaji M. Familial occurrence of HTLV-I associated myelopathy and adult T-cell leukemia. In: Roman GC, Vernant JC, Osame M, eds. *HTLV-I and the nervous system.* New York: Alan R. Liss, 1989:307–309.
24. Barkhaus PE, Morgan O. Jamaican neuropathy: an electrophysiological study. *Muscle Nerve* 1988;11:380–385.
25. Nakamura T, Shirabe S, Matsuo H, Tsujihata M, Nagataki S. Proviral DNA studies in an HTLV-I isolate from a patient with HTLV-I associated myelopathy. In: Roman GC, Vernant JC, Osame M, eds. *HTLV-I and the nervous system.* New York: Alan R. Liss, 1989:39–42.
26. Sarin PS, Rodgers-Johnson P, Sun DK, et al. Comparison of human T-cell lymphotropic virus type I strain from cerebrospinal fluid of a Jamaican patient with tropical spastic paraparesis with a prototype human T-cell lymphotropic virus type I. *Proc Natl Acad Sci USA* 1989;86:2021–2025.
27. Osame M, Igata A, Matsumoto M. HTLV-I associated myelopathy (HAM) revisited. In: Roman GC, Vernant JC, Osame M, eds. *HTLV-I and the nervous system.* New York: Alan R. Liss, 1989:213–223.
28. Robertson WB, Cruickshank EK. Jamaican (tropical) myeloneuropathy. In: Minckler J, ed. *Pathology of the nervous system, vol. 3.* New York: McGraw Hill Book Co., 1972:2466–2476.
29. Piccardo P, Ceroni M, Rodgers-Johnson PEB, et al. Pathological and immunological observations on tropical spastic paraparesis in patients from Jamaica. *Ann Neurol* 1988;23(suppl):156–160.
30. Liberski P, Rodgers-Johnson P, Char G, Piccardo P, Gibbs CJ, Gajdusek DC. HTLV-I–like particles in spinal cord cells in Jamaican tropical spastic paraparesis patients. *Ann Neurol* 1988;23(suppl):S185–S187.
31. Hugon J, Giordano C, Dumas M, et al. HIV-2 antibodies in African with spastic paraplegia. *Lancet* 1988;1:189.
32. Morgan OStC, Montgomery RD, Rodgers-Johnson PEB. The myeloneuropathies of Jamaica: an unfolding story. *Q J Med* 1988;252:273–281.
33. Usuku K, Sonoda S, Osame H, et al. HLA haplotype-linked high immune responsiveness against HTLV-I in HTLV-I-associated myelopathy: comparison with adult T-cell leukemia/lymphoma. *Ann Neurol* 1988;23(suppl):S143–S150.

Human Retrovirology: HTLV,
edited by William A. Blattner.
Raven Press, Ltd., New York © 1990.

Multiple Sclerosis Clinical and MRI Characteristics: Is There a Link Between HAM/TSP and MS?

Donald W. Paty

Division of Neurology, University of British Columbia, Vancouver, British Columbia

INTRODUCTION

Multiple sclerosis (MS) is a disease characterized pathologically by inflammation and focal areas of demyelination in the central nervous system. It presents in the majority of patients with episodes called relapses. Seventy percent of patients with MS begin with spontaneous relapses and remissions. The other 30% begin with a slowly evolving neurological deficit. The slowly evolving patients tend to be those with an older age of onset.

The myelopathy associated with HTLV-I virus overlaps in age distribution and in clinical characteristics with the slowly evolving spinal cord form of MS. Because HTLV-I–associated myelopathy (HAM/TSP) patients can also have multifocal abnormalities on the magnetic resonance imaging (MRI) scan and can have oligoclonal banding in the cerebrospinal fluid, the differential diagnosis between MS and the HAM/TSP syndrome can be difficult.

In addition, there have been recent reports suggesting a direct link between the HTLV-I virus and multiple sclerosis. The HTLV-I virus is endemic in the tropics and southern Japan. As noted above, in a small percentage of the HTLV-I positive population in endemic regions, there is a slowly progressive spinal cord disorder (HAM/TSP) that could be confused with spinal MS. In contrast, the geographic distribution of MS is primarily in temperate climates, and the genetic susceptibility factors in MS are primarily found in Europeans. The geographic distribution, and therefore the populations at risk, of the HAM/TSP syndrome and MS are almost mutually exclusive. For this reason, this author thinks that any link between the HTLV-I virus and MS, even though potentially very exciting, is highly unlikely.

CLINICAL CHARACTERISTICS OF MULTIPLE SCLEROSIS

Multiple sclerosis is a disease of young adults, with a peak age of onset of 25 yr and a mean age of onset of 31 yr (1). The ratio of females to males is 2:1.

Genetic studies have shown that there is a link between major histocompatibility complex (MHC) antigens DR2 and DQW1 and susceptibility. However, the MHC antigens that are most associated with MS tend to vary from population to population. For example, in most of Europe and the southern British Isles, patients with MS have a higher-than-normal frequency of DRW2. In Scotland, where MS is more common than in England, the unaffected population has a high frequency of DRW2 and the MS population has a frequency that is the same as in the general population.

Family studies have shown that ~22% of patients with MS have other family members with the disease (2). The lifelong risk for developing MS in a first-degree relative of someone already diagnosed with the disease is ~3%–5%. This frequency is ~30 times higher than in persons without a family history.

Twin studies, when done carefully on unselected populations, have shown that identical twins have almost a 30% chance of being concordant for MS when the first twin already has MS (3). Non-identical (paternal) twins have the same risk for developing MS as do siblings of someone with MS (~3%). These studies have provided strong support for the concept of a strong genetic susceptibility factor. However, the fact that >70% of the identical twins of persons diagnosed as having MS do not have clinical MS themselves is strong evidence against MS being an inherited disorder. It is interesting that MRI studies have shown that as many as 30% of clinically unaffected second identical twins have multiple abnormalities on the scan compatible with MS, even though they have never had neurological symptoms (4).

The above-mentioned twin studies, and the MRI studies, taken in conjunction with pathological studies in unaffected individuals (5), have suggested that there is a reservoir of undiagnosed MS in the susceptible community.

The diagnosis of MS is made by the demonstration of multiple abnormal areas in the white matter of young adults from the genetically susceptible population, disseminated in both time and space (6). However, because many other neurological and medical conditions can mimic MS, investigations must be extensive enough to identify clearly other possible causes of the syndrome such as cerebral vasculitis, sarcoidosis, Lyme disease, neurosyphilis, cerebrovascular disease, and multifocal tumors such as cerebral lymphomas.

The use of MRI has helped enormously in the diagnosis of difficult cases (7); however, the abnormalities seen on the MRI scan are not specific for multiple sclerosis (8). The interpretation of the MRI scan and clinical findings therefore requires very critical clinical judgment before making the diagnosis. It is the pattern of abnormalities that is highly suggestive of MS rather than the presence of any particular abnormality.

PATHOLOGY

Specific demyelination is the most characteristic pathological abnormality in MS. The areas of demyelination tend to be scattered throughout the central nervous system, particularly in the white matter, but there is a peculiar tendency for the chronic areas of demyelination to be clustered around the cerebral ventricles. Recent studies have shown that intense inflammation is an invariable characteristic of the acute lesion (9) and that remyelination can occur with considerable intensity in the early lesions.

Computerized tomography (CT) (10) and MRI studies (11) have now shown very dramatically that breakdown in the blood-brain barrier is an early finding in the evolution of MS lesions. In addition, serial studies with MRI have shown that the lesions of MS can come and go with considerable frequency (12–14). It is now thought that the early, and perhaps most important, lesions of MS involve inflammation and breakdown in the blood-brain barrier. Demyelination may occur only secondarily and late in the course of the evolution of an individual lesion.

MRI has also provided a window on the brain for the visualization of the extent and the activity of MS lesions over time. Therefore, MRI is now being used as one measure of the evolution of MS pathology in order to better describe the natural history and to detect changes in lesion activity in the adjudication of new therapies (15).

MRI IN THE EVALUATION OF DISEASE ACTIVITY AND EXTENT

When MRI was shown to visualize the lesions of MS with such accuracy, many investigators began to look at the evolution of the disease process, as it could be described by MRI, over time. Figure 1 shows a typical MRI scan in MS. Measurement of the extent of disease on the MRI scan has shown that there is an increase in the "burden of MS," as detected by MRI, of ~20% over 2 yr (16,17). This fact means that MRI can be used as a measure of the extent of disease. Clinical trials of new drugs in recent years have been using MRI as one measure of therapeutic outcome.

Unfortunately, there is not very good correlation between the extent of disease as measured by MRI and the disability seen in the patients (18). Neuropathologists have been telling us of this lack of correlation for years, but we can now see dynamic evidence of lesion changes on the MRI without clinical symptoms and vice versa. Such nonagreement should not really be a surprise. The location of lesions in the nervous system should determine, for the most part, what symptoms will be produced. Therefore, lesions in the spinal cord, optic nerves, and brain stem are much more likely to produce symptoms than are lesions in the cerebral hemispheres. The majority of lesions that are seen in MS, however, occur in the cerebral hemispheres; therefore, the ability to detect the presence of those lesions by standard neurological examination is limited.

Recent studies of neuropsychological function have shown that there is a greater

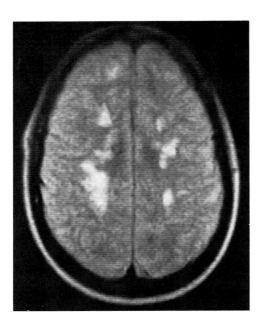

FIG. 1. This MRI slice shows a number of the typical MS lesions. MS lesions are white on spin echo sequences and typically are concentrated in the supra and periventricular areas of the cerebral hemispheres, although they can be found in any part of the central nervous system, including the gray matter. This photograph courtesy of Dr. David Li, Director of MRI Program, University Hospital, University of British Columbia.

correlation between cognitive abnormalities and cerebral MRI scans than can be found by standard neurological examination (19).

Early in our experience with the evaluation of MRI in MS, we were interested in the identification of disease activity in order to understand the fluctuations in immune function that we were observing in our patients (20). Immune function fluctuation is well known to occur in MS, but the correlation between these fluctuations and clinical symptoms was not very good (21).

We therefore began to use serial and frequent MRI scans of the head as a method of identifying new and active areas of pathology in order to see if we could improve our correlation (20). In 1984 we identified several patients with new lesions that then disappeared in follow-up (22). We therefore set out to do systematic studies comparing immune functions with both neurological and MRI monitoring. We now have completed several studies comparing these functions and have shown that new lesions are quite frequent. New and/or enlarging MRI lesions can be identified in patients at a frequency of at least five times the frequency of clinical symptoms. Patients with both the relapsing form and the chronic progressive form of MS show evidence for new and enlarging lesions considerably more frequently than would be predicted by symptoms alone. In fact, in most of the 24 patients that we have studied so far, there were many more new and enlarging lesions than would have been predicted by symptoms. In fact, almost all of the new lesions seen were not associated with new neurological symptoms.

We think that these new lesions are fundamental to the evolution of MS pathology. They probably represent the primary lesion in MS. We also think that they are probably inflammatory rather than demyelinating (13).

Gadolinium studies using MRI have also shown that the majority of new lesions also show MRI enhancement (11,23). These findings suggest that the new inflammatory lesions are very closely associated with breakdown in the blood-brain barrier.

Long-term follow-up of enhancing and changing MRI lesions has also shown that, even though many of these lesions continue to be active (they change in size, disappear and reappear), ~20% of new lesions go on to become permanent and merge with their neighbors with the MRI appearance of confluence (24).

CONCLUSIONS CONCERNING THE DIAGNOSIS OF HAM/TSP AND MS

Multiple sclerosis is a disease of young adults seen in temperate climates that has a strong genetic susceptibility component. Thirty percent of patients with MS have a slowly evolving clinical course that could be confused clinically with the HAM/TSP syndrome. Both MS and HAM/TSP can be associated with disseminated MRI lesions in the head and with oligoclonal banding in the cerebrospinal fluid. Therefore, if patients with chronic progressive myelopathy show evidence for elevated titers of the HTLV-I virus, they should not be considered to have MS until we know more about the specificity of the diagnostic studies.

At this date, there is nothing to suggest that the HAM/TSP syndrome occurs with any significant frequency outside of the endemic regions. Most of the few cases of HAM/TSP that have been seen in temperate climates have occurred in patients who have originated from or who have strong links with the endemic areas. In addition, the patients with HAM/TSP seen in temperate climates also come primarily from the racial populations at risk.

It would seem prudent, therefore, that until such time as a specific differential feature can be established, that patients with MS-like syndromes, particularly chronic progressive spinal disorders, who come from non-Caucasian populations should be screened for HTLV-I antibody. Those persons found to have elevated titers to the HTLV-I virus should not be considered as having MS until more is known concerning specific differentiating features. Certainly such patients should not be used in clinical trials of new drugs designed for the treatment of MS.

ACKNOWLEDGMENTS

Thanks go to my colleagues in the MS Research Program at UBC, especially Dr. David Li, the Director of our MRI program. Studies cited in this chapter were supported by the British Columbia Health Care Research Foundation, the

Medical Research Council, the MS Society of Canada, and the Jacob W. Cohen Fund for Research in MS.

REFERENCES

1. Hashimoto SA, Paty DW. Multiple sclerosis. *Dis Mon* 1986;32.
2. Sadovnick AD, MacLeod PMJ. The familial nature of multiple sclerosis: empiric recurrence risks for first, second, and third-degree relatives of patients. *Neurology* 1981;31:1039–1041.
3. Ebers GC, Bulman DE, Sadovnick AD, et al. A population based study of multiple sclerosis in twins. *N Engl J Med* 1986;315:1638–1642.
4. McFarland HF, Patronas NJ, McFarlin DE, et al. Studies of multiple sclerosis in twins using nuclear magnetic resonance. *Neurology* 1985;35(suppl 1):137.
5. Gilbert IJ, Sadler M. Unsuspected multiple sclerosis. *Arch Neurol* 1983;40:533–537.
6. Schumacher GA, Beebe G, Kibler RF, et al. Problems of experimental trials of therapy in multiple sclerosis. *Ann NY Acad Sci* 1965;122:552–568.
7. Paty DW, Hashimoto S, Hooge J, et al. Magnetic resonance imaging in the diagnosis of multiple sclerosis (MS): a prospective study of comparison with clinical evaluation, evoked potentials, and oligoclonal banding. *Neurology* 1988;38:180–185.
8. Paty DW, Asbury AK, Herndon RM, et al. Use of magnetic resonance imaging in the diagnosis of multiple sclerosis: policy statement. *Neurology* 1986;36:1575.
9. Prineas JW, Kwon LR, Chu E-S. Massive early remyelination in acute multiple sclerosis. *Neurology* 1987;37(suppl 1):109.
10. Aita JF, Bennett DR, Anderson RE, Ziter F. Cranial CT appearance of acute multiple sclerosis. *Neurology* 1978;28:251–255.
11. Miller DH, Rudge P, Johnson G, et al. Serial Gadolinium enhanced magnetic resonance imaging in MS. *Brain* 1988;111:927–939.
12. Isaac C, Li DKB, Genton M, et al. Multiple sclerosis: a serial study using MRI in relapsing patients. *Neurology* 1988;38:1511–1515.
13. Willoughby E, Grochowski E, Li D, Oger J, Kastrukoff L, Paty D. Serial magnetic resonance scanning in multiple sclerosis: a second prospective study in relapsing patients. *Neurology* 1989;25:43–49.
14. Koopmans RA, Li DKB, Oger JJF, et al. Chronic progressive multiple sclerosis: serial magnetic brain imaging over six months. *Ann Neurol* 1989;26:248–256.
15. Paty DW. Magnetic resonance imaging in assessment of disease activity in multiple sclerosis. *Can J Neurol Sci* 1988;15:266–272.
16. Kastrukoff LF, Hashimoto SA, Oger JJ, et al. A trial of namalway interferon in the treatment of chronic progressive (CP) multiple sclerosis (MS). I. Clinical and MRI analysis. *Can J Neurol Sci* 1987;14:242.
17. The Multiple Sclerosis Study Group. The efficacy of cyclosporin immunosuppression in multiple sclerosis: a preliminary report of a randomized blinded, placebo-controlled clinical trial. Program and Abstracts, American Neurological Association. *Ann Neurol* 1988;24:174.
18. Paty DW, Bergstrom J, Palmer M, MacFadyen J, Li D. A quantitative magnetic resonance image of the multiple sclerosis brain. *Neurology* 1985:35(suppl 1):137.
19. Franklin GM, Heaton RK, Nelson LM, Filley CM, Seibert C. Correlation of neuropsychological and MRI findings in chronic progressive multiple sclerosis. *Neurology* 1988;38:1826–1829.
20. Oger J, Kastrukoff LF, Li DKB, Paty DW. Multiple sclerosis: in relapsing patients, immune functions vary with disease activity as assessed by MRI. *Neurology* 1988;38:1739–1744.
21. Kastrukoff LF, Oger J, Paty DW. Multiple sclerosis: correlation of peripheral blood lymphocyte phenotype and natural killer cell activity with disease assessed clinically and by MRI. *Ann Neurol* 1986;20:164.
22. Li D, Mayo J, Fache S, Robertson WD, Paty D, Genton M. Early experience in nuclear magnetic resonance imaging of multiple sclerosis. *Ann NY Acad Sci* 1984;436:483–486.
23. Grossman RI, Braffman BH, Brorson JR, Goldberg HI, Silberberg DH, Gonzalez-Scarano F. Multiple sclerosis: serial study of Gadolinium-enhanced MR imaging. *Radiology* 1988;169:117–122.

24. Koopmans RA, Li DKB, Oger JJF, Mayo J, Paty DW. The lesion of multiple sclerosis: imaging of acute and chronic stages. *Neurology* 1989;39:959–963.

DISCUSSION

One discussant was asked what the risk was for the HLA antigen DR2 in MS. In this discussant's experience DR2 was four times more common in MS than in the general population.

Another discussant had found only atrophy of the spinal cord in TSP, and increased signal had not been found by MRI. This discussant inquired whether the Vancouver group had studied the spinal cord. It was explained that although other groups probably had, the equipment in Vancouver did not allow it.

Another speaker brought up the importance of Northern European stock in MS, a point that had been stressed during the presentation. In this speaker's opinion, the evidence indicated an increased risk as a function of geography. Another participant did not doubt that environmental factors were important and thought that both points—that there are genetic and environmental factors in MS—have to be made.

Human Retrovirology: HTLV,
edited by William A. Blattner.
Raven Press, Ltd., New York © 1990.

HTLV-I and Tropical Spastic Paraparesis in Caribbean Migrants in Britain: Clinical and Familial Studies

J. K. Cruickshank, *Jennifer Richardson, *Anne Newell,
*A. G. Dalgleish, and †P. Rudge

*Department of Epidemiology, †Department of Neurology, Northwick Park Hospital;
and *Retrovirus Research Group, Clinical Research Centre,
Harrow, HA1 3UJ, United Kingdom*

INTRODUCTION

HTLV-I has become closely associated worldwide with the chronic myelopathy currently named tropical spastic paraparesis (TSP), a condition first recognized in Jamaica (1).

This short chapter outlines the current experience of TSP in Britain among Caribbean-born patients, mainly Jamaican, who had migrated many years before their disease onset. Details of this work are available in references 2–9. To April 1989, 29 patients with TSP have been studied, with all 27 tested being positive for HTLV-I antibodies. The most recent case is the first patient (age 22) to have been born in Britain of Jamaican parents. The clinical features which define TSP in our cases are those described in the original large series from Jamaica (1).

DIAGNOSTIC STUDIES

Magnetic resonance imaging (MRI) of the central nervous system demonstrates important differences between multiple sclerosis and TSP. When images of the brain of patients with equal disability and duration of disease are compared, the abnormalities are in general much less marked in the TSP patients. Some of the TSP patients have normal MRI of the brain; the remainder typically have isolated supratentorial white matter lesions and a smooth rim of periventricular increased signal with sparing of the posterior fossa structures. The cervical cord is normal, whereas the dorsal cord is characteristically atrophied and of uniformly increased signal return. This contrasts with multiple sclerosis (MS), where focal lesions of the cerebrum, brain stem, and cervical cord are often seen, and the periventricular abnormality is irregular and prominent (10). In the dorsal cord, which is also of decreased volume, the increased signal return is less common, and, when it occurs, it is of a focal rather than a diffuse nature.

Serologic Analysis

Quantitative and qualitative aspects of the antibody response to HTLV-I were examined using a range of serological assays (5). HTLV-I antibody levels and virus neutralizing ability were considerably higher in TSP patients than in asymptomatic relatives or ATL patients. All HTLV-I sera mediated antibody-dependent cell-mediated cytotoxicity (ADCC), and there was poor correlation between overall antibody level (as determined by ELISA and particle agglutination assays) and the ability of individual sera to neutralize or mediate ADCC. In the CSF, some but not all of the IgG oligoclonal bands that are found are directed against HTLV-I (4); in two patients with recent disease onset, CSF oligoclonal IgM bands were also eluted out by HTLV-I.

Family Studies

In a study of viral prevalence among family members of HTLV-I–infected individuals, 60 of 66 first-degree relatives of the Jamaican-born patients in our series were traced in the UK and Jamaica; 20%–30% of those born in the Caribbean had antibodies to HTLV-I, irrespective of their present country of residence, whereas none of those born in the UK, who were the children of the patients and a generation younger than the Caribbean-born relatives, had antibodies (7,8). The absence of HTLV-I antibodies in the UK-born offspring (mean age 20 yr), all of whom had been breast-fed, points to sexual transmission as the major route of infection in the older Jamaican-born relatives. The mother and father of the UK-born 22-yr-old case were seropositive, and this patient had been breast-fed. An alternative explanation—that some people infected in infancy do not seroconvert until later in life—has been tested by the use of polymerase chain reaction (PCR) to look for evidence of HTLV-I DNA in peripheral blood. Viral DNA sequences could be amplified from the TSP cases in all the seropositive relatives, but PCR failed to demonstrate HTLV-I DNA in any seronegative relative. PCR and sequence analysis of DNA obtained from our TSP patients have revealed minor sequence variation in the *pol* region of HTLV-I (9) and more recently in the *env* gene (C. Bangham, in preparation).

CONCLUSION

Review of the original pathological material (it remains very scarce worldwide), together with data such as the above and animal models (particularly visna in sheep, which shares both clinical and pathological features with TSP), suggest that TSP is due to an HTLV-I–associated lymphocyte/macrophage–mediated inflammatory response in the spinal cord.

REFERENCES

1. Montgomery R, Cruickshank EK, Robertson WB, McMenemey WH. Clinical and pathological observations in Jamaican myelopathy. *Brain* 1964;87:425–462.
2. Newton M, Cruickshank JK, Miller D, et al. Antibodies to HTLV-I in West Indian-born, UK-resident patients with spastic paraparesis. *Lancet* 1987;1:415–416.
3. Lancet Editorial. HTLV-I comes of age. *Lancet* 1988;1:217–219.
4. Cruickshank JK, Rudge P, Dalgleish AG, et al. TSP and HTLV-I in the United Kingdom. *Brain* 1989;112:1051–1090.
5. Dalgleish AG, Richardson JH, Matutes E, et al. HTLV-I infection in TSP: lymphocyte culture and serological response. *AIDS Hum Retroviruses* 1988;4:475–485.
6. Sinclair A, Habeshaw JA, Muir L, Chandler P, Forster S, Cruickshank K, Dalgleish AG. Antibody-dependent cell-mediated cytotoxicity: comparison between HTLV-I and HIV-1 assays. *AIDS* 1988;2:465–472.
7. Cruickshank JK, Knight J, Morgan O, et al. Prevalence of HTLV-I antibodies in relatives of UK-resident Jamaican-born patients with TSP in Britain and Jamaica. *Clin Sci* 1988;78:31.
8. Richardson JH, Cruickshank JK, Newell AL, et al. Prevalence of HTLV-I antibodies and HTLV-I DNA among relatives of UK TSP patients. *J Cell Biochem* 1989; suppl. 13B:288.
9. Bangham CRM, Daenke S, Phillips RE, Cruickshank JK, Bell JI. Enzymic amplification of exogenous and endogenous retroviral sequences from DNA of patients with tropical spastic paraparesis. *EMBO J* 1988;7:4179–4184.
10. Ormerod IEC, Miller DH, MacDonald WI, et al. The role of NMR imaging in the assessment of multiple sclerosis and isolated neurological lesions: a quantitative study. *Brain* 1987;110: 1579–1660.

Human Retrovirology: HTLV,
edited by William A. Blattner.
Raven Press, Ltd., New York © 1990.

Peripheral Neuropathies and Myositis Associated to HTLV-I Infection in Martinique

*J.-C. Vernant, *R. Bellance, *G. G. Buisson,
*S. Havard, †J. Mikol, and ‡G. Roman

*Service de neurologie Hôpital la Meynard, 97200-Fort de France, Martinique;
†Service de neuropathologie, Hôpital Lariboisière, 2 rue Ambroise Paré,
75010 Paris, France; ‡7 Brentwood Circle, Lubbock, Texas 19406

Until 1985 the only clinical manifestations related to HTLV-I infection were hematological. Since 1985 we know that the virus may be responsible for neurological disturbances as well (1,2), and there is now sufficient clinical data to allow us to conclude that there is no direct relation between hematological and neurological diseases.

It is also well established that the picture of a spastic paraplegia is the most common neurologic expression of HTLV-I infection. Nevertheless peripheral features may also be observed. They may coexist with spastic paraplegia (3), sometimes simulating amyotrophic lateral sclerosis (ALS) (4,5), or they may be isolated and evolve as a peripheral neuropathy (5). In addition, HTLV-I–associated polymyositis has been reported (6). Thus muscular, peripheral nerve, and central neurologic disturbances (7,8) are linked to HTLV-I.

We now report clinical and neuropathological data of six patients who illustrate some of these features of HTLV-I–associated neurological disease.

CASE REPORTS

Case 1

This 75-yr-old black man was referred to the Department of Neurology in 1989. At age 25, he developed a disturbance of gait, which was diagnosed as syphilitic myelitis. His clinical symptoms progressively worsened, and in 1982 the patient became bedridden. Physical examination currently reveals a striking atrophy of the patient's hands, legs, and thighs (Fig. 1). Deep tendon reflexes are absent in his upper and lower limbs. There is no sensory disturbance, no bulbar manifestation, nor Babinski sign. Laboratory data are normal, syphilitic serology

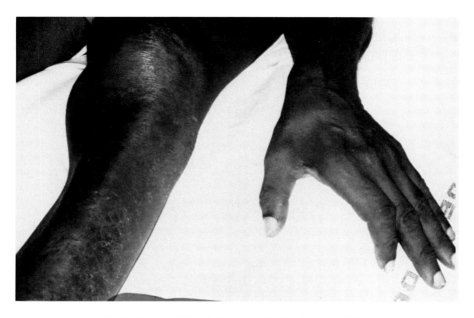

FIG. 1. Case 1: marked atrophy of the hand, leg, and thigh.

in particular is negative, but HTLV-I serology is positive (ELISA and Western Blot). A peroneal muscular biopsy shows neurogenic atrophy, suggesting anterior horn disease, and a deltoid biopsy shows mononuclear cells infiltrating interstitial tissue (Figs. 2 and 3).

Conclusion: Clinical history evoking a chronic lower motor neuron disease evolving over 50 yr. Muscular biopsy in keeping with clinical data, with additional features of a myositic process.

Case 2

A 51-yr-old black patient complained from 1979 on of weakness of his lower limbs progressing slowly to his upper limbs. In 1984, at the age of 56, he was examined for the first time in the department of neurology. Examination was remarkable for a proximal weakness of the four limbs. In fact, there was a typical waddling gait, the patient could not stand up without use of the upper limbs, and the pectoral girdle was weak and atrophic. There were mild distal sensory disturbances of the lower limbs; all tendon reflexes were normal. Serum creatine phosphokinase (CPK) activity was threefold greater than normal, and a first muscle biopsy showed inflammatory features. A diagnosis of polymyositis was established, and a corticosteroid therapy was therefore started with a good responsiveness but with corticodependence. In 1987, at the age of 59, the patient was reexamined. Deep tendon reflexes were abolished at the lower limbs; serum CPK activity was still elevated. HTLV-I serology performed for the first time was positive (ELISA and Western Blot). A second muscle biopsy showed lymphoplasmocytoid interstitial infiltrates consistent with an inflammatory muscular process. Beside these lesions, there was atrophy or disappearance of clustered muscular fibers characteristic of a neurogenic process.

Conclusion: Clinical and neuropathological features, characteristic both of intersitial myositis and peripheral neuropathy in an HTLV-I seropositive patient.

Case 3

A 29-yr-old black woman complained in 1972 of slowly progressive weakness of the four limbs. She was examined for the first time in the Department of Neurology in 1976. One clinician noticed at the time: "obvious clinical picture of muscular disease with waddling gait. The patient can't walk up stairs; there are no sensory disturbances, no Babinski sign, and deep tendon reflexes are normal." Serum CPK activity was tenfold greater than normal value, sedimentation rate was 40 mm at first hour, and there was a positive Latex agglutination (Rheumatoid Factor) test. The patient wasn't seen again until 1979, when she was 36 yr of age. For 2 yr she hadn't been able to walk down stairs, to stand up alone, or to raise her arms. She complained of muscular pains. On examination, there was prominent atrophy and weakness of the four limbs and pectoral girdle. An electromyogram showed diffuse peripheral-type abnormalities. A muscular biopsy was performed with the following results: "At the quadriceps level there are mainly myositic- and necrotizing-type lesions, but the fascicular distribution is that of a neurogenic process. At the peroneal level there is a typical neurogenic atrophy evoking lesions of the medullary anterior horn. I reexamined the biopsies with Professor Fardeau; he thinks that clinical data alone can determine the diagnosis. There is a striking neurogenic atrophy with inflammatory lesions around capillaries. I hope that future data will help us to understand this difficult problem " (Prof. J. Mikol) (Figs. 4 and 5).

This woman was seen last in November 1988. She was completely bedridden; there was a striking diffuse atrophy. She could no longer move the lower limbs. Some movements of flexion-extension could be performed with the forearms and fingers and at the level of the upper arms. The muscles of the face were intact, there was no bulbar abnormality, and the neuropsychiatric evaluation was normal. HTLV-I serology was performed and was strongly positive (ELISA and Western Blot).

Conclusion: Chronic disease with clinical and anatomical features of polymyositis coexisting with clinical, electromyographical, and anatomical features of a peripheral neuropathy in an HTLV-I–positive patient.

(text continues on page 233)

FIG. 2. Case 2: mononuclear cell infiltrates with perivascular distribution in the deltoid muscle.

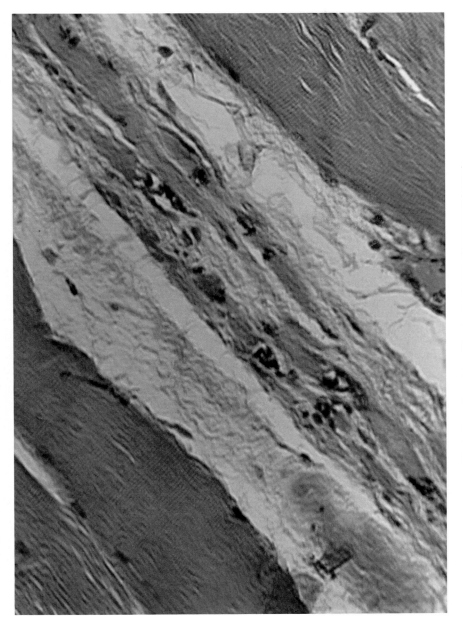

FIG. 3. Case 2: peripheral fascicular atrophy of the peroneal muscle.

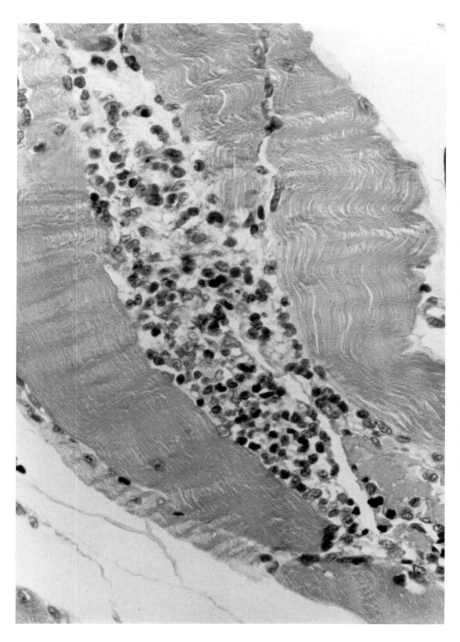

FIG. 4. Case 3: mononuclear cellular infiltrates in the quadriceps.

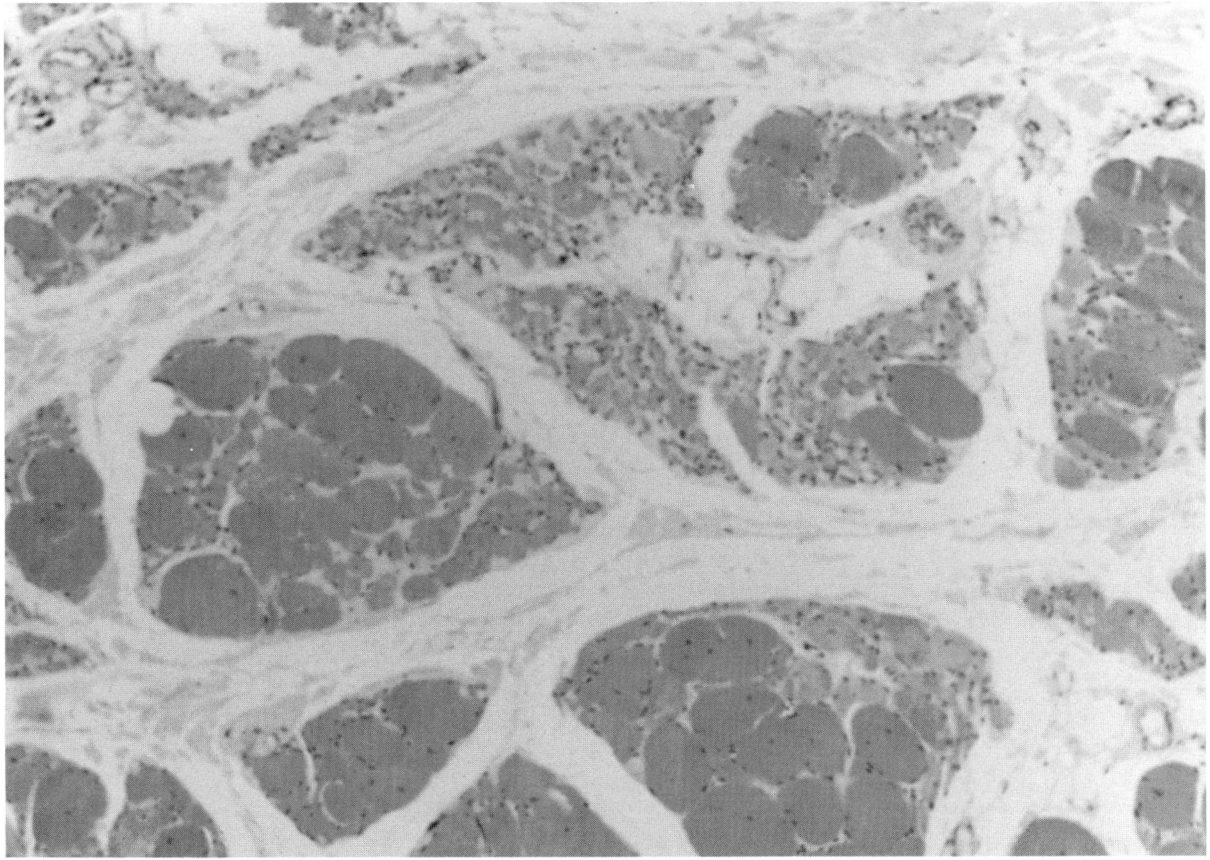

FIG. 5. Case 3: typical peripheral neurogenic atrophy evoking chronic anterior horn disease.

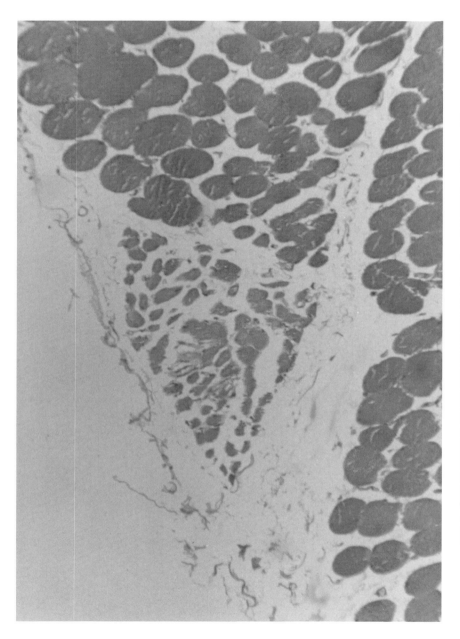

FIG. 6. Case 5: peroneal atrophy evoking a peripheral process by a HAM/TSP evolving over 32 yr.

Case 4

A 48-yr-old black man complained of muscular weakness and of pains of the four limbs. He had lost 10 kg in the course of 1984. One year later circular, squamous, cutaneous lesions of the trunk and upper limbs appeared. Clinical examination was remarkable because of proximal weakness of the four limbs and musculature was painful to pressure. The deep tendon reflexes were normal and there were no sensory disturbances nor Babinski sign, but he complained of a dryness of the mouth and irritation of the eyes. Sedimentation rate was 40 mm, serum CPK activity was tenfold greater than normal, latex test positive, and HTLV-I serology was positive. Schirmer's test showed a sicca syndrome, and biopsy of the accessory salivary glands histologically consistent with a Gougerot Sjögren syndrome stage 1. A muscular biopsy showed lymphoplasmocytoid interstitial infiltrates mainly around small vessels with features of neurogenic atrophy.

Corticosteroid therapy was started and led to dramatic improvement. Within 2 wk the patient became free of any symptoms, and by 3 mo he returned to his normal previous weight. On reexamination of the patient in March 1988, tendon reflexes and bilateral Babinski sign were now present.

Conclusion: Clinical, biological, and anatomical features of dermatomyositis, anatomical features of peripheral neuropathy, and secondary appearance of a pyramidal syndrome.

Case 5

In 1957 a 45-yr-old black man developed progressive paraplegia and remained paraplegic from that time on. In 1987, when he was 75 yr old, the diagnosis of typical HTLV-I-associated myelopathy was established. A peroneal muscular biopsy was performed because the patient showed atrophy of the left leg and hand. The biopsy showed clustered atrophy characteristic of a peripheral neuropathy (Fig. 6).

Conclusion: clinical and anatomical features of peripheral neuropathy in the course of a typical HAM/TSP.

Case 6

A 53-yr-old black man developed weakness of the lower limbs over a period of 4 mo. On neurologic examination, he exhibited a total paralysis of the left lower limb, where only static contractions were possible. The strength of the right lower limb was less severely impaired. There was striking atrophy of the left thigh and leg. Deep tendon reflexes were abolished in the left lower limb but were normal elsewhere. The patient complained of paresthesias of both legs, where we noticed some sensory disturbances. There was no Babinski sign. A myelography was performed and was normal. In the CSF there was hyperproteinemia (2.3 g/L) and an intrathecal synthesis of immunoglobulins. The HTLV-I serology was positive (ELISA and Western Blot).

Conclusion: a peroneal muscular biopsy showed features of neurogenic atrophy coexisting with signs of interstitial myositis.

DISCUSSION

The six patients presented in this report demonstrate that some cases with typical "HTLV-I–associated myelopathy/tropical spastic paraplegia (HAM/TSP)" (case 5) or, independent of this syndrome (cases 2–4, 6), have peripheral neuropathy or polymyositis. Furthermore, as seen for case 1, HTLV-I–associated paraplegia is readily distinguishable from typical HAM/TSP in the sense that the syndrome most resembles chronic anterior horn disease. It is also striking to notice that when the clinical picture is that of a polymyositis, the biopsy always shows features of peripheral neuropathy coexisting with those of the polymyositis. For the latter, the most constant anatomical feature is that of perivascular mononuclear cellular infiltrates in interstitial areas, whereas necrotic lesions are absent contrary to classic polymyositis.

Our cases are closely similar to those reported by Goudreau et al. (7) and Tarras et al. (8) and demonstrate that, when the HTLV-I infection expresses itself, it leads to anatomical lesions stretching over the whole nervous system, even if the spastic paraplegia remains the most frequent expression. Furthermore we have reported elsewhere that more than two-thirds of our patients suffering from HAM/TSP have features of a latent lymphocytic alveolitis (9). All these data confirm our concept that the spastic paraplegia associated to HTLV-I is but one expression of a generalized disease which may be linked to the antigenicity of the virus. Thus HAM/TSP may be more frequently recognized because of the nature of clinical signs while the broader spectrum of features may be as or more frequent but less frequently diagnosed.

REFERENCES

1. Gessain A, Barin F, Vernant J-C, et al. Antibodies to human T-lymphotropic virus type-I in patients with tropical spastic paraparesis. *Lancet* 1985;2:407–411.
2. Rodgers-Johnson P, Gajdusek DC, Morgan OStC, et al. HTLV-I and HTLV-III antibodies and tropical spastic paraparesis. *Lancet* 1985;2:1247–1248.
3. Saïd G, Goulon-Goeau C, Lacroix C, Fève A, Descamps H, Fouchard M. Inflammatory lesions of peripheral nerve in a patient with human T-lymphotropic virus type I-associated myelopathy. *Ann Neurol* 1988;24:275–277.
4. Arimura K, Nakashima H, Matsumoto W, et al. HTLV-I associated myelopathy (HAM) presenting with ALS-like features. In: Roman G, Vernant J-C, Osame M (eds). *HTLV-I and the nervous system. Neurology and neurobiology,* vol. 51, New York: Alan R. Liss, 1989;367–370.
5. Vernant J-C, Buisson G, Bellance R, François MA, Madkaud O, Zavaro O. Pseudo-amyotrophic lateral sclerosis, peripheral neuropathy and chronic polyradiculoneuritis in patients with HTLV-

I associated paraplegias. In: Roman G, Vernant J-C, Osame M (eds). *HTLV-I and the nervous system. Neurology and neurobiology,* vol. 51. New York: Alan R. Liss, 1989;361–365.

6. Mora CA, Garruto RM, Brown P, et al. Seroprevalence of antibodies to HTLV-I in patients with chronic neurological disorders other than tropical spastic paraparesis. *Ann Neurol* 1988;23(suppl): 192–195.

7. Goudreau G, Karpati S, Carpenter S. Inflammatory myopathy in association with chronic myelopathy in HTLV-I seropositive patients. *Neurology* 1988;38(suppl 1):206.

8. Tarras S, Sheramata WA, Snodgrass S, Ayyar DR. Polymyositis and chronic myelopathy associated with presence of serum and cerebrospinal fluid antibody to HTLV-I. In: Roman G, Vernant J-C, Osame M (eds). *HTLV-I and the nervous system. Neurology and neurobiology,* vol. 51. New York: Alan R. Liss, 1989;435–441.

9. Couderc LJ, Caubarrère I, Venet A, et al. Bronchoalveolar lymphocytosis in patients with tropical spastic paraparesis associated with human T-cell lymphotropic virus type 1 (HTLV-I). *Ann Intern Med* 1988;109:625–628.

Human Retrovirology: HTLV,
edited by William A. Blattner.
Raven Press, Ltd., New York 1990.

The Clinical Spectrum of HTLV-I Infection in Trinidad and Tobago

*Courtenay Bartholomew, †Farley Cleghorn,
and ‡William Blattner

*The Department of Medicine, University of the West Indies, General Hospital, Port of Spain, Trinidad; †The Caribbean Epidemiology Centre, Port of Spain, Trinidad; and ‡The Viral Epidemiology Section, National Cancer Institute, Bethesda, Maryland

Trinidad and Tobago are the two southernmost islands in the Caribbean basin with a population of 1.2 million comprising people of African origin (41%), people of Asian origin from India (41%), people of mixed race (16%), Caucasians (1%), and Chinese (1%). The people of African descent came to Trinidad mainly from the west coast of Africa via the Portuguese slave trade from 1680 onward, whereas those of Indian origin came after the abolition of slavery as indentured laborers, beginning in the year 1845. There have not been any reports of studies of HTLV-I seroprevalence from India; however, recent studies of the seroprevalence of HTLV-I antibodies in the west coast of Africa have shown a seroprevalence of 4% in Ghana (1) and survey samples in the Gabon ranged from 5% in urban areas to 12% in rural areas (southern Cameroon, Equatorial Guinea, Gabon) (2). This higher rural seroprevalence was also observed in studies in Jamaica (3).

In Trinidad, out of 1578 people randomly tested for HTLV-I antibodies, 37 were positive as determined by an enzyme-linked immunosorbent assay (ELISA) and confirmed by competition assay (4,5). Of those who tested positive, 16 were males and 21 were females. Thirty-one of 807 (3.9%) Trinidadians of African ancestry and 5 of 218 (2.4%) people of mixed African ancestry were positive, whereas only 1 out of 148 (0.2%) people of Indian descent was positive. In this latter respect, although Trinidad is a very cosmopolitan island, intermarriage between people of Indian and African ancestry is still relatively uncommon. This restriction of the virus almost exclusively to the African population lends support to the hypothesis of Gallo et al. that HTLV-I came to the Caribbean via the African slave trade (6).

The fact that, after three centuries of HTLV-I presence in this island community, only 4% of Afro-Trinidadians are seropositive bespeaks the low infectivity of this virus. The same can be said of the low seroprevalence rates in west Africa. Still to be explained, however, is the relative paucity of reports of HTLV-I–associated clinical diseases in west Africans.

Studies of the neighboring eastern Caribbean islands of Barbados and Grenada have shown similar seroprevalence rates. In Barbados a rate of 4.2% was observed and in Grenada, 5.6% (7,8). On the other hand, a study of a small rural village in the island of Tobago gave a surprisingly high seroprevalence of 12%, a figure that, coincidentally, is similar to the rural seroprevalence rate observed in west Africa (2).

ADULT T-CELL LEUKEMIA IN TRINIDAD AND TOBAGO

After the recognition by Catovsky et al. (9) that six Afro-Caribbean people living in the United Kingdom were found to have adult T-cell leukemia (ATL) and were HTLV-I–antibody positive, the first case of ATL associated with HTLV-I seropositivity was diagnosed in Trinidad in 1982. Since then, in a prospective study of hematologic malignancies in Trinidad and Tobago from October 1, 1985, to June 30, 1989, 48 cases of ATL have been diagnosed. In concert with the Afro-Trinidadian reservoir of infection as identified by the serosurvey, all the cases of ATL have so far been found only in people of African origin. The details of these cases have been reported by Cleghorn et al.

TROPICAL SPASTIC PARAPARESIS IN TRINIDAD AND TOBAGO

In a series of 100 neurological patients seen in a 3-yr period at the University of the West Indies in Jamaica, Cruickshank described a clinical entity which he called "Jamaican Neuropathy," of whom 80% had a predominantly spastic syndrome. The others presented with a mixture of spasticity and ataxia (10). A nutritional, infectious, and toxic cause was sought to explain the syndrome. In fact, in their paper entitled "Clinical and Pathological Observations on Jamaican Neuropathy" published in 1964, Montgomery and Cruickshank stated that "The possibility of a viral infection playing a part in the Jamaican neuropathic syndrome has been considered but laboratory investigations have given no support to this idea" (11). Likewise, in her 1965 report on "The Clinical Features and Aetiology of the Neuropathic Syndrome in Jamaica," Rodgers stated that "viral infections must also be considered, particularly as it is almost impossible to rule them out at this stage. They may play a part in this syndrome, or in some patients may be the sole cause. . . ." (12).

However, it was not until 1985, 30 yr later, that Gessain et al. showed evidence suggesting that the retrovirus HTLV-I was etiologically linked to cases of tropical spastic paraparesis (TSP) in Martinique (13). After this historic recognition, the first cases of TSP in association with HTLV-I were recognized in Trinidad in 1986, and, to date, 32 such cases have been diagnosed. Also in 1986 Osame et al. (14) reported cases of HTLV-I–associated myelopathy (HAM) in Japan, and— although initial differences were noted between the myelopathies in the sub-tropical Japanese archipelago and those in the tropical Caribbean basin—with studies of additional cases, discrepancies have narrowed and now HAM and TSP are considered to be the same disorder (15).

CONCOMITANT ATL AND TSP

Although it has also been demonstrated that the viruses isolated from ATL and HAM have identical genomic compositions (16), cases of patients suffering from both ATL and HAM/TSP have been very rare. In this respect, there is a similarity to the murine leukemia virus, in that this type C retrovirus has been shown to cause both a lymphoma and a hind-leg paralysis in wild mice (*Mus musculatum*) in California. However, in one study, only 2% of these infected mice had both paralysis and lymphoma (17).

Although ATL was first described as a specific clinical entity in 1977 by Takatsuki et al. (18), the first report of a case of ATL and TSP in the same patient was by Bartholomew et al. in 1986 (19). This was in a 49-yr-old man in Trinidad who was HTLV-I–antibody positive in both serum and CSF and who developed ATL 16 yr after the onset of his spastic paraparesis. Since then two other cases of concomitant ATL and TSP have been recognized in Trinidad and Tobago. In one case the patient developed a spastic paraparesis one month after being treated with immunosuppressive therapy (MOPP) for what was first thought to be Hodgkin's disease but which on later histological review was considered to be ATL. After an unusual remission of close to 2 yr, the patient developed an acute aggressive relapse of his lymphoma and died. The third patient had TSP of 4 yr duration before developing a smoldering T-cell lymphoma with a generalized dermatitis in which ATL-like cells were shown on histological examination of the skin. This patient is still alive.

FAMILIAL CASES OF ATL

We have only seen one instance of ATL occurring in two brothers. One brother, age 31, had a classical acute aggressive adult T-cell leukemia and died on March 26, 1984. The other sibling, age 28, presented with a smoldering type of ATL initially misdiagnosed as mycosis fungoides in 1982; 2 yr later there was a transformation of his clinical picture to the classical aggressive ATL. He died on March 5, 1984, 3 wk before his older brother.

FAMILIAL CASES OF TSP

In addition to familial occurrence of ATL, familial cases of HAM have been recently reported in Japan (20). We now report here for the first time familial occurrence of HTLV-I–associated TSP in Trinidad. In this family of six, the father (age 51), mother (age 48), a son (age 22), and a daughter (age 19) were all seropositive for HTLV-I antibodies. Two other daughters (ages 24 and 8) were seronegative. The seropositive mother, son, and daughter all had spastic paraparesis (Fig. 1).

HTLV-I–ASSOCIATED FACIAL NERVE PALSY

We first suspected that HTLV-I might be neurotropic in 1984 when we saw two patients with ATL, a 24-yr-old man and a 30-yr-old woman who also had unexplained unilateral facial nerve palsies. One of these patients also complained of pain and paresthesiae in the fingers and feet. Yet another patient with aggressive ATL had bilateral facial nerve palsies but was also seropositive for HIV as well as for HTLV-I (19). Since then several cases of HTLV-I–associated facial nerve palsies have been reported from Japan; thus, in addition to the spastic syndrome caused by pyramidal tract lesions in the spinal cord, HTLV-I is also associated with peripheral neuropathies, either through a direct neurovirulent effect or— much more likely—via an indirect autoimmune phenomenon.

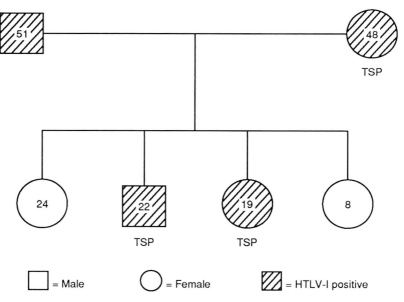

FIG. 1. Family tree of members showing positive and negative HTLV-I antibody and Tropical Spastic Paraparesis.

HTLV-I–ASSOCIATED GUILLAIN-BARRÉ SYNDROME

More recently we have recognized three cases of the Guillain-Barré syndrome associated with HTLV-I antibodies in the serum. This syndrome, which was first reported by Guillain, Barré, and Strohl from Paris in 1916, is an acute inflammatory demyelinating polyradiculoneuropathy characterized by muscular weakness and mild sensory loss and usually following a banal infection.

The syndrome has been associated with the cytomegalovirus, the Epstein-Barr virus, and smallpox vaccinia and is probably also associated with measles, varicella-zoster, hepatitis, and mumps. In this respect, it is noteworthy that all viruses linked to the Guillain-Barré syndrome are—at least to a limited extent—neurotropic and are enveloped viruses like HTLV-I (22). There is considerable evidence to suggest that this syndrome represents an aberrant immune response manifested by segmental demyelination and mononuclear cellular infiltration of the peripheral nerves.

Characteristically, it was described as "un syndrome de radiculo-névrite avec hyperalbuminose du liquid cephalo-rachidien sans réaction cellulaire." This albumin-cytologic dissociation or the absence of a pleocytosis in the cerebrospinal fluid (CSF) in association with a very high CSF protein is a characteristic finding in this syndrome. The clinical signs and symptoms of these three cases are listed in Table 1. It will be observed that bilateral facial palsies were found in two of the cases.

A cell-free CSF was seen in all three cases, but in one case the CSF protein was normal. This CSF tap was done on the fourth day of symptoms, and it is well known that the protein rise may not occur until after several days of illness— peak protein values occurring ~4–6 wk after the onset of clinical symptoms. Significantly, in contrast to most cases with HTLV-I–associated TSP, HTLV-I antibodies were not found in the CSF in these three patients. This is probably

TABLE 1. *HTLV-I–associated Guillain-Barré syndrome*

Clinical signs/symptoms	Case 1 (JJ)	Case 2 (DM)	Case 3 (DB)
Low-grade fever	−	+	+
Flaccid paresis/upper limbs	+	+	+
Flaccid paresis/lower limbs	+	+	+
Hyporeflexia/upper limbs	+	+	+
Hyporeflexia/lower limbs	+	+	+
Peripheral sensory deficit	+	+	+
Numbness/paresthesiae	+	+	+
Dysphagia	−	+	+
Dyspnea	−	+	+
Confusion	−	−	+
Bilateral facial palsy	−	+	+

not an unexpected finding because the Guillain-Barré syndrome is believed to be an autoimmune reaction to a virus directed against peripheral nerves and not against the spinal cord. In addition, unlike cases of TSP, when high HTLV-I titers are usually found, these three cases of HTLV-I–associated Guillain-Barré syndrome had titers of 1806, 2838, and 2706, respectively with a mean of 2450. In our series of cases of TSP the mean titer of HTLV-I antibodies has been 25 129.

Light microscopy of sural nerve biopsies on two patients showed mild to moderate reduction of myelin and axons in the one and focal mononuclear inflammatory infiltrates in the other. These two findings were consistent with the Guillain-Barré syndrome; however, electron microscopic studies were more discerning, and, in the former biopsy, electron microscopy showed that there was a reduction of myelin and axons. The latter biopsy showed mononuclear cell infiltrates, mainly lymphocytes, in the endoneurium with a perivascular inflammatory cuff also composed of mononuclear cells. There was no significant myelin or axonal loss. However, this sural nerve biopsy was taken 4 yr after the diagnosis of the Guillain-Barré syndrome was made and remyelination could have occurred. On the other hand, the mononuclear inflammatory process in the peripheral nervous system has been shown to persist in low-grade fashion for months and even years after clinical recovery (23).

These findings suggest that the Guillain-Barré syndrome should be added to the list of peripheral nerve diseases linked to HTLV-I.

ACKNOWLEDGMENTS

We wish to thank Dr. Clarence Gibbs Jr., National Institute of Neurological Diseases and Stroke, Bethesda, Maryland, for electron microscopy studies on the sural nerve biopsies and Mrs. Anne Blache-Fraser for secretarial assistance.

REFERENCES

1. Biggar RJ, Saxinger C, Gardiner C, et al. Type-1 HTLV antibody in urban and rural Ghana, West Africa. *Int J Cancer* 1984;34:215–219.
2. Delaporte E, Peeters M, Durand JP, et al. Infection by HTLV-I among populations of four countries in Western Central Africa. Abstract presented at III International Conference on AIDS and Associated Cancers in Africa. September 1988.
3. Murphy EL, Blattner WA. HTLV-I associated leukaemia: a model for chronic retroviral disease. *Am J Neurol* 1988;23(Suppl):S174.
4. Saxinger C, Gallo RC. Application of the indirect enzyme linked immunosorbent assay microtest to the detection and surveillance of human T-cell leukaemia/lymphoma virus. *Lab Invest* 1983;49: 371–377.
5. Saxinger C, Blattner WA, Levine T, et al. HTLV-I antibodies in Africa. *Science* 1984;225:1473–1476.
6. Gallo RC, Sliski A, Wong-Staal F. Origin of human T-cell leukaemia lymphoma virus. *Lancet* 1983;2:963.
7. Reidel DA, Evans AS, Saxinger C, Blattner WA. A historical study of human T lymphotropic virus type 1 transmission in Barbados. *J Infect Dis* 1989;159(4):603–609.

8. Bartholomew C, Hull B, Cleghorn F. Seroprevalence of HTLV-I in Grenada, West Indies. Unpublished data, 1988.
9. Catovsky D, Greaves MR, Rose M, et al. Adult T-cell lymphoma-leukaemia in blacks from the West Indies. *Lancet* 1982;1:639–643.
10. Cruickshank EK. A neuropathic syndrome of uncertain origin. *West Indian Med J* 1956;5:147.
11. Montgomery RD, Cruickshank EK. Clinical and pathological observations on Jamaican neuropathy. A report on 206 cases. *Brain* 1964;87:425.
12. Rodgers PEB. The clinical features and aetiology of the neuropathic syndrome in Jamaica. *West Indian Med J* 1965;14:36.
13. Gessain A, Barin F, Vernant JC, et al. Antibodies to human T lymphotropic virus type 1 in patients with tropical spastic paraparesis. *Lancet* 1985;2:407.
14. Osame M, Osuku K, Izumo F. HTLV-I associated myelopathy—A new clinical entity. *Lancet* 1986;1:1031.
15. Roman GC, Osame M. Identity of HTLV-I associated tropical spastic paraparesis and HTLV-I associated myelopathy. *Lancet* 1988;1:651.
16. Yoshida M, Osame M, Osuku K, et al. Viruses detected in HTLV-I associated myelopathy and adult T cell leukaemia are identical on DNA blotting. *Lancet* 1987;1:1085.
17. Gardner MB, Henderson BE, Officer JE, et al. A spontaneous lower motor neuron disease apparently caused by indigenous Type C RNA virus in wild mice. *J Natl Cancer Inst* 1973;51:1243.
18. Takatsuki K, Uchiyama T, Sagawa K, Yodoi J. Adult T-cell leukaemia in Japan. In: Seno S, Takaku F, Irino S, eds. *Topics in haematology.* Amsterdam: Excerpta Medica, 1977:73–77.
19. Bartholomew C, Cleghorn F, Charles W, et al. HTLV-I and tropical spastic paraparesis. *Lancet* 1986;2:99.
20. Shoji H, Kuwasaki N, Natori H, Kaji M. Familial occurrence of HTLV-I associated myelopathy and adult T-cell leukaemia. In: Roman G, Vernant J, Osame M, eds. *HTLV-I and the nervous system.* New York: Alan R. Liss, 1988:307–309.
21. Guillain G, Barré JA, Strohl A. Sur un syndrome de radiculonevrite avec hyperalbuminose du liquide cephalo-rachidien sans reaction cellulare. Remarques sur les caracters cliniques et graphiques des reflexes tendineux. *Bull Soc Med Hop Paris* 1916;40:1462.
22. Arnason BGW. Acute inflammatory demyelinating polyradiculoneuropathy. In: Dyck PJ, Thomas PK, Lambert EH, Bunge R, eds. *Peripheral neuropathy.* Philadelphia, W. B. Saunders, 1984;2050–2100.
23. Asbury AK, Arnason BG, Adams RD. The inflammatory lesion in idiopathic polyneuritis. *Medicine* (Baltimore) 1969;48:173.

DISCUSSION

A discussant asked about the definition of ATL used in surveillance in Trinidad, suggesting that the definitions used here and elsewhere might be helpful in determining the best definition to use for surveillance of ATL in the United States. Another discussant indicated that a combination of clinical, serological, and immunological parameters is used to define ATL in Trinidad. A participant asked whether the cases of Guillain-Barré syndrome (GBS) associated with HTLV-I infection reported by the previous discussant are typical of GBS in terms of resolution of neurologic symptoms. It was explained that one of the patients recovered quickly, the second has shown gradual improvement over a period of several weeks and the third has shown little improvement over two to three months. These clinical courses are typical of GBS. Another participant commented that these GBS cases would probably fit Asbury's criteria for the diagnosis of GBS but suggested the need for additional investigation of the neuropathology in such cases of GBS. Apparently what is available is consistent with GBS, but the association of the syndrome with TSP raises the question of whether the pathology is in the central or the peripheral nervous system. One discussant commented that GBS is quite common following sero-conversion to HIV. It was stated that GBS can also be seen later in the course of HIV infection, and it should not be a surprise that the syndrome is seen in association with HTLV-I as well. A discussant questioned whether Borrelia infection and/or Lyme disease have been investigated in association with the facial paralysis reported and was informed that this possibility has not been investigated.

Human Retrovirology: HTLV,
edited by William A. Blattner.
Raven Press, Ltd., New York © 1990.

HTLV-I, Tropical Spastic Paraparesis, and Other Neurological Diseases in South India

*A. G. Dalgleish, *J. H. Richardson, †P. K. Newman,
*A. L. Newell, ‡G. Rangan, and ‡K. S. Mani

*MRC Clinical Research Centre, Harrow, Middlesex; †Department of Neurology,
General Hospital, Middlesbrough, Cleveland, UK; and ‡Neurological Clinical,
1 Old Veterinary Hospital Road, Basavanagudi, Bangalore, India

INTRODUCTION

HTLV-associated myelopathy/tropical spastic paraparesis (TSP) (HAM/TSP) has been recognized as a rare form of spastic paraplegia in many tropical countries following the early descriptions from Jamaica (1,2) and South India (3). The manifestations may show some variation, but typically patients have a slowly progressive spastic paraparesis with spasticity more prominent than weakness, often with back pain and cramps and frequently having sensory features, particularly paraesthesia, which may be of lesser clinical significance. Optic atrophy and evidence of cerebral involvement are uncommon, but bladder dysfunction and male impotence often develop as the disease progresses. It is difficult to be confident of the diagnosis until myelography has excluded spinal cord compression. Clinical features may overlap with multiple sclerosis (MS) in some cases.

A seroepidemiological survey in Martinique revealed an association between HAM/TSP and antibodies to the HTLV-I virus (4), an association which has since been confirmed in many other countries (5–9), including Japan (10). The present study has investigated the hypothesis that HTLV-I or similar viruses may be associated with HAM/TSP and other neurological diseases in South India. A group of patients who had previously been seen in a private neurological practice and diagnosed by two of us (KSM and GR) as having definite or possible TSP and MS were reviewed by KSM, GR, and PKN. Details of these patients are given in Table 1. Blood was obtained from the 14 patients and 10 household contact relatives for serological analysis and DNA extraction.

SEROLOGICAL RESULTS: THE IMPORTANCE OF CONFIRMATORY ASSAYS

In primary antibody screens using immunofluorescence (IF) and DuPont ELISA assays (11), HTLV-I reactivity was observed in six patients and four

TABLE 1. *Clinical and serological details*

Patient	Age	Sex	Years since onset	Diagnosis	IF	ELISA	WB
1	76	M	30	TSP (fever at onset, slowly progressive painful myelopathy).	+ (W)	+ (1/20)	–
2	35	F	5	TSP (slowly progressive painful myelopathy with some improvement following steroid therapy).	+ (W)	–	–
3	35	M	8	TSP (slowly progressive painful myelopathy but still walking).	–	–	NT
4	65	M	5	TSP (initially progressive painful myelopathy, now static, still walking).	+ (W)	–	–
5	58	F	15	? TSP (progressive painful myelopathy with nystagmus and optic atrophy).	+ (W)	+ (1/20)	–
6	73	M	3	? TSP (progressive painful myelopathy in an elderly hypertensive).	–	–	NT
7	39	F	13	? MS (seizures and L. hemi, then quadriparesis and optic atrophy; nonprogressive).	+ (W)	+ (1/2000)	+
8	46	F	6	? MS (acute transverse myelitis, later seizures, not progressive).	–	–	NT
9	46	M	10	? MS (nonprogressive myelopathy, nystagmus, and diplopia; also thyrotoxicosis).	–	–	NT
10	48	F	6	MS (relapsing and remitting myelopathy).	–	–	NT
11	37	F	5	MS (myelopathy with ataxia and visual impairment).	–	–	–
12	55	M	20	MS (myelopathy, cerebellar signs, and visual impairment).	–	–	–
13	28	F	2	MS (relapsing and remitting myelopathy with fits).	–	–	–
14	29	F	6	MS (relentlessly progressive myelopathy with nystagmus and dysarthria).	–	+ (1/20)	–

W, weak; NT, not tested.

controls (Tables 1 and 2). Only one of these reactive sera could be confirmed by Western blot analysis (Fig. 1). The remaining nine sera showing weak activity in IF or ELISA assays were tested for antibodies to HTLV-II. None reacted against the HTLV-II–producing cell line Mo on Western blotting, and all were negative in an HTLV-II syncytial inhibition assay (12). Sera from patients 7, 8, 9, and 14 displayed extremely weak (VSV) HTLV-II pseudotype neutralizing activity (13), but none were clearly positive for HTLV-II antibodies. Gene amplification (PCR) confirmed the presence of HTLV-I DNA in the peripheral blood of patient 7. The HTLV-I antibody titer in this patient was considerably lower than that in a control Caribbean TSP patient (not shown).

The high frequency of false positive reactions seen in the HTLV-I IF and ELISA assays underscores the necessity of performing confirmatory assays and also the need for better primary screening assays, perhaps using recombinant HTLV-I antigens. HTLV-I positive results have been reported in up to 37% of African sera (14) although only 1%–6% give true positive results using specific confirmatory assays (15,16). A similarly high frequency of false positive results has been reported in New Guinea, where up to 26% of the screened population were positive for HTLV-I antibodies using ELISA or particle agglutination kits (17,18). Few of these results could be confirmed by Western blot (19,20), a fact consistent with the apparent absence of TSP and the T-cell malignancies associated with HTLV-I in New Guinea. It is important, however, not to disregard weakly positive sera as they may represent cross-reaction with other, so far unidentified, infectious agents.

NON-HTLV-I–ASSOCIATED TSP IN SOUTH INDIA

The clinical and neurological features of South Indian spastic paraplegia are identical to the HAM/TSP reported from the Caribbean and Japan (2,3,10). Most cases of HAM/TSP tested to date have antibodies to HTLV-I, and, in one

TABLE 2. *Serological status of relatives of neurological patients*

Subject	Relationship to patient	IF	ELISA	WB
1	Son of 1	−	−	NT
2	Husband of 11	+ (W)	+ (1/50)	−
3	Son of 12	−	−	−
4	Husband of 2	−	−	NT
5	Brother of 2	−	−	NT
6	Husband of 5	−	−	NT
7	Daughter of 5	−	+ (1/20)	−
8	Wife of 4	−	+ (1/50)	−
9	Son-in-law of 6	−	+ (1/20)	−
10	Daughter of 6	−	−	−

NT, not tested; W, weak.

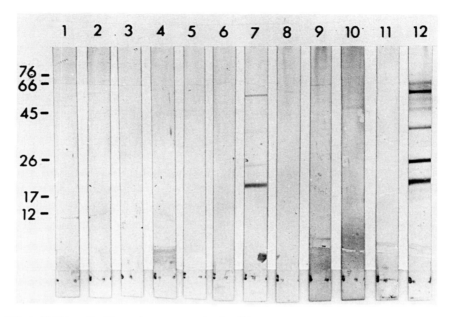

FIG. 1. ELISA and/or IF reactive sera examined on Western blot. Lane 1, patient 1; lane 2, relative 2; lane 3, patient 2; lane 4, patient 13; lane 5, relative 7; lane 6, patient 5; lane 7, patient 7; lane 8, relative 4; lane 9, relative 8; lane 10, patient 14; lane 11, patient 13; lane 12, TSP patient. The control (TSP) serum was used at 1/200 dilution and all other sera at 1/20.

series of patients seen in the UK, this association has been absolute (5,21). In the small series from South India described here, none of the TSP cases were clearly HTLV-I positive. A similar negative finding has been reported in Bombay (22). The absence of HTLV-I in four otherwise classical TSP patients suggests that agents other than HTLV-I may be etiologically associated with this disease.

The single patient in this series with positive serology did not have the typical features of HAM/TSP. HTLV-I antibody titers are extremely high in HTLV-I–associated HAM/TSP (10,11), whereas in this patient the titer was comparatively low and the PCR signal weak. The clinical presentation was with focal motor seizures and coma, which was followed by hemiparesis and dysarthria as improvement occurred. The next year bulbar symptoms and quadriparesis developed but slowly recovered following steroid treatment. Occasional seizures with reactive depression continued, but at review thirteen years after onset the patient was fully independent, with urgency of micturition and occasional incontinence. She was euphoric, had pallor of the optic discs, mild spastic quadriparesis with extensor plantar responses, and loss of vibration sense at the ankles. The diagnosis is probable multiple sclerosis. However, there are unusual features, and this subacute encephalomyelopathy may have an infective etiology. The association with HTLV-I in this case must remain speculative. It is interesting that this patient is a member of the Brahmin community originating from the west coast of India, who are fair with blue eyes and are thought to have Portuguese ancestry.

This small study suggests that HTLV-I does exist in India but not in the patient group in which it has been identified elsewhere. The clinical similarity between South Indian spastic paraplegia and TSP from other parts of the world appears not to be matched by serological correlation, and if the South Indian variety has an infective etiology, then other agents must be sought. Further studies in India are needed to define the local epidemiology of HTLV-I and to examine specific clinical entities, particularly those patients with unexplained encephalo-myelopathy.

SUMMARY

Tropical spastic paraparesis (TSP), indistinguishable from that described else-where, is prevalent in South India. In contrast to TSP patients from other coun-tries, HTLV-I antibodies were not detectable in the South Indian patients. Mul-tiple sclerosis patients were also negative, but HTLV-I was identified in a patient with a subacute encephalomyelopathy without features of TSP. Weak, uncon-firmable HTLV seroreactivity was found in some patients and some controls.

ACKNOWLEDGMENTS

Thanks are due to Dr. P. Clapham and Professor Robin Weiss of the Chester Beatty Laboratories for the HTLV-II antibody testing, and Dr. R. C. Gallo for the C91/PL and Mo cell lines.

REFERENCES

1. Cruickshank EK. A neuropathic syndrome of uncertain origin—review of 100 cases. *West Ind Med J* 1956;5:147–158.
2. Montgomery RD, Cruickshank EK, Robertson WB, McMenemey WH. Clinical and pathological observations in Jamaican myelopathy. *Brain* 1964;87:425–462.
3. Mani KS, Mani AJ, Montgomery RD. A spastic paraplegic syndrome in South India. *J Neur Sci* 1969;9:179–199.
4. Gessain A, Barliu F, Vernant TC, et al. Antibodies to HTLV-I in patients with tropical spastic paraparesis. *Lancet* 1985;2:407–410.
5. Newton M, Cruickshank EK, Miller D, et al. Antibody to HTLV-I in West Indian born UK residents with spastic paraparesis. *Lancet* 1987;1:415–416.
6. Roman GC, Spencer PS, Schoenberg BS, et al. Tropical spastic paraparesis in the Seychelles Islands: a clinical and case-control neuroepidemiologic study. *Neurology* 1987;37:1323–1328.
7. Rodgers-Johnson P, Garruto RM, Gajdusek DC, Mora C, Zaninovic V, Morgan OSC. Tropical spastic paraparesis: evidence for the etiological role of human T-cell lymphotropic virus type I in endemic foci in Jamaica and Colombia. *Ann Neurol* 1987;22:116–117.
8. Bartholomew C, Cleghorn F, Charles W, et al. HTLV-I and tropical spastic paraparesis. *Lancet* 1986;2:99–100.
9. Gessain A, Francis H, Sonan T, et al. HTLV-I and tropical spastic paraparesis in Africa. *Lancet* 1986;2:698.

10. Osame M, Matsumoto M, Usuku K, et al. Chronic progressive myelopathy associated with elevated antibodies to HTLV-I and adult T-cell leukaemia-like cells. *Ann Neurol* 1987;21:117–122.
11. Dalgleish AG, Richardson J, Matutes ES, et al. HTLV-I infection in tropical spastic paresis: lymphocyte culture and enhanced serological response. *AIDS Res Hum Retroviruses* 1988;4:475–485.
12. Nagy K, Clapham P, Cheingsong-Popow R, Weiss RA. HTLV-I: induction of syncytia and inhibition by patients sera. *Int J Cancer* 1983;32:321–328.
13. Clapham P, Nagy K, Weiss RA. Pseudotypes of HTLV-I and HTLV-II: neutralisation by patients sera. *Proc Natl Acad Sci USA* 1981;3083–3086.
14. Ben Ishai Z, Haas M, Triglia D, et al. HTLV-I antibodies in Falashas and other ethnic groups in Israel. *Nature* 1985;315:665–666.
15. Weiss RA, Cheingsong-Popov R, Clayden S, et al. Lack of HTLV-I antibodies in Africans. *Nature* 1986;319:794–795.
16. Fleming AF, Yamamoto N, Bhusnurmath SR, et al. Antibodies to HTLV in Nigerian blood donors and patients with chronic lymphatic leukaemia or lymphoma. *Lancet* 1983;2:334–335.
17. Kazura JW, Saxinger WC, Wenger J, et al. Epidemiology of HTLV-I in East Sepik Province, Papua New Guinea. *J Infect Dis* 1987;155:1100–1107.
18. Babona DV, Nurse GI. HTLV-I antibodies in Papua New Guinea. *Lancet* 1988;2:1148.
19. Constantine NT, Fox E. Need to confirm HTLV-I screening assays. *Lancet* 1989;1:108–109.
20. Hardy DB, Carlson JR, Lee JL, Armstrong MYK. Need to confirm HTLV-I screening assays. *Lancet* 1989;1:109.
21. Cruickshank JK, Rudge P, Dalgleish AG, et al. Tropical spastic paraparesis and HTLV-I in the United Kingdom. *Brain* 1989;112:1057–1090.
22. Lalkaka JA, Savant CV, Singhal BS. HTLV-I antibody study in non-compressive myelopathies—Indian Experience Abstract. Neurological Society of India, 1988.

Human Retrovirology: HTLV,
edited by William A. Blattner.
Raven Press, Ltd., New York 1990.

Epidemiology of HTLV-I and Associated Diseases

William A. Blattner

Viral Epidemiology Section, National Cancer Institute, National Institutes of Health, Bethesda, Maryland 20892

INTRODUCTION

The 1980 report of the discovery of the first human retrovirus, human T-lymphotrophic virus type-I (HTLV-I) by Gallo, Poiesz, and colleagues was the culmination of a search for such an agent that began in the early days of the twentieth century (1). The isolation of HTLV-I allowed for the development of reagents suitable for application in epidemiologic surveys. Since 1981 the Viral Epidemiology Section of the National Cancer Institute has pursued a series of epidemiologic investigations to characterize the nature of associated diseases linked to HTLV-I as well as to broaden our understanding of the epidemiology for transmission of the virus and its role in the pathogenesis of leukemia/lymphoma. In this report are summarized data relating HTLV-I to the pathogenesis of various malignant and nonmalignant conditions.

HTLV-ASSOCIATED DISEASES

The convergence of virologic discoveries first made in the laboratory of Dr. Robert C. Gallo with the clinical observations of Kyoshi Takatsuki (2) led to the recognition that HTLV-I was associated with a unique form of adult T-cell leukemia/lymphoma termed ATL (3,4). The ATL syndrome, as first reported, encompassed a spectrum of mature T-cell lymphomas, including the presence of leukemic involvement with cells of unusual morphology, and the frequent occurrence of cutaneous involvement as well as hypercalcemia, sometimes with lytic bone lesions. This frequently aggressive form of leukemia/lymphoma has been further subdivided as described in other chapters of this monograph and summarized in Fig. 1. In particular, studies by Hanchard and Gibbs in Jamaica have helped to independently corroborate the classification first proposed by Takatsuki and colleagues of ATL classic type, chronic ATL (a syndrome resembling T-chronic lymphocytic leukemia), smoldering ATL (a form of disease resembling mycosis fungoides/Sézary syndrome), and ATL lymphoma type, here

251

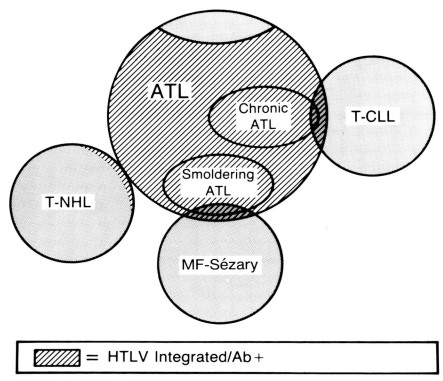

FIG. 1. Spectrum of HTLV-associated diseases. HTLV-I is etiologically linked to adult T-cell leukemia/lymphoma (ATL). This syndrome has a spectrum of clinical phenotypes represented diagrammatically in this figure and discussed in the text. T-NHL represents T-cell non-Hodgkin's lymphoma; MF represents mycosis fungoides; T-CLL represents T-cell chronic lymphocytic leukemia. Reproduced with permission from Blattner W and Gallo R, *Proceedings of the XIIth Symposium of the International Association of Comparative Research on Leukemia and Related Diseases* (F. Deinhardt, ed.), pp. 361–382, Hamburg, 1986.

labeled on Fig. 1 as T-non-Hodgkin's lymphoma (5,6). This spectrum of clinical presentations has made it difficult to define ATL for the epidemiologic purpose of quantifying the risk for ATL in HTLV-I–infected populations. Studies in Jamaica document that over 50% of incident non-Hodgkin's lymphomas are HTLV-I associated, with a diversity of presentations encompassing the spectrum of disease as originally described in Japan (7). As in Japan, survival is poor and the presence of hypercalcemia is associated with a particularly poor prognosis (6). The peak occurrence of ATL occurs between the ages of 30 and 50, with a more or less equal male-to-female ratio of cases. The age-specific curve of ATL occurrence in Jamaica is similar to that reported in Japan, albeit with a slightly younger age of peak occurrence, presumably reflecting the different underlying distribution of age in the population at risk (3,8). A preliminary analysis of an ongoing case-control study of non-Hodgkin's lymphoma in Jamaica documents a 20- to 35-fold increased risk for HTLV-I positivity among patients classified as ATL on the basis of adult onset T-cell non-Hodgkin's lymphoma (A. Manns, E. Murphy, and B. Hanchard, unpublished data). Nonetheless there is some significant portion of cases who have some features of ATL but lack antibody to HTLV-I. Such HTLV-I–negative ATL cases are the focus of ongoing molecular epidemiologic studies to determine whether sequences of HTLV-I might be present in tumor tissues but without detectable antibody.

In the original survey of lymphoreticular neoplasia undertaken in Jamaica, several cases of HTLV-I–associated chronic lymphocytic leukemia were noted. These cases at first were thought to represent examples of HTLV-associated T-chronic lymphocytic leukemia. However, cell sorter analysis showed that these cases were B-cell chronic lymphocytic leukemia which, on Southern blot analysis, were HTLV-I–virus negative. However, when peripheral blood from these cases was cultured, virus-positive lines with T-cell markers and polyclonally integrated HTLV-I virus were detected. Dr. Dean Mann and colleagues employed human hybridoma technology to express the immunoglobulin genes of the B-cell leukemia cells (9). In two cases studied by this technique, the immunoglobulin produced by the tumor cells had specific reactivity to HTLV-I antigens. In one case the antibody related to the p24 core protein, and in the other an envelope-related reactivity was identified (9). Furthermore, in one case, direct binding of p24 to the surface of the chronic lymphocytic leukemia cells could be documented. It is hypothesized that HTLV-I might play an indirect pathogenic role with HTLV-I infection of T-cells resulting in chronic antigenic stimulation of B-cell lines with altered immune regulation resulting in a B-cell malignancy (9).

A third HTLV-associated condition is the HTLV-associated myelopathy/tropical spastic paraparesis syndrome (HAM/TSP) first shown to be HTLV associated by Gessain, de Thé, and colleagues and summarized in several chapters of this monograph (10–12). The HAM/TSP syndrome is characterized by symptoms due to demyelination of long motor tracts in the spinal cord, resulting in spastic paraparesis. This condition has a similar age distribution to that of ATL but with a predominance of female cases. The geographic distribution of this syndrome overlaps that of adult T-cell leukemia, and there are occasional cases of coincident occurrence of both ATL and HAM/TSP in the same patient (13). Findings from Japan have documented an excess occurrence of HAM/TSP among recipients of blood transfusion (14). The Japanese transfusion data suggest that HAM/TSP has a shorter latency from exposure to disease than does ATL, where a transfusion association has not been documented.

Recently HTLV-I has also been linked to some cases of polymyositis and a number of other neurologic syndromes, including some cases with features indistinguishable from multiple sclerosis (15).

Possible links of HTLV-I to clinical and subclinical immunosuppression come from clinical and laboratory observations. Cases from Japan of patients with AIDS-like illnesses associated with HTLV-I (in the absence of underlying malignancy) have been reported (16). The association of HTLV-I with some parasitic infestations has also been interpreted to suggest that HTLV-I may have immunosuppressive effects (17). Recent findings document that HTLV-I positives express in vitro spontaneous proliferation (18). This phenomenon, first reported by Jacobson and McFarlin in patients with HAM/TSP, may provide an insight as to how HTLV-I may alter the immunologic milieu of the infected host (19). It has recently been postulated that HTLV-I envelope may act as a T-cell mitogen and such ongoing immune stimulation may adversely affect T-cell regulatory processes (20). This may also provide an explanation for the observed accelerated progression to AIDS among HTLV-I and HIV-1 co-infected individuals (21). It is postulated that spontaneous proliferation of HTLV-I could facilitate the killing effects of HIV-1 by stimulating HIV-1 cells to undergo cell division, resulting in cell death, and to activate T-cells to make them more susceptible to HIV infection, thus amplifying the rate of HIV accumulation (18).

Although some of the clinical conditions putatively linked to HTLV-I are not fully characterized, it is clear that there is a broad spectrum of clinical outcomes that appear to be HTLV-associated. In this regard, HTLV-I provides a useful model for considering HTLV-related viruses in the pathogenesis of a variety of idiopathic conditions, as suggested by recent data possibly linking HTLV or a related virus to multiple sclerosis (22,23).

EPIDEMIOLOGIC PATTERN AND MODES OF TRANSMISSION OF HTLV-I

The epidemiology of HTLV has been defined through the use of tests which detect antibodies to the virus. Because there is no antigen test or other method for detecting the virus save via virus culture—and, more recently, via polymerase chain reaction (PCR)—detection, the epidemiologic profile observed could potentially be distorted because an immune response to the virus is required for its detection. Nonetheless, small-scale surveys employing PCR have not detected large numbers of virus-positive, antibody-negative individuals, although some instances have been anecdotally reported (22). This is a subject of continued investigation. Furthermore, epidemiologic studies of HTLV-I are complicated by the inability of current serologic assays to distinguish HTLV-I from the closely related HTLV-II virus (22).

The patterns of antibody to HTLV-I in populations are striking. First, there is the marked geographic variation in HTLV-I positivity initially noted in surveys in Japan and subsequently in other areas of the world. In broad terms HTLV-I has been documented to have an endemic pattern of occurrence in the southern areas of Japan: Kyushu, Shikoku, and the islands of the Ryukyu chain, including Okinawa (23). Other areas of the Far East, including China, Korea, and Taiwan, have been extensively surveyed, and endemic foci have not been identified (23). In Papua New Guinea there are reports of high rates of seropositivity, but the specificity of the assays employed and the epidemiologic pattern of occurrence is not consistent with that observed in other viral endemic areas, and the nature of the reactivity is still the subject of dispute. Nonetheless, there may be areas of true HTLV-I positivity in this region, although virus isolation is warranted to search for variant viruses (24).

A second major focus of HTLV-I infection has been identified in the Caribbean region, including Jamaica, Trinidad and Tobago, Martinique, Guadeloupe, Barbados, and Haiti (23). In Trinidad and Tobago, seropositivity is restricted almost exclusively to persons of African descent, despite sharing a common environment with the Indo-Asian ethnic subpopulation for over 100 yr (25,26). Although rates of positivity are fairly uniform throughout the region, in Jamaica there are unexplained variations in seroprevalence by altitude with the highest rates observed in low-altitude, high-rainfall areas (E. Maloney and E. Murphy, unpublished).

In Central and South America areas of HTLV-I positivity are reported. For example, in Panama, rates of positivity appear highest in the northwestern region adjacent to Costa Rica (27). In Colombia, HTLV-I clusters along the Pacific Coast in an area with an unusually high rate of HAM/TSP. The highest rates are observed in persons of predominantly African or mixed African ancestry, and, controlling for race, the highest rates are observed at sea level (28). Other areas of South America with documented foci of HTLV-I include Brazil, Venezuela, Surinam, and Guyana.

It has been postulated that HTLV-I originated in Africa. However, numerous reports of HTLV-I seropositivity in the African continent have tended to exaggerate the true rate of seropositivity because of high rates of nonspecific reactivity (29). Other surveys have documented that rates of HTLV-I seropositivity are more similar to those in the Caribbean region. In a recent survey from Zaire, a microgeographic cluster of seropositivity was noted in the Equateur Province, an area also reported with clinical cases of HAM/TSP (30). HTLV-I tends to occur broadly in the equatorial regions of Africa, including the Ivory Coast, Nigeria, Zaire, Kenya, and Tanzania (31). Rates of positivity, based on limited data, are low in the North African countries bordering the Mediterranean. Further work, utilizing more precise and specific assay systems, is needed to define the pattern of HTLV-I.

Studies of migrant populations have documented that HTLV-I is often acquired early in life and can be carried with the individual to a nonendemic area with subsequent disease occurrence decades after putative infection. This pattern was first recognized in Japan and among migrant populations from Okinawa to Hawaii and from the Caribbean to the United States and the United Kingdom (32–34). In the Hawaiian migrant population, rates of positivity in Japanese Americans are the same in persons born in Okinawa but residing in Hawaii most of their lives and in their first-generation, Hawaii-born offspring (33). However, the rate of positivity in the migrant population residing in Hawaii from a young age was lower than that of lifelong residents of Okinawa matched for age. This finding raised the possibility that some environmental factors in the viral endemic area amplified the probability of infection. Supporting this concept are more recent data from further follow-up of the Hawaii-Okinawa cohort showing an apparent diminution of rates of positivity in third-generation, Hawaii-born Japanese Americans residing in Hawaii (G. Ho and A. Nomura, unpublished).

HTLV-I by Age

There is a characteristic age-dependent rise in HTLV-I seroprevalence typical for viral endemic areas. This rise in seroprevalence becomes prominent in the adolescent years, with a more rapid rise in females compared with males occurring throughout life, with the marked female predominance further exacerbated by a plateauing of rates in males occurring around age 40 and onward (8,35). Comparing the curves for Okinawa and Jamaica, the sharper rise at a younger age in Jamaica may reflect lifestyle differences between the two locales. The higher rate of positivity in Okinawa most likely results from differences in sampling approaches, because the Okinawa survey included family units where the virus is known to cluster. The explanation for this age-dependent rise in HTLV-I seroprevalence has been limited by the fact that such data derive from cross-sectional surveys. A cohort effect is one possibility, with improvements in standard of living over time causing a decline in the rate of positivity in successive birth cohorts because of the disappearance of yet-to-be-identified cofactors that promote infection (23). Data in favor of this hypothesis have been reported from prospective studies in Japan (36) and are supported by the recent data from among Okinawan migrants residing in Hawaii (G. Ho and A. Nomura, unpublished). However, in one study from Barbados, the rate and pattern of seroprevalence in a historic cohort were indistinguishable from that of rates observed in the same region some 15–20 yr later (37).

A second possible explanation is that the rise in seroprevalence results from cumulative exposure to a poorly infectious agent over the lifetime of the individual, resulting in a permanent index of infection, antibody positivity which is persistent and detectable because of the integration of this virus into the genetic material of the host, and its persistent expression over the lifetime of the individual (23). A third possibility is that infection as detected by the antibody test is insensitive and that substantially more individuals are virus positive than are appreciated by the presence of antibody positivity alone. In this case significant numbers of persons may carry the virus in a latent stage, and the age-dependent rise in seroprevalence represents the reactivation of latently acquired infection (23).

These three possibilities are the current subject of ongoing research studies aimed at defining the modes of transmission and in particular to look at patterns of infection over the lifetime of prospectively followed individuals.

TABLE 1. *Determinants of HTLV infection*

I. Route
  A. Sexual
    1) Male-to-female, female-to-male, and male-to-male
  B. Parenteral
    1) Transfusion—cellular components
    2) IV Drug abuse
  C. Mother-to-child—breast-feeding
II. Cofactors
  A. Sexual
    1) Large number of male sexual partners
    2) Steady sexual relationship with seropositive
    3) Coincident sexually transmitted diseases—males and females
  B. Environmental
    1) Birthplace/ancestry in viral endemic area
    2) Chronic antigenic stimulation?
III. Infectivity
  A. Virus replication
    1) High antibody titer
    2) Spontaneous lymphocyte proliferation
  B. Immune status—activated T-cell targets

Modes of HTLV Transmission

Summarized in Table 1 are the routes, cofactors, and viral characteristics associated with transmission of HTLV-I. The basic modes of transmission of HTLV-I are quite analogous to those of the human immunodeficiency virus (HIV-1).

Sexual Transmission

Sexual transmission from male-to-female and female-to-male as well as from male-to-male has been documented (23,38,39). The receptive partner, be this the result of male-to-female or male-to-male contact, appears more susceptible than the insertive partner, undoubtedly reflecting the relative inefficiency of transmission of this highly cell-associated virus. The contrast between the routes of infection for HTLV-I and HIV-1 in a male homosexual cohort in Trinidad reinforces the fact that HIV-1—which can be transmitted cell free, whereas HTLV-I is cell-associated—appears to be at least an order of magnitude more infectious than HTLV-I (38).

An analytic study of HTLV-I shows that, for females, a large number of lifetime male partners is associated with increased likelihood of being seropositive (39). However, there appear to be other factors involved, as suggested by recent data from Mueller and colleagues, who prospectively followed discordant spouse pairs in a viral-endemic area of Japan (40). In that study, the likelihood of a female spouse becoming seropositive from her seropositive husband was associated with higher antibody titer of the infectious spouse. As the discordant couple became older, the effect of rising antibody titer on transmission suggests that, with advancing years, the ability of the host immune system to hold the virus in check may decline, with HTLV-I antibody titer acting as a surrogate for virus load and infectivity. In this regard there is some resemblance to data from the HIV-1 studies of Goedert et al., where HIV-1 antigenemia and decreased T-cell subsets are a marker of increased virus load, which in turn is associated with heightened virus transmission (41). Recent data from Ho, Nomura, and colleagues in Hawaii (unpublished) report that the likelihood of a female spouse being seropositive is directly correlated with elevated antibody titer of the husband, also supporting the concept that antibody titer is a surrogate for virus load and transmission.

Another cofactor for sexual transmission is the coincidence of other sexually transmitted diseases, particularly ulcerative genital lesions such as occur in syphilis (39). In this case the coincidence of other sexually transmitted diseases seems to amplify the likelihood of transmission for both males and females, presumably by a break in the normal mucosal barrier but also perhaps by an increase in lymphocytic infiltration resulting in more target cells for infection. In this regard, infrequent female-to-male transmission may be amplified under this circumstance—as suggested by a recent study from Murphy et al., which identified this as a major independent risk factor for HTLV-I infection in males (39). It is also possible that the lifestyle associated with ulcerative sexually transmitted disease may play a role, although, in the Murphy et al. study, ulcerative genital lesions (but not number of partners) were the only independent risk factors for male seropositivity.

Mother-to-Child Transmission

The second major route of transmission is from mother to child. Breast-feeding, as documented from Japanese studies, is more efficient than perinatal transmission. Thus whereas 20% of breast-fed infants seroconvert, 1–2% of bottle-fed infants become infected (42). In this regard HTLV-I differs from HIV because perinatal transmission of HIV-1 appears to be associated with up to 30% of neonatal infections (43). The rate of breast milk–associated HIV-1 transmission is unknown because most HIV-1 positive mothers in the United States are discouraged from breast-feeding. Antibody titer in the mother is an important determinant of HTLV-I transmission. Mothers who have a high antibody titer are more likely to transmit the virus to their offspring, a finding supporting the concept that titer is a surrogate marker for virus load and infectivity (42). Duration of breast-feeding, however, was not associated with increased transmission, and this may reflect the fact that the window for transmission by this route may be open only in the early postpartum period.

Parenteral Transmission

A third major route of transmission is parenteral, either via transfusion or intravenous drug abuse. In the case of transfusion transmission, Okochi and colleagues were the first to document that cellular components of blood are associated with transmission (44). More recently, A. Manns and colleagues, in a study in Jamaica, have corroborated this finding and have also shown that, similar to the reports from Japan, approximately one-half of recipients of cellular blood components experience a seroconversion (A. Manns, R. Wilks, et al., unpublished). In the study of A. Manns and colleagues, antibodies to both envelope and core proteins appear to develop almost synchronously. No acute clinical syndrome has been linked to seroconversion. An interesting finding from Osame and colleagues is the fact that transfusion may be an associated risk for TSP (14). These data have been used to suggest that the latency for TSP may be shorter than that for ATL, because transfusion does not appear to be associated with ATL.

Parenteral drug abuse has also been associated with transmission of HTLV virus (45). However, recent data from New Orleans among a cohort of black drug abusers document that by PCR, contrary to expectation, the majority of positives are HTLV-II and not HTLV-I (22). This curious finding raises the possibility that the efficiency of transmission of HTLV-II is favored in the case of parenteral drug abusers, in contrast to HTLV-I, which may be more efficiently transmitted by other routes.

Environmental Cofactors

Other factors that appear to contribute to transmission of HTLV-I include poorly defined environmental factors that have not been fully characterized. A number of studies show that lower socioeconomic status may be associated with heightened risk of being antibody positive (26). Some have argued that this finding supports a role for insect vector transmission since altitude, markers of hygiene, and poor housing quality are also associated with higher rates but are difficult to disentangle because these markers are all highly correlated with each other (25). Insect transmission, however, seems unlikely because HTLV-I is not associated with seropositivity for other anthropod-borne viruses (45) and because HTLV-I is of such low titer and is so highly cell associated that even whole-blood transfusion only results in infection in one-half of cases (44). Thus insects do not appear to be a likely vector for transmitting such a poorly infectious virus. An alternate explanation for the association of HTLV-I with markers of poor social status is the possibility that living in such an environment predisposes an individual to infection through indirect immune-mediated mechanisms. In this scenario, susceptibility to infection resulting from immune activation of the host, as well as the likelihood of transmitting the virus through amplification of virus load, may be coupled, resulting in increased rates of virus positivity (26).

MODEL FOR INFECTION AND DISEASE OCCURRENCE

In Fig. 2 is summarized the cumulative prevalence of HTLV-I by transmission route from males (2A) and females (2B). In this model, over the lifetime of the host, the age-dependent rise in seroprevalence results from the cumulative exposure of the individual to the virus through a variety of different routes of infection. Early in life, transplacental and breast milk account for a small pro-

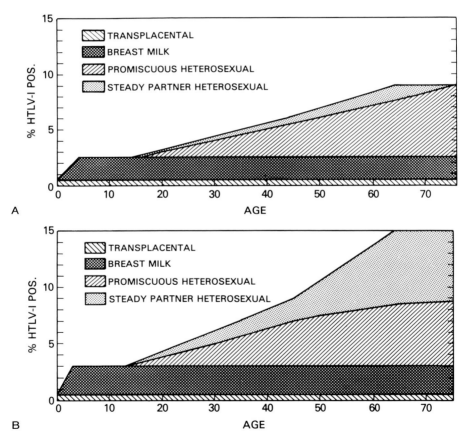

FIG. 2. The cumulative prevalence of HTLV-I as a function of age is portrayed for males (**A**) and females (**B**) to result from different routes of transmission. The bulk of transmission results from sexual transmission, which accounts for the disparity between males and females because male-to-female transmission is more efficient than female-to-male. (See text for further discussion).

portion of infection. Subsequently, adult-acquired infection occurs almost exclusively through sexual transmission, at least in viral endemic areas where intravenous drug abuse is not a significant factor. In this circumstance, the efficiency of transmission from male-to-female versus female-to-male results in the disproportionate rise among females compared with males. Early in the sexually active years, promiscuous sexual exposure—especially if associated with other sexually transmitted diseases—may account for the majority of HTLV-I infections. Later in life sustained steady exposure to a single partner associated with a rise in HTLV antibody titer linked to increased infectivity may explain the later disproportionate rise in females compared to is a plateauing for males in this age group. Thus in this model the cumulative seroprevalence rises largely because of sexual transmission of the virus later in life. This pattern also fits with that reported for HIV, where the major mode of transmission of the epidemic is in the sexually active age group.

Figure 3, A and B, models disease occurrence based on the patterns of infection and disease occurrence. ATL differs from HAM/TSP both in overall incidence and the fact that ATL occurs equally in males and females, whereas HAM/TSP occurs more frequently among females. For ATL we postulate that early life exposure, perhaps at birth, is the major risk factor for subsequent disease occurrence (8). As documented from migrant studies, ATL occurs with a 20-yr or more latent period between exposure and disease occurrence (34). This hypothesis would best explain the pattern observed for equal male-to-female incidence of disease observed in Jamaica and Japan, because it is unlikely that females are more resistant than males to developing ATL (46). It is possible that some ATL cases result from adult exposure, again with long latency from exposure to disease, but the proportion of these cases is unlikely to be significant, given the equal male-to-female ratio of cases.

The pattern for HAM/TSP shown in Fig. 3B contrasts with that of ATL. In this model, a significant portion of HAM/TSP patients also results from early-life exposure, but perhaps an equal or larger proportion also results from adult-acquired infection. A shorter latency, as suggested from transfusion data, would favor this as well as the female predominance of cases (14). Based on available data it is likely that the incidence of HAM/TSP is at least twice that of ATL, although more precise data on this issue are needed—especially given the marked differences in survival between ATL and HAM/TSP, which favor the more ready detection of the neurologic syndrome.

CONCLUSION

As summarized in this report and other papers in this monograph, leukemogenesis resulting from HTLV-I is a long-term process that is an infrequent outcome of exposure. This has led to the hypothesis that ATL has a multistep etiology, where the virus may play a role in initiation of leukemogenesis but

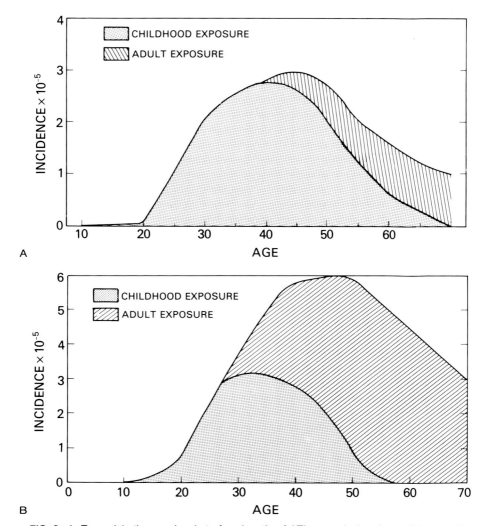

FIG. 3. **A:** To explain the equal male-to-female ratio of ATL cases in Jamaica and Japan, given the higher rates of HTLV-I seropositivity in females, we postulate that prepubescent nonsexual transmission, mainly in the perinatal infant period, is most important for ATL occurrence. (See text for discussion.) **B:** In contrast to ATL, HAM/TSP occurs disproportionately more frequently in females, which suggests that adult-acquired sexual exposure, associated with a shorter incubating period from infection to disease, results in a higher incidence of cases. (See text for discussion.)

subsequent events are necessary for conversion to ATL (47). In contrast, we have recently postulated that HAM/TSP may involve a mechanism of indirect pathogenesis, with virus exposure resulting in a cascade of immunologic events that may help explain the shorter latency from exposure to disease (17). If direct viral infection of the nervous system plays a role for HAM/TSP, the number of events required for neurologic disease pathogenesis would appear to result from a more direct effect of the virus, with a limited number of "events" needed for disease to become evident.

These contrasting patterns of disease latency and occurrence point out that HTLV-I is an extremely useful model for dissecting out the role of viruses of long latency in the pathogenesis of diverse human diseases. Furthermore, results of these studies provide a useful context for considering other (including yet to be discovered) viruses of this class in the pathogenesis of diseases of unknown etiology, including multiple sclerosis and other cancers and leukemias that are suggested from retrovirus animal models.

REFERENCES

1. Poiesz BJ, Ruscetti FW, Gazdar AF, Bunn PA, Minna JD, Gallo RC. Detection and isolation of type-C retrovirus particles from fresh and cultured lymphocytes of patients with cutaneous T-cell lymphoma. *Proc Natl Acad Sci USA* 1980;77:7415–7419.
2. Uchiyama T, Yodoi J, Sagawa K, Takatsuki K, Uchino H. Adult T-cell leukemia: clinical and hematologic features of 16 cases. *Blood* 1977;50:481–492.
3. The T- and B-cell Malignancy Study Group. Statistical analyses of clinicopathological, virological and epidemiological data on lymphoid malignancies with special reference to adult T-cell leukemia/lymphoma: a report of the second nationwide study of Japan. *Jpn J Clin Oncol* 1985;15:517–535.
4. Blattner WA, Kalyanaraman VS, Robert-Guroff M, et al. The human type-C retrovirus, HTLV, in Blacks from the Caribbean region, and relationship to adult T-cell leukemia/lymphoma. *Int J Cancer* 1982;30:257–264.
5. Yamaguchi K, Nishimura H, Kawano F, et al. A proposal for smoldering adult T-cell leukemia-diversity in clinical pictures of adult T-cell leukemia. *Jpn J Clin Oncol* 1983;13:189–199.
6. Gibbs WN, Lofters WS, Campbell M, et al. Non-Hodgkin lymphoma in Jamaica and its relation to adult T-cell-lymphoma. *Ann Intern Med* 1987;106:361–368.
7. Blattner WA, Gibbs WN, Saxinger C, et al. Human T-cell leukaemia/lymphoma virus–associated lymphoreticular neoplasia in Jamaica. *Lancet* 1983;2:61–64.
8. Murphy EL, Hanchard B, Figueroa JP, et al. Modelling the risk of adult T-cell leukemia/lymphoma in persons infected with human T-lymphotropic virus type I. *Int J Cancer* 1989;43:250–253.
9. Mann DL, DeSantis P, Mark G, et al. HTLV-I-associated B-cell CLL: indirect role for retrovirus in leukemogenesis. *Science* 1987;236:1103–1106.
10. Gessain A, Barin F, Vernant JC, et al. Antibodies to human T-lymphotropic virus type-I in patients with tropical spastic paraparesis. *Lancet* 1985;2:407–410.
11. Rodgers-Johnson P, Morgan OS, Mora C, et al. The role of HTLV-I in tropical spastic paraparesis in Jamaica. *Ann Neurol* 1988;23(Suppl):S121–S126.
12. Osame M, Matsumoto M, Usuku K, et al. Chronic progressive myelopathy associated with elevated antibodies to human T-lymphotrophic virus type I and adult T-cell leukemia-like cells. *Ann Neurol* 1987;21:117–122.
13. Cleghorn F, Charles W, et al. HTLV-I and tropical spastic paraparesis. *Lancet* 1986;2:99–100.
14. Osame M, Izumo S, Igata A, et al. Blood transfusion and HTLV-I associated myelopathy. *Lancet* 1986;2:104–105.
15. Morgan OS, Char G, Mora C, Rodgers-Johnson P. HTLV-I and polymyositis in Jamaica. *Lancet* 1989;2:8673:1184–1186.
16. Kobayashi M, Yoshimoto S, Fujishita M, et al. HTLV-positive T-cell lymphoma/leukaemia in AIDS patients. *Lancet* 1984;1:1361–1362.
17. Krämer A, Blattner WA. The HTLV-I model and chronic demyelinating neurological diseases. In: Notkins AL, Oldstone MBA, eds. *Concepts in viral pathogenesis.* New York: Springer-Verlag. 1989:204–214.

18. Krämer A. Spontaneous lymphocyte proliferation in symptom-free HTLV-I positive Jamaicans. *Lancet* 1989;2:923–924.
19. Jacobson S, Zaninovic V, Mora C, et al. Immunologic findings in neurological diseases associated with antibodies to HTLV-I: activated lymphocytes in tropical spastic paraparesis. *Ann Neurol* 1988;23(suppl):S196–S200.
20. Gazzolo L, Duc Dodon M. Human T-lymphotrophic virus type I is a direct activator of resting T lymphocytes. *Nature* 1987;326:714–717.
21. Bartholomew C, Blattner W, Cleghorn F. Progression to AIDS in homosexual men co-infected with HIV and HTLV-I in Trinidad. *Lancet* 1987;2:1469.
22. Lee H, Swanson P, Shorty V, et al. High rate of HTLV-II infection in seropositive IV drug abusers in New Orleans. *Science* 1989;244:471–475.
23. Blattner WA. Retroviruses. In: Evans AS, ed. *Viral infections of humans: epidemiology and control, 3d ed.* New York: Plenum Medical Book Co. 1989:545–592.
24. Kazura JW, Saxinger WC, Wenger J, Forsyth K, et al. Epidemiology of human T cell leukemia virus type I infection in East Sepik Province, Papua New Guinea. *J Infect Dis* 1987;155(6):1100–1107.
25. Miller GJ, Pegram SM, Kirkwood BR, et al. Ethnic composition, age and sex, together with location and standard of housing as determinants of HTLV-I infection in an urban Trinidadian community. *Int J Cancer* 1986;38:801–808.
26. Blattner W, Saxinger C, Riedel D, et al. A study of HTLV-I and its associated risk factors in Trinidad and Tobago. (submitted)
27. Reeves WC, Saxinger C, Brenes MM, et al. Human T-cell lymphotropic virus type I (HTLV-I) seroepidemiology and risk factors in metropolitan Panama. *Am J Epidemiol* 1988;127(3):532–539.
28. Maloney EM, Ramirez H, Levin A, Blattner WA. A survey of the human T-cell lymphotropic virus type I (HTLV-I) in south-western Colombia. *Int J Cancer* 1989;44:419–423.
29. Saxinger WC, Blattner WA, Levine PH, et al. Human T-cell leukemia virus (HTLV-I) antibodies in Africa. *Science* 1984;225:1473–1476.
30. Wiktor SZ, Mann JM, Nzilabmi N, et al. Human T-cell lymphotrophic virus type I (HTLV-I) among female prostitutes in Kinshasa, Zaire. *J Infect Dis (in press)*.
31. de Thé G, Giordano C, Gessain A, et al. Human retroviruses HTLV-I, HIV-1 and HIV-2 and neurological diseases in some equatorial areas of Africa. *J AIDS* 1989;2:550–556.
32. Hinuma Y, Komoda H, Chosa T, et al. Antibodies to adult T-cell leukemia-virus–associated antigen (ATLA) in sera from patients with ATL and controls in Japan: a nation-wide seroepidemiologic study. *Int J Cancer* 1982;29:631–635.
33. Blattner WA, Nomura A, Clark JW, et al. Modes of transmission and evidence for viral latency from studies of HTLV-I in Japanese migrant populations in Hawaii. *Proc Natl Acad Sci USA* 1986;83:4895–4898.
34. Greaves MF, Verbi W, Tilley R, et al. Human T-cell leukaemia virus (HTLV) in the United Kingdom. *Int J Cancer* 1984;33:795–806.
35. Kajiyama W, Kashiwagi S, Ikematsu H, et al. Intrafamilial transmission of adult T-cell leukemia virus. *J Infect Dis* 1986;154:851–857.
36. Ueda K, Kusuhara K, Tokugawa K, et al. Cohort effect on HTLV-I seroprevalence in southern Japan. *Lancet* 1989;2:979.
37. Riedel DA, Evans AS, Saxinger C, Blattner WA. A historical study of human T lymphotropic virus type I transmission in Barbados. *J Infect Dis* 1989;159(4):603–609.
38. Bartholomew C, Saxinger WC, Clark JW, et al. Transmission of HTLV-I and HIV among homosexuals in Trinidad. *JAMA* 1987;257:2604–2608.
39. Murphy EL, Figueroa JP, Gibbs WN, et al. Sexual transmission of the HTLV-I. *Ann Intern Med* 1989;111:555–560.
40. Mueller N, Tachibana N, Essex M. Natural history of HTLV-I infection. In: *Proceedings IV International Conference on AIDS.* 1988;2:225.
41. Goedert JJ, Eyster ME, Biggar RJ, et al. Heterosexual transmission of human immunodeficiency virus: association with severe depletion of T-helper lymphocytes in men with hemophilia. *AIDS Res Hum Retroviruses* 1987;3:335–361.

42. Hino S, Yamaguchi K, Katamine S, et al. Mother-to-child transmission of human T-cell leukemia virus type-I. *Gann* 1985;76:474–480.
43. Goedert JJ, Mendez H, Drummond JE, et al. Mother-to-infant transmission of human immunodeficiency virus type 1: association with prematurity and low anti-gp120. *Lancet* 1989;2:1351–1355.

DISCUSSION

One speaker was asked to expand on the observation that the frequency of seropositivity in the female population increased with age. The speaker replied that this may be due to different patterns of transmission or because the virus was present, and viral replication increased as the patient became older, thus inducing an increased antibody response that appeared as a new infection, but which was actually latent in the individual. A discussant asked if breast-feeding habits in the Hawaiian and Okinawan populations were investigated where the frequency of HTLV-I seropositivity appeared to decrease in successive generations. The speaker replied that the data were being gathered but were not yet analyzed. Another discussant pointed out that the study had only considered breast milk and the transplacental spread of viruses from mother to infant and that with many viruses, infection occurred when the fetus passed through the birth canal. The speaker agreed that this could be a route whereby the virus was transmitted. A discussant raised the question of occupational exposure versus place of birth as a contributing factor to HTLV-I in prostitutes in the Kinshasha study. The question was referred to another participant, who replied that the association was with place of birth and not with current residence or duration of living in Kinshasha. One discussant asked if PCR had been used to identify viruses in the population where infection might be expected and antibody activity not detected. The speaker replied that these studies were underway in the West Indies.

Another discussant made the following comments related to the diagnosis of hairy cell leukemia in HTLV-II–infected patients: hairy cell leukemia is a B-cell disorder and represents a specific disease entity; finding T-cell markers indicated the patient had another disorder. This discussant also suggested that HTLV-II might be associated with other B-cell proliferative disorders such as B-CLL, which was found to be associated with HTLV-I infection in Jamaica.

Human Retrovirology: HTLV,
edited by William A. Blattner.
Raven Press, Ltd., New York © 1990.

Prospective Studies of HTLV-I and Associated Diseases in Japan

*Kazuo Tajima, †Shin-Ichiro Ito, and
‡Tsushima ATL Study Group

*Division of Epidemiology, Aichi Cancer Center Research Institute, Chikusa-ku,
Nagoya 464; †Director, Tsushima Izuhara Hospital, Shimoagata-gun, Nagasaki 817;
‡This group comprises: Kami-Tsushima Hospital, National Tsushima Hospital,
Nagasaki Health Center, National Nagasaki Central Hospital,
and Atomic Disease Institute of Nagasaki University, Japan

INTRODUCTION

Adult T-cell leukemia (ATL) was first reported in 1976 in Japan (1). In 1981, it was revealed that most patients with ATL had serum antibody to ATL-associated antigens (ATLA) (2,3), and also some healthy people in the ATL-endemic areas had antibody to ATLA (4,5). A causative relation between ATL-associated virus (ATLV) and ATL was clarified in 1982 (6). A human T-cell lymphoma virus (HTLV) was isolated from cultivated tumor cells of cutaneous T-cell lymphoma in the United States in 1980 (7). HTLV and ATLV were classified as human oncoviruses, found to be identical in virological homology checks (8,9), and named human T lymphotropic virus type I (HTLV-I) in 1984. Anti-HTLV-I antibody in sera from patients with tropical spastic paraparesis (TSP) in the Caribbean basin was detected in 1985 (10). Independently of TSP, many cases of HTLV-I–associated myelopathy (HAM) in the ATL-endemic areas of Japan were reported in 1986 (11). It was suggested that TSP in the Caribbean basin and HAM in Japan would be in the same disease entity. Subsequently, two HTLV-I–associated diseases—ATL, a neoplastic disease, and HAM/TSP, a reactive disease—have been detected in the end of this century and have become a new target for comprehensive studies on human oncovirus-associated diseases.

Carriers of HTLV-I and patients with its associated diseases (ATL & HAM) are highly clustered in some areas of Japan. Currently in Japan, the following studies for the primary prevention of these mysterious diseases have been initiated: 1) Continuous nationwide surveillance of HTLV-I carriers and patients with HTLV-I–associated diseases; 2) a long-term follow-up study of HTLV-I carriers; and 3) intervention trials for vertical transmission of HTLV-I. Continuous nationwide surveillance of patients with ATL will give detailed information on the geographical variation of ATL patients in Japan. To clarify the leukemogenic steps of ATL after HTLV-I infection, it is necessary to calculate the average onset time of ATL in men and women in relation to the distribution of HTLV-I carriers in both sexes. A long-term follow-up study of HTLV-I carriers will give information on the risk of ATL manifestation among HTLV-I carriers. And from the seroepidemiological studies on HTLV-I carriers, the risk of vertical transmission from mother to child and horizontal transmission from husband to wife can be evaluated. Intervention trials for vertical transmission of HTLV-I by quitting breast-feeding can prevent the spread of HTLV-I in the ATL endemic areas and, subsequently, can prevent ATL in future. And the natural history of HTLV-I infection from mother to children could be clarified from these preventive cohort studies. In the present paper, ongoing prospective studies on these subjects are summarized.

NATIONWIDE SURVEILLANCE

Geographical Comparison

It has been shown that ATL patients are distributed widely throughout the world, i.e., among Japanese in Asia (1,12–14) and blacks in central Africa (15), the southern part of the United States (16), and the Caribbean Basin (17,18). It is estimated that approximately 1.2 million HTLV-I carriers exist in Japan and 700 new cases of ATL occur annually among these adult HTLV-I carriers (Table 1). ATL patients cluster in Kyushu, in the southern part of Japan, and more than 50% of ATL cases in Japan are from Kyushu. An important epidemiological feature of ATL in Japan is the geographical distribution of the birthplaces of ATL patients. Most ATL patients in metropolitan areas such as Tokyo and Osaka were born and reared in Kyushu (1,13,14,19). Recently, however, ATL patients have been observed throughout Japan, mainly in the coastal areas along

TABLE 1. *Estimated number of HTLV-I carriers and incidence of ATL by district*

District (location)	Population (1000)	Carrier rate of HTLV-I*	HTLV-I carriers†	ATL‡ patients (%)
Hokkaido and Tohoku (north)	15 330	0.01	108 000	65 (9.3)
Kanto (around Tokyo)	36 650	0.005	128 300	77 (11.0)
Osaka and Hyogo§	13 520	0.015	141 900	85 (12.2)
Kyushu (south)	14 460	0.06	607 300	364 (52.2)
Other districts	40 660	0.005	175 000	106 (15.2)
Total	120 720	—	1 160 500	697 (100)

* Estimated from positivity rate of HTLV-I antibody in blood donors, screened by immunofluorescence assay.
† Estimated number of HTLV-I carriers in adults older than 20 yr.
‡ Calculated from annual incidence rate as 0.6/1000 adult carriers.
§ Northern country areas in Hyogo prefecture were excluded.

the Pacific Ocean and the Japan Sea (13,14,20). The seroepidemiological evidence shows that HTLV-I carriers are found mainly among people in the ATL endemic areas, e.g., among Japanese and Melanesians in Papua New Guinea (21) in southeast Asian countries, and blacks in central Africa (15) and the Caribbean Basin (18). It is interesting that HTLV-I carriers are also found among immigrant Japanese in Hawaii (22), Brazil, and Bolivia (23).

From the anthroepidemiological point of view, the worldwide distribution of ATL patients and/or HTLV-I carriers presents very important evidence as to the history of the movement of the human race, however, there is no unequivocal explanation as to the reasons for this mysterious distribution. Recently it was revealed that HTLV-I is highly prevalent among Japanese aborigines in Hokkaido, who have been long isolated from other Japanese people (24). Furthermore, Hokkaido is the northernmost part of Japan and is very far from the ATL-endemic Kyushu district. It is well known that HTLV-I is transmissible only through lymphocytes infected with HTLV-I and is not very contagious under natural conditions. Considering these things, it is very difficult to explain how HTLV-I could have spread rapidly throughout Japan in modern ages and been preserved only in limited areas. It could be speculated that HTLV-I was introduced into Japan in an earlier time, at least several thousand years ago (25).

Prospective Changes Over Time in HTLV-I Carriers

The most interesting finding on HTLV-I infection is that the carrier rate of HTLV-I increased remarkably with age in both men and women. A cross-sectional seroepidemiological survey on HTLV-I among healthy people in the ATL endemic areas showed a characteristic age- and sex-specific distribution of HTLV-I carriers (4,5,26). A detailed analysis of the age-specific positivity rate of HTLV-I antibody in Tsushima Island is shown in Fig. 1. Age-specific rates of positivity (by 5-yr group) in men parallel those in women—until age 40, at which the male rate tends to plateau, whereas the female rate rises more steeply after 40 yr of age. Several ideas have been proposed to explain the age distribution characteristics of HTLV-I carriers: 1) the rate of carriers in the ATL endemic areas might depend on individuals' birth cohort, with earlier generations having a higher infection rate; 2) people at risk are continuously exposed to HTLV-I and infections increase with age, especially in females; and 3) anti-HTLV-I antibody is expressed later in life after infection early in life. Putting these ideas together, the characteristics of the age distribution of HTLV-I carriers could be clarified.

It had been suggested that antibody expression after HTLV-I infection is delayed by immunotolerant condition and, therefore, anti-HTLV-I antibody could not be detected among child and young adult HTLV-I carriers. It was reported, however, that even a young baby HTLV-I carrier expressed antibody to HTLV-I in sera (27). A retrospective-prospective seroepidemiological study in the Goto Islands in which serum samples from school children were collected more than

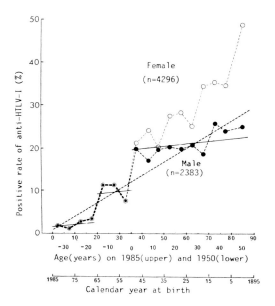

FIG. 1. Age- and sex-specific positive rate for anti-HTLV-I antibody in Tshushima Island. Solid and open circles indicate males and females, respectively.

10 years ago and stored in a freezer, revealed that the positive rate of anti-HTLV-I antibody in sera among school children (primary, junior high, and senior high school students) has decreased in recent years (28). Another prospective study (29) in Okinawa prefecture revealed that the positive rate of anti-HTLV-I antibody among 3-yr-old children of HTLV-I-carrier mothers did not show an increase during the 15-yr follow-up period (corresponding to 18-yr-old children). This epidemiological evidence strongly supports the first hypothesis mentioned above, that the characteristic age distribution of HTLV-I carriers depends mainly on birth cohort; that is, the natural infection rate of HTLV-I, such as from mother to child and between men and women, has decreased in the more recent periods.

FOLLOW-UP STUDY OF HTLV-I CARRIERS

Possible Host Factors for ATL

It is known that the age of onset of ATL in Japan is over 25, commonly over 40, and there has been no case of a child with ATL—even though patients with ATL are often infected with HTLV-I by their mother in childhood, probably through breast milk within 1 yr after birth. It remains unclear why ATL becomes overt after passing a long latent period (averaging longer than 50 yr). There is a possibility that an unknown host-related factor can regulate or control the progression of T cells transformed by HTLV-I integration. Because ATL does not become overt in children and young adults, it is speculated that the age of onset of clinical ATL could be related to age-dependent loss of the immune regulation of T cells, which is part of the normal aging process.

As ATL patients were clustered in limited areas and among family members (30), some host factors for ATL may exist (12,31). To approach this hypothesis, HLA type as a genetic marker was compared between HTLV-I carriers and noncarriers (32) and between ATL patients and healthy controls (33). No consistent result was obtained, which would indicate the existence of genetic susceptibility for HTLV-I infection and ATL manifestation. Recently, however, it was pointed out that the frequency of phenotype BW52 is lower in patients with ATL and HTLV-I carrier relatives compared with that in general controls and with noncarrier relatives (34), and that some haplotypes might be correlated to the risk for ATL manifestation (35). These results can neither support nor refute the possibility of genetic susceptibility to HTLV-I infection and risk for ATL.

Natural Transmission of HTLV-I

Familial clustering of HTLV-I carriers in the ATL-endemic areas suggests two natural transmission routes (5). One is vertical transmission from mother to child, and the other is transmission between men and women. For HTLV-I infection, cell-cell contact is required, and HTLV-I is not easily passed under natural conditions. Detailed seroepidemiological studies showed that breast-feeding posed a risk for HTLV-I transmission from mother to child, particularly from mothers with HTLV-I antigen-positive lymphocytes in their breast milk (36–41). The recent overall infection rate from carrier mothers to their children younger than 19 yr is estimated at 10–30% (5,29,38,41) (Table 2). However, retrospective studies showed ~50% of children of HTLV-I-carrier mothers older than 20 yr generated antibody to HTLV-I in sera (5). There is no evidence that can clearly explain the discrepancy in positivity rates between the two different generation groups. The same ideas mentioned for the age-specific distribution of HTLV-I carriers can be suggested here. One factor that might have affected

TABLE 2. *Risk for HTLV-I transmission from carrier mother to children*

Reporter (year)	Children Positives/Total	Transmission rate (%)	Age (yr) of children
Hino (1985)	5/30	17	1–9
Tajima (1986)	2/19	11	6–19
Kinoshita (1987)	8/32	25	3
Kussuhara (1987)	10/65	15	3–18*
Total	25/146	15	1–19

* Followed up for 15 yr from 3- to 18-yr-old.

the chance of vertical transmission of HTLV-I is the duration of breast-feeding, that is, the relative period of breast-feeding has decreased remarkably recently, especially during the past 30–40 yr. In Japan, alternative milk has come into wide use since the early 1950s and recently mothers have been advised to stop breast-feeding babies earlier than before. As shown in Fig. 2, the average period of breast-feeding increased with age, especially in age groups older than 50. That is, older people from the ATL-endemic areas were more exposed to breast milk that might have been contaminated by lymphocytes infected with HTLV-I. From this indirect evidence, it is suggested that the risk for transmission of HTLV-I by breast milk has decreased recently, which may be associated with the remarkable decrease of HTLV-I carriers in the young generation.

The second natural transmission route is horizontal transmission between men and women through sexual contact (5,27). A study of married couples strongly suggested that HTLV-I is transmitted mainly from husband to wife (5) through semen (37). Among wives with carrier husbands, the carrier rate (positive rates of anti-HTLV-I antibody) was remarkably high (5,27). However, the age-specific carrier rate among wives with noncarrier husbands was not very high and was the same as (or a little lower than) that for all husbands. On the other hand, the carrier rate of husbands with noncarrier wives decreased with age, because most couples with positive husband and negative wife changed to couples with positive husband and positive wife after transmission from the carrier husband. Consequently, the positive rate of anti-HTLV-I antibody in husbands with

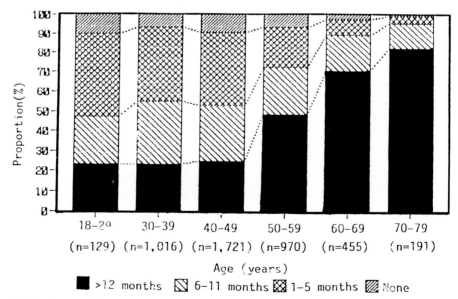

FIG. 2. Age-specific proportional distribution of average lactating period in women who visited Aichi Cancer Center Hospital in 1988.

TABLE 3. *Estimated incidence rate of ATL among carriers of HTLV-I in Japan*

Reporter (year)	Annual rate/1000		Cummulative rate (70 yr)*	
	Male	Female	Male %	Female %
Tajima (1985)	1.4	0.5	4.2	1.5
Tajima (1987)	2.2	0.8	6.6	2.4
Kondo (1987)†	0.8–2.0	0.4–0.7	5.0	1.5
Average	1.7	0.6	5.3	1.8

* Other causes of death were neglected.
† Age-specific incidence rate was calculated.

noncarrier wives decreased with age. Several possible explanations were suggested concerning the increasing (wives with carrier husbands) or decreasing (husbands with noncarrier wives) trends in the positive rates for anti-HTLV-I antibody with age among married couples. The transmission rate increased with age with the cumulative exposure to HTLV-I through repeated sexual contact between spouses. This hypothesis seems more likely for natural transmission routes. However, because sexual activity diminished in the older age group, it is not very easy to explain the continuous increase with age in the carrier rate of HTLV-I in the older age group. A time lag of antibody expression after sexual transmission cannot be excluded; however, what seems most important is that the transmission rate between spouses after marriage has changed over time, decreasing in recent birth cohort groups.

Risk for ATL in HTLV-I Carriers

Seroepidemiological studies of HTLV-I in the ATL-endemic areas show there are more female than male carriers of HTLV-I in the older age group, which corresponds to the one-way transmission of HTLV-I from men to women after

TABLE 4. *Age-specific positive rate* * of anti-HTLV-I antibody in Tsushima Island in 1985–1987*

Age (yr)	Male (%) Positive/Total	Female (%) Positive/Total	Total (%) Positive/Total
1	1/308 (0.3)	5/316 (1.5)	6/624 (0.9)
3	3/130 (2.3)	5/140 (3.5)	8/270 (2.9)
1–4	8/365 (2.1)	7/306 (2.2)	15/671 (2.2)
5–9	3/246 (1.2)	3/196 (1.5)	6/444 (1.4)
10–14	3/153 (1.9)	8/178 (4.4)	11/331 (3.3)
15–19	1/47 (2.1)	3/83 (3.6)	4/130 (3.1)

* Positives by PA and IFA methods.

TABLE 5. *Changes in antibody positivity on 1.5- and 3-yr-old children in Tsushima Island in 1985–1987*

Age at examination				
1.5-yr		3-yr		Number of children
PA	IFA	PA	IFA	n = 130 (%)
−	−	−	−	120 (92.3)
+	−	−	−	2 (1.5)
−	−	+	−	4 (3.1)
−	−	+	+	3 (2.3)
+	+	+	+	1 (0.8)

PA: Particle agglutination test.
IFA: Immunofluorescence assay.

marriage, as mentioned above. Still, more male ATL patients were observed, as the sex ratio (male to female) for the estimated incidence rate of ATL, 1.5, shows (42–44). No clear cases of ATL developed after transfusion of blood from HTLV-I carrier donors; furthermore, most patients with ATL and carriers of HTLV-I who were detected in the metropolitan areas were born in the ATL-endemic areas and were probably infected with HTLV-I in childhood by their mothers (13,14,44). From this evidence, it was suggested that horizontal transmission routes of HTLV-I rarely, if ever, lead to ATL manifestation. Infection with HTLV-I at a young age (probably during the first few years) is very important for ATL manifestation after growing up, which would mean that vertical transmission of HTLV-I from mother to child poses a higher risk of ATL manifestation.

Epidemiological observation and virological findings show that HTLV-I infection is a main causative agent for ATL manifestation (4,6,7). However, many healthy HTLV-I carriers live in the ATL-endemic areas in Japan, and so HTLV-I infection alone does not necessarily cause the onset of ATL. The annual incidence rate of ATL among HTLV-I carriers older than 40 yr can be estimated at 0.6–1.7 (27,42,43) (Table 3). Therefore, the cumulative (life span of 70 yr) incidence rate of ATL among HTLV-I carriers in Japan is estimated at 4–5% in males and 1–3% in females if competing risks for other diseases are neglected.

INTERVENTION TRIAL

Latent Period of Seroconversion

The age-specific positivity rate of anti-HTLV-I antibody in cross-sectional surveys increases remarkably with age (Fig. 1), however, the increment is not significant in children and young adults. One-year-old children have a low rate of HTLV-I antibody (Table 4) but the rate of positivity rose to 3% in 3-yr-old children, and it remained stable until young adulthood. These findings suggest that seroconversion after vertical transmission of HTLV-I may occur during the first 3 yr, as reported previously (27). In the follow-up studies of seroconversion against HTLV-I in young children (Table 5), only 1 baby among 130 was positive among 1.5-yr-olds; however, 3 had seroconverted among 3-yr-olds. From these observations, the average time lag of seroconversion after vertical transmission of HTLV-I from the mother was estimated at 1–3 yr. Therefore, follow-up studies to clarify the risk for HTLV-I infection from mother to child through breast milk should be continued at least 3 yr.

Prevention Trials for Vertical Transmission

The most important measure for primary prevention of ATL is to control vertical transmission of HTLV-I. The detailed mechanisms of HTLV-I infection from mother to child have not been clarified, but breast-feeding by an HTLV-I-carrier mother is the most conceivable route of HTLV-I transmission (36–41). Therefore, the use of breast milk substitutes should be recommended if expectant mothers are HTLV-I carriers. To evaluate the preventive effect of physicians advising pregnant women who are HTLV-I carriers not to breast-feed their new-born babies, several intervention trials of breast-feeding control are now being carried out in the ATL-endemic areas (27,39,40). Since 1985 in Tsushima Island, more than 100 pregnant women with HTLV-I antibody in sera were advised during the terminal stage of pregnancy not to breast-feed but to bottle-feed (Table 6). More than 70% of these women followed this recommendation, but the remainder refused and continued breast-feeding. For 43 babies, HTLV-I antibody in sera at 18 mo for all 43 babies and at 2 yr for some was checked by particle agglutination test (PA), immunofluorescence assay (IFA), and Western blotting method (WB) (Table 7). No baby showed positive antibody by IFA; however, one in the group fed bottle milk showed positive antibody at 2 yr by both PA and WB (IgG to p24). Very fortunately, no baby in the group of those breast-fed has yet seroconverted. These interim results cannot evaluate the effect of an intervention trial for the risk reduction of vertical transmission of HTLV-I. To

TABLE 6. *Positive rate for anti-HTLV-I antibody* among pregnant women in Tsushima Island in 1985–1987*

Hospital	Subjects	Positives	Rate (%)
National Hospital	853	67	7.8
Nii Hospital	514	36	7.0
Sasuna Clinic	348	21	6.0
Total	1715	121	7.1

* By particle agglutination test (PA) and immunofluorescence assay (IFA).

TABLE 7. *Seroconversion to HTLV-I after quitting breast milk among 18-mo-old children from carrier mother*

Testing method	Children		
	Quitted	Not quitted	Total
PA(+) IFA(+)	0	0	0
PA(+) IFA(−)	3*	0	3
PA(−) IFA(−)	28	12	40
Total	31	12	43

* One case showed positive antibody (IgG and IgM) by Western blotting method and two cases became negative by PA test on 2-year old.

discuss the effect of this prevention trial properly, a further follow-up period of 2–3 yr is necessary.

Issues of Prevention Trials

To prevent HTLV-I infection from an HTLV-I-carrier mother to her child, carrier mothers should be advised not to breast-feed their children. To do this it is important to identify HTLV-I carriers among pregnant women and, therefore, an accurate positivity check for HTLV-I antibody in pregnant women is indispensable. To inform pregnant women of the risks of being an HTLV-I carrier, physicians in charge of intervention trials should know the basic information concerning the risk of HTLV-I transmission from carrier mother to baby and the risk for manifestation of HTLV-I–associated diseases after infection with HTLV-I. Generally, obstetricians first advise expectant mothers who are infected with HTLV-I to refrain from breast-feeding, and pediatricians care for and monitor these children to evaluate the effect of this prevention effort. For this effort to succeed, therefore, continuous teamwork care of expectant women and mothers by several medical departments is necessary. In some cases, the cooperation of family members, especially of the husband, will become the chief support of pregnant women and mothers, who often suffer from confusion and anxiety about the future after being informed of their HTLV-I infection. In the same way, keeping the confidentiality of HTLV-I-carrier mothers and their babies is indispensable.

SUMMARY

HTLV-I carriers and patients with ATL are clustered in limited areas in the world for unknown reasons. Especially in Japan, many HTLV-I carriers and patients with its associated diseases are found. Four nationwide studies on ATL have been conducted, and the fundamental clinicoepidemiological features of

ATL in Japan have been clarified. A sequential surveillance is now going on to clarify the geographical and chronological variation of ATL in Japan, which can give important information on HTLV-I–associated diseases to both the fields of epidemiology and basic science.

To evaluate the risk for ATL and its related diseases among HTLV-I carriers, several large-scale prospective studies have been developed and are now being carried out in the ATL-endemic areas in Japan. It will take several years to follow up and to get results from these prospective studies.

The detailed manifestation mechanisms of ATL are not yet clarified; however, it is certain that HTLV-I infection is the main cause of ATL. Therefore, a study designed to prevent the vertical transmission of HTLV-I from mother to child through breast milk by getting carrier mothers to refrain from breast-feeding has been started. The results of these preventive cohort studies can contribute to the future prevention of HTLV-I–associated diseases in the ATL-endemic areas, not only in Japan but throughout the world.

ACKNOWLEDGMENT

This work was supported by a Grant-in-Aid for Cancer Research from the Ministry of Education, Science and a Grant-in-Aid from the Ministry of Health and Welfare for the Comprehensive 10-Year Strategy for Cancer Control, Japan.

REFERENCES

1. Uchiyama T, Yodoi J, Sagawa K, Takatsuki K, Uchino H. Adult T-cell leukemia. Clinical and hematologic features of 16 cases. *Blood* 1977;50:481–492.
2. Miyoshi I, Kubonishi M, Sumida S, et al. Characteristics of a leukemic T-cell line derived from adult T-cell leukemia. *Jpn J Clin Oncol* 1979;9:485–494.
3. Hinuma Y, Nagata K, Hanaoka M, et al. Adult T-cell leukemia: antigen in a ATL cell line and detection of antibodies to the antigen in human sera. *Proc Natl Acad Sci USA* 1981;78:6476–6480.
4. Hinuma Y, Komoda H, Chosa T, et al. Antibodies to adult T-cell leukemia-virus–associated antigen (ATLA) in sera from patients with ATL and controls in Japan: a nation-wide seroepidemiologic study. *Int J Cancer* 1982;29:631–635.
5. Tajima K, Tominaga S, Suchi T, et al. Epidemiological analysis of the distribution of antibody to adult T-cell leukemia-virus–associated antigen (ATLA): possible horizontal transmission of adult T-cell leukemia virus. *Gann* 1982;73:893–901.
6. Yoshida M, Miyoshi I, Hinuma Y. Isolation and characterization of retrovirus (ATLV) from cell lines of human adult T-cell leukemia and its implication in the disease. *Proc Natl Acad Sci USA* 1982;79:2031–2035.
7. Poiesz BJ, Ruscetti FW, Gazdar AF, Bunn PA, Minna JD, Gallo RC. Detection and isolation of type-C retrovirus particles from fresh and cultured lymphocytes of patients with cutaneous T-cell lymphoma. *Proc Natl Acad Sci USA* 1980;77:7415–7419.
8. Popovic M, Reitz MS, Sarngadharan MG, et al. The virus of Japanese adult T-cell leukemia is a member of the human T-cell leukemia virus group. *Nature* 1982;300:63–66.
9. Watanabe T, Seki M, Yoshida M. ATLV (Japanese isolated) and HTLV (US isolated) are the same strain of retrovirus. *Virology* 1984;133:238–241.
10. Gessain A, Barin F, Vernant JC, et al. Antibodies to human T-lymphotropic virus type-I in patients with tropical spastic paraparesis. *Lancet* 1985;2:407–409.

11. Osame M, Usuku K, Izumo S, et al. HTLV-I–associated myelopathy. A new clinical entity. *Lancet* 1986;1:1031–1032.
12. Tajima K, Tominaga S, Kuroishi T, Shimizu H, Suchi T. Geographical features and epidemiological approach to endemic T-cell leukemia/lymphoma in Japan. *Jpn J Clin Oncol* 1979;9(Suppl. 1):495–504.
13. The T- and B-cell Malignancy Study Group. Statistical analysis of immunologic, clinical and histopathologic data on lymphoid malignancies in Japan. *Jpn J Clin Oncol* 1981;11:15–38.
14. The T- and B-cell Malignancy Study Group. Statistical analyses of clinico-pathological, virological and epidemiological data on lymphoid malignancies with special references to adult T-cell leukemia/lymphoma: a report of the second nationwide study of Japan. *Jpn J Clin Oncol* 1985;15: 517–535.
15. Fleming AF, Maharajan R, Abraham M, et al. Antibodies to HTLV-I in Nigerian blood-donors, their relatives and patients with leukaemias, lymphomas and other diseases. *Int J Cancer* 1986;38: 809–813.
16. Blayney DW, Jaffe ES, Blattner WA, et al. The human T-cell leukemia/lymphoma virus associated with American adult T-cell leukemia/lymphoma. *Blood* 1983;62:401–405.
17. Catovsky D, Greaves MF, Rose M, et al. Adult T-cell lymphoma-leukemia in blacks from the West Indies. *Lancet* 1982;1:639–643.
18. Blattner WA, Kalyanaraman VS, Robert-Guroff M, et al. The human type-C retrovirus, HTLV, in blacks from the Caribbean region, and relationship to adult T-cell leukemia/lymphoma. *Int J Cancer* 1982;30:257–264.
19. Shimoyama M, Minato K, Tobinai K, et al. Anti-ATLA (antibody to the adult T-cell leukemia cell associated antigen)–positive hematologic malignancies in the Kanto district. *Jpn J Clin Oncol* 1982;12:109–116.
20. Shibata A, Aoyagi Y, Aoki A. Adult T-cell leukemia observed at Niigata University Hospital. *Jpn J Clin Oncol* 1983;13:657–666.
21. Brindle RJ, Eglin RP, Parsons AJ, Hill AVS, Selkon JB. HTLV-I, HIV, hepatitis B and hepatitis delta in the Pacific and south-east Asia: a serological survey. *Epidemiol Infect* 1988;100:153–156.
22. Blattner WA, Clark JW, Gibbs WN, et al. Epidemiology and relationship to disease. In: *Retrovirus in human lymphoma/leukemia. Proceedings of the 15th International Symposium of the Princess Takamatsu Cancer Fund.* Tokyo: Japan Scientific Society Press, 1984;23–24.
23. Tsugane S, Watanabe S, Sugimura H, et al. Infectious states of human T lymphotropic virus type I and hepatitis B virus among Japanese immigrants in the republic of Bolivia. *Am J Epidemiol* 1988;128:1153–1161.
24. Ishida T, Yamamoto K, Omoto K, Iwanafa M, Osata T, Hinuma Y. Prevalence of a human retrovirus in native Japanese: evidence for a possible ancient origin. *J Infect* 1985;ii:153–157.
25. Hinuma Y. Seroepidemiology of adult T-cell leukemia virus (HTLV-I/ATLV): origin of virus carriers in Japan. *AIDS Res* 1986;2:s17–21.
26. Maeda Y, Fukuhara M, Takehara Y, et al. Prevalence of possible adult T-cell leukemia virus carriers among healthy volunteer blood donors in Japan: a nation-wide study. *Int J Cancer* 1984;33: 717–720.
27. Tajima K, Kamura S, Ito S, et al. Epidemiological features of HTLV-I carriers and incidence of ATL in an ATL-endemic island: a report of the community-based co-operative study in Tsushima, Japan. *Int J Cancer* 1987;40:741–746.
28. Miyamoto T, Hino S, Muneshita T. Seroepidemiological studies of antibody against antigens expressed on adult T-cell leukemia cells in Nagasaki area. In: Rich MA, ed. *Leukemia Reviews International,* vol. 1. New York and Basel: Marcel Dekker, Inc. 1983:105–106.
29. Kusuhara K, Sonoda S, Takahashi K, Tokugawa K, Fukushige J, Ueda K. Mother-to-child transmission of human T-cell leukemia virus type I (HTLV-I): a fifteen-year follow-up study in Okinawa, Japan. *Int J Cancer* 1987;40:755–757.
30. Kinoshita K, Kamihira S, Yamada Y, et al. Adult T-cell leukemia-lymphoma in Nagasaki district. *GANN Monogr Cancer Res* 1982;28:167–184.
31. Tajima K, Tominaga S, Shimizu H, Suchi T. A hypothesis on the etiology of adult T-cell leukemia/lymphoma. *Gann* 1981;72:684–691.
32. Tajima K, Akaza T, Koike K, Hinuma Y, Suchi T, Tominaga S. HLA antigens and adult T-cell leukemia virus infection: a community based study in the Goto Islands, Japan. *Jpn J Clin Oncol* 1984;14:347–352.
33. Tanaka K, Sato H, Okochi K. HLA antigens in patients with adult T-cell leukemia. *Tissue Antigens* 1984;23:81–83.

34. The T- and B-cell Malignancy Study Group. The third nation-wide study on adult T-cell leukemia/lymphoma (ATL) in Japan: characteristic patterns of HLA antigens and HTLV-I infection in ATL patients and their relatives. *Int J Cancer* 1988;41:505–512.
35. Usuku K, Sonoda S, Osame M, et al. HLA haplotype-linked high immune responsiveness against HTLV-I in HTLV-I–associated myelopathy: comparison with adult T-cell leukemia/lymphoma. *Ann Neurol* 1988;Suppl. 23:s143–150.
36. Kinoshita K, Hino S, Amagasaki T, et al. Demonstration of adult T-cell leukemia virus antigen in milk from three seropositive mothers. *Jpn J Cancer Res* 1984;75:103–105.
37. Nakano S, Ando Y, Ichijo M, et al. Search for possible routes of vertical and horizontal transmission of adult T-cell leukemia virus. *Jpn J Cancer Res (Gann)* 1984;75:1044–1045.
38. Hino S, Yamaguchi K, Katamine S, et al. Mother-to-child transmission of human T-cell leukemia virus type-I. *Jpn J Cancer Res* 1985;76:474–480.
39. Hino S, Doi H, Yoshikuni H, et al. HTLV-I carrier mothers with high-titer antibody are at high risk as a source of infection. *Jpn J Cancer Res* 1987;78:1156–1158.
40. Ando Y, Nakano S, Saito K, et al. Transmission of adult T-cell leukemia retrovirus (HTLV-I) from mother to child: comparison of bottle- with breast-fed babies. *Jpn J Cancer Res* 1987;78:322–324.
41. Kinoshita K, Amagasaki T, Hino S, et al. Milk-borne transmission of HTLV-I from carrier mothers to their children. *Jpn J Cancer Res* 1987;68:674–680.
42. Tajima K, Kuroishi T. Estimation of incidence rate of ATL among ATLV-carriers in Kyushu, Japan. *Jpn J Clin Oncol* 1985;15:423–430.
43. Kondo T, Kono H, Nonaka N, et al. Risk of adult T-cell leukemia/lymphoma in HTLV-I carriers. *Lancet* 1987;2:159.
44. Tajima K, Tominaga S, Suchi T, Fukui H, Komoda H, Hinuma Y. HTLV-I carriers among migrants from an ATL-endemic area to ATL non-endemic metropolitan areas in Japan. *Int J Cancer* 1986;37:383–387.

DISCUSSION

Asked to address the issue of male:female ratio of ATL in Japan, one participant stated that it was 1.2:1 and was thus lower than the previous reports of 1.4:1.

One discussant raised the argument that, since maternal-to-child transmission apparently accounts for only a 3% seroprevalence rate among children up to the age of 19 years, such transmission might not pose a public health problem significant enough to justify the stoppage of breast feeding. Another discussant countered that southwestern Japan is one of the places where maternal-to-child transmission of HTLV-I is most important in the world and that intervention studies should be performed to investigate the effect of cessation of breast feeding. This discussant commended the Japanese for their early intervention efforts.

Human Retrovirology: HTLV,
edited by William A. Blattner.
Raven Press, Ltd., New York © 1990.

Epidemiologic Perspectives of HTLV-I

*Nancy Mueller, ‡Nobuyoshi Tachibana, *Sherri O. Stuver,
†‡Akihiko Okayama, ‡Junzo Ishizaki, ‡Eiichi Shishime,
‡Koichi Murai, ‡Shigemasa Shioiri, and ‡Kazunori Tsuda

*Departments of Epidemiology and †Department of Cancer Biology, Harvard School of
Public Health, Boston, Massachusetts 02115; ‡Second Department of Medicine,
Miyazaki Medical School, Miyazaki 889-16, Japan

INTRODUCTION

The human T-cell leukemia virus (HTLV-I) provides a unique challenge to the epidemiologist. Although its causal connection to the occurrence of adult T-cell leukemia (ATL) and to HAM/TSP is not questioned, little is known at present of the natural history of this infection nor of the role of cofactors in the induction of disease. Analytic studies now underway (1–3), involving the prospective observation of infected populations, are likely to clarify these issues, providing an unprecedented window on viral oncogenesis. However, more fundamental questions related to the descriptive epidemiology of this retroviral infection remain. These include the curious age-prevalence curve of antibodies against HTLV-I and the highly restricted geographic distribution of infected populations. The purpose of this chapter is to discuss, from an epidemiologic perspective, these two related puzzles.

AGE-SPECIFIC SEROPREVALENCE OF HTLV-I ANTIBODY

The seroprevalence of antibodies against HTLV-I has been most extensively described for populations living in endemic areas of Japan. A typical example is that derived from our prospective cohort study based in two highly endemic villages in southern Miyazaki prefecture (Fig. 1). These data from 1533 persons (918 women and 615 men), who were seen in conjunction with a free government-sponsored annual health examination, represent 69% of 2221 adults (aged 40 or older) in the study area. The data are much less representative for younger persons. Blood samples were collected from 1984 through 1987 and tested by the particle agglutination test (Fuji Rebio, Tokyo, Japan) with confirmation by indirect immunofluorescent staining with acetone fixed Hut102 cell antigen, or by immunoblot and radioimmunoprecipitate assay. In this population, there is a gradual increase of seroprevalence between age 30 and 55 with a slight female predom-

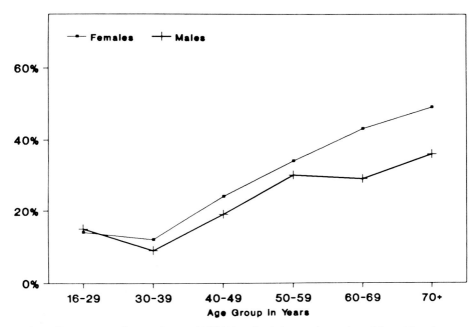

FIG. 1. The age-specific prevalence of HTLV-I antibody by sex in southern Miyazaki prefecture study population, 1984–1987. Age and antibody status at first screen. The number of positive females/males for each age group is as follows: 16–29 yr, 30/26; 30–39 yr, 77/43; 40–49 yr, 180/110; 50–59 yr, 306/206; 60–69 yr, 203/133; 70+ yr, 122/97.

inance. Among older persons, the seroprevalence among men tends to plateau but continues to rise among women.

This same pattern is apparent among 2582 adults sampled in Tsushima Island off Kyushu (1) (Fig. 2). In this population, screening was also done on blood specimens from 1093 children less than 15 yr of age. Among these children, there is a seroprevalence of 1–2%.

These two cross-sectional surveys give us a snapshot of the effect of the cumulative incidence of infection across the age and sex groups. The snapshot is in effect a collage, a cutting and pasting together of the net effect both of age- and time-related transmission patterns. The fact that these two geographically separated populations have similar shapes suggests that they are a valid representation of what characterizes the highly endemic regions of southern Kyushu. For the sake of this discussion, we will term this age-seroprevalence curve a "mature pattern."

Mature Endemic Pattern

As discussed by Blattner elsewhere in this volume, transmission of HTLV-I occurs primarily by two routes, perinatal and sexual exposure. However, this

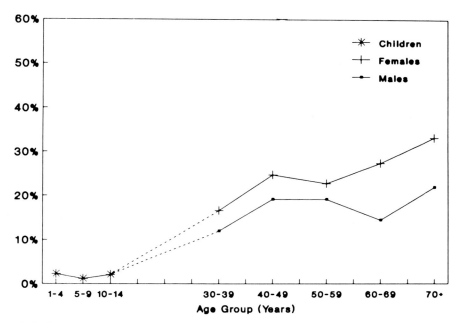

FIG. 2. The age-specific prevalence of HTLV-I antibody, Tsushima Island, 1984–1985. The number of positive children or females/males for each age group is as follows: 1–4 yr, 226; 5–9 yr, 273; 10–14 yr, 201; 30–39 yr, 157/83; 40–49 yr, 311/167; 50–59 yr, 500/291; 60–69 yr, 398/234; 70+ yr, 268/173. Adapted from Tajima et al. (1).

mature pattern differs substantially from the age-seroprevalence curve of antibodies against other viruses with similar transmission patterns. A comparison can be made with the related retrovirus, the human immunodeficiency virus-type 1 (HIV-1), in populations where intravenous drug abuse is quite rare and transfusion infrequent. Such a population would be represented in parts of Central Africa.

Among children, the age-seroprevalence curves are similar for HTLV-I and HIV-1. Both viruses are transmitted perinatally. Because HTLV-I is highly cell-associated, most mother-to-child infections are attributed to exposure via infected lymphocytes in breast milk (4,5). The transmission rate among children born to HTLV-I–infected mothers is ~10–20% (1,6,7). Kusuhara et al. (7) have found in a study of 311 mother/child pairs followed over 18 yr that no children seroconverted by age 18 yr other than those infected perinatally (i.e., with HTLV-I antibodies by age 3). Thus, there is no evidence of horizontal transmission of HTLV-I among children, which is also true of HIV-1.

However, beginning in young adulthood, a very slow increase in prevalence becomes evident with HTLV-I carrier state. This observation contrasts dramatically with what is seen in HIV-1 infection, where there is a steep "shoulder of increase" for seroprevalence beginning in the 20s and 30s (8), as is commonly seen in most sexually transmitted infections (9).

An explanation of this puzzling difference is suggested by emerging data on transmissibility of HIV-1 in relation to the time course of its natural history. As observed by Goedert et al. (10), the sexual transmission of HIV-1 infection by seropositive hemophiliac husbands to their wives was associated with progression of the husbands' immunosuppression and T-cell depletion. Because this progression is accompanied by a reappearance of serologically detectable p24, it would appear that the presence of antigen is a good marker of infectivity. In HIV-1 infection, this occurs at two time points in the natural history: a burst during the initial viremia after primary infection, and a reappearance several years later in the presymptomatic phase, which is sustained into disease (11). By analogy, this suggests that a similar cycle of infectivity holds for HTLV-I, but with a much more extended time course. This time course coupled with the lack of cell-free HTLV-I virus could explain the extremely low transmission rate by sexual exposure.

Several observations are consistent with this explanation. Hino et al. have

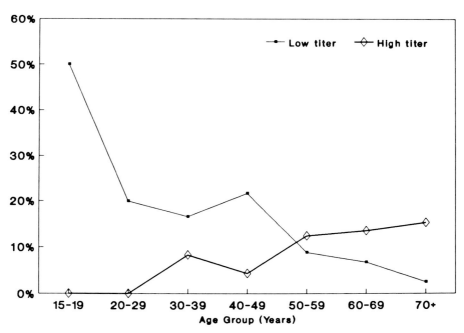

FIG. 3. The distribution of HTLV-I carriers by low titer (1:4) and high titer (>1:160) by age among 193 positive blood donors in Miyazaki prefecture. Antibody titers were determined by fixed membrane antigen method. The number of positive blood donors in each age group is as follows: 15–19 yr, 14; 20–29 yr, 5; 30–39 yr, 12; 40–49 yr, 23; 50–59 yr, 56; 60–69 yr, 44; 70+ yr, 39. Adapted from Okayama et al. (12).

TABLE 1. *Distribution of couples by concordance for antibody status (by sex) and by age-group of husband*

Age	M+/F+	M+/F−	M−/F+	M−/F−	Total
≤39*	1 (0.03)†	3 (0.10)	2 (0.06)	25 (0.81)	31
40–49	16 (0.15)	10 (0.10)	9 (0.09)	69 (0.66)	104
50–59	34 (0.20)	12 (0.07)	26 (0.15)	102 (0.59)	174
≥60	37 (0.30)	3 (0.02)	16 (0.13)	68 (0.55)	124
Total	88 (0.20)	28 (0.06)	53 (0.12)	264 (0.61)	433

* Age at first screen.
† Proportion.
Note: Status defined before seroconversion.

shown that the risk of perinatal transmission of HTLV-I increases with maternal antibody titer (6). As reported by Blattner at this meeting, among HTLV-I–positive Japanese men who migrated to Hawaii, titer level was correlated with seroprevalence among their wives. We have found that among 193 seropositive blood donors from Miyazaki City, the proportion with elevated titers increases with age (Fig. 3) (12). Taken together, these observations suggest that the titer level of antibody may be proportional to the number of infected lymphocytes and concurrent transmissibility, and that this increases with length of infection (age). Thus the average risk of infection by sexual exposure is likely quite low during the most sexually active years, accounting for the very gradual increase in seroprevalence in young adulthood.

An additional feature of note is the divergence of seroprevalence by sex among older persons. Although it is often stated that sexual transmission of HTLV-I occurs predominantly in one direction, i.e., male to female, the parallel increase of seroprevalence for men and women during early and middle adulthood implies that sexual transmission occurs in both directions. However, after age 50, it appears that sexual transmission occurs almost solely from men to women. This shift may reflect the effect of physiologic changes in the genital tract of women following menopause, changes which reduce the probability of transmission to a male partner during sexual intercourse. An alternative explanation is that after menopause, women are much more susceptible to infection.

Evidence to support the validity of the assertion that sexual transmission accounts for the "growth" of the curve among adults comes from the prospective follow-up of our study population. In the first four years of observation, a total of 433 married couples were identified within the study population (Table 1). Similar to the reports of Tajima (13) and Kajiyama (14), we found a high concordance of antibody-status within couples. Cross-sectionally, the portion of discordant couples with only the husband infected decreases from 10% for couples in which the husband is less than 40 yr of age to 2% after age 60. For discordant couples with only the wife infected, the reverse is true, increasing from 6% to 13% over the age groups.

TABLE 2. *Characteristics of subjects seroconverting for HTLV-I antibody by particle agglutination assay during prospective follow-up*

Case	Sex	Age*	Length of marriage†	1984	1985	1986	1987
					Antibody status		
1	Female	61	40 yr	−	−	ND‡	+
	(husband	59		+	+	ND	+)
2	Female	47	25 yr	−	−	−	+
	(husband	51		+	+	ND	ND)
	Note: Both of husband's parents are positive.						
3	Male	54	25 yr	ND	−	+	+
	(wife	48		ND	+	+	+)
	Note: daughter, aged 24 yr, is positive.						
4	Female	62	39 yr	ND	ND	−	+
	(husband	62		ND	ND	+	ND)
5	Female	59	2 yr	ND	−	+	+
	(husband	75		ND	−	−	−)
6§	Female	52	32 yr	ND	−	−	+
	(husband's status is unknown)						

* Age at first screen.
† Length of marriage at seroconversion.
‡ ND = not done.
§ Not confirmed. Both 1985 and 1986 sera were positive by ELISA and radioimmunoprecipitation assay.

By the end of 1987, after four rounds of screening in one village and three in the other, six seroconversions had been observed: five of these occurred among study couples (Table 2). All negative sera of these six subjects were retested by immunoblot, radioimmunoprecipitation assay, and ELISA (Cambridge Bioscience, Cambridge, MA) by Dr. Yi-ming A. Chen (HSPH). All were confirmed negative except those of subject number 6 which were positive on ELISA and radioimmunoprecipitation assay. Among the 433 couples, there were 609 persons

TABLE 3. *Distribution of subjects for risk factors for seroconversion and the relative risk associated with each factor among study couples*

Risk factor	Number seroconverting (%)	Number at risk*	RR†	p-value‡
Carrier spouse				
Yes	4 (6.6)	61	24.8	0.002
No§	1 (0.3)	378		
Sex of carrier spouse				
Male	3 (14.3)	21	5.7	0.11
Female§	1 (2.5)	40		

* Seronegative spouses with more than one annual blood screen.
† Relative risk.
‡ Fisher's exact p-value.
§ Referent category.

TABLE 4. *Incidence rate of seroconversion of seronegative wives of carrier husbands by age of husband at last annual screen and the relative risk associated with husband's age*

Age of husband	Number seroconverting	Women-years of observation	Incident rate per year	Relative risk
≥60 yr	2	8	0.25	8.5
<60 yr*	1	34	0.03	(p = 0.10)†

* Referent category.
† Fisher's exact p-value.

without antibody. Of these, 439 had been seen for more than one screen, providing the opportunity for seroconversion to be observed. As shown in Table 3, the relative risk (RR) of seroconversion among those "exposed" to a carrier spouse compared with those with a negative spouse was more than 20-fold. This was five times more likely to occur among wives of carriers than among husbands. To estimate the RR, we then examined the risk of seroconversion among all seronegative women with carrier husbands in relation to the husband's age (Table 4). We considered each interval between annual screenings as person units of observation for seroconversion at the end of the interval. Each interval was categorized by the husband's age at the start of the interval. In the case where an intervening screen was missed, we assumed that the woman's serostatus at the previous screen held for the missed observation. Based on this analysis, we found that the uninfected wives were more than eight times as likely to seroconvert if their infected husband were at least 60 years of age than if he were younger. Note that the only apparent female-to-male transmission involved a woman in her late 40s. These observations are consistent with sexual transmission of the infection particularly by older infected men.

In summary, the mature age-specific seroprevalence curve for HTLV-I antibodies begins at less than 5% from birth to adolescence, gradually increases from age 20 to the mid 50s to ~25–30% for both sexes, then continues to increase to 35–50% among elderly women (Figs. 1 and 2).

Evolving Endemic Pattern

In contrast, an "evolving" curve can be seen in less endemic areas of Japan, where a slow diffusion of the infection appears to be occurring. Data from Miyazaki Prefecture illustrate this effect. Data were collected from 7055 consecutive persons who came to Miyazaki City Health Promotion Center from September 1983 through December 1984. HTLV-I antibody seropositivity was determined with the use of the indirect immunofluorescence assay using acetone-fixed HTLV-I–producing cells. As this test is somewhat less sensitive than the particle agglutination assay (see Okayama et al., this volume), the seroprevalence in this population may be underestimated. These data are shown in Fig. 4.

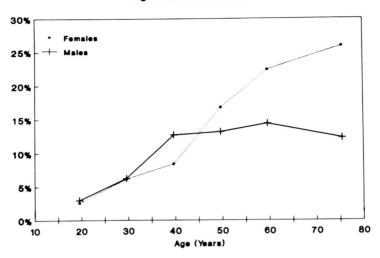

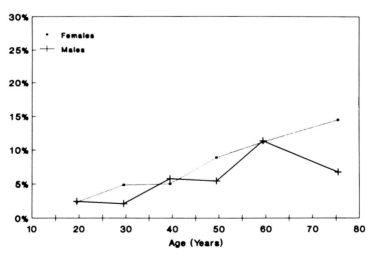

FIG. 4. The age-specific prevalence of HTLV-I antibody by sex in southern (a) and northern (b) Miyazaki prefecture general population by sex, September 1983–December 1984 (S. Stuver, unpublished data). In (a) southern Miyazaki prefecture, the numbers of females/males tested in each age group were as follows: 15–24 yr, 157/169; 25–34, 195/224; 35–44, 333/173; 45–54, 422/167; 55–64, 340/118; 65+, 58/57. In (b) northern Miyazaki prefecture, the numbers were: 15–24 years, 297/284; 25–34, 329/427; 35–44, 716/450; 45–54, 774/311; 55–64, 553/220; 65+, 159/118.

In the southern (more endemic) half of the prefecture, a mature curve is evident, although the level of saturation is about one-half that in the highly endemic populations noted above. However, in the northern half of the prefecture, a less clearly defined curve is apparent with a lower overall magnitude. This likely reflects the slow diffusion of the infection brought by migrants from the southern area. Given the evidence that suggests that HTLV-I has been present in Kyushu since prehistoric times (15), the time required to establish a stable, mature age curve of seroprevalence at a relatively high saturation is exceedingly long, likely requiring many generations.

Gallo et al. (16,17) have proposed that HTLV-I infection was brought to Japan by Africans aboard Portuguese ships visiting southern Kyushu in the 16th century. This hypothesis is based on the assumption that the initial focus of HTLV-I infection was in Africa. If this assumption is true, then we would expect that the seroprevalence of HTLV-I infection in African populations would be characterized by a mature age curve with relatively high saturation. However, this is not evident in available data.

De Thé and Gessain (18) have recently summarized the seroprevalence data from adult populations in Africa; the overall prevalence ranges from 0 to 10%. Because these estimate the crude seroprevalence in each of the populations sampled, they may be heavily weighted by younger adults and include relatively fewer old persons in comparison with Japanese data. Even so, the estimates are surprisingly low. Delaporte et al. (19) surveyed nearly 2000 adults in Gabon, identified by representative sampling (Fig. 5). Based on their report, it appears that the infection is more established in the rural population than in the urban, where it may have been more recently introduced by population migration. In the rural group, there is no evidence of an increase among the oldest group, as would be expected from Japanese data; however, the number, sex, and age range of subjects in this group are not given and are probably not comparable. Therefore, comparing the age seroprevalence curve of those younger than age 40 with the Japanese data presented here, that seen in the rural Gabon population is more similar to the southern half of Miyazaki Prefecture (Fig. 4) than to a fully mature, highly saturated curve. This is despite the fact that the "shoulder of increase" among young adults is much sharper than in Japan, suggesting that more sexual transmission occurs in Gabon at younger ages.

These observations, in conjunction with the lack of evidence of endemic infection in Portugal (20,21), support the hypothesis that HTLV-I "was already present in Japan long before the first Portuguese came to Japan in the 16th century" (16).

INTERPRETATION OF ENDEMIC PATTERNS

It appears that establishment of a relatively saturated, mature endemic HTLV-I infection in a population takes an exceedingly long time. In Japan, the highest

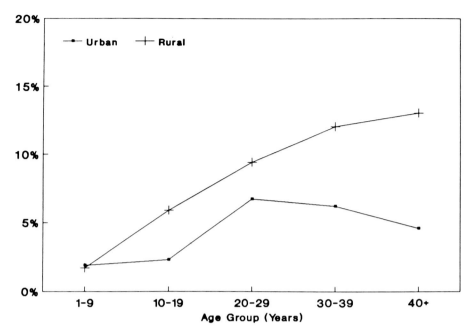

FIG. 5. The age-specific prevalence of HTLV-I antibody for 1115 rural and 383 urban adults in Gabon. Adapted from Delaporte et al. (19).

rates of seropositivity are characteristically seen in "island" populations; that is, geographically restricted groups. The reasons for this microepidemicity likely reflect the difficulty of transmitting this infection other than by perinatal exposure. To illustrate this, consider the following. Assume that the perinatal transmission rate observed in present-day Japan has been true over the lifetime of the virus. Simply to maintain the number of infected people as established by perinatal transmission, each infected woman would have to produce at least one infected female child and one infected male child. With an average transmission rate of 20%, each infected woman would then have to produce 10 children surviving to reproduction. However, this number can be reduced if the proportion of women who are carriers increases because of sexual transmission by or during the childbearing years. In fact, the curve is driven primarily by the proportion of women who are infected (and infectious) during their reproductive years. It may be that women who become infected shortly before childbearing are more infectious than those who had been infected as infants. The level of infection among older people has little effect on the endemicity of infection, but rather appears to be an informative indicator of the stability of the curve and the degree of saturation. If this scenario is true, then the relative amount of increase in seroprevalence within a generation is likely to be very small.

Evidence to challenge the validity of our interpretation of the present data could include both epidemiologic and biologic investigation. The prospective follow-up of endemic populations with particular focus on discordant couples would provide data on the validity of our observed seroconversions. Systematic pedigree studies could evaluate the relative transmission rate between women who were likely infected perinatally (i.e., whose own mothers are positive) and those who were likely sexually infected (i.e., whose mothers are negative). Screening of population-based samples from Africa which are enriched with older people could verify whether the infection in Africa is low as presently thought. In addition, attempts to isolate virus from genital tissues and fluids from healthy carriers over a range of ages, combined with data on virus status (such as antibody titer), could provide insight concerning the natural history of transmissibility of HTLV-I.

SUMMARY AND CONCLUSIONS

In summary, looking at HTLV-I from an epidemiologic perspective suggests that the establishment and spread of the infection in a population involve a complex mixture of effects of behavior and biology over an extended time period. Elements of this mixture include normative sexual behavior, fertility patterns, and factors influencing survival to reproduction. Our hindsight cannot illuminate what prevailed in endemic areas in the past, or what accommodation has occurred between virus and host to result in the present seroprevalence. However, we should be able to gain a clearer understanding of present transmission risks to guide the development of sensitive, minimally disruptive public health policies to reduce the transmission of infection among future generations.

ACKNOWLEDGMENTS

This work is supported by a grant from the National Institutes of Health (USA), 2 R37 CA38450. Dr. Mueller is a recipient of an American Cancer Society Faculty Research Award. Mrs. Stuver is supported by an Institutional National Research Service Award in Environmental Epidemiology, 1 T32 ES07069.

REFERENCES

1. Tajima K, Kamura S, Ito S-i, et al. Epidemiological features of HTLV-I carriers and incidence of ATL in an ATL-endemic island: a report of the community-based co-operative study in Tsushima, Japan. *Int J Cancer* 1987;40:741–746.
2. Tachibana N, Okayama A, Ishizaki J, et al. Suppression of tuberculin skin reaction in healthy HTLV-I carriers from Japan. *Int J Cancer* 1988;42:829–831.
3. Murphy EL, Hanchard B, Figueroa JP, et al. Modelling the risk of adult T-cell leukemia/lymphoma in persons infected with the human T-lymphotropic virus type I. *Int J Cancer* 1989;43:250–253.

4. Kinoshita K, Hino S, Amagasaki T, et al. Demonstration of adult T-cell leukemia virus antigen in milk from three sero-positive mothers. *Gann* 1984;75:103–105.
5. Hino S, Yamaguchi K, Katamine S, et al. Mother-to-child transmission of human T-cell leukemia virus type 1. *Jpn J Cancer Res* 1985;76:474–480.
6. Hino S, Sugiyama H, Doi H, et al. Breaking the cycle of HTLV-I transmission via carrier mothers' milk. *Lancet* 1987;2:158–159.
7. Kusuhara K, Sonoda S, Takahashi K, Tokugawa K, Fukushige J, Ueda K. Mother-to-child transmission of human T-cell leukemia virus type I (HTLV-I): a fifteen-year follow-up study in Okinawa, Japan. *Int J Cancer* 1987;40:755–757.
8. Quinn TC, Mann JM, Curran JW, Piot P. AIDS in Africa: an epidemiologic paradigm. *Science* 1986;234:955–963.
9. Aral SO, Holmes KK. Epidemiology of sexually transmitted disease. In: Holmes KK, Mardh P-A, Sparling PF, Wiesner PJ, eds. *Sexually transmitted diseases.* New York: McGraw-Hill, 1984;126–141.
10. Goedert JJ, Eyster ME, Biggar RJ, Blattner WA. Heterosexual transmission of human immunodeficiency virus: association with severe depletion of T-helper lymphocytes in men with hemophilia. *AIDS Res Hum Retroviruses* 1987;3:355–361.
11. Redfield RR, Burke DS. HIV infection: the clinical picture. *Sci Am* 1988;259:90–99.
12. Okayama A, Tachibana N, Ishizaki J, Yokota T, Shishime E, Tsuda K. An immunological study of human T-cell leukemia virus type (HTLV-I) infection (I). Detection of low-titer anti-HTLV-I antibodies by HTLV-I associated membrane antigen (HTLV-MA) method. *J Jpn Assoc Infect Dis* 1987;61:1363–1368. (In Japanese)
13. Tajima K, Tominaga S, Suchi T, et al. Epidemiological analysis of the distribution of antibody to adult T-cell leukemia-virus-associated antigen (ATLA): possible horizontal transmission of adult T-cell leukemia virus. *Gann* 1982;73:893–901.
14. Kajiyama W, Kashiwaga S, Ikematsu H, Hayashi J, Nomura H, Okochi K. Intrafamilial transmission of adult T-cell leukemia virus. *J Infect Dis* 1986;154:851–857.
15. Ishida T, Yamamoto K, Omoto K, Iwanaga M, Osata T, Hinuma Y. Prevalence of a human retrovirus in native Japanese: evidence for a possible ancient origin. *J Infect Dis* 1985;11:153–157.
16. Gallo RC, Sliski A, Wong-Staal F. Origin of human T-cell leukemia-lymphoma virus. *Lancet* 1983;2:962–963.
17. Gallo RC, Sliski A. Origins of human T-lymphotropic viruses. *Nature* 1986;320:219.
18. de Thé G, Gessain A. HTLV-I and associated diseases in Europe and Africa: a sero-epidemiological survey. Working paper, Scientific group on HTLV-I Infections and Its Associated Diseases, World Health Organization, Regional Office for the Western Pacific, Kagoshima, Japan, December 10–15, 1988.
19. Delaporte E, Dupont A, Peeters M, et al. Epidemiology of HTLV-I in Gabon (western equatorial Africa). *Int J Cancer* 1988;42:687–689.
20. Larouze B, Schaffar-Deshayes L, Blesonski S, et al. Antibodies to HTLV-I p24 in African and Portuguese populations. *Cancer Res* (suppl) 1985;45:4630S–4632S.
21. Cardoso EA, Robert-Guroff M, Franchini G, et al. Seroprevalence of HTLV-I in Portugal and evidence of double retrovirus infection of a healthy donor. *Int J Cancer* 1989;43:195–200.

DISCUSSION

One speaker suggested that it is difficult to account for the marked increase in HTLV-I seroprevalence in women over the age of 50 on the basis of sexual transmission. This speaker pointed out that work performed in Gabon suggests that sexual transmission is highest in younger age groups, as might be expected. Another participant responded that of the three phases of the HTLV-I seroprevalence curve—the low stable seroprevalence in children up to the age of 19, the parallel increase in both sexes between 20 and 50 years of age, and then a further increase in females beyond the age of 50—the latter is the most difficult to explain, and further research will be necessary in several endemic areas to explain this selective increase in females over the age of 50. A discussant questioned whether the interval between exposure and seroconversion in sexual transmission of HTLV-I might account for the observed age-specific rates. The previous discussant responded that the time to seroconversion in sexual transmission of HTLV-I is unknown. The question of whether observed seroprevalence might be affected by a diminished antibody response caused by HTLV-I–induced immunosuppression was raised. It was stated that there are some studies concerning immunosuppression associated with HTLV-I infection, in particular a study of skin-test responsiveness published recently in the *International Journal of Cancer,* which indicates that the highest rates of low response or anergy are in adults over the age of 60. A discussant commented that hormonal changes in both males and females may be responsible for intermittent activation of the virus and intermittency or a delay in the antibody response. Such hormonal effects could explain seroconversion many years following infection; in vitro studies support this hypothesis. A question was asked whether, in the rabbit model of HTLV-I infection, females demonstrate a greater antibody response than males. One of the participants indicated that this is in fact the case.

Human Retrovirology: HTLV,
edited by William A. Blattner.
Raven Press, Ltd., New York © 1990.

The Epidemiology of HTLV-I: Modes of Transmission and Their Relation to Patterns of Seroprevalence

Edward L. Murphy

Departments of Laboratory Medicine and Medicine, University of California at San Francisco, San Francisco, California 94110

INTRODUCTION

Human T-lymphotropic virus type I (HTLV-I) was discovered by Poiesz and Gallo in 1980 (1) and was subsequently found to be identical to adult T-cell leukemia/lymphoma virus independently isolated by Hinuma et al. (2). The virus has been etiologically linked to adult T-cell leukemia/lymphoma (ATL), a malignancy of mature CD4 lymphocytes. HTLV-I has also been associated with a progressive demyelinating myelopathy known variously as tropical spastic paraparesis or HTLV-I–associated myelopathy (HAM/TSP) (3,4). Endemic areas for the virus include southern Japan, most Caribbean islands, areas of continental North and South America bordering the Caribbean, and sub-Saharan Africa. The prevalence of HTLV-I antibodies increases linearly with age and is higher in women than in men. HTLV-I is transmitted from mother to child (5), by sexual intercourse (6,7), by contaminated blood transfusion (8,9), and by sharing of hypodermic needles (10). This chapter will review the modes of transmission of HTLV-I, and attempt to explain the measured contemporary age- and sex-specific seroprevalence of the virus in Jamaica by a model that combines these infection rates.

GEOGRAPHIC DISTRIBUTION OF HTLV-I

Tajima was the first to describe the geographically limited distribution of HTLV-I seroprevalence. He found high levels (30% in those aged over 40) of HTLV-I antibodies in residents of the Goto islands in southern Japan, an area with high incidence of ATL (11). There was an almost linear increase in HTLV-I seroprevalence with age, and women had higher prevalence than men over the age of 40. Maeda et al. tested for HTLV-I antibodies among blood donors at numerous blood centers across Japan (12). He found the highest seroprevalence

(3%) in the southern district of Kyushu and confirmed the age-related increase in seroprevalence, but also noted variation in HTLV-I prevalence among neighboring blood centers.

In Okinawa, antibody to HTLV-I has been found at high levels in patients with ATL and other hematologic malignancies (13). In the general population of Okinawa, prevalence of antibodies to HTLV-I increased linearly with age from 3% (age 0–9) to 22% (age over 70) in men and from 2% (age 0–9) to 36% (age over 70) in women (14). The pattern of age-specific antibody prevalence is different in Okinawa than in Kyushu, with a female excess of seroprevalence being apparent at a younger age, namely in the 20- to 30-yr-old age group.

HTLV-I was described initially in patients from the Caribbean region who had immigrated to the United Kingdom (15) and to the United States (16). Similar patterns of seroprevalence, at differing absolute magnitudes, have been observed in countries bordering the Caribbean Sea. Prevalence of HTLV-I at levels of 1% or greater has been reported from Barbados (17), Trinidad (18), Panama (19), Venezuela (20), Colombia (21), and Brazil (22). Of interest was the fact that HTLV-I antibodies were concentrated in residents of African rather than Oriental or European origin (18). In the southeastern United States, HTLV-I antibodies have also been described in samples referred to state virology laboratories at frequencies of 1/189 (Florida) and 2/95 (Georgia) (23).

Immigrants from Japan or the Caribbean living in Hawaii, New York, and the United Kingdom also have elevated HTLV-I seroprevalence. Finally, serologic reactivity to the virus has also been detected in Amerindian populations in Panama, Alaska, and Greenland, although the results of more specific confirmatory HTLV-I assays have been variable in this group, suggesting the possibility of a cross-reactive virus.

AGE- AND SEX-SPECIFIC SEROPREVALENCE

The most consistent feature of HTLV-I epidemiology has been the distinctive age- and sex-specific variation in seroprevalence. Prevalence is low and similar in both sexes during childhood. Beginning in adolescence, prevalence increases linearly with age to maximal values in those over the age of 60, and may decline slightly after this age. However, female prevalence increases at a greater rate with age than does male prevalence, leading to an almost twofold excess of female prevalence by middle age. In Okinawa and in the Caribbean, the female excess is apparent as early as the 20- to 29-yr age group, whereas surveys from the Kyushu district of Japan do not show a female excess until middle age.

Among patients at a hospital clinic in Kingston, Jamaica, an age-specific increase in HTLV-I seroprevalence from 2% in ages 0–19 to 17–18% in ages 60–69 was seen, with no significant difference between the sexes (24). However, in a much larger survey of healthy Jamaicans throughout the island, HTLV-I seroprevalence rose from 1% in both sexes for ages 0–19 to 9% (ages over 70) in men and to 19% (ages over 70) in women (25).

Explanation of this age-sex pattern has included at least three hypotheses: 1) Accumulation of seroconversion due to newly acquired infections leads to the age-related increase, with females more susceptible to sexual infection; 2) a cohort effect, with declining seroprevalence in more recent generations due to improvement in general hygiene or changes in breastfeeding and sexual habits; and 3) if a condition of latent HTLV-I infection exists (viral integration into host genome without antibody production), the age-related increase in antibody prevalence could be due to a time-dependent expression of antibodies to latent infections, perhaps in relation to lymphocyte activation by other infectious agents.

MODES OF TRANSMISSION

HTLV-I has several known modes of transmission. Mother-to-child transmission, sexual intercourse, blood transfusion, and re-use of contaminated needles are the major routes by which the virus is spread (see Table 1). Cross-sectional epidemiologic studies from Japan and the Caribbean have provided most of the data in support of the various transmission modes. I shall attempt to summarize the current state of knowledge in this section.

Mother to Child

Mother-to-child transmission was postulated early for HTLV-I as the result of family studies that showed unusually high rates of HTLV-I seropositivity in parents, siblings, and offspring of ATL patients (11,26,27). Subsequent studies confirmed that children of seropositive mothers had an approximate 20% prevalence of antibodies as compared with 1–2% among unselected control children (7). Small prospective studies resulted in conflicting reports on the ability to isolate HTLV-I from the blood of infants born to seropositive mothers (28,29). Passive IgG antibody from the mother may be present for at least 6 mo after birth, and a minority of children will show a subsequent reappearance of antibody after 1 yr of age (5). These data, along with observations of transmission of HTLV-I by breast milk in the marmoset model, have led to the implication of breast-feeding as an important route of transmission (30). The extent to which transplacental or intrapartum blood-borne infection occurs will require larger prospective studies.

Sexual Transmission

Sexual transmission was postulated to be important because HTLV-I seropositivity is more prevalent in the spouses of infected individuals in Japan. Cross-

TABLE 1. *Modes of transmission of HTLV-1*

Mother-to-Child
20% of offspring of seropositive mothers are seropositive
Breast-feeding most likely route
Transplacental and intrapartum transmission may also occur
Sexual
Male to female during vaginal intercourse
Higher number of sexual partners is risk factor for women
Female-to-male sexual transmission less efficient
Penile lesions increase a man's risk of infection
Homosexual anal intercourse is risk factor in endemic areas
Blood Transfusion
40–60% risk of seroconversion per positive blood unit
Seroconversion occurs average of 2 mo post-transfusion
Fresh frozen plasma carries low risk, platelets high risk
Longer shelf life decreases infectivity
Intravenous Drug Abuse and Needle Sharing
HTLV-II may be prevalent in U.S. drug abusers
This virus cross-reacts with HTLV-I antibody assays
? Vectors
Increased HTLV-I seroprevalence in patients with parasitic infection
Healthy HTLV-I seropositives do not have higher prevalence of antibodies
    to Strongyloides or arboviruses

---

sectional analysis of the number of discordant husband positive/wife negative and husband negative/wife positive pairs has led to the conclusion that HTLV-I is transmitted by vaginal intercourse, and that the efficiency of transmission is more than 100-fold greater from male to female than from female to male (7,11). In Jamaica, the author has observed an elevated HTLV-I seroprevalence among women, but not men, who were treated at two sexually transmitted disease clinics (compared with a reference group ascertained through occupation). Having an increased lifetime number of sexual partners was associated with an elevated risk of HTLV-I seropositivity in women but not in men, but penile sores or ulcers increased a man's chances of seropositivity (6).

Blood Transfusion

Blood transfusion provides an optimal means for transmission of HTLV-I–infected lymphocytes. Okochi et al. have shown that cellular blood products from HTLV-I–infected donors produce infection in ∼60% of transfusions (8). Fresh frozen plasma carries a much lower, but not zero, risk of infection. This result is consistent with the laboratory observation that HTLV-I particles are predominantly found in close association with lymphocytes, and that infection of target cells requires cocultivation with infected cells rather than free virus particles (31). Increased time between donation and transfusion of blood appears to decrease infectivity, suggesting the viability of HTLV-I–infected lymphocytes is a crucial factor.

Screening for antibodies to HTLV-I in donated blood has been instituted in both Japan and the United States. It is clear that such measures are warranted in areas of high HTLV-I prevalence because of the relatively high risk of transmitting a carcinogenic and neuropathic virus. In countries with low prevalence, the cost versus benefit of such screening has been debated because of the apparent low risk of disease among carriers. In the absence of long-term studies, however, such estimates are relatively uncertain, and it is probably prudent to decide in favor of screening.

Intravenous Drug Abusers

Reactivity on HTLV-I antibody assays has been demonstrated at reasonably high prevalence in intravenous drug abusers in several areas of the United States (10,32). Prevalence has been circa 20% of drug abusers in several studies, in contrast to HIV prevalence, which ranges from 10 to 50% in these same populations. Prevalence is higher in the older age groups and in subjects of black race. There has been concern, however, because patterns of reactivity on the Western blot confirmatory have been different from those observed with samples from Japan or Jamaica (33). Polymerase chain reaction analysis of DNA from drug abusers has suggested that a large proportion of this serologic reactivity may in fact be due to human T-lymphotropic virus type II (HTLV-II) (34).

Vectors

Although high rates of antibody to HTLV-I have been demonstrated in patients with clinical strongyloidiasis (35) and in subjects with high titers of anti-filarial antibodies (36), the exact nature of the relationship between HTLV-I and parasites has not been determined. In contrast, there was no difference in the prevalence of antibody to Strongyloides or to five arboviruses between HTLV-I–positive and HTLV-I–negative groups of healthy Jamaicans (37,38). Because HTLV-I is generally more cell-associated than HIV, it is felt that vector-borne transmission of HTLV-I is even less likely than such spread of HIV.

MODELING TRANSMISSION

The possibility of latent HTLV-I infection without antibody production has been raised to account for some portion of the age-related rise in seroprevalence of the virus. There is little evidence to suggest that this mechanism is biologically important for HTLV-I; however, studies using the polmerase chain reaction technique have demonstrated HIV provirus sequences in homosexual men as long as 2 yr before HIV seroconversion. Additional virologic studies will be necessary to determine the existence of such a phenomenon in HTLV-I infection.

However, current knowledge about the modes of transmission of HTLV-I may allow the construction of a mathematical model that will estimate the proportion of prevalent seropositivity attributable to known routes of transmission. The difference between observed age- and sex-specific seroprevalence and that predicted by the model would represent an upper limit to the proportion of seropositivity attributable to latent infection.

The working hypothesis is that the proportion of seropositivity due to latent infection is equal to the difference between observed seroprevalence and seropositivity attributable to known sources of infection. Several assumptions are necessary to test this hypothesis. The first is that age- and sex-specific seroprevalence rates have been stable over time (lack of a cohort effect). A corollary of this is that an individual's risk of infection as he/she ages may be approximated by current age-specific seropositivity rates. The second assumption is that no other major routes of infection exist. If seroconversion resulting from latent infection is to be estimated by the observed minus expected seroprevalence, an unknown route of transmission would result in an overestimate of latent infection. Finally, the usefulness of this model is limited by the accuracy (or inaccuracy) of its parameters. For transfusion, these are relatively well determined, but the parameters for sexual transmission are poorly characterized at this time and may distort the model.

I shall now estimate the numerical parameters for each mode of transmission to be included in the model. The example of HTLV-I infection in Jamaica will be used to illustrate the discussion, and parameters based on Jamaican data will be used whenever possible.

Mother-to-Child Parameters

The seroprevalence of HTLV-I among Jamaican women of childbearing age is ~4%. The efficiency of transmission from a seropositive mother to her child has been estimated at 20% by various authors. Therefore, one would expect a seroprevalence of 0.8% among babies resulting from this route of transmission. In fact, a survey of 500 children less than 2 yr of age found 4 seropositives, which corresponds well with the estimate (Murphy and Ramlal, unpublished data). Of 50 000 births per year in Jamaica, an estimated 0.8%, or 400 infants, will be infected by mother-to-child transmission.

Sexual Transmission Parameters

The risk of HTLV-I infection in a woman per episode of intercourse with an infected man is unknown. Estimates for HIV vary according to the clinical status of the male partner, but 1 per 1000 is a representative estimate. If the same risk is used for HTLV-I, the annual risk per woman per year may be estimated: .04 (HTLV-I prevalence among sexually active men) times .001 risk per intercourse times 50 episodes of intercourse per year (from survey data), or .002 (0.2% per year). For 570 000 Jamaican women in the sexually active age groups, this leads to an estimate of 1140 new infections annually.

Published and unpublished data on female-to-male sexual transmission indicate that the risk may be substantially lower than that from male to female. My best estimate is that the risk is only one-tenth of the male-to-female risk per intercourse. The annual risk of sexual infection per man may be estimated: .08 (HTLV-I prevalence among sexually active women) times 60 episodes of intercourse per year (from survey data) times .0001 risk per intercourse, or 0.0005 (0.05% per year). For 550 000 Jamaican men in the sexually active age groups, this leads to an estimate of 275 new infections annually.

Blood Transfusion Parameters

Our estimates of the risk of HTLV-I infection resulting from blood transfusion are based on the firmest data; ironically, this route is numerically the least important. Risk of blood-borne infection per person per year may be estimated: 0.003 (average annual probability of receiving a blood transfusion in Jamaica), times 2 (average units of blood per transfusion), times .024 (seroprevalence among Jamaican blood donors), times 0.5 (risk of infection per positive blood unit), or 0.000072 (0.007 percent per year). For two million Jamaicans, leads to an estimate of 144 new infections per year.

Formulation of Equations

The above parameters may be formulated into equations that express the age-specific prevalence of HTLV-I seropositivity as the sum of cumulative infection by each mode of transmission. The male and female equations differ only in the sexual term:

Males: Model Seroprev. = 0.8 + 0.05 (Age − 15) + 0.007 (Age)
(Percent)

            Mother-   Sexual   Transfusion
            child

Females: Model Seroprev. = 0.8 + 0.2 (Age − 15)   + 0.007 (Age)
(Percent)

These equations yield the results depicted in the lower (solid) portion of the bar graphs in Fig. 1. Seroprevalence data by sex and by 10-yr age group are available from Jamaica (25). These data are represented in Fig. 1 by the total height of each bar.

The difference between observed and model seroprevalence is represented by the upper (striped) portion of each bar graph. According to the hypothesis stated

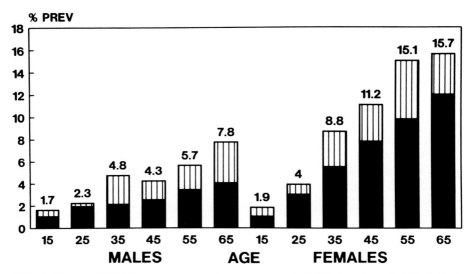

FIG. 1. Observed HTLV-I seroprevalence from a survey of 13 000 Jamaicans (total height of each bar) and HTLV-I seroprevalence predicted by the transmission model (solid portion of each bar). For males (left side) and females (right side), ages shown are the midpoints of 10-yr age groups. The striped portion of each bar represents the difference between these two values, and may be interpreted as the upper limit of other causes of seroconversion (i.e., seroconversion from latent infection).

earlier, this difference may represent the proportion of seroconversion because of delayed antibody response to latent infection. In numerical terms, seroconversions resulting from latent infection (i.e., the difference between observed and the model) occur at a rate approximately 0.06% per year. This is lower than seroconversion from male-to-female sexual transmission, but 10-fold greater than seroconversion because of blood transfusion. In relative terms, seroconversion from latent infection may account for up to 20% (females) or up to 50% (males) of the age-related increase in seroprevalence.

CONCLUSIONS

Most of our knowledge on transmission of HTLV-I is drawn from cross-sectional studies. In the 9 yr since the discovery of the virus, substantial progress has been made in determining the main modes of transmission and in estimating their relative importance. However, there are still puzzling aspects to the epidemiology of HTLV-I, in particular its unusual pattern of age- and sex-specific seroprevalence. At this stage of investigation, more exact parameters on the risk of transmission are needed to understand the dynamics of HTLV-I in human populations and the reasons for the unusual geographical distribution of the virus, and to plan strategies to combat the infection.

By definition, cross-sectional epidemiological studies do not allow separation of causes from effects. For this reason, prospective studies of the transmission of HTLV-I are needed. Good prospective studies of HTLV-I transmission by blood transfusion have already been accomplished in Japan and Jamaica and have led to the institution of HTLV-I–antibody screening of blood donors in several countries. Studies of mother-to-child transmission are currently underway in both Jamaica and Japan and should yield important data on this route. Sexual transmission of HTLV-I and transmission of HTLV-I or HTLV-II by contaminated needles are areas ripe for additional prospective study. In addition, the new polymerase chain reaction technique, which detects HTLV-I–specific nucleic acid sequences in human cells, will allow direct, as opposed to serologic, diagnosis of infection.

Finally, I believe that the numerical estimates of risk and transmission rate obtained from cross-sectional and prospective studies may be combined in mathematical models of the epidemiology of HTLV-I. The initial aim of such modeling is to form hypotheses that can be tested in new studies. In an iterative fashion, the parameters from the new studies may then be used to construct new models with incrementally better description of reality. "Holes" in our current understanding may be revealed—for example, the possibility that seroconversion from latent infection may account for 20%–50% of the age-related increase in seroprevalence. Further investigation of this "hole" may confirm latent infection or even reveal an unsuspected mode of transmission.

ACKNOWLEDGMENTS

The author thanks Drs. J. P. Figueroa, A. Manns, B. Hanchard, W. N. Gibbs, R. Wilks, and Mrs. B. Cranston, P. McKay, P. Robinson, I. Thomas, and C. Hamilton for fruitful collaboration; as well as Dr. William A. Blattner and Dr. Girish N. Vyas for advice and guidance. Supported in part by NCI Research Contract NO1-CP-31006 and by NHLBI Program Project Grant 5-PO1-HL-36589 and Research Contract 1-NO1-HB-6-7024. Dr. Murphy is a Charles E. Culpeper Foundation Scholar.

REFERENCES

1. Piesz BJ, Ruscetti FW, Gazdar AF, et al. Detection and isolation of type-C retrovirus particles from fresh and cultured lymphocytes of patients with cutaneous T-cell lymphoma. *Proc Natl Acad Sci USA* 1980;77:7415–7419.
2. Hinuma Y, Nagata K, Hanaoka M, et al. Adult T-cell leukemia antigen in an ATL cell line and detection of antibodies to the antigen in human sera. *Proc Natl Acad Sci USA* 1981;78:6476–6480.

3. Gessain A, Barin F, Vernant JC, et al. Antibodies to human T-lymphotropic virus type I in patients with tropical spastic paraparesis. *Lancet* 1985;2:407–409.
4. Rodgers-Johnson P, Gajdusek DC, Morgan WStC, et al. HTLV-I and HTLV-III antibodies and tropical spastic paraparesis. *Lancet* 1985;2:1247–1248.
5. Hino S, Yamaguchi K, Katamine S, et al. Mother-to-child transmission of human T-cell leukemia virus type I. *Jpn J Cancer Res (Gann)* 1985;76:474–480.
6. Murphy EL, Figueroa JP, Gibbs WN, et al. Sexual transmission of human T-lymphotropic virus type I. *Annals Int Med* 1989;111(7):555–560.
7. Kajiyama W, Kashiwagi S, Ikematsu H, et al. Intrafamilial transmission of adult T-cell leukemia virus. *J Infect Dis* 1986;154(5):851–857.
8. Okochi K, Sato H, Hinuma Y. A retrospective study on transmission of adult T-cell leukemia virus by blood transfusion: seroconversion in recipients. *Vox Sang* 1984;46:245–253.
9. Manns A, Murphy EL, Wilks R, et al. Prospective cohort study of transfusion-mediated HTLV-I transmission in Jamaica. Abstract number 5568, *IV International Conference on AIDS,* Stockholm, 1988.
10. Weiss SH, Saxinger WC, Ginzburg HM, et al. Human T-cell lymphotropic virus type I (HTLV-I) and HIV prevalences among U.S. drug abusers. Abstract 16, *Proceedings of the American Society of Clinical Oncology.* Atlanta, GA, May 1987.
11. Tajima K, Tominaga S, Suchi T, et al. Epidemiological analysis of the distribution of antibody to adult T-cell leukemia-virus–associated antigen: possible horizontal transmission of adult T-cell leukemia virus. *Gann* 1982;73:893–901.
12. Maeda Y, Furukawa M, Takehara Y, et al. Prevalence of possible adult T-cell leukemia virus-carriers among volunteer blood donors in Japan: a nationwide study. *Int J Cancer* 1984;33:717–720.
13. Clark JW, Robert-Guroff M, Ikehara O, et al. Human T-cell leukemia-lymphoma virus type I and adult T-cell leukemia-lymphoma in Okinawa. *Cancer Res* 1985;45:2849–2852.
14. Kajiyama W, Kashiwagi S, Nomura H, et al. Seroepidemiologic study of antibody to adult T-cell leukemia virus in Okinawa, Japan. *Am J Epidemiol* 1986;123(1):41–47.
15. Catovsky D, Greaves MF, Rose M, et al. Adult T-cell lymphoma-leukemia in blacks from the West Indies. *Lancet* 1982;1:639–643.
16. Blattner WA, Kalyanaraman VS, Robert-Guroff M, et al. The human type-C retrovirus, HTLV, in blacks from the Caribbean region, and relationship to adult T-cell leukemia/lymphoma. *Int J Cancer* 1982;30:257–264.
17. Riedel DA, Evans AS, Saxinger C, Blattner WA. A historical study of human T-lymphotropic virus type I transmission in Barbados. *J Infect Dis* 1989;159:603–609.
18. Blattner WA, Saxinger C, Cleghorn F, et al. HTLV-I: associated risk factors in Trinidad and Tobago. Abstract number 4658, *IV International Conference on AIDS,* Stockholm, 1988.
19. Reeves WC, Saxinger C, Brenes M, et al. Human T-cell lymphotropic virus type I (HTLV-I) seroepidemiology and risk factors in metropolitan Panama. *Am J Epidemiol* 1988;127:532–539.
20. Merino F, Robert-Guroff M, Clark J, et al. Natural antibodies to human Tcell leukemia/lymphoma virus in healthy Venezuelan populations. *Int J Cancer* 1984;34:501–506.
21. Maloney EM, Ramirez H, Levin A, Blattner WA. A survey of the human T-cell lymphotrophic virus type I (HTLV-I) in southwestern Colombia. *Int J Cancer (in press).*
22. Cortes E, Detels R, Aboulafia D, et al. HIV-1, HIV-2, and HTLV-I in high-risk groups in Brazil. *N Engl J Med* 1989;320:953–958.
23. Blayney DW, Blattner WA, Robert-Guroff M, et al. The human T-cell leukemia-lymphoma virus in the southeastern United States. *JAMA* 1983;250(8):1048–1052.
24. Clark J, Saxinger C, Gibbs WN, et al. Seroepidemiologic studies of human T-cell leukemia/lymphoma virus type I in Jamaica. *Int J Cancer* 1985;36:37–41.
25. Murphy EL, Riedel D, Figueroa JP, et al. Demographic determinants of human T-lymphotropic virus type I (HTLV-I) seroprevalence in healthy Jamaicans. Abstract. *Proceedings of the American Society for Clinical Oncology* Volume 8, May 1989.
26. Robert-Guroff M, Kalyanaraman VS, Blattner WA, et al. Evidence for human T-cell lymphoma-leukemia virus infection of family members of human T-cell lymphoma-leukemia virus positive T-cell leukemia-lymphoma patients. *J Exp Med* 1983;157:248–258.
27. Kawano F, Tsuda H, Yamaguchi K, et al. Unusual clinical courses of adult T-cell leukemia in siblings. *Cancer* 1984;54:131–134.

28. Komuro A, Hayami M, Fujii H, et al. Vertical transmission of adult T-cell leukemia virus. *Lancet* 1983;1:240.
29. Nakano S, Ando Y, Ichijo M, et al. Search for possible routes of vertical and horizontal transmission of adult T-cell leukemia virus. *Gann* 1984;75:1044–1045.
30. Yamanouchi K, Kinoshita K, Moriuchi R, et al. Oral transmission of human T-cell leukemia virus type I into a common marmoset (*Callithrix jacchus*) as an experimental model for milk-borne transmission. *Jpn J Cancer Res (Gann)* 1985;76:481–487.
31. Miyamoto K, Tomita N, Ishii A, et al. Transformation of ATLA-negative leukocytes by blood components from anti-ATLA–positive donors in vitro. *Int J Cancer* 1984;33:721–725.
32. Robert-Guroff M, Weiss SH, Giron JA, et al. Prevalence of antibodies to HTLV-I, -II, and -III in intravenous drug abusers from an AIDS endemic region. *JAMA* 1986;255(22):3133–3137.
33. Agius G, Biggar RJ, Alexander SS, et al. Human T-lymphotropic virus type I antibody patterns: evidence of difference by age and risk group. *J Infect Dis* 1988;158(6):1235–1244.
34. Lee H, Swanson P, Shorty VS, et al. High rate of HTLV-II infection in seropositive IV drug abusers in New Orleans. *Science* 1989;244(4903):471–475.
35. Nakada K, Kohakura M, Kmoda H, Hinuma Y. High incidence of HTLV antibody in carriers of *Strongyloides stercoralis. Lancet* 1984;1:633.
36. Tajima K, Fujita K, Tsukidate S, et al. Seroepidemiological studies on the effects of filarial parasites on infestation of adult T-cell leukemia virus in the Goto islands, Japan. *Gann* 1983;74:188–191.
37. Neva FA, Murphy EL, Gam A, et al. Antibodies to *Strongyloides stercoralis* in healthy Jamaican carriers of HTLV-I. *N Engl J Med* 1989;320(4):252–253.
38. Murphy EL, Calisher CH, Figueroa JP, et al. HTLV-I infection and arthropod vectors. *N Engl J Med* 1989;320(17):1146.

DISCUSSION

A discussant was asked whether there are plans to use PCR to elucidate further the prevalence of HTLV-I infection in the study population in Jamaica. The discussant replied that such studies are in progress in persons most likely to exhibit seronegative infection, such as children born to seropositive mothers. The discussant was also asked whether the mothers of seropositive children in the study are seropositive, and responded that all of them are seropositive. The discussant was further asked whether changes in sexual behavior might be occurring in response to HTLV-I infections in Jamaica. The discussant responded that some of the educational programs initiated in response to the AIDS epidemic might have the effect of reducing transmission of HTLV-I in addition to HIV. Another participant asked about the duration of residence in the various altitude zones. The discussant responded that analysis by length of residence is pending. When analyzed by birthplace, the results did not show an association between HTLV-I seroprevalence and altitude. A speaker commented that this conference presented an opportunity to bring together data from various countries and to have statisticians produce better models to assess the contribution of latency to the age-specific rise in seroprevalence.

Human Retrovirology: HTLV,
edited by William A. Blattner.
Raven Press, Ltd., New York © 1990.

Stochastic Analysis of the Carcinogenesis of Adult T-Cell Leukemia-Lymphoma

*Takashi Okamoto, *Shigehisa Mori, †Yuko Ohno,
†Shoichiro Tsugane, †Shaw Watanabe, ‡Masanori Shimoyama,
§Kazuo Tajima, *Masanao Miwa, and *Kunitada Shimotohno

*Virology Division, †Epidemiology Division, ‡Hematology-Oncology and
Clinical Cancer Chemotherapy Division, National Cancer Center, Tokyo;
and §Epidemiology Division, Aichi Cancer Center, Aichi, Japan

INTRODUCTION

Human T-cell leukemia virus type 1 (HTLV-I) is a primary etiologic agent of adult T-cell leukemia-lymphoma (ATL) based on the seroepidemiological studies (1–4) as well as in vitro studies (5–7). It has been demonstrated that HTLV-I infection (5,6), and possibly a viral gene, *tax,* alone (7) could immortalize human cord blood T cells. It is proposed that *tax* gene product of HTLV-I is responsible for immortalization of T cells and that expression of both components of an autocrine circuit, interleukin-2 and its own receptor, is stimulated (8–10). However, viral infection alone could not explain the entire processes of ATL leukemogenesis such as, for example, the monoclonal origin of the leukemic cells. In this study we attempted to clarify possible natures of the events between HTLV-I infection and clinical manifestation of ATL.

The multistep theory of carcinogenesis as a mathematical model, in which it is assumed that a cell must go through a sequence of specific changes to become malignant, was advanced to account for the fact that, in certain types of cancer, the logarithm of incidence rate of cancers increases in direct proportion to the logarithm of age (11,12). Similarly, it was shown that continuous-carcinogenesis experiments could be analyzed by fitting appropriate Weibull distributions, in which the "shape parameter" could be regarded as the number of "hits" or "steps" in which healthy tissue would give rise to a cancer (13–15).

PATIENTS AND METHODS

Study Population

The ages of disease onset of 357 ATL cases, 203 males and 154 females, collected during the nationwide surveys from 31 institutions in Japan from 1982 through 1985 were analyzed (16,17). These data were considered to cover approximately one-eighth of all new ATL patients during the observed period (16,18). The average ages of ATL onset for male and female patients were 55.1 (±12.3 SD) and 55.8 (±12.5 SD), respectively. In the following study we assumed the onset age of ATL as an incubation period of the disease because epidemiological studies demonstrated that in most Japanese ATL cases the disease developed from those who acquired HTLV-I infection during infancy, probably through breast-feeding (19–22). We treated these cross-sectional data as cohort data. The ages of death from other hematopoietic malignancies, including Hodgkin's or non-Hodgkin's lymphomas, acute myelocytic leukemia (AML), and chronic myelocytic leukemia (CML), were taken from the vital statistics report of Japan, 1984–1986.

Data Analysis

A cumulative incidence of ATL occurrence was analyzed by fitting to various distribution models using a computer program (the LIFEREG procedure, SAS Institute) (23) based on the age distribution of onset ages of the patients. Log-likelihood ratio tests were calculated to examine the model-fitting (24).

RESULTS

Figure 1 shows the age-specific distribution of disease onset for these 357 ATL patients. ATL development started from the age of 24 yr, peaked at 50–54 yr and then decreased rapidly. This rapid decrease of ATL occurrence after 55 yr could not be attributed to the competitive deaths by more common causes, because this trend was still observable after the correction for the competing effects of deaths resulting from other causes according to a life table from the national population survey in Japan conducted in 1985 (Fig. 1). A slight decrease in the age rank of 55–59 yr was noted in both of two independent surveys (one from 1982 through 1983 and the other from 1984 through 1985, with no overlapping cases) (not shown) and was considered to be due to a cohort effect rather than a sampling fluctuation. From this distribution of onset ages of ATL patients, a cumulative percentage incidence curve was obtained (Fig. 1).

The cumulative percentage of ATL occurrence thus obtained was analyzed by fitting the data to various distribution models using the LIFEREG computer program package. On the basis of high log-likelihood values, both the Weibull and the gamma distribution models were accepted (log-likelihood values for these two models were 19.35 and 20.83, respectively, whereas that for the log normal distribution model was 6.30). We adopted the Weibull model because of its simplicity as a mathematical model (in fact, the Weibull model can be conceptually included in the gamma distribution model). Thus estimated, the

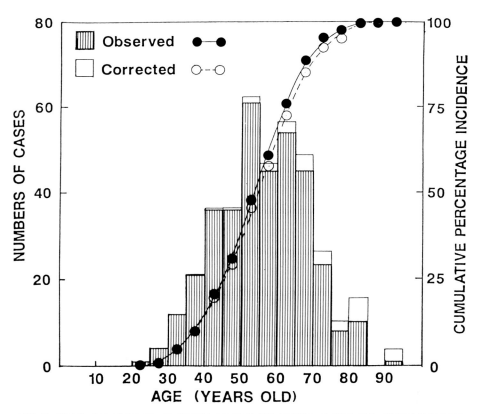

FIG. 1. Distribution of the age at disease onset of ATL patients studied. The original data were categorized into 5-yr ranks. The observed number of patients in each rank is represented by a shaded bar. The number of cases in each category was corrected with the accumulated death rate for each age rank according to data from the national population survey in Japan (1985). The cumulative incidence, both observed and corrected, is shown by circles and lines. Reprinted with permission from Okamoto et al. (17).

probability density function for the present Weibull model, f(t), expressed as a function of time (t; years), was described as:

$$f(t) = a/b \cdot t^{a-1} \cdot \exp(-t^a/b)$$

where a "shape parameter" was 5.03 and b "scale parameter" was 8.00×10^8. Therefore, the hazard rate, $\lambda(t)$, expressed as a function of time could be written as:

$$\lambda(t) = \frac{f(t)}{\displaystyle\int_t^\infty f(x)dx} = a/b \cdot t^{a-1}$$

Thus, it is shown that the risk of ATL development at each age is approximately proportionate to the fourth power of the age, suggesting that at least five events, each of which occurs spontaneously (randomly) and independently, might be required for development of the disease. Figure 2a shows that a cumulative ATL occurrence could be expressed as a single linear line on the Weibull plot. The distributions for both sexes showed nearly identical (confirmed by log-rank and Wilcoxon's rank tests). The age distributions of other commoner hematopoietic malignancies were similarly analyzed with Weibull plots. Because the ages of disease onset were not available, the ages of death from these malignancies were studied (Fig. 2b). These malignancies did not fit the Weibull model or any others. In some of these diseases for which the ages of disease onset were available, the patterns were almost identical (not shown) with those obtained with the ages of death.

DISCUSSION

We demonstrated that the age-dependent occurrence of ATL can be simply described by a Weibull model (typical tear-off type), which is a feasible mathematical model for multistep carcinogenesis, as indicated by previous investigators (13–15). This model suggested to us that ATL leukemogenesis might be the result of accumulation of a number of critical events, most likely somatic mutations. The above assumption might be attainable considering the presence of unusually high incidence of chromosomal aberrations in ATL cells (25,26).

The number of critical events, or mutations, within the cell to develop ATL is not predictable. However, according to the interpretation of the previous investigators, it can be expressed as a shape parameter in the Weibull distribution function, supposing that each event occurs randomly (therefore, a Poisson distribution can be applicable for each case) and independently. Thus, the putative number of independent leukemogenic events involved in ATL could be estimated to be around five. This number is similar to the numbers estimated for other malignancies, although such a well-matched fit to a single Weibull distribution function has not been reported.

It then appears that, although HTLV-I infection plays a primary role in the pathogenesis of ATL as an "initiator," it may be only a prerequisite for accumulation of later events and may not be sufficient for ATL leukemogenesis— unless HTLV-I is subsequently shown to possess a mutagenic potential in T cells.

Moreover, because the lifelong rate of ATL occurrence among HTLV-I carriers has been estimated to be not more than a few percent (20,27,28), it is considered that the rates of critical ATL-prone mutations are very low or that quite a large number of the HTLV-I–immortalized cells are required as a reservoir for ATL to develop within the normal human life span. We are currently unable to speculate further on the natures of the target genes for the somatic "mutations" involved in ATL leukemogenesis. These may include abrogation of some tumor

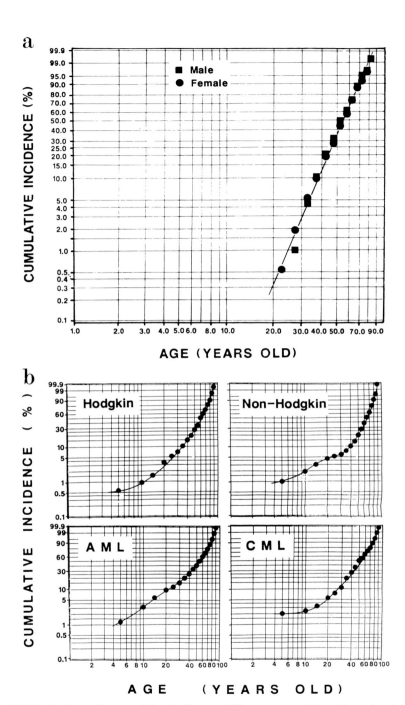

FIG. 2. Weibull plots of the cumulative incidence of ATL occurrence (a) and that of various hematopoietic malignancies (b) by age. Identical lines for both males and females are noted for ATL. The original data of the ages of death from Hodgkin's (544 cases) and non-Hodgkin's (12 743 cases), acute myelocytic leukemia (AML) (2248 cases), and chronic myelocytic leukemia (CML) (3089 cases) were obtained from the vital statistics report of Japan between 1984 and 1986. Ordinates indicates log log $[1/\{1 - F(t)\}]$ and abscissa, log t.

suppressor genes, as well as activation of some oncogenes (29,30), considering the presence of multiple gross chromosomal abnormalities, although locations of the critical abnormalities have not been determined in ATL.

SUMMARY

A multistep model has been proposed to explain carcinogenic processes of human cancers. However, only a few human cancers could be explained by this model. Here we reported that adult T-cell leukemia-lymphoma (ATL) could be described by a simple multistep model. This simulation is considered feasible because the latency period between the acquisition of the cancer-causing agent, human T-cell leukemia virus type 1 (HTLV-I), and clinical onset of ATL can be easily estimated unlike other malignancies. In this study, the cumulative incidence rate throughout the lifespan was analyzed for 357 ATL cases reported from 31 institutions in Japan. It has been demonstrated that the age distribution of ATL onset can be described by a single Weibull distribution function (tear-off type), which is a feasible mathematical model for multistep carcinogenesis. Assumption of preclinical period was not needed for this approximation, and there was no sex difference with regard to the present model. Based on this stochastic model, it is assumed that age-dependent accumulation of leukemogenic events, which occur randomly and most likely are somatic mutations, within the target cell might be required before the development of the disease. Comparison with other hematopoietic malignancies is discussed.

ACKNOWLEDGMENTS

We thank Drs. Takashi Sugimura, Hiroshi Tanooka and Sunao Adachi for encouragement and helpful discussions. This work is supported by grants-in-aid from the Ministry of Health and Welfare and the Ministry of Education, Science and Culture, Japan.

REFERENCES

1. Hinuma Y, Nagata K, Hanaoka M, et al. *Proc Natl Acad Sci USA* 1981;78:6476–6480.
2. Blattner WA, Kalyanaraman VS, Robert-Guroff M, et al. *Int J Cancer* 1982;30:257–264.
3. Robert-Guroff M, Nakao Y, Notake K, Ito Y, Sliski A, Gallo RC. *Science* 1982;215:975–978.
4. Catovsky D, Greaves MF, Rose M, et al. *Lancet* 1982;1:639–643.
5. Miyoshi I, Kubonishi I, Yoshimoto S, et al. *Nature* 1981;294:770–771.
6. Popovic M, Sarin PS, Robert-Guroff M, et al. *Science* 1983;219:856–859.
7. Grassmann R, Dengler C, Muller-Fleckenstein I, et al. *Proc Natl Acad Sci USA* 1989;86:3351–3355.
8. Green WC, Leonard WJ, Wano Y, et al. *Science* 1986;232:877–880.
9. Inoue J, Seiki M, Taniguchi T, Tsuru S, and Yoshida M. *EMBO J* 1986;5:2883–2888.
10. Cross SL, Feinberg MB, Wolf JB, Holbrook NJ, Wong-Staal F, Leonard WJ. *Cell* 1987;49:47–56.

11. Armitage P, Doll R. *Br J Cancer* 1954;8:1–12.
12. Pike MC. *Biometrics* 1966;22:142–161.
13. Burch PRJ. *The Biology of cancer.* Lancaster, England: MTP Press, 1976.
14. Peto R, Lee P. *Biometrics* 1973;29:457–470.
15. Hakama M. *Int J Cancer* 1971;7:557–564.
16. The T- and B-cell Malignancy Study Group. *Int J Cancer* 1988;41:505–512.
17. Okamoto T, Ohno Y, Tsugane S, et al. *Jpn J Cancer Res* 1989;80:191–195.
18. The T- and B-cell Malignancy Study Group. *Int J Cancer* 1988;41:505–512.
19. Tajima K, Tomonaga S, Suchi T, et al. *Jpn J Cancer Res (Gann)* 1982;73:893–901.
20. Tajima K, Kamura S, Ito S, et al. *Int J Cancer* 1987;40:741–746.
21. Hino S, Yamaguchi K, Katamine S. *Jpn J Cancer Res* 1985;76:474–480.
22. Kusuhara K, Sonoda S, Takahashi K, Tokugawa K, Fukushige J, Ueda K. *Int J Cancer* 1987;40: 755–757.
23. SAS Institute. *The LIFEREG procedure, SAS user's guide, Version 5 Edition.* Cary, N.C.: SAS Institute, 1985.
24. Cox DR, Oakes D. *Analysis of survival data* London: Chapman and Hall, 1984.
25. Shimoyama M, Abe T, Miyamoto K, et al. *Blood* 1987;69:984–989.
26. The Fifth International Workshop on Chromosomes in Leukemia-Lymphoma. *Blood* 1987;70: 1554–1564.
27. Kondo T, Nonaka H, Miyamoto N, et al. *Int J Cancer* 1985;35:749–751.
28. Tajima K, Kuroishi T. *Jpn J Clin Oncol* 1985;15:423–430.
29. Farber E. *Cancer Res* 1984;44:4217–4223.
30. Knudson AG. *Cancer Res* 1985;45:1437–1443.

Human Retrovirology: HTLV,
edited by William A. Blattner.
Raven Press, Ltd., New York © 1990.

Genetic and Immunologic Determinants of HTLV-I–Associated Diseases

Shunro Sonoda

Department of Virology, Faculty of Medicine, Kagoshima University, Kagoshima 890 Japan

INTRODUCTION

Adult T-cell leukemia (ATL) and the newly defined neurological disease, HAM/ TSP (HTLV-I associated myelopathy/tropical spastic paraparesis), are caused by the same species of HTLV-I (1) and are prevalent in HTLV-I–endemic populations of southwestern Japan, the Caribbean, and other tropics (2–5). ATL and HAM are found to occur independently among family members carrying HTLV-I (6). It is thus suggested that there might be some host factors involved in the segregated incidence of ATL and HAM. The clinical entities of ATL and HAM are characterized by the distinct immunological features, the former being an immunocompromised state with a lowered cellular immunity (7) and the latter being a hyperimmune state with an increased titer of antibody to HTLV-I in serum and cerebrospinal fluid (8,9), as well as an increased T-cell response to HTLV-I (10,11). The immunopathological backgrounds of the two diseases are probably associated with a genetically determined host factor(s). Reality of the genetic and immunologic determinants of ATL and HAM may be studied in three issues: 1) ethnic background of ATL and HAM in the southern Japanese population; 2) HLA haplotype-linked immune responsiveness to HTLV-I in ATL and HAM; and 3) immunopathological effectors involved in disease development of ATL and HAM.

ETHNIC BACKGROUND OF ATL AND HAM IN THE SOUTHERN JAPANESE POPULATION

Geographic peculiarity of HTLV-I carriers in Japan, the Caribbean, and other tropical countries has been considered with the specific mode of HTLV-I transmission among these ethnic groups (12–15). Although the origin of HTLV-I has been controversially debated, some ancestral ethnic groups should have been permissive to HTLV-I infection and have carried the virus, with resultant development of ATL or HAM/TSP in their descendants, as those postulated (16,17).

Provided that the disease-prone genetic factor might have associated with a particular ethnic group and determined the type of HTLV-I–associated diseases, it would be possible to define the genetic factors associated with ATL and HAM ethnics. To investigate the ethnic factors, we analyzed HLA types of ATL and HAM in Kagoshima, southern Kyushu, Japan. To this end, peripheral blood lymphocytes (PBL) were collected from the ATL and HAM patients and their family members and frozen in liquid nitrogen until their HLA antigens were typed by the microcytotoxicity test (18). When HLA antigens of ATL were analyzed, the frequency of A26 was significantly increased in the patient group with acute- and lymphoma-type ATL (cp < 0.004). Several other HLA antigens revealed an increased or decreased tendency of the antigen frequency among ATL and HAM patients. These results suggested that A26 and the associated antigens may be associated with the ethnic group of ATL and the other with HAM ethnics. To confirm the specific HLA association, we further analyzed HLA haplotypes of ATL and HAM with the use of the familial collection of PBL. Acute- and lymphoma-type ATL were revealed to possess one specific haplotype, A26Cw3Bw62DR5DQw3 and the related ones, whereas HAM had five major haplotypes commonly observed in the general Japanese population (Table 1). When HLA haplotypes of asymptomatic HTLV-I carriers were compared with those of ATL and HAM in Kagoshima, they shared all of the representative HLA haplotypes associated with ATL and HAM. It was thus suggested that ATL and HAM might have developed from HTLV-I carrier population in which "ATL-associated" and "HAM-associated" haplotypes segregated to each family trait of ATL or HAM. By reviewing HLA haplotypes inherited from paternal and maternal traits, it was evident that ATL patients originated from

TABLE 1. *HLA haplotype frequency in HAM, ATL, and asymptomatic HTLV-I carriers*

HLA haplotypes	HAM (N = 112)	ATL[a] (N = 56)	AC[b] (N = 50)	Control[c] (N = 200)
A24Cw7B7 DR1 DQw1	9.8*%	0%	2.0%	4.0%
A24Cw–Bw52DR2 DQw1	8.9	0	12.0	9.5
A11Cw1Bw54DR4 DQw4	6.3*	0	0	1.0
A24Cw1Bw54DR4 DQw4	5.4	0	8.0	5.5
A2 Cw7Bw60DRw8DQw–	2.7	0	0	0
Sum	33.1	0	22.0	20.0
A26Cw3Bw62DR5 DQw3	0	7.1*	2.0	0.5
A26Cw1Bw54DR4 DQw4	0	7.1	0	1.5
A26Cw3B35 DR2 DQw1	0	5.4	0	1.5
A26Cw–Bw61DR9 DQw3	0	3.6	0	0.5
Sum	0	23.2	2.0	4.0

Differs from control: *p < 0.01.
[a] Acute ATL patients.
[b] Asymptomatic HTLV-I carriers.
[c] Kagoshima population.

TABLE 2. *HLA haplotypes of ATL patients with three different clinical types*

Acute ATL

Haplotype

#			/	
1	KAG–071	A26 Cw3B35 DR2 DQw1	/	A26Cw3Bw62DR5 DQw3
2	KAG–001	A2 Cw3Bw61DRw9DQw3	/	A26Cw3Bw62DR5 DQw3
3	KAG–004	A2 Cw–B44 DRw6DQw–	/	A26Cw3Bw62DR5 DQw3
4	KAG–037	A2 Cw–Bw61DR4 DQw4	/	A26Cw3Bw62DR5 DQw3
5	KAG–040	A26 Cw–Bw– DR2 DQw1	/	A26Cw3Bw62DR2 DQw1
6	KAG–011	A11Cw–Bw61DRw9DQw3	/	A26Cw3Bw62DRw–DQw3
7	KAG–027	A31Cw3Bw61DR4 DQw4	/	A26Cw3B35 DR2 DQw1
8	KAG–018	A11Cw–B51 DR5 DQw3	/	A26Cw–Bw61DRw9DQw3
9	KAG–002	A2 Cw–Bw60DRw8DQw1	/	A26Cw1Bw54DR4 DQw4
10	KAG–020	Aw33Cw–B44DRw6DQw1	/	A26Cw1Bw54DR4 DQw4
11	KAG–019	A24Cw–Bw55DRw–DQw–	/	A24Cw3Bw48DR2 DQw1
12	KAG–035	A31Cw–B51 DRw8DQw–	/	A24Cw7B39 DR4 DQw3
13	KAG–064	A24Cw3Bw62DRw9DQw3	/	A24Cw–B39 DR4 DQw3
14	KAG–038	A24Cw1Bw48DR2 DQw1	/	A11Cw1Bw54DRw8DQw1
15	KAG–036	A– Cw–Bw62DR4 DQw3	/	A2 Cw3Bw60DRw8DQw1
16	KAG–051	A24Cw–Bw– DR4 DQw3	/	A2Cw11Bw46DRw8DQw1

ATL–Lymphoma

#			/	
1	KAG–043	A– Cw–Bw59DRw–DQw–	/	A26Cw3Bw62DR1 DQw1
2	KAG–052	A2 Cw–Bw– DRw6DQW1	/	A26Cw3Bw61DRw9DQw3
3	KAG–029	A11Cw3B35 DR4 DQw–	/	A26Cw3B35 DR2 DQw1
4	KAG–033	A31Cw3B35 DR2 DQW1	/	A26Cw1Bw54DR4 DQW4
5	KAG–046	A31Cw3B35 DR2 DQw1	/	A26Cw1Bw54DR4 DQw4
6	KAG–069	A24Cw3Bw61DRw9DQw3	/	A26Cw–Bw52DR2 DQW1
7	KAG–042	A26 Cw–Bw61DRw9DQw3	/	A26Cw–B13 DR5 DQw3
8	KAG–084	A31Cw–B27 DR4 DQw4	/	A11Cw–B51 DR5 DQw3
9	KAG–066	A– Cw3Bw– DR4 DQw–	/	A24Cw–Bw48DRw8DQw–
10	KAG–041	A31Cw–B51 DRw9DQW3	/	A2 Cw–Bw61DRw9DQw3
11	KAG–085	Aw33Cw–B17DRw6DQw1	/	A2 Cw–Bw52DR2 DQW1
12	KAG–034	A24Cw7B7 DRw9DQw3	/	A2 Cw1Bw– DRw8DQW–

Chronic ATL

#			/	
1	KAG–012	A24Cw1Bw54DR4 DQw4	/	A24Cw–B51 DRw–DQw–
2	KAG–067	A24Cw1Bw54DR4 DQw4	/	A31Cw3Bw61DR2 DQW1
3	KAG–039	A24Cw7B7 DR1 DQw1	/	A31Cw–Bw52DR2 DQw1
4	KAG–068	A24Cw–Bw52DR2 DQw1	/	A24Cw4Bw60DR2 DQw1

TABLE 3. *HLA haplotypes of HAM patients*

Patients	Haplotype		
1 HAM-03710	A11Cw1Bw54DR4 DQw4	/	A31Cw4Bw60DRw9DQw3
2 HAM-07710	A11Cw1Bw54DR4 DQw4	/	A31Cw3Bw60DR1 DQw1
3 HAM-05710	A11Cw1Bw54DR4 DQw4	/	A26Cw3Bw61DR1 DQw1
4 HAM-06710	A11Cw1Bw54DR4 DQw4	/	A2 Cw3Bw61DR5 DQw3
5 HAM-01710	A11Cw1Bw54DR4 DQw4	/	A2 Cw1Bw- DR5 DQw3
6 HAM-05210	A11Cw1Bw54DR4 DQw4	/	A- Cw3Bw- DRw8DQw3
7 HAM-03510	A24Cw1Bw54DR4 DQw4	/	A31Cw-B51 DR5 DQw3
8 HAM-07610	A24Cw1Bw54DR4 DQw4	/	A26Cw3B35 DR4 DQw3
9 HAM-08310	A24Cw1Bw54DR4 DQw4	/	A2 Cw-B35 DRw8DQw-
10 HAM-09010	A24Cw1Bw54DR4 DQw4	/	A- Cw3Bw60DR2 DQw1
11 HAM-04910	A24Cw1Bw54DR4 DQw4	/	A- Cw-Bw- DRw-DQw-
12 HAM-07510	A31Cw1Bw54DR4 DQw4	/	A2 Cw4Bw60DR5 DQw3
13 HAM-03210	A31Cw1Bw54DR4 DQw4	/	A- Cw-Bw62DR2 DQw1
14 HAM-01810	A2 Cw7B7 DR1 DQw1	/	Aw33Cw-B44DRw6DQw1
15 HAM-03810	A2 Cw7B7 DR1 DQw1	/	A31Cw3Bw62DR4 DQw3
16 HAM-08210	A24Cw7B7 DR1 DQw1	/	A26Cw3Bw62DR2 DQw1
17 HAM-00710	A24Cw7B7 DR1 DQw1	/	A26Cw3Bw48DRw9DQw3
18 HAM-07210	A24Cw7B7 DR1 DQw1	/	A26Cw3Bw- DR5 DQw3
19 HAM-08110	A24Cw7B7 DR1 DQw1	/	A26Cw3B35 DR4 DQw3
20 HAM-00910	A24Cw7B7 DR1 DQw1	/	A24Cw1Bw42DR4 DQw3
21 HAM-11710	A24Cw7B7 DR1 DQw1	/	A24Cw1Bw- DRw8DQw1
22 HAM-11010	A24Cw7B7 DR1 DQw1	/	A24Cw-Bw52DRw6DQw1
23 HAM-08810	A24Cw7B7 DR1 DQw1	/	A11Cw1Bw- DR5 DQw3
24 HAM-09010	A24Cw7B7 DR1 DQw1	/	A- Cw3Bw60DR2 DQw1
25 HAM-10810	A24Cw7B7 DR1 DQw1	/	A- Cw-Bw55DRw-DQw-
26 HAM-07110	A24Cw7B7 DRw8DQw-	/	A- Cw7B7 DRw6DQw1
27 HAM-04410	A- Cw1Bw- DRw8DQw-	/	A2 Cw4Bw62DR4 DQw3
28 HAM-13110	A24Cw1Bw- DRw8DQw-	/	A2 Cw3Bw- DR5 DQw3
29 HAM-07310	A2 Cw7Bw60DRw8DQw-	/	A24Cw-Bw60DRw-DQw-
30 HAM-07810	A2 Cw7Bw60DRw8DQw-	/	A11Cw1Bw55DR5 DQw3
31 HAM-05010	A2 Cw7Bw60DRw8DQw-	/	A26Cw3B39 DR4 DQw3
32 HAM-04010	A2 Cw7Bw60DRw8DQw1	/	A31Cw-B51 DR5 DQw3
33 HAM-04310	A31Cw7Bw60DRw8DQw1	/	A2 Cw-B51 DR5 DQw3
34 HAM-04210	A24Cw-Bw52DR2 DQw1	/	A31Cw-B51 DRw9DQw3
35 HAM-04810	A24Cw-Bw52DR2 DQw1	/	A26Cw3Bw60DR4 DQw3
36 HAM-02410	A24Cw-Bw52DR2 DQw1	/	A26Cw-Bw- DRw-DQw-
37 HAM-01410	A24Cw-Bw52DR2 DQw1	/	A24Cw7B7 DR4 DQw3
38 HAM-01310	A24Cw-Bw52DR2 DQw1	/	A24Cw7B7 DR1 DQw1
39 HAM-00310	A24Cw-Bw52DR2 DQw1	/	A24Cw1Bw- DRw8DQw1
40 HAM-05310	A24Cw-Bw52DR2 DQw1	/	A24Cw-Bw52DRw9DQw3
41 HAM-09510	A24Cw-Bw52DR2 DQw1	/	A11Cw1Bw54DR4 DQw4
42 HAM-02610	A24Cw-Bw52DR2 DQw1	/	A2 Cw1Bw- DRw8DQw1
43 HAM-03910	A24Cw-Bw52DR2 DQw1	/	A- Cw3Bw60DR4 DQw3
44 HAM-00810	A24Cw-Bw52DR1 DQw1	/	A11Cw7Bw60DR4 DQw3
45 HAM-00410	A24Cw-Bw52DRw9DQw3	/	A2 Cw1Bw- DRw8DQw-
46 HAM-01210	A24Cw-Bw52DRw9DQw3	/	A26Cw3Bw62DRw8DQw-
47 HAM-03610	A24Cw-Bw52DRw9DQw3	/	Aw33Cw-B44DRw6DQw1
48 HAM-10610	A2 Cw1Bw- DR4 DQw-	/	A26Cw3Bw52DR2 DQw1
49 HAM-04510	A11Cw1Bw- DRw6DQw1	/	A- Cw3B51 DR5 DQw3

Boxes refer to a cluster of HLA haplotypes.

TABLE 3. *Continued.*

Patients	Haplotype		
50 HAM-12510	A24Cw3Bw- DR4 DQw3	/	A2 Cw1B35 DRw9DQw3
51 HAM-07410	A24Cw3B35 DR2 DQw1	/	A26Cw-B51 DR5 DQw3
52 HAM-00110	A31Cw4B35 DR4 DQw3	/	Aw33Cw-B44DRw6DQw1
53 HAM-09510	A2 Cw1Bw59DR4 DQw-	/	A26Cw3Bw61DRw-DQw-
54 HAM-13610	A24Cw-Bw59DR4 DQw3	/	A26Cw3Bw- DR2 DQw1
55 HAM-01610	A24Cw-Bw61DRw8DQw3	/	A26Cw3B51 DRw-DQw-
56 HAM-00510	A24Cw3Bw61DRw6DQw1	/	A26Cw3Bw61DR5 DQw3

a particular ethnic group of ATL-associated haplotypes which was closely mated, and HAM patients derived from two ethnic groups of ATL-associated and HAM-associated haplotypes (Tables 2 and 3). In view of HLA haplotypes associated with ATL and HAM, those of chronic ATL were remarkable for the dissimilarity to those of acute- and lymphoma-type ATL and the similarity to those of HAM (Tables 2 and 3). The HAM-associated HLA haplotypes may be correlated with the clinical manifestation of chronic ATL, whose prognosis is quite different from those of acute- and lymphoma-type ATL but similar to those of HAM in the aspect of chronic progression.

HLA HAPLOTYPE-LINKED IMMUNE RESPONSIVENESS TO HTLV-I IN ATL AND HAM

Because the immune response capacity is genetically linked to MHC genes, the difference in HLA haplotypes may associate with a different level of immune response to HTLV-I. This was true in the T-cell proliferative response to HTLV-I in the culture of PBL obtained from ATL and HAM patients (Fig. 1). HTLV-I–specific T-cell response was estimated by subtracting the products of the autologous proliferation of T cells, which was invariably induced in in vitro culture with PBL of HTLV-I carriers, as documented previously (10). T-cell response of ATL was very much lower than those of HAM and asymptomatic HTLV-I carriers (Fig. 1). The difference in the average of responses between ATL and HAM was statistically significant. Some patients with HAM showed an extraordinarily high T-cell response compatible with their hyperimmune state, as expected from the increased titer of HTLV-I antibody (9). The response of asymptomatic HTLV-I carriers was comparable with that of HAM.

To correlate the T-cell response with HLA haplotypes, we investigated the HTLV-I–specific T-cell responsiveness among normal donors who possessed either ATL-associated or HAM-associated HLA haplotypes. The donors of ATL-associated haplotypes were found to be low responders to HTLV-I, whereas those of HAM-associated haplotypes were high responders (Fig. 2). All of these T-cell responses were demonstrated by use of HTLV-I virion antigen. The HLA haplotype-linked T-cell responsiveness could be reproduced by the use of a re-

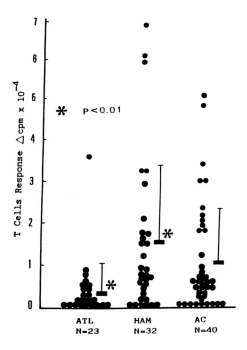

FIG. 1. HTLV-I–specific T-cell immune response in ATL, HAM, and asymptomatic HTLV-I carriers (AC). 5×10^4 PBL of ATL, HAM, or AC were incubated for 6 d in 96-well microtrays (Corning, Round-bottom #25850), with or without an appropriate amount of purified HTLV-I virion antigen in 0.2 ml of RPMI-1640 medium supplemented with 10% pooled human serum (HTLV-I seronegative) and antibiotics (streptomycin 50 μg/ml, penicillin 50 IU/ml). On day 6, ^{3}H-thymidine (1 μCi) was incorporated to each well to pulse the cultured cells for 16 h and harvested the cells for estimation of ^{3}H-uptake. T-cell response (Δcpm) was measured by subtracting ^{3}H-uptake in PBL cultured without HTLV-I antigen from that with HTLV-I antigen as described (10).

combinant protein, HTLV-I (*gag-env*), which was kindly provided from KYOWA MEDEX Co., Ltd., Tokyo (data not shown). These results indicate that the low immune response in ATL is not ascribed to the immunocompromised state of the patients but rather to the genetically determined factor(s) involved in the low immune responsiveness to HTLV-I that was linked to ATL-associated haplotypes. The high immune response in HAM, similarly, was determined by the HLA haplotype-linked factor whose immune response capacity was irrelevant to the history of HTLV-I carrier state and the amounts of memory T cells reactive to HTLV-I antigens. Thus, HLA haplotype-linked immune responsiveness appeared to be the primary genetic determinant for HTLV-I permissiveness and disease susceptibility to ATL and HAM.

IMMUNOPATHOLOGICAL EFFECTORS INVOLVED IN DISEASE DEVELOPMENT OF ATL AND HAM

When T cells are infected with HTLV-I, they are activated to induce Tac/IL-2 receptor (19) and express the altered HLA antigens (20). These phenotypic alterations were detectable in peripheral blood T lymphocytes which expressed "gain" or "loss" of HLA antigens on their cell surface (21). The altered antigens may be an effective stimulator for autologous mixed lymphocyte reaction (MLR)

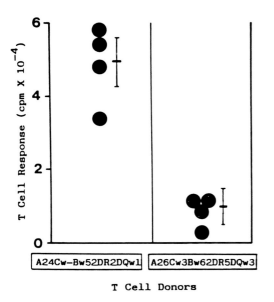

FIG. 2. HLA haplotype-linked immune responsiveness to HTLV-I in T cells of normal donors. PBL isolated from normal donors possessing "ATL-associated" and "HAM-associated" HLA haplotypes were tested for HTLV-I–specific T-cell response as was done in Fig. 1.

by presenting the altered HLA (equivalent to alloantigens), which is complexed with HTLV-I viral antigens. This possibility was tested in vitro with the use of PBL isolates from HTLV-I carriers. When PBL were cultured in vitro, T cells proliferated spontaneously without any addition of mitogens, the so-called "autologous proliferative response" (APR) as previously described (10). Among these T-cell responses, some PBL donors proliferated CD4 T cells and the others proliferated CD8 T cells preferentially (Fig. 3). The PBL isolated from ATL patients were found to proliferate only CD4 cells, whereas those from HAM were revealed to proliferate CD8 cells predominantly (23). Therefore, differential proliferation of CD4 and CD8 cells was significant in the light of T-cell effectors which might have been involved in the immunopathology of ATL and HAM.

To realize the immunologically significant roles of CD4 and CD8 cells in HTLV-I infection, we investigated the virological and immunological functions of CD4 and CD8 cells in the propagation of HTLV-I and HTLV-I–specific immune effector function, respectively. CD4 and CD8 cells were separated by the panning method with monoclonal antibodies (22) and infected with HTLV-I produced by MT-2 cells. CD4 cells infected with HTLV-I allowed the growth of HTLV-I antigen-positive cells in 20%–30% of the cultured CD4 cells after 14 d incubation. In contrast, when CD8 cells were added back to CD4 cell culture, the growth of HTLV-I antigen-positive cells was remarkably suppressed in the cases of donors possessing the HAM-associated HLA haplotypes. However, no suppression was produced by the CD8 cells of ATL-associated HLA haplotype donors (Fig. 4). It was thus suggested that CD8 responders of HAM-associated

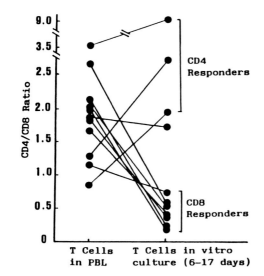

FIG. 3. Differential proliferative response of CD4 and CD8 T-cell subsets in the culture of asymptomatic HTLV-I carrier's PBL. Five million PBL isolated from asymptomatic HTLV-I carriers were cultivated for 6–17 d in a culture flask (Falcon #3013) containing 10 ml of RPMI-1640 medium supplemented with 10% pooled human serum and the antibiotics. The freshly isolated and cultured PBL were doubly stained with T11/T4 or T11/T8 antibodies (Coulter Immunology) and analyzed by Coulter EPICS-C flowcytometer (Coulter Electronics, Hialeah, Florida).

haplotypes might develop an immune effector to lyse or suppress the growth of HTLV-I–infected CD4 cells, and those of ATL-associated haplotypes might lack the effector function. This notion was confirmed by the evidence that CD8 cells isolated from HAM patients exerted a potent immune suppression to the autologous CD4 cells expressing HTLV-I antigens (23). Therefore, the CD8 cells in HAM are likely autoimmune effectors of immunopathology, as those implicated for other autoimmune diseases associated with HTLV-I infection (24–29).

CONCLUSION

Genetic background of ATL and HAM was determined by HLA haplotype. ATL patients possessed a unique HLA haplotype, A26Cw3Bw62DR2DQw3 and the related ones (ATL-associated HLA haplotypes). HAM patients had heterozygous HLA haplotypes that were composed of one from the ATL associated haplotypes and the other from the common HLA haplotypes of the general Japanese population (HAM-associated HLA haplotypes). The former HLA haplotypes were a low immune responder to HTLV-I and the latter HLA haplotypes were a high immune responder to HTLV-I. The low and high immune responders were similarly segregated among normal donors whose T cells had either ATL-associated or HAM-associated HLA haplotypes.

CD4 and CD8 T cells of ATL, HAM, and the asymptomatic HTLV-I carriers proliferated in vitro after transformation or responding to HTLV-I–induced altered self-antigens. CD8 cells of HAM-associated HLA haplotypes showed a potent immune suppression against HTLV-I antigen-positive CD4 cells, whereas

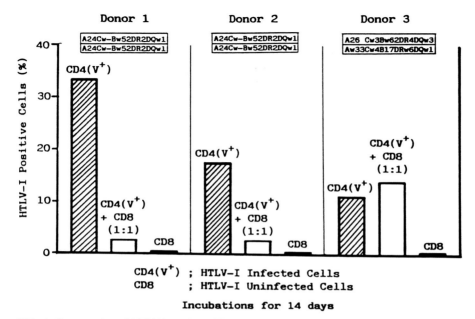

FIG. 4. Suppression of HTLV-I–positive CD4 cells by autologous CD8 cells. Five million PBL isolated from three normal donors possessing different HLA haplotypes were rendered to "Panning" for CD4 or CD8 cells as described (22). Two hundred thousand CD4 cells (or CD8 cells), or the mixture of CD4 and CD8 cells at the ratio indicated, were co-cultivated with 2×10^5 MT-2 cells (HTLV-I producer, mitomycin-C treated) for 14 d in 24-well multiplate (Nunclon #143982) containing 2 ml of RPMI-1640 medium supplemented with 10% pooled human serum, 1 U/ml recombinant IL-2 (Takeda Pharm. Co.), 2 μg/ml PHA-P (Difco) and the antibiotics. The cultured cells were stained with anti-HTLV-I monoclonal antibodies (30,31) GIN-14 (anti-p19) and F-10 (anti-gp21), and the numbers of HTLV-I–positive cells were counted under a fluorescence microscope.

no suppression was produced by CD8 cells of ATL-associated HLA haplotypes. The similar immune effectors were demonstrated in CD8 cells isolated from HAM patients.

It was thus concluded that the HTLV-I immune response genes were linked to HLA haplotypes which determined the genetic and immunologic background of ATL and HAM. The high and low immune response genes were segregated among HTLV-I carriers to develop the different entities of diseases such as those exemplified by ATL and HAM. The similar immunogenetic factors may be postulated for other autoimmune diseases associated with HTLV-I infection, as depicted in Fig. 5.

ACKNOWLEDGMENT

The author thanks Dr. M. Osame for his valuable advice and support to this investigation; Dr. M. Tara for his kind provision of the clinical specimens from

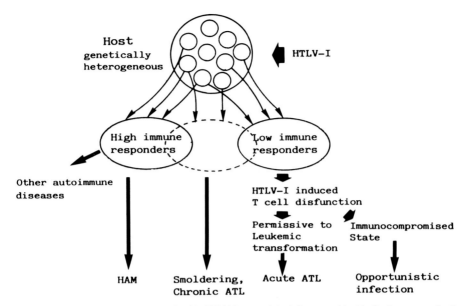

FIG. 5. Immunogenetic factors involved in HTLV-I–associated diseases. Host indicates a genetically heterogeneous population which includes low- and high-immune responders to HTLV-I, and develops the different types of leukemias and autoimmune diseases, including HAM.

ATL and HAM patients; and Drs. S. Yashiki, N. Eiraku, S. Ijichi, and K. Machigashira for their excellent laboratory works and data analyses.

This investigation was supported in part by the Grant-in-Aid for Cancer Research (Nos. 62015069 and 63015075) from the Ministry of Education, Science, and Culture of Japan.

REFERENCES

1. Yoshida M, Osame M, Usuku K, et al. Viruses detected in HTLV-I–associated myelopathy and adult T-cell leukemia are identical on DNA blotting. *Lancet* 1987;i:1085–1086.
2. The T- and B-Cell Malignancy Study Group. Statistical analyses of clinico-pathological, virological and epidemiological data on lymphoid malignancies with special reference to adult T-cell leukemia/lymphoma: a report of the second nationwide study of Japan. *Jpn J Clin Oncol* 1985;15:517–535.
3. Blattner WA, Kalyanaraman VS, Robert-Guroff M, et al. The human type-C retrovirus, HTLV, in blacks from the Caribbean region, and relationship to adult T-cell leukemia/lymphoma. *Int J Cancer* 1982;30:257–264.
4. Osame M, Usuku K, Izumo S, et al. HTLV-I–associated myelopathy, a new clinical entity. *Lancet* 1986;i:1031–1032.
5. Gessain A, Barin F, Vernant JC, et al. Antibodies to human T-lymphotropic virus type-I in patients with tropical spastic paraparesis. *Lancet* 1985;ii:407–409.
6. Shoji H, Kuwasaki N, Natori H, et al. HTLV-I–associated myelopathy and adult T-cell leukemia cases in a family. *Eur Neurol* 1989;29:33–35.

7. Shaw GM, Broder S, Essex M, Gallo R. Human T-cell leukemia virus: its discovery and role in leukemogenesis and immunosuppression. *Adv Intern Med* 1984;30:1–27.

8. Osame M, Matsumoto M, Usuku K, et al. Chronic progressive myelopathy associated with elevated antibodies to human T-lymphotropic virus type I and adult T-cell leukemia-like cells. *Ann Neurol* 1987;21:117–122.

9. Ceroni M, Piccardo P, Rodgers-Johnson P, et al. Intrathecal synthesis of IgG antibodies to HTLV-I supports an etiological role for HTLV-I in tropical spastic paraparesis. *Ann Neurol* 1988;23(suppl): S188–191.

10. Usuku K, Sonoda S, Osame M, et al. HLA haplotype-linked high immune responsiveness against HTLV-I in HTLV-I–associated myelopathy: comparison with adult T-cell leukemia/lymphoma. *Ann Neurol* 1988;23(suppl):S143–150.

11. Itoyama Y, Minato S, Kira J, et al. Spontaneous proliferation of peripheral blood lymphocytes increased in patients with HTLV-I–associated myelopathy. *Neurology* 1988;38:1302–1307.

12. Hino S, Yamaguchi K, Katamine S, et al. Mother-to-child transmission of human T-cell leukemia virus type-I. *Jpn J Cancer Res (Gann)* 1985;76:474–480.

13. Kajiyama W, Kashiwagi S, Ikematsu H, et al. Intrafamilial transmission of adult T cell leukemia virus. *J Infect Dis* 1986;154:851–857.

14. Blattner WA, Nomura A, Clark JW, et al. Modes of transmission and evidence for viral latency from studies of human T-cell lymphotrophic virus type I in Japanese migrant populations in Hawaii. *Proc Natl Acad Sci USA* 1986;83:4895–4898.

15. Ueda K, Kusuhara K, Tokugawa K. Transmission of HTLV-I. *Lancet* 1988;i:1163–1164.

16. Gallo RC, Sliski AH, de Noronha CMC, et al. Origins of human T-lymphotropic viruses. *Nature* 1986;320:219.

17. Ishida T, Hinuma Y. The origin of Japanese HTLV-I. *Nature* 1986;322:504.

18. Terasaki PI, Bernoco D, Park MS, et al. Microdroplet testing for HLA-A, -B, -C and -D antigens. *Am J Clin Pathol* 1978;69:103–120.

19. Uchiyama T, Hori T, Tsudo M, et al. Interleukin-2 receptor (Tac antigen) expressed on adult T cell leukemia cells. *J Clin Invest* 1985;76:446–453.

20. Mann DL, Popovic M, Murray C, et al. Cell surface antigen expression in newborn cord blood lymphocytes infected with HTLV. *J Immunol* 1983;131:2021–2024.

21. Sonoda S, Yashiki S, Takahashi K, et al. Altered HLA antigens expressed on T and B lymphocytes of adult T-cell leukemia/lymphoma patients and their relatives. *Int J Cancer* 1987;40:629–634.

22. Wysocki LJ, Sato VL. "Panning" for lymphocytes: a method for cell selection. *Proc Natl Acad Sci USA* 1978;75:2844–2848.

23. Eiraku N, Ijichi S, Machigashira K, et al. (*submitted*).

24. Vernant JC, Buisson GG, Sobesky G, et al. Can HTLV-I lead to immunological diseases? *Lancet* 1987;ii:404.

25. Sugimoto M, Nakashima H, Watanabe S, et al. T-lymphocyte alveolitis in HTLV-I–associated myelopathy. *Lancet* 1987;ii:1220.

26. Ohba N, Sameshima M, Uehara F, et al. Ocular manifestations in patients infected with human T-lymphotropic virus type 1 (HTLV-I). In: Ferraz de Oliveira LN, ed. *Ophthalmology today.* Location: Elsevier Science Publishers B.V., 1988;385–386.

27. Couderc LJ, Caubarrere I, Venet A, et al. Bronchoalveolar lymphocytosis in patients with tropical spastic paraparesis associated with human T-cell lymphotropic virus type I (HTLV-I). *Ann Intern Med* 1988;15:625–628.

28. Dixon AC, Kwock DW, Nakamura JM, et al. Thrombotic thrombocytopenic purpura and human T-lymphotrophic virus, type I (HTLV-I). *Ann Intern Med* 1989;110:93–94.

29. Maruyama I, Sakashita I, Mori S, et al. Pulmonary involvement in HTLV-I-associated myelopathy: T-cell accumulation in the lung and cellular hyperimmune responsiveness against HTLV-I. In: Roman GC, Vernant JC, Osame M, eds. *HTLV-I and the nervous system.* New York: Alan R. Liss, 1989;471–478.

30. Tanaka Y, Koyanagi Y, Chosa T, et al. Monoclonal antibody reactive with both p28 and p19 of adult T-cell leukemia virus-specific polypeptides. *Gann* 1983;74:327–330.

31. Sugamura K, Fujii M, Ueda S, et al. Identification of a glycoprotein, gp21, of adult T cell leukemia virus by monoclonal antibody. *J Immunol* 1984;132:3180–3184.

DISCUSSION

A discussant noted that, after many years of investigation, a genetic factor was found to be associated with nasopharyngeal carcinoma. This discussant mentioned that, with regard to ATL, it would be useful to look at the families of ATL patients and to study siblings with and without the disease. Another discussant indicated that such studies are in progress in four ATL families. One participant mentioned that in last year's neuro-pathology meetings a patient with TSP with a predominance of CD8 cells had been reported; is it possible that the CD8 cells that are activated in TSP might be lysing CD4 cells? The previous discussant responded that CD8 cells isolated from HAM patients have an inhibitory effect on HTLV-I antigen-positive CD4 cells. Therefore, it is possible that CD8 cells activated in HAM may be lysing CD4 cells expressing HTLV-I or related antigens. Such questions raise the issue of the role of the activated CD8 cells in the pathogenesis of TSP.

Human Retrovirology: HTLV,
edited by William A. Blattner.
Raven Press, Ltd., New York © 1990.

Multiple Sclerosis in Cubans: A Preliminary Study of HLA Antigens

*William A. Sheremata, †Joseph G. Montes,
and ‡Violet Esquenazi

*Multiple Sclerosis Center, School of Medicine, University of Miami, Florida;
†Department of Biophysics, School of Medicine, University of Maryland, Baltimore,
Maryland; and the ‡Histocompatibility Laboratory, University of Miami/Jackson
Memorial Hospital and Veterans Administration Medical Center, Miami, Florida.

INTRODUCTION

The prevalence of multiple sclerosis (MS) in Florida exceeds the low rates predicted by published epidemiological studies carried out elsewhere (1–3). In 1984 a study by the Department of Health and Rehabilitative Services, State of Florida, established a prevalence rate approaching 80 per 100 000 for the state (4). This unexpectedly high rate may reflect migration from high-prevalence areas. However, southern Florida is also home to a large number of immigrants from Central America and the Caribbean, especially from Cuba. In the 1980 census year, 324 976 Cuban-born individuals were enumerated in Dade County, the most populous county in Florida (5).

NEUROLOGICAL DISEASE IN YOUNG AND MIDDLE-AGED CUBANS RESIDING IN SOUTH FLORIDA

The immigrant Cuban population living in Miami is afflicted by a chronic neurological illness recognized and diagnosed by experienced neurologists as MS. The prevalence rates have yet to be established for this subpopulation.

The occurrence of MS in a population emigrating from a Caribbean island, where tropical spastic paraparesis (TSP) might have been expected (6), suggests that genetic factors may play a role in their predisposition to MS. Cubans are categorized as "Hispanics" by the United States government, but we have observed that they differ in physical characteristics from most other Latin American populations living in south Florida, including those of Mexican and Central American origin. MS appears to be rare in these other populations.

Whereas HLA antigens and haplotypes appear to be associated with an increased risk of HTLV-I–associated myelopathy, a TSP equivalent in Japan (7), extensive investigation has clearly implicated such factors in several MS populations (8–10). Given HLA antigens have been shown predominant in certain geographic areas (11,12), and certain antigens that occur commonly in northern Europe (principally A1, A3, B7, B8, DR2, and DR3) are increased in most MS populations (8–10). Our observations that many Cuban MS patients have fair complexions and that a larger than the usual 10%–15% have a progressive myelopathy (13) prompted us to study the HLA antigens of this population.

Thirty-two Cuban patients, 23 females and 9 males, diagnosed as MS were randomly selected and referred for HLA testing. Eye color, hair color, and skin tone were subjectively assessed by both patients and independent observers. "Light eyes" was recorded if their eyes were blue, green, or light hazel. Blond and light brown hair was characterized as "light hair" (vs. "dark hair"). Skin tone was recorded as "light" or "dark". "Dark" included one black patient with progressive myelopathy (PM).

A diagnosis of MS was accepted if patients met the Schumacher criteria (14). Presence or absence of antibody to HTLV-I was not a criterion for inclusion nor exclusion. Of the total group, relapsing-remitting MS (R/R) was identified in 19 of the 32 patients (60%) and PM in 13 (40%). Of the R/R group, 14 were female and 5 were male; of the PM group, 9 were female and 4 were male. The female-to-male ratio in the R/R disease was similar to the PM group. R/R disease was identified if improvement or stabilization occurred between episodes of neurological deficit in patients with at least two attacks separated by 1 mo or longer. Illness was characterized as PM if the predominant feature of their illness was a progressive myelopathy of at least 6 mo duration. Patients who had retrobulbar neuritis (2 patients) but who then progressed to a spastic paraparesis, or who after many years developed extraocular movement abnormalities (1 patient), also were characterized as PM.

HLA typing was carried out in the regional transplantation histocompatibility laboratory at the University of Miami. Data from 226 unrelated Cuban-born individuals within studied families with a potential organ transplant recipient served as a control population. HLA-A and HLA-B antigens were identified using the standard Terasaki microlymphocytotoxicity test (15). Antisera to 19 antigens were used for HLA-A and 23 antisera for HLA-B typing. DR determinants were identified on B lymphocytes using nine antisera and standardized procedures. HLA-A and -B antigens were identified in 252 subjects and DR antigens in 166. Data analysis was performed using χ^2 testing with correction for numbers of antigens for each locus, applied according to the method of Svejgaard et al. (16,17).

PHYSICAL CHARACTERISTICS, HLA ANTIGENS, AND ETHNIC ORIGIN OF CUBANS

Review of the physical characteristics, i.e., eye color, hair color, and skin tone is shown in Table 1. No control data was available for these physical features.

TABLE 1. *Relationship between eye, hair, and skin color in Cubans related to type of multiple sclerosis*

Color	Total MS	R/R	PM	p value
Eyes: light/dark	50/50	68/32	22/78	<0.01
Hair: light/dark	15/85	21/79	8/92	NS
Skin: light/dark	42/58	63/37	22/78	<0.01

Abbreviations: MS = Multiple sclerosis, R/R = relapsing remitting MS, PM = progressive myelopathy. All numbers indicate percentage.

Of the R/R group, 68% had light eye color, 21% had light hair, and 63% had light skin. Of the PM group, these features were 22%, 8%, and 22%, respectively. Light eye color and skin tone were significantly more common in the R/R group compared with the PM group of patients (p = < 0.01 level).

HLA data were arranged for each of the antigens for controls, R/R, and PM (Fig. 1). χ^2 analysis was carried out, with correction for the number of antigens (P_c). Preliminary characterization of HLA haplotypes was determined by inspection of individual HLA antigens and reference to linkages with the control Cuban-born population. Selected data for antigens A-1, A-3, B-7, B-8, and B-15, and complete DR antigen data, are presented in Fig. 1. Antigen frequencies found in the control population were 18.1% for A-1, 19% for A-3, 15% for B-7, 10.6% for B-8, and 8.4% for B-15. In the MS patients, A-1 was increased in frequency, being present in 34.4% of all patients (26.3% of R/R patients compared with 46% of PM patients) (p = 0.002; P_c = 0.03). A-3 was present in only 12.5%

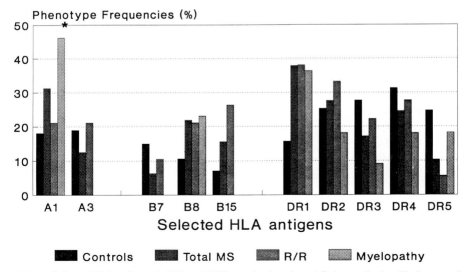

FIG. 1. Selected HLA antigens in R/R and PM in randomly selected Cuban patients with diagnosed MS. B-15 is significantly increased (p < 0.01). DR-1 is also increased (p = 0.01).

of MS patients ($p > 0.05$). Similarly, B-7 was not increased, being present in only 6.3% of the patients (10.5% of the R/R group but none of the PM group). B-8 was increased for the MS patients as a whole but did not reach significance. B15 occurred in 26.3% of the R/R group ($p = 0.001$; $P_c = 0.01$) but not in the PM group. However, B17 did not occur in the R/R group, but the frequency in the PM group did not reach significance when compared with the controls. The only DR determinant to be increased was DR1 ($p < 0.001$; $P_c = 0.01$). A similar increase was seen in the R/R and PM groups (38.1% vs. 36.4%).

Anticipated elevated frequencies of A-3, B-7, and DR-2 were not observed in Cuban MS patients. However, HLA A-1 and B-8 tended to be somewhat increased in these patients, as is seen in most MS populations (10). In the study in northern France by Madigand et al., a similar low frequency in the general population was seen, together with a relative increase in "progressive MS," a population not limited to progressive myelopathies. Our twofold increase might have reached significance in a larger sample size. The antigens and probable haplotypes seen in the Cuban-born population are generally typical of a northwestern European population, an important observation agreeing with the clinical features of the patients. The 10.6% frequency of B-8 in Cubans is lower than the 22.4% frequency in northern France (10) but slightly higher than the 11.1% frequency in Madrid (11). However, the frequency of B-8 in Cuban MS (21.9%) and in French MS (22.4%) patients is remarkably similar. B-8 is distributed in an increasing south-north fashion in Europe and appears to be a better marker for European populations than B-7 (11). B-15 occurred in controls with a frequency of 7.1%. In MS this antigen was found only in R/R disease, with an incidence of 26.3%, comparable with that of the antigen in central Norway and Sweden. Conversely, B-17 occurred only in those with PM, but the increase in comparison with controls fell short of significance. Nevertheless, this datum is particularly interesting because an A2 B17 linkage occurs in west Africa (16). This suggests that genetic factors of Mediterranean origin might be important in the predisposition to myelopathy in the Cuban population. However, most of the B-17 in the control group appeared to be linked to A-1: the A-1 B-17 haplotype is essentially absent in pure African populations. Serological data using reagents for "splits" of B17 (B57 and B58) should distinguish between the predominantly Northern European haplotype (A-1 B-57) and Mediterranean variant (A-1 B-58) of A1 B17. Haplotype data in the Cuban population will also be necessary to elucidate these issues by showing whether an A1 B17 linkage is present, as in the European population.

Interview and questioning of the unaffected Cubans and MS patients reveals interesting and relevant information. The majority are aware that their forebears originated from regions in northwestern Spain, principally Galicia and Asturia. These areas were settled by successive legions of invaders, including the Celts (18). Local traditions include wearing of kilts and playing of bagpipes (18). A few indicate ancestry from the Canary Islands (which are situated off the coast of Africa), where cases of MS have been reported (19).

CONCLUSIONS

Cubans are afflicted by classical relapsing-remitting MS. Our preliminary findings indicate that affected individuals appear to possess HLA antigens typically associated with northern and western European origins but not those most commonly identified in other studies. We have also identified an association of B-15 with this form of disease. Although the sample size is small, the findings are strengthened by the evidence of physical characteristics of the patient population and consideration of their ancestry (12,13).

The finding of an apparently specific association of B-17 with the progressive myelopathic presentation was not predicted, but we had expected a possible association of HLA antigens indicating an association of Mediterranean origin. Should further serological study reveal an association with B-57 or B-58, this could facilitate understanding of the relationship of R/R and PM as well as the role of immune responses and environmental factors such as viruses (16). Further study is clearly indicated.

Our definition of progressive myelopathic disease allowed inclusion of other neurological signs and events, but the major characteristic of their illness was that of a progressive disease of the spinal cord from the outset of the disease. While falling within the Schumacher criteria, the clinical definition could have been more stringent. The decisions for this definition were solely based on the perception that such illness could be not characterized as typical relapsing-remitting illness. The findings suggest that such a classification is valid for the purpose of study. The conclusion that genetic factors play an important role in determining the risk for MS and are probably more important than geographic factors cannot be avoided. However, the role of environmental factors is certainly not excluded. The myelopathic form of illness could represent two or more different diseases, i.e., MS and TSP.

TSP could have been diagnosed in 9 or 10 patients on clinical grounds. In accepting a diagnosis of the progressive myelopathic form of demyelinating illness, we did not segregate patients with myelopathy associated with HTLV-I infection. We wished to avoid classifying patients on the basis of antibody to HTLV-I, because this might obscure a possible role for genetic differences in the ability to respond to such infection with levels of antibody required to obtain positive Western blot assays. Indeed we have found diagnostic antibody in only three patients with progressive disease. At present acute adult T-cell leukemia (ATL) is not diagnosed on the presence or absence of detectable antibody to HTLV-I, nor should there be a need to redefine our patients as having TSP on the basis of such criteria. A survey of the total Cuban MS patient population and controls using more sensitive antibody assays, in situ hydrization, and polymerase chain reaction assays will be required to address these issues.

REFERENCES

1. Kurtzke JF. Epidemiology of multiple sclerosis. In: Hallpike JF, Adams CWM, Tourtelotte WW, eds. *Multiple sclerosis.* Baltimore, MD: Williams and Wilkins, 1983;47–95.
2. Stazio A, Paddison RM, Kurland LT. Multiple sclerosis in New Orleans, Louisiana, and Winnipeg, Manitoba, Canada: follow-up of a previous survey in New Orleans, and comparison between the patient populations in the two communities. *J Chronic Dis* 1967;20:311–332.
3. Chipman M. Multiple sclerosis in Houston, Texas, 1954–1959. A study of the methodology used in determining a prevalence rate in a large Southern city. *Acta Neurol Scand* 1966;42(Suppl 19): 77–82.
4. Health and Rehabilitative Services, State of Florida. *An analysis of the adequacy of services provided to multiple sclerosis patients.* 1984.
5. United States Census, 1980. General social and economic characteristics, Table 59. Persons by Spanish origin, race, and sex.
6. Roman G. Retrovirus associated myelopathies. *Arch Neurol* 1987;44:659–663.
7. Usuku K, Sonada S, Osame M, et al. HLA haplotype-linked high immune responsiveness against HTLC-I in HTLV-I associated myelopathy: comparison with adult T-cell leukemia/lymphoma. *Ann Neurol* 1988(Suppl):S143–S150.
8. Batchelor JR, Compston DAS, McDonald WI. HLA and multiple sclerosis. In: HLA System, *Br Med Bull* 1978;34:279–284.
9. Compston A. Genetic factors in the aetiology of multiple sclerosis. In: McDonald I, Silberberg DH, eds. *Multiple sclerosis.* Kent, England: Butterworth, 1986;56–73.
10. Madigand M, Oger JJ-F, Fauchet R, et al. HLA profile in multiple sclerosis suggests two forms of the disease and existence of protective haplotypes. *J Neurol Sci* 1982;53:519–529.
11. Ryder LP, Andersen E, Svejgaard A. An HLA map of Europe. *Hum Hered* 1978;28:171–200.
12. Moreno ME, Kreisler JM. HLA phenotype and haplotype frequencies in a sample of the Spanish population. *Tissue Antigens* 1977;9:105–110.
13. As discussed by Paty DW (this volume).
14. Schumacher GA, Beebe GW, Kibler RF, et al. Problems of experimental trials of therapy in multiple sclerosis. *Ann NY Acad Sci* 1965;49:253–271.
15. Terasaki PI, McClelland JD. Microdroplet assay of human serum cytotoxicity. *Nature* 1964;206: 998–2001.
16. Svejgaard A, Stabb Nielsen L, Bodmer WF. HLA antigens and disease: statistical and genetical considerations. *Tissue Antigens* 1974;4:95–105.
17. Baur ND, Danilous JA. Joint Report. Population analysis of HLA -A, B, C, DR and other genetic markers. In: *Histocompatibility testing.* Copenhagen: Munksgaard, 1980;955–993.
18. Waldron D. The Celts who live in Spain. *Geogr Mag* 1983;9:118–205.
19. Garcia JR, Rodriguez S, Henriquez S, et al. Prevalence of multiple sclerosis in Lanzarote (Canary Islands). *Neurology* 1989;39:265–267.

DISCUSSION

A participant was asked whether the MS patients described in the presentation were seropositive for HTLV-I. The participant responded that three of the patients with progressive disease, but none of those with typical remitting disease, were seropositive by Western blot. Another participant mentioned that in regard to racial origin, skin color is a particularly poor indicator of race; might it be better to use grandparental origin? It was explained that the origin of the patients could be traced to northwestern Spain and the Canary Islands.

Human Retrovirology: HTLV,
edited by William A. Blattner.
Raven Press, Ltd., New York © 1990.

HTLV-I Associated Myelopathies in Western Europe and Tropical Africa: Clinical, Epidemiological, and Immunovirological Observations

*Guy de Thé, †Olivier Gout, *Antoine Gessain,
and †Olivier Lyon-Caen

*CNRS Laboratory of Epidemiology and Immunovirology of Tumors, Faculty
Medicine Alexis Carrel, 69372 Lyon Cedex 8, France; †Neurology
and Neuropsychology Clinic, Pitié-Salpêtrière Hôpital 43,
Bd de l'Hôpital, 75013 Paris, France

INTRODUCTION

HTLV-I (human T-cell leukemia-lymphoma virus type I), discovered in 1980 by Poiesz et al. (1), represents the prototype human oncoretrovirus having a multipathological potential. First associated with severe cases of adult T lymphoproliferations in Japan (2,3) and the Caribbean (4,5), HTLV-I recently has been found to be closely associated with a subgroup of chronic progressive myelopathies (CPM), as we first observed in Martinique, French West Indies, in 1985 (6,7). We are presenting herewith the situation, as observed in France and in some tropical African countries.

The comparative investigations of neurological manifestations of the two human retroviral subfamilies (the oncoretroviruses HTLV-I, and the lentiretroviruses HIV-1 and HIV-2) (8) should clarify the mechanisms linked to neuropathogenesis. The fact that HIVs are new agents for nonimmune populations leads to great severity of associated diseases, contrasting with the expected low pathogenicity of HTLV-I, a suspected old parasite for certain ethnic groups and geographical areas.

THE MURINE MODEL

The relationship existing between the wild mouse population of *Mus musculus* and the naturally infecting murine leukemia virus (MuLV) has been studied in the Los Angeles area since 1969 by Murray Gardner and may represent a pertinent model for HTLV-I (see Gardner, pages 1–14). This is because MuLV has both hematological and neurological pathogenic potential. The epidemiology of the ecotropic MuLV, in a way similar to that of HTLV-I, shows regional clustering with infection limited to certain families and evidence of mother-to-offspring breast-feeding transmission. This wild strain of MuLV has a very low recombination rate with endogenous host-cell sequences and a low pathogenicity, with a long latent period leading either to hematological or to neurological disorders (9, and Gardner, this volume). The neuropathogenicity of MuLV appears to be related to specific nucleotide sequences situated either at the 3' end of the polymerase gene or throughout the envelope gene region (10). It appears as if certain *env* determinants control the interactions between the viral envelope glycoproteins and the brain cell receptors. The wild mice population susceptible to MuLV would segregate a dominant gene named FV-4, representing a defective endogenous provirus encoding an *env* glycoprotein, closely related to the ecotropic MuLV, and competing with this virus for the virus attachment and entry in target cells (11). Molecular epidemiological studies in HTLV-I–endemic areas should allow researchers, in the near future, to see if similar events take place in the human species.

CHRONIC PROGRESSIVE MYELOPATHIES (CPM), AS SEEN IN WESTERN EUROPE

Chronic progressive myelopathies (CPM) or spastic paraparesis-paraplegia (SPP) seen in Europe are defined by a gradual onset of a paraplegia and a lack of any sign of compression of the spinal cord, with a lack of evidence of motor neuron involvement and lack of supramedullary disseminated lesions. In the last century, Charcot and Erb (12,13) thought that these syndromes could represent an entity that they named primary lateral sclerosis (PLS). However, with such a strictly clinical definition, the CPM group includes miscellaneous diseases. For example, McAlpine et al. (14) described the progressive SPP as a form of possible multiple sclerosis (MS). Paraclinical evaluations, such as evoked potentials, CT scan, and nuclear magnetic resonance imaging (MRI), recently helped to distinguish SPP from MS.

In Western Europe, to make a diagnosis within the CPM group, one must take into account the patient's origin. This is because, in temperate countries of the northern hemisphere, about 60% of CPM are considered as laboratory-supported definite (MS), according to the following internationally accepted criteria: one attack, clinical evidence of one lesion, paraclinical evidence of another separate lesion by EP or MRI, and IgG oligoclonal bands in cerebrospinal fluid (15,16). On the other hand, in tropical countries where MS is rare, most of the CPM correspond to endemic tropical spastic parapareses (TSP) (17). The evidence for a close relationship between HTLV-I and some of these TSP (6) opened a new era in the diagnosis of chronic myelopathies. This observation was rapidly confirmed in other tropical countries, where both HTLV-I and TSP were endemic, and was extended to a nontropical area of southern Japan, where the term HTLV-I–associated myelopathies (HAM) was proposed (18). These developments urged us to study a series of chronic progressive myelopathies in France, in relation to HTLV-I.

The Pitié-Salpêtrière Hospital in Paris houses the largest neurological department in France. HTLV-I antibodies were systematically searched for in all patients with chronic progressive myelopathies (CPM) during a period of 36 mo (1986–1988). Forty-five cases of CPM were observed, and among those, 14 (representing a relatively homogeneous group) exhibited anti-HTLV-I antibodies in both their sera and CSF, as reported in part elsewhere (19).

Table 1 presents the demographic and clinical features of these 14 patients: 7 originated from the French West Indies, 2 from French Guyana, 3 from equatorial Africa, and 2 patients were born in France. Worth noting here is that patients 3 and 4 were transfused 7 and 2 years, respectively, before the onset of their disease. The mean age at onset of these 14 patients was 44 yr, and the mean duration of illness was 6 yr prior to diagnosis. They all had a progressive spastic paraparesis or paraplegia, with brisk reflexes of the upper limbs and bladder dysfunction. Paresthesias and minimal sensory loss were present at diagnosis in most of the patients (9 of 14). Four patients had signs or symptoms suggesting supraspinal involvement, and one exhibited peripheral nerve lesions. Progression of the disease was highly variable. In cases 1 and 7, the course was subacute, and the patients became severely disabled in less than 1 yr. In cases 5, 6, 8, 9, and 10, the progression was slow, while in cases 2, 3, 4, and 11, a stable evolution was noted over a mean observation period of 4 yr (range: 3 to 6 yr). Treatment by corticosteroids had no effect on the clinical progression of the disease. Spinal cord imaging (myelography or spinal MRI) was normal in all cases. Evoked potential (EP) studies showed an abnormal central conduction in 11 patients out of the 13 tested. Visual EP was abnormal in 5, brain stem auditory EP in 7 and somatosensory EP in 9. Brain MRI showed high signal lesions on T2 sequences in 7 cases out of the 12 tested. They consisted of either large lesions in the white matter contiguous to the ventricular system, most often the frontal or occipital horn, or referred to small isolated lesions scattered through the white matter without contact with the ventricular system.

Table 2 presents the immunovirological data of these 14 HTLV-I–associated CPM as seen in Paris. No difference could be observed in the two continental French cases (nos. 7 and 8), when compared to cases originating from HTLV-I–endemic tropical areas. The serum HTLV-I titers varied from 1/1280 to 1/10 240, and the CSF titers from 1/20 to 1/320, as determined by ELISA. An increase in intrablood brain barrier IgG synthesis rate could be observed in every case, except in cases 2 and 3. Pleiocytosis in CSF (more than 5 white cells/ml^3) was present in 10 of 14 cases. Viral isolation was achieved in 10 patients from PBMC or CSF cells, and results are described separately (20, and Gessain et al., this volume, pp. 129–142).

Apart from these chronic progressive myelopathies, 152 definite cases of MS,

TABLE 1. Clinical features of the 14 patients with HTLV-I associated chronic progressive myelopathies seen in Paris

Patient no.	Sex/Age (yr)	Geographical origin	Illness duration (yr)	Progression duration (yr)	Spasticity	Handicap level*	Sensory disturb.	Bladder dysfunct.	Other signs	EP†	Brain MRI‡
1	M/60	Martinique	1	1	+	2	+	+	Optic atrophia	ND	ND
2	F/51	Martinique	12	6	+	2	−	+	Recurrent conjunctivitis	AbN	N
3	F/44	Martinique	6	2	+	1	+	+	Ulnar neuropathy, transient pulm. infiltrate	N	N
4	F/62	Guadeloupe	9	5	+	2	−	+	Recurrent conjunctivitis	AbN	AbN
5	F/34	French Guyana	4	2	+	2	−	+		AbN	N
6	M/44	Senegal	8	7	+	2	+	+		AbN	ND
7	F/52	France	1	1	+	3	+	+	Tinnitus	AbN	AbN
8	F/59	France	3	2	+	1	+	+	Recurrent conjunctivitis	AbN	AbN
9	F/52	Martinique	3	2	+	2	+	+		AbN	N
10	F/54	French Guyana	5	3	+	1	−	+	Cerebellar ataxia, Tinnitus	AbN	AbN
11	F/49	Guadeloupe	11	8	+	3	+	+	Optic neuropathy	AbN	AbN
12	F/56	Zaire	4	4	+	1	−	+		N	N
13	F/34	Central Africa	3	2	+	3	−	+		AbN	AbN
14	F/38	Martinique	13	13	+	2	+	+	Recurrent conjunctivitis	AbN	AbN

* Handicap level: 1 = ability to walk; 2 = ability to walk with two sticks; 3 = bedridden.
† EP: evoked potentials.
‡ Brain MRI: brain magnetic resonance imaging.
ND: not done; AbN: abnormal; N: normal; +: present; −: absent.

TABLE 2. *Immunovirological data of the 14 patients with HTLV-I–associated chronic progressive myelopathies as seen in Paris*

| Patient | HTLV-I antibodies | | ELISA titers | | HTLV-I Ab index* | IBBB IgG synt. mg/d† | IgG olig. bands | CSF white cells N/mm³ | Polycl. int. PBL‡ | HTLV-I in cult. PBL‡ |
	IFI	WB	Serum	CSF						
1	+	+	1/10 240	1/320	7.7	54.5	+	30.4	+	+
2	+	+	1/1280	1/20	3.8	3.0	+	9.8	+	+
3	+	+	1/1280	1/20	3.9	−2.7	−	15	NA	NA
4	+	+	1/2560	1/40	5.2	15.3	+	3.6	+	+
5	+	+	1/2560	1/20	3.0	4.4	−	3	NA	NA
6	+	+	1/1280	1/40	4.2	26.9	+	20.4	NA	NA
7	+	+	1/5120	1/160	4.8	35.1	+	36	+	+
8	+	+	1/2560	1/80	7.5	9.2	+	26.3	NA	NA
9	+	+	1/10 240	1/160	5.6	19.9	+	0.2	+	+
10	+	+	1/10 240	1/160	6.0	17.1	+	11	+	+
11	+	+	1/5120	1/320	4.7	59.7	+	3	+	+
12	+	+	1/2560	1/80	5.2	30.4	+	14	+	−
13	+	+	1/10 240	640	ND	30.6	+	31.2	+	+
14	+	+	1/1280	1/40	4.3	7.8	+	35.2	+	+

* Cutoff value of 2.
† Upper normal reference in our laboratory: 3.5 mg/day.
‡ See details in Gessain et al, this volume, pp. 129–142; NA: not available.

38 cases of Creutzfeldt-Jakob disease, and 14 cases of amyotrophic lateral sclerosis were tested for HTLV-I antibodies. All were negative (21,22). Of interest was that, among 64 unselected patients with "other neurological diseases," three patients originating from the French West Indies and presenting with dementia, stroke, or algoneurodystrophy exhibited HTLV-I antibodies in their sera.

MULTIPLE SCLEROSIS AND HTLV-I/TSP

Supramedullary involvement in HAM/TSP, frequently suggested clinically and by EP or brain MRI investigations, has been proven by necropsy findings in patients from Japan, Jamaica, and Chile (23,24,25). This leads to the unavoidable question of the relationship between the HTLV-I positive CPM, such as HAM/TSP, and multiple sclerosis (MS).

As shown in Table 3, there exist major clinical differences between HAM/TSP and MS, with the unresolved problem of the chronic spinal form of MS. In such cases the paraclinical investigations do not easily differentiate these conditions, since in both diseases, pleiocystosis, IgG oligoclonal bands, and abnormal central conduction time on EP study can be detected. Brain MRI often allows one to differentiate the two conditions, since MS exhibits more extensive abnormality in contrast to the punctate lesions, without contact with the ventricules of HAM/TSP (26,27). Furthermore, in chronic progressive spinal forms of MS, sequential MRI indicates that, with time, some of the abnormal MRI-detected lesions disappear, whereas some others appear, contrasting with the lesions of HAM-I/TSP, which, at least in the case reported by Gout et al. (19), remained stable during a three-year period. Thus, as shown in Table 3, HTLV-I serology, for the time being, remains a most valuable tool to differentiate the two diseases. As already noted, among 152 definite French MS cases, none exhibited HTLV-I seropositivity.

RARITY OF THE NEUROLOGIC SYNDROMES ASSOCIATED WITH HTLV-I IN EQUATORIAL AFRICA

Some reports suggest that equatorial Africa represents an endemic area for HTLV-I, with significant differences according to geographic areas and ethnic groups (28–34). In 1986 we began to investigate the prevalence of HTLV-I infection among neurological patients in some equatorial African countries, as well as to search for the presence of HTLV-I-associated myelopathies.

Tropical neuromyelopathies include different clinical entities, such as tropical ataxic neuropathies (TAN), tropical polyneuritis, and chronic pyramidal syndromes (CPS), with the frequent observation of overlapping syndromes (17). The CPS group, of special interest, here comprises two different epidemiological entities, namely the epidemic spastic paraparesis (ESP) and the sporadic tropical spastic paraparesis (TSP). If the pyramidal tract involvement is the major feature

TABLE 3. *Comparative characteristics of the two main groups of chronic progressive myelopathies (CPM)*

	TSP/HAM	Multiple Sclerosis (MS)
Geographical origin	HTLV-I–endemic areas (tropical countries, Japan)	Western countries (Europe, USA)
Clinical and Epidemiological Aspects		
Sex ratio F/M	3:1	1.4:1
Prevalence/100 000	20–100 in endemic foci	30–80
Ethnics	Black, Asian (low in white)	White (low in Black and Asian)
Age of onset	45 yr (30–60)	30 yr (15–45)
Evolution	Progressive	Remission and attack (15% of progressive form)
Major clinical features	Paraplegia or paraparesis with sphincter involvement and minimal sensory loss	Great polymorphism characterized by various signs or symptoms of CNS involvement, which cannot be explained by a single lesion
Visual evoked potential	Abnormalities in 20–60%	Abnormalities in 85% of definite MS
Brain MRI	Abnormalities in 60%	Abnormalities in 90% of definite MS
Pathological findings	1. Vasculitis of the blood vessels, mostly in the spinal cord and in the sub-arachnoid space 2. Loss of myelin and axons in the lateral and anterior columns of the spinal cord	Multiple foci of demyelinization (plaques) in particular locations (optic nerves and chiasma, periventricular white matter, corticomedullary junction)
Laboratory and Virological Features		
CSF cells	Mild pleocytosis 50/mm^3 (90%)	Mild pleocytosis 50 mm^3 (25%)
ATL-like cells, blood and CSF	+	Absence
IgG CSF oligoclonal bands	>95%	>90% in definite MS
Specificity of some olig. bands	HTLV-I	Unknown
Intrathecal IgG synthesis	95%	90% in definite MS
HTLV-I serology	++ High titers in serum and CSF	neg. or partial reactivities
HTLV-I DNA hybridization in PBL	Southern blot + PCR ++	Southern blot neg. PCR +/−
HTLV-I viral isolation	Blood + CSF +	neg. neg.

in both ESP and TSP, they differ by their etiopathological and epidemiological characteristics. The ESP cases are related to the consumption of cyanide-containing cassava and are observed during famines. They occur with an acute clinical onset, mainly in children and women, and are not associated with HTLV-I (35–37). In contrast, the sporadic cases of TSP occur in adults and have an insidious onset and a slow progressive evolution. Among putative etiological factors (malnutrition, lathyrism, vitamin deficiency, etc.), HTLV-I represents the major factor in the Caribbean (6) and South America (38) but not in South India (39), where TSP may represent yet another etiopathological entity.

Table 4 summarizes data on the three groups of African patients with tropical neurologic syndromes that we were able to test, and gives the results of the serological testing (ELISA and Western blots) for HTLV-I, and HIV-1 and -2. Detailed information and results are to be published elsewhere (40).

Ninety-four cases of ESP collected in 1985 and 1987 in central and eastern Africa were tested and found negative for HTLV-I reactivities, as well as for HIV reactivities.

Among 47 cases of tropical neuromyelopathies observed during an 18-mo survey (1986–1988) in the only neurological department in Abidjan (Ivory Coast), twenty-six sporadic cases of chronic pyramidal syndromes were detected, and among those, four patients exhibited antibodies to HTLV-I (Table 5). None of 21 TAN cases were positive.

Four of 195 other unselected neurological patients from Abidjan and 1 of 58 neurological patients at the Kilimanjaro Christian Medical Center in Moshi, Tanzania, exhibited HTLV-I antibodies.

Clinical and immunovirological data of these nine HTLV-I–positive neurological patients are seen in Table 5. Of the three TSP cases from Ivory Coast, none had received blood transfusion, and their initial symptoms, involving primarily lower-limb paresthesias and abnormal gait, were noticed 2 to 4 yr prior to consultation. At diagnosis, weakness of both legs (with spasticity of both upper

TABLE 4. *HTLV-I and HIVs serological reactivities in African neurological patients* *

	Geographic area	Yr	Specimen	HTLV-I N+/N tested	HIV N+/N tested	Collaborators
Epidemic spastic paraparesis (ESP)	Zaïre	1985	Serum	0/10	0/10	Dr. Carton
			CSF	0/10	0/10	
	Zaïre	1987	Serum	0/15	0/15	Dr. Rösling
	Tanzania	1985	Serum	0/61	0/61	Dr. Howlett
Tropical neuromyelo-pathies						
Chronic pyramidal syndromes (CPS)	Ivory Coast	1986/1988	Serum	3/26	3/26	Pr. Giordano
			CSF	4/21	3/21	
Tropical ataxic neuropathy and polyneuritis	Ivory Coast	1986/1988	Serum	0/21	0/21	Dr. Sonan
Unselected neurological patients	Ivory Coast	1986	Serum	4/195	34/195	Dr. Akani
			CSF	4/45	13/45	
	Congo	1985	Serum	0/62	5/62	Dr. Mouanga
	Tanzania	1987	Serum	1/58	7/58	Dr. Howlett

* From de Thé et al. (40).

TABLE 5. *Characteristics of HTLV-I-positive African neurological patients* *

Age/sex	Geographical origin	Clinical diagnosis	Serum HTLV-I/Ab	CSF/HTLV-I-related data				
				Ab titers	IBBB IgG synth.	IgG/OB	CSF white cells	HIV serology
33/M	Ivory Coast	Typical TSP	1/500	1/20	278 mg/d	+	22	HIV-1+, HIV-2+ in serum and CSF
34/M	Ivory Coast	Typical TSP	1/8000	1/320	68.5 mg/d	+	NA	Negative
26/M	Ivory Coast	Typical TSP	1/2000	1/80	7.5 mg/d	+	28	Negative
52/M	Ivory Coast	Chronic pyramidal syndrome	NA	1/320	NA	NA	NA	Negative
25/M	Ivory Coast	Rhombencephalitis	+	+	NT	NT	NT	HIV-1+, HIV-2+ in serum
27/M	Ivory Coast	Peripheral facial palsy	+	NA	NA	NA	NA	HIV-1+, HIV-2+ in serum
31/F	Ivory Coast	Meningeal hemorrhage	+	NA	NA	NA	NA	Negative
38/M	Ivory Coast	Vascular stroke—hemiparesis	+	+	NT	NT	NT	HIV-2+ in serum
16/M	Tanzania	Coma	+	NA	NA	NA	NA	Negative

* See details in de Thé et al (40).
NA: not available; NT: not tested.

and lower limbs) and bladder dysfunction were obvious, with absence of sensory signs. As discussed elsewhere (40), these Ivory Coast TSP cases were clinically undistinguishable from the TSP cases from Martinique (41). The fourth case represented a chronic pyramidal syndrome, most probably a TSP, but final diagnosis could not be achieved.

Their serum HTLV-I antibody titers varied from 1/500 to 1/8000 and their CSF titers from 1/20 to 1/320, with complete profile of HTLV-I reactivities in Western blot. Evidence of intrathecal HTLV-I–specific antibodies was obtained for the first three patients, with elevated HTLV-I antibody index and the presence of HTLV-I–specific oligoclonal bands in the CSF (42). Besides these three typical TSP cases, and a possible fourth one, five further neurological patients, presenting various unconnected diagnoses, were found to have HTLV-I antibodies (see Table 5).

In summary, only 9 out of 448 (2%) tested African neurological patients exhibited HTLV-I antibodies. Among 26 CPS diagnosed during an 18-mo period in Abidjan, 3 were found to be typical TSP, undistinguishable from TSP seen in the French West Indies (6,7,41). This apparent rarity of typical HTLV-I/TSP in Africa must be taken with some caution, since 3 further African cases were diagnosed during the same period at the Pitié-Salpêtrière Hospital in Paris (see above).

Tropical neuromyelopathies, as noted above, comprise different diseases, and it is of importance to have established that HTLV-I/TSP or HAM does exist, as an etiopathological entity, in tropical Africa. However, there is little doubt that the African situation is different from that observed in the black populations of the Caribbean, where 5% of unselected neurological patients were HTLV-I positive from 1983 to 1986 (22).

Is the rarity of HAM/TSP in equatorial Africa related to a low level of HTLV-I infection in the areas investigated here, or to an underreporting of these cases to the main hospitals? In Ivory Coast during the same period, Ouattara et al. (34) observed a 1% to 2.7% prevalence rate of confirmed HTLV-I infection in the populations of the four regions of Ivory Coast. Based on this, underreporting of such slowly progressive diseases, with no known curative treatment is a likely explanation for this discrepancy between seropositivity and disease occurrence. The recent report by Carton et al. (communication at the International Symposium on Retrovirus in Multiple Sclerosis and Related Diseases, Copenhagen, September 8–9, 1988), of 25 HTLV-I–positive HAM/TSP out of 32 patients presenting with chronic spastic paraparesis in a village of northern Zaïre supports the concept that the disease occurs in significant numbers when looked for. In my opinion the most likely hypothesis concerning equatorial Africa and HTLV-I–related diseases is that an ancient and low endemic virus results in a low pathogenic potential, except in certain geographical areas and ethnic groups, where the viral infection may be higher or more recent and therefore results in more obvious clinical manifestations.

In contrast with the relative rarity of HTLV-I infection observed in unselected neurological African patients, infection by HIV-1, and to a lesser extent by HIV-2, was much more frequent. As seen in Table 4, HIV antibodies were detected in 14.6%, as compared with 1.6% antibodies to HTLV-I (for details, see ref. 40).

HTLV-I: THE CAUSAL AGENT OF HAM/TSP?

It thus appears that HAM/TSP represents an etiopathological and clinical entity that occurs infrequently in HTLV-I nonendemic areas, such as France or Italy (19,43); with low frequency in HTLV-I–low-endemic countries of equatorial Africa (40,44); and with high prevalence in HTLV-I–endemic areas, such as Japan (18,45), the Caribbean (41,46,47), or South America (38,48). A major unanswered question is whether HTLV-I represents the etiological agent of a subgroup of TSP, with the suggested label of HAM/TSP.

In fact, a major piece of the puzzle is still missing, i.e., the evidence that primary infection by HTLV-I, proven by a seroconversion, leads, within a determined period of time, to the development of typical TSP. Japanese authors (49) have shown that HAM occurs with increased frequency in recipients of contaminated blood, which is highly suggestive of a causal relationship, but seroconversion followed by development of the disease has not yet been reported.[1] If, as we believe, HTLV-I is the causative agent of HAM/TSP, the next question to investigate will be the molecular and immunological mechanisms involved in the pathogenesis of the disease. Another question will be whether the limited but existing diversity of HTLV-I isolates can be linked to specific diseases, such as T-cell leukemia-lymphomas (3) or HAM/TSP, as appears to be the case in the murine system (see above). Molecular epidemiology of HTLV-I, now possible by various techniques such as polymerase chain reaction, could demonstrate whether or not genetic variability of the virus is associated with specific diseases in Ivory Coast, where both HAM/TSP and ATL are present (50). The role of environmental factors in the development of one or the other disease also merits investigation. In this context, the three different EBV-associated diseases, namely infectious mononucleosis in adolescents of industrialized countries, Burkitt's lymphoma in African children, and nasopharyngeal carcinoma in southern Chinese adults, represent an interesting model (51).

SUMMARY

Out of 45 patients with CPM seen during a 36-mo period at the neurological department of the Pitié-Salpêtrière Hospital in Paris, 14 exhibited antibodies to HTLV-I and were diagnosed as HAM/TSP. Whereas 12 patients originated from

---

[1] Note added in proofs: The recent case reported by Gout et al. (52) of a subacute syndrome identical to HAM/TSP and developing after an HTLV-I contaminated blood donation for a cardiac transplantation brings the up-to-now missing evidence that HTLV-I is the etiological agent of HAM/TSP syndromes.

HTLV-I–endemic overseas areas, 2 were born in continental France, without known risk factor. Immunovirological characterization and clinical evolution of these 14 patients were indistinguishable from those of HAM/TSP as observed in the Caribbean or Japan. Although differential diagnosis between certain spinal forms of MS and HAM/TSP might be difficult, HTLV-I serology appears to be a useful tool to separate them.

The apparent rarity of HTLV-I–positive TSP cases in Africa (3 out of 47 tropical neuromyelopathies diagnosed during 18 mo in Ivory Coast's only neurology department) may be related to underreporting, since 3 black-African cases were also diagnosed during the same period in Paris at the Pitié-Salpêtrière Hospital. The African cases were identical to the original Caribbean cases. In contrast to the relatively low prevalence of HTLV-I antibodies (1.6% of unselected African neurological patients), HIV antibodies were detected in 14.6% of the same patients.

Thus, HAM/TSP represents an etiopathological entity existing in many different parts of the world, allowing the initiation of molecular epidemiology, with the aim of determining whether minor genetic diversity of HTLV-I may be linked to different pathologies, as is the case for MuLV in the wild murine population of southern California.

ACKNOWLEDGMENTS

We are indebted to Pr. Giordano and his collaborators, Dr. Akani and Dr. Sonan, at Cocody University Hospital, Abidjan, for allowing us to discuss their patients, already described in Giordano et al. (44) and de Thé et al. (40); to C. Caudie for her participation in CSF investigation; to F. Stenger for serological testing; and to B. Maret for her assistance in preparing this manuscript. Work reported here benefitted from the financial support of NIH-NCI (contract no. NO1-CO-74102), the Mérieux Foundation, the ARC Villejuif (contract 6670), and the World Laboratory (project MCD-2 309-6).

REFERENCES

1. Poiesz BJ, Ruscetti FW, Gazdar AF, Bunn PA, Minna JD, Gallo RC. Detection and isolation of type C retrovirus particles from fresh and cultured lymphocytes of a patient with cutaneous T-cell lymphoma. *Proc Natl Acad Sci USA* 1980;77:7415–7419.
2. Miyoshi I, Kubonishi I, Yoshimoto S, et al. Type C virus particles in a cord T-cell line derived by co-cultivating normal human cord leukocytes and human leukaemic T cells. *Nature* 1981;294: 770–771.
3. Yamamoto N, Hinuma Y. Viral aetiology of adult T-cell leukaemia. *J Gen Virol* 1985;66:1641–1660.
4. Catovsky D, Greaves MF, Rose M, et al. Adult T-cell lymphoma-leukaemia in Blacks from the West Indies. *Lancet* 1982;1:639–643.
5. Blattner WA, Gibbs WN, Saxinger C, et al. Human T-cell leukaemia/lymphoma virus-associated lymphoreticular neoplasia in Jamaica. *Lancet* 1983;2:61–64.
6. Gessain A, Barin F, Vernant JC, Gout O, Maurs L, Calender A, de-Thé G. Antibodies to human

T lymphotropic virus type I in patients with tropical spastic paraparesis. *Lancet* 1985;2:407–410.

7. Vernant JC, Gessain A, Gout O, et al. Paraparésies spastiques tropicales en Martinique. Etude clinique. Haute prévalence d'anticorps anti-HTLV-I. *Presse Med* 1986;15(9):419–422.

8. Gabzuda DH, Hirsch MS. Neurological manifestations of infection with HIV: clinical features and pathogenesis. *Ann Int Med* 1987;107:383–391.

9. Gardner MB. Naturally occurring leukaemia viruses in wild mice: how good a model for humans? *Cancer Surveys* 1987;6(1):56–71.

10. Rassart E, Nelbach L, Jolicoeur P. The cao-BR-E MuLV: sequencing of the paralytogenic regions of its genome and derivation of specific probes to study its origin and the structure of its recombinant genomes in leukemic tissues. *J Virol* 1986;60:910–919.

11. Gardner B. Retroviral infection of the nervous system in animals and man. In: *Neuroimmune Networks: Physiology and Disease,* New York: Alan R. Liss, Inc, 1989:24–37.

12. Charcot JM. Sclérose des cordons latéraux de la moelle épinière chez une femme hystérique atteinte de contracture permanente des quatre membres. *Bull Soc Med Hop (Paris)* 1865;2:24.

13. Erb WA. Über einen wenig bekannten spinalen Symtomen-Komplex. *Klin Wochenschr* 1875;12:357–359.

14. McAlpine D, Lumsden CE, Acheson D. *Multiple Sclerosis. A Reappraisal,* 2nd ed. Baltimore, MD: Williams and Wilkins, 1972.

15. Paty DW, Blume WT, Brown WB, Jaabul N, Kentesz A, McInnis W. Chronic progressive myelopathy: investigation with CSF electrophoresis, evoked potentials and CT scan. *Ann Neurol* 1979;6:419–429.

16. Poser CM, Paty DW, Scheinberg L, et al. New diagnostic criteria for multiple sclerosis: guidelines for research protocols. *Ann Neurol* 1983;13:227–231.

17. Roman GC, Spencer PS, Schonberg BP. Tropical neuromyelopathies: the hidden endemias. *Neurology* 1985;35:1158–1170.

18. Osame M, Usuku K, Izumo S, et al. HTLV-I associated myelopathy, a new clinical entity. *Lancet* 1986;1:1031–1032.

19. Gout O, Gessain A, Bolgert F, et al. Chronic myelopathies associated with HTLV-I. A clinical, serological and immuno-virological study of 10 patients seen in France. *Arch Neurol* 1989;46:255–260.

20. Gessain A, Saal F, Morozov V, et al. Characterization of HTLV-I isolates and T lymphoid cell lines derived from French West Indian patients with tropical spastic paraparesis. *Int J Cancer* 1989;43:327–333.

21. Hauser SL, Aubert C, Bruks JS, Kerr C, Lyon-Caen O, de-Thé G, Brahic M. Analysis of human T-lymphotropic virus sequences in multiple sclerosis tissue. *Nature* 1986;322:176–178.

22. Gessain A, Gout O, Caudie C, et al. Chronic progressive myelopathies associated with HTLV-I in French West Indies, Africa, and France: a clinical, serological and immuno-virological update. In: Roman G, Vernant JC, Osame M, eds. *HTLV-I and the Nervous System,* Neurology and Neurobiology, vol. 51. New York: Alan R. Liss, Inc, 1989:103–122.

23. Akizuki S, Nakazato O, Higuchi Y, et al. Necropsy findings in HTLV-I associated myelopathy. *Lancet* 1987;1:156–157.

24. Piccardo P, Ceroni M, Rodgers-Johnson P, et al. Pathological and immunological observations on tropical spastic paraparesis in patients from Jamaica. *Ann Neurol* 1988;23(suppl):S156–S160.

25. Cartier L, Araya F, Verdugo R. Progressive spastic paraparesis in Chile. In: Roman G, Vernant JC, Osame M, eds. *HTLV-I and the Nervous System,* Neurology and Neurobiology, vol. 51. New York: Alan R. Liss, Inc, 1989:167–173.

26. Kira J, Minato S, Itoyama Y, Goto I, Kato M, Hasuo K. Leukoencephalopathy in HTLV-I-associated myelopathy: MRI and EEG data. *J Neurol Sci* 1988;87:221–232.

27. Cruickshank JK, Rudge P, Dalgleish AG, et al. Tropical spastic paraparesis and human T-cell lymphotropic virus type 1 in the United Kingdom. *Brain* 1989;112:1057–1090.

28. Biggar RJ, Johnson BK, Oster C, et al. Regional variation of prevalence of antibody against human T-lymphotropic virus types I and III in Kenya, East Africa. *Int J Cancer* 1985;35:763–767.

29. De-Thé G, Gessain A, Gazzolo L, et al. Comparative seroepidemiology of HTLV-I and HTLV-III in the French West Indies and some African countries. *Cancer Res* (suppl) 1985;45:4633s–4636s.

30. Fleming AF, Maharajan R, Abraham M, et al. Antibodies to HTLV-I in Nigerian blood donors,

their relatives and patients with leukemias, lymphomas and other diseases. *Int J Cancer* 1986;38: 809–813.

31. Weiss RA, Cheingsong-Popov R, Clayden S, Pegram S, Tedder RS, Barzilai A, Rubenstein E. Lack of HTLV-I antibodies in Africans. *Nature* 1986;319:794–795.

32. Delaporte E, Dupont A, Peeters M, et al. Epidemiology of HTLV-I in Gabon (Western Equatorial Africa). *Int J Cancer* 1988;42:687–689.

33. Delaporte E, Peeters M, Durand JP, et al. Sero-epidemiological survey of HTLV-I infection among randomized populations of Western Central African countries. *JAIDS* 1989;2:410–413.

34. Ouattara SA, Gody M, de-Thé G. Prevalence of HTLV-I as compared to HIV-1 and HIV-2 antibodies in different groups in Ivory Coast (West Africa). *JAIDS* 1989;2:481–485.

35. Carton H, Kayembe K, Kabeya, Odio, Billiau A, Maertens K. Epidemic spastic paraparesis in Bandundu (Zaire). *J Neurol Neurosurg Psych* 1986;49:620–627.

36. Rosling H, Gessain A, de-Thé G. Tropical and epidemic spastic parapareses are different. *Lancet* 1988;1:1222.

37. Howlett WP, Brubaker GR, Mingi N, Rosling H. Konzo, an epidemic upper motor neuron disease studied in Tanzania. *Brain* (in press).

38. Zaninovic V, Arango C, Biojo R, et al. Tropical spastic paraparesis in Colombia. *Ann Neurol* 1988;23(suppl):S127–S132.

39. Richardson JH, Newell AL, Newman PK, Mani KS, Rangan G, Dalgleish AG. HTLV-I and neurological disease in South India. *Lancet* 1989;1:1079.

40. de-Thé G, Giordano C, Gessain A, et al. Human retroviruses HTLV-I, HIV-1, HIV-2 and neurological diseases in some equatorial areas of Africa. *JAIDS* 1989;2:550–556.

41. Vernant JC, Maurs L, Gessain A, et al. Endemic tropical spastic paraparesis associated with human T-cell leukemia virus type I. A clinical and sero-epidemiological study of 25 cases. *Ann Neurol* 1987;21(2):123–130.

42. Link H, Cruz M, Gessain A, Gout O, de-Thé G, Kam-Hansen S. Chronic progressive myelopathy associated with HTLV-I: oligoclonal IgG and anti-HTLV-I antibodies in cerebrospinal fluid and serum. *Neurology* 1989;39:1566–1572.

43. Annunziata P, Fanetti G, Giarratana M, Aucone AM, Guzazi GC. HTLV-I associated spastic paraparesis in an Italian woman. *Lancet* 1987;2:1393–1394.

44. Giordano C, Dumas M, Hugon J et al. Neuromyélopathies tropicales africaines: 61 cas étudiés en Côte d'Ivoire. *Rev Neurol* 1988;144(10):578–585.

45. Osame M, Igata A, Matsumoto M. HTLV-I associated myelopathy (HAM) revisited. In: Roman GC, Vernant JC, Osame M, eds. *HTLV-I and the nervous system,* Neurology and Neurobiology, vol 51. New York: Alan R. Liss, Inc, 1989:213–223.

46. Rodgers-Johnson P, Gajdusek DC, Morgan OSC, Zaninovic V, Sarin PS, Graham DS. HTLV-I and HTLV-III antibodies and tropical spastic paraparesis. *Lancet* 1985;2:1247–1248.

47. Rodgers-Johnson P, Garruto M, Gajdusek C. Tropical myelopathies—a new aetiology. *TINS* 1988;11:526–532.

48. Arango C, Concha M, Zaninovic V, et al. Epidemiology of tropical spastic paraparesis in Colombia and associated HTLV-I infection. *Ann Neurol* 1988;23(suppl):S161–S165.

49. Osame M, Izumo S, Igata A, et al. Blood transfusion and HTLV-I associated myelopathy. *Lancet* 1986;2:104–105.

50. Baurmann H, Miclea JM, Ferchal F, et al. Adult T-cell leukemia associated with HTLV-I and simultaneous infection by human immunodeficiency virus type 2 and human herpesvirus 6 in an African woman: a clinical, virologic and familial serologic study. *Am J Med* 1988;85:853–857.

51. De-Thé G, Zeng Y. Population screening for EBV markers toward improvement of nasopharyngeal carcinoma. In: Epstein MA, Achong BG, eds. *The Epstein-Barr Virus: Recent Advances.* London: William Heinemann Medical Books, 1986:236–249.

52. Gout O, Baulac M, Gessain A, et al. Rapid development of myelopathy after HTLV-I infection acquired by transfusion during cardiac transplantation. *N Engl J Med* 1990;322:383–388.

DISCUSSION

A speaker commented that the epidemiological studies under discussion were well carried out. A prevalence of seropositivity to HTLV-I of 10% has been established in

Gabon. Study of sexual transmission is, however, not a simple matter, because the transmission efficiency is low. Sex is frequent in Africans, but study of prostitutes is a very different matter.

Another speaker suggested that, as regards laboratory procedures used to support the diagnosis of TSP, magnetic resonance imaging (MRI) was not a useful tool. In that respect TSP is very different from multiple sclerosis.

One participant was surprised that African prostitutes had lower rates of HTLV-I antibody than might have been expected, based on other studies. This participant asked how long it took prostitutes to seroconvert and cited researchers who have presented evidence that it can take 5 to 20 years. Since the situation with intravenous drug use in addicts is different, this participant wondered if the lower rates might reflect age and limited length of exposure in these women. Another participant agreed that this was a very important point and that it probably took a very long time to seroconvert but indicated that data were not available for African prostitutes. This participant also agreed that limited length of occupational service as a prostitute was a reasonable explanation for the observed rates of seroconversion, because intravenous drug abuse and transfusion were not important factors in this population. Lack of knowledge when an immune response can be found following HTLV-I infection in this population was also mentioned.

Human Retrovirology: HTLV,
edited by William A. Blattner.
Raven Press, Ltd., New York 1990.

HTLV-I/II and Blood Transfusion in the United States

*Alan E. Williams, †Chyang T. Fang, and ‡S. Gerald Sandler

*Transmissible Diseases Laboratory and †National Reference Laboratory for Infectious Diseases, American National Red Cross, Jerome H. Holland Laboratory, Rockville, Maryland 20855; and ‡Blood Services, American National Red Cross, National Headquarters, Washington, D.C. 20006

RATIONALE FOR HTLV-I SCREENING OF BLOOD DONORS

The decision to screen donated blood for antibodies to the human T-cell leukemia virus type I (HTLV-I) in the United States was made from considerations based on epidemiologic observations about HTLV-I in Japan and the Caribbean, as well as direct evidence of the presence of HTLV-I/II[1]–infected blood donors in the United States. Although HTLV-I was the first human retrovirus to be identified, only limited data about the natural history, epidemiology, and importance of HTLV-I, and its closely cross-reacting relative HTLV-II, have been available for the United States. Despite the preliminary nature of HTLV-I epidemiologic data in the U.S., the decision to test all donated blood intended for transfusion was firmly based on five key observations about the transmissibility, distribution, and pathologic manifestations of this agent. We describe below the rationale of this decision, the steps which led to HTLV-I test kit licensure by the U.S. Food and Drug Administration, and the early epidemiologic findings from blood donor screening by the American Red Cross.

HTLV-I Association with ATL and HAM/TSP

Evidence continues to accumulate documenting the etiological association of HTLV-I and adult T-cell leukemia/lymphoma (ATL) (1). Although few natural history data are available, two separate reports have estimated the risk for ATL development in Japan as 2.0% over the lifetime of an individual infected at birth (2,3), whereas a third report has estimated a 4.0% risk of ATL for Jamaican natives infected before the age of 20 (4). The risk of ATL development in an

---

[1] Throughout this paper, the designation of HTLV-I/II is used to indicate the presence of infection when the infecting agent is not known. HTLV-I or HTLV-II is used when the identity of the infecting agent has been determined or when referring to a licensed diagnostic test.

individual who acquires HTLV-I/II infection later in life is unknown. The fact that ATL is rare in the U.S., yet recognized more frequently in HTLV-I–endemic areas of the world, is a direct reflection of both the etiologic relationship of this disease with HTLV-I, and the perinatal modes of HTLV-I transmission that occur in an area of high endemicity (5). ATL is rare in the United States (6) and, as of September 1989, has not been reported in association with transfusion-transmitted HTLV-I infection. Despite the clinical severity of ATL, it is questionable whether the association of ATL and HTLV-I infection alone would have led to testing all donated blood in the United States, given the 20- to 30-yr incubation period between infection and ATL development, the advanced age of most U.S. blood recipients, and the observation that as many as 60% of transfusion recipients may die within ~3 yr of transfusion because of their underlying disease (7).

Recent studies from Japan estimate the incidence of neurological disease after HTLV-I infection to be less than 1.0% (8). Although this figure is lower than the 2–4% lifetime incidence currently estimated for ATL, HTLV-I–associated myelopathy (HAM)—also known as tropical spastic paraparesis (TSP)—appears to have an incubation period that is well within the lifespan of younger transfusion recipients (9). Although reports of HAM/TSP occurring subsequent to HTLV-I infection by transfusion have not been published to date, several retrospective studies from Japan and the Caribbean have defined an increased relative risk for TSP associated with a history of transfusion, and several unpublished prospective observations of posttransfusion HAM/TSP have been made both in Japan and the U.S. (9–12).

Transmission of HTLV-I Infection by Cellular Blood Components

The transmission of HTLV-I infection by cellular blood components was first described and quantitated in 1984 as a result of a large retrospective study conducted by the Kyushu University Hospital and the Fukuoka Red Cross Blood Center in southern Japan (13). This study began in 1981, before the routine testing of Japanese blood donors for HTLV-I. Serial serum specimens from recipients of HTLV-I seropositive blood products demonstrated a 62% rate of HTLV-I seroconversion. These early data, as well as later reports from this study, established that HTLV-I could be transmitted by certain blood components, and also showed that posttransfusion HTLV-I infection is primarily associated with the receipt of cellular components. It is also of interest that the efficacy of transmission appeared to decrease with advancing product age (14,15).

HTLV-I screening became mandatory for Japanese blood donors in November 1986. Two subsequent reports, one from Nagasaki (16) and one from Kumamoto (17), have confirmed that the transfusion-associated transmission of HTLV-I can be greatly reduced or eliminated by removing blood components that test positive for HTLV-I antibodies. Longitudinal studies conducted by Manns et al. in Jamaica have also confirmed these basic observations (18).

Cross-sectional seroprevalence studies of multitransfused patients in New York City have shown seroprevalence rates of 2.3–6.0% (19,20). Although no disease was observed in direct association with these infections, these data provided the first indication that transfusion-transmitted HTLV-I infection was occurring in the United States with an approximate per-unit risk of 0.023%. Remarkably similar findings have also been recently reported from a large cross-sectional study of transfused cardiac surgery patients (21).

HTLV-I/II Infection in U.S. Intravenous Drug Users

Several cross-sectional studies of sera collected from intravenous (IV) drug users in the U.S. have demonstrated that HTLV-I/II infection is endemic in this population, particularly on the East Coast. Cross-sectional seroprevalence studies conducted in IV drug users have shown marked geographical clustering, with rates ranging from 2 to 9% in New Haven and New York, respectively (22,23). Focal rates of infection as high as 49.3% have been observed in black IV drug users in New Orleans (24). Because of needle-sharing activities, it can be assumed that the prevalence of HTLV-I/II infection in the U.S. IV drug–using population is growing, and is likely to continuously seed the spread of infection to the general population via perinatal and sexual transmission routes.

HTLV-I/II Seroprevalence in U.S. Blood Donors

In 1986–1987, the American Red Cross conducted a large cross-sectional/case-control study to evaluate the magnitude of HTLV-I/II infection in U.S. blood donors. Of 39 898 donors in eight geographically diverse regions, 10 (0.025%) were identified as HTLV-I seropositive. Six of seven donors participating in the identity-linked portion of the study revealed risk factors compatible with HTLV-I exposure during postdonation interviews.[2] These risk factors included intravenous drug use and sexual contact with an intravenous drug user or person of Caribbean or Japanese origin. These findings were compatible with current knowledge about the epidemiology of HTLV-I, and indicated that the number of HTLV-I–seropositive blood donors in the U.S. in 1986–1987 was at least comparable with that for HIV-1–infected donors in 1985 and subsequently (25,26).[3]

---

[2] The original report of this study indicated the presence of risk factors in 5/7 donors. An additional individual was identified with risk on subsequent questioning.

[3] HIV-1 seroprevalence in the donor population currently averages 0.017%.

Risk of Blood Transfusion as a Means of HTLV-I Transmission

Estimates of the seroprevalence of HTLV-I/II infection in blood donors and transfusion recipients have made it possible to project the total impact of blood transfusion as a means of HTLV-I/II transmission within the United States in the absence of blood donor screening.

Assuming a seroprevalence of 0.025% in blood donors (25) and extrapolating this figure to 8 million blood donors per year (representing 12 million different donations), we predicted the identification of ~2000 donors who will need to be excluded and counseled in the first year of HTLV-I antibody screening. We also estimated that in the absence of screening, HTLV-I/II–infected donors would be capable of transmitting infection to approximately 2800 recipients of homologous cellular products in a single year in the U.S. (25).

HTLV-I Serological Screening and Confirmation Tests

Serological testing for HTLV-I antibodies in Japan has been conducted primarily by indirect immunofluorescence assay (IFA) and gelatin particle agglutination. In the U.S., research level assays for HTLV-I antibody have utilized a variety of methods, including fluorescence, competitive enzyme-linked immunosorbent assay (EIA), and radioimmunoassay. EIA is the screening test of choice for most infectious disease markers screened in a high-volume blood bank setting because of its ease of use, reproducibility, automation potential, shelf life, and lack of reliance on radioisotopes. Early assays for HTLV-I antibodies lacked adequate specificity because of the difficulties inherent in the separation of viral antigen from cellular proteins. Blood donor prevalence estimates of 1–3% based on these early HTLV assays (27) have now proven to be invalid. As the need for blood donor screening became increasingly apparent, several commercial HTLV-I screening assays were developed which have improved sensitivity and specificity. EIA test kits from three commercial manufacturers (Abbott Laboratories, Dupont/Biotech, and Cellular Products Inc.) were licensed by the U.S. Food and Drug Administration in November 1988 for testing donated blood and plasma.

Four years of experience in the screening and confirmation of HIV-1 antibody in blood donors provided a valuable foundation on which to partially structure HTLV-I antibody confirmation testing. However, although protein immunoblot (Western blot) using purified viral proteins provides reliable diagnostic information for most cases of true HIV-1 infection, the confirmation of HTLV-I antibody is somewhat more complex because of the close cellular association of several HTLV-I viral proteins. Before licensure of test kits, the issue of HTLV-I antibody confirmation was the subject of an FDA working group on confirmation composed of representatives from key virological laboratories in the field. After comparing the results obtained from these individual laboratories on a large representative serum panel, this working group agreed that HTLV-I antibody confirmation requires the presence of antibody to the p24 *gag* gene product, as well one of the HTLV-I *env* gene products, gp46 or gp61. Sera with no reactivity to viral protein bands are considered negative, whereas sera with partial reactivity, but not meeting the confirmation definition, are considered indeterminate (28).

Following the model established for HIV testing, the Western blot assay is the initial test in the American Red Cross HTLV-I antibody confirmation strategy. Although the Western blot assay readily detects antibody reactivities directed toward *gag*-encoded viral proteins, additional tests, such as the radioimmunoprecipitation assay (based on virus-infected whole cells) are often necessary for the demonstration of antibody to *env*-encoded virus components, which tend to be highly cell-associated. Although not readily available to the blood bank, advanced methods of viral nucleic acid detection and characterization, such as the polymerase chain reaction (PCR), can also be used to generate important virologic data at the research level. Similarly, radioimmunoprecipitation using particular HTLV-I–infected cell lines such as SLB-1 can also detect antibody against the p40x *tax* protein, although the role of this reactivity in the diagnosis of infection remains to be determined (29).

Although the initial strategy for serological confirmation has provided a working basis for donor notification, it is increasingly apparent that this strategy may need to be modified to address new virological information related to HTLV-II infection.

Positive HTLV-I Test Results May Reflect HTLV-II Infection

HTLV-II (human T-cell leukemia virus type II) cross-reacts with current HTLV-I test systems. At the present time, distinction between HTLV-I and HTLV-II infection requires nucleic acid hybridization studies based on viral isolates or specific segments of proviral genome amplified by the polymerase chain reaction (30). It is likely, however, that serological distinction of these agents will be possible in the future by recombinant polypeptides specific to the dissimilar portions of the viral genomes. The extent to which HTLV-I seropositivity in U.S. blood donors actually reflects HTLV-II infection is currently unknown. As discussed below, intravenous drug use or sexual contact with an intravenous drug user (IVDU) are major risk factors for HTLV-I/II infection in the U.S. blood donor population. There are now preliminary data showing that HTLV-II may be more common than HTLV-I in the U.S. IVDU population. Viral isolates from two seropositive donors participating in the 1986–1987 American Red Cross HTLV-I/II seroprevalence study were characterized by Southern blot hybridization analysis. The first isolate, characterized as HTLV-I, was derived from a 61-yr-old white male donor whose only identifiable risk factor was sexual contact with Japanese prostitutes in 1947–1949. The second isolate, from a 37-yr-old female residing in Washington D.C.—who had used IV drugs, as well as had numerous sexual contacts with male IV drug users—was characterized as HTLV-II (25). A recent virological study of 23 IV drug users from New Orleans revealed that 21/23 were infected with HTLV-II, whereas two were infected with HTLV-I. It is important to note that, of the 23 donors in this latter study found to carry HTLV-I or HTLV-II provirus genome identifiable by PCR, 3 individuals were serologically indeterminate, and 2 individuals were serologically negative (30). As recently described for HIV-1 infection (31), there is epidemiological evidence that HTLV-I may remain quiescent in the human host without the production of a full spectrum of specific antibodies (32). For these reasons, an exact definition of the sensitivity and specificity of currently licensed HTLV-I screening assays in relation to the detection of all HTLV-I/II infection remains to be determined.

LICENSURE AND IMPLEMENTATION OF HTLV-I SCREENING

On November 4, 1988, a joint statement on "Interim Guidelines for Testing Donated Blood for HTLV-I Antibodies" was issued by the American Association of Blood Banks, the American Red Cross, and the Council of Community Blood Centers, formalizing the commitment to screen all blood donations for HTLV-I antibodies. This statement also provided guidelines for the notification and counseling of seropositive blood donors, and notification and investigation of blood recipients exposed to prior donations of currently seropositive individuals ("lookback").

Licensure of test kits for HTLV-I antibodies occurred in November 1988. Performance of all three assays was generally excellent with sensitivity and specificity greater than 99% (based on patient clinical diagnosis). It was recognized throughout the licensure process that the clinical basis for determination of test performance would not necessarily provide an accurate measure of sensitivity, specificity, and predictive value for the screening of asymptomatic blood donors, but that the relationship of the screening tests to HTLV-I clinical disease represented the only means by which test performance could be assessed in a systematic and timely manner.

Immediately after the licensure of HTLV-I antibody screening assays, U.S. blood donor services began to phase in testing of donated blood, including the inventories of blood components collected before test licensure. By March 1, 1989, all blood components issued by the American Red Cross were tested and found to be negative for HTLV-I antibodies by EIA.

Test Performance

Beginning in December 1989, the American Red Cross began screening blood donors using the Abbott HTLV-I EIA. Initial screening results for Red Cross

Blood Services are indicated in Table 1. For the first 1.6 million blood collections, the systemwide seroprevalence was 0.02%, a figure remarkably consistent with the estimates of infectivity determined by recipient surveillance (19–21) and by the earlier Red Cross seroprevalence/case-control study (25). Despite the use of specialized Western blot techniques that are capable of detecting gp46 reactivity, it has been our experience that demonstration of antibodies to *env*-encoded glycoproteins, and the subsequent confirmation of HTLV-I/II seropositivity, requires the use of radioimmunoprecipitation procedures as often as 60% of the time (33).

Specificity of Abbott Laboratories' EIA screening assay has proven to be very high for blood donor samples, with approximately one-third of repeatedly reactive specimens confirmed positive and one-third negative. Comparative testing of confirmed positive sera with other licensed test kits revealed differences in sensitivity between the kits of different manufacturers (36). These differences are currently under investigation. Because HTLV-II may be the infecting agent in at least a portion of seropositive blood donors, differences in test sensitivity could be due to a differential ability to detect HTLV-II antibodies. Further information will require molecular analyses such as genome amplification to distinguish HTLV-I and HTLV-II.

The remaining one-third of HTLV-I, EIA-reactive donor sera have reactivity on the Western blot and/or radioimmunoprecipitation confirmation assays, but do not meet the current confirmation criteria. The infection status of these individuals is unknown and requires further study.

Donor Notification and Counseling

Although there is a growing consensus regarding the need to test donated blood for HTLV-I/II, the notification of seropositive donors about their test results and provision of adequate counseling remains controversial. Guidelines issued by the U.S. Public Health Service provide an outline of the epidemiology and clinical associations of HTLV-I (34); however, in the absence of information on the natural history of HTLV-I/II infections, these guidelines were unable to quantitate the risk of an infected individual for adult T-cell leukemia or neurologic disease. Although the guidelines do provide general counseling recommendations,

TABLE 1. *American Red Cross HTLV-I/II screening and confirmation test experience*

Three months (1.6 million donations) tested by Abbott HTLV-I EIA
Repeat reactive rate   0.050%
Repeat reactive sera confirmed   35.8%
Overall confirmed positive rate   0.02%
HTLV-I positive donors found in 40/56 ARC regions

the area of notification and counseling of seropositive individuals remains developmental, particularly with respect to the recommended modification of sexual activities to prevent further virus transmission.

The current notification and counseling programs of the American Red Cross are based on the US Public Health Service guidelines, with the additional provision that serological testing of donor family and sexual contacts be conducted, whenever possible, to allow more specific guidance about risks of exposing others by sexual contact.

Concurrent with the implementation of blood donor screening, the American Red Cross began a program to collect epidemiologic data and blood samples from all HTLV-I/II seropositive donors, referred sexual and family contacts, and potentially exposed ("lookback") recipients. Each of these programs has been designed so as to fulfill the operational needs of the regional blood center, as well as to address the need for additional natural history data with which to further refine permanent operational procedures.

I. OBJECTIVES OF THE AMERICAN RED CROSS SYSTEMWIDE HTLV-I/II INVESTIGATIONAL PROGRAMS:
   A. HTLV-I/II testing of contacts and recipients to assist counseling.
   B. Demography and risk status of the infected donors.
   C. Demography and risk status of potentially exposed recipients.
   D. Incidence of recipient HTLV-I/II infection
      1. vs. age of product at transfusion
      2. vs. type of product (including FFP and cryoprecipitate)
      3. vs. characteristics of donor infection
   E. Distinguish HTLV-I from HTLV-II . . . (others?)
   F. Support future operational decisions regarding screening, notification, and counseling of donors and recipients.

Characteristics of Seropositive Donors

The American Red Cross's earlier investigation of HTLV-I/II infection in blood donors (25) indicated that most seropositive individuals had a defined characteristic in their medical history that placed them at an increased risk of HTLV-I/II infection. Red Cross has now identified more than 300 HTLV-I/II seropositive donors. A compilation of data from 28 Blood Services Regions that conducted standardized in-person interviews with 105 HTLV-I/II seropositive donors at the time of result notification is presented in Table 1. A significant overrepresentation of black and female donors is seen in the HTLV-I/II seropositive cohort (Table 2). A risk history compatible with HTLV-I/II exposure was ascertained for 76 individuals (72%), showing the strong influence of sexual exposure to infected parenteral drug abusers or individuals originating from the Caribbean Basin. (Table 3). Considered hierarchically, 24 donors (22.8%) had a history of birth in the Caribbean or sexual contact with a Caribbean native. Nine

TABLE 2. *Demographic characterization of HTLV-I/II seropositive donors: (28 ARC regions)*

	n	%
Black females	39/105	37.1
White females	22/105	21.0
Black males	14/105	13.3
White males	9/105	8.6
Hispanic males	8/105	7.6
Hispanic females	8/105	7.6
Asian females	3/105	2.9
Asian males	1/105	1.0
Native American females	1/105	1.0
Native American males	0/105	0.0

males and one female (9.5%) reported a history of IV drug use. Twenty-four female donors (22.8%) reported sexual contact with IV drug–using males. Of 46 donors without identified risks related to sexual contact or IVDU, 2 were of Japanese descent and 11 reported a history of blood transfusion (10.5%). Twenty-nine donors (83% of whom were female) had no risk identified by questionnaire. Many of the reported risk events, such as IV drug use and sexual contact with persons from endemic areas, occurred as long as 20–30 yr previously, supporting observations from Japan of long periods of asymptomatic infection. When stratified by gender (Table 4), sexual contact with a current or former IVDU was the principal hierarchical risk factor for females, accounting for 33% of the infections. A history of IVDU was the primary hierarchical risk factor for male donors, accounting for 28% of the infections. Odds ratios (OR) for individual risk variables (Table 5) were computed by comparison with overall estimated system demographics. Infection was strongly associated with black race (OR = 9.2, Puerto

TABLE 3. *Evaluation of HTLV-I/II seropositive donors*

	BF	WF	HF	OF	BM	WM	HM	OM	TOTAL
Hierarchical risk of infection (n = 105)									
Birth or sex/c in Caribbean	2	0	1	0	4	0	1	0	8 (7.6%)
IV drug use	0	0	1	0	0	3	6	0	10 (9.5%)
Sex/c born or lived in Caribbean	8	2	0	0	3	2	1	0	16 (15.2%)
Sex/c with IV drug user	12	7	5	0	0	0	0	0	24 (22.8%)
Birth or sex/c in Japan	0	0	0	0	1	0	0	0	1 (1.0%)
History of blood transfusion	2	4	0	1	2	2	0	0	11 (10.5%)
Japanese descent	0	0	0	1	0	0	0	1	2 (1.9%)
No overt identified risk	15	7	1	1	4	1	0	0	29 (27.6%)
Other risk	0	2	0	1	0	1	0	0	4 (3.8%)

TABLE 4. *Hierarchical risk factors for HTLV-I/II infection stratified by gender (all regions)*

Females (n = 73)	
Sex/c with an IV drug user	24/73 (32.9%)
No identified risk	24/73 (32.9%)
Sex/c born or lived in Caribbean	10/73 (13.7%)
History of blood transfusion	7/73 (9.6%)
Born or lived in Caribbean	3/73 (4.1%)
Other risk	3/73 (4.1%)
IV drug use	1/73 (1.4%)
Japanese descent	1/73 (1.4%)

Males (n = 32)	
IV drug use	9/32 (28.1%)
Sex/c born or lived in Caribbean	6/32 (18.8%)
Born or lived in Caribbean	5/32 (15.6%)
No identified risk	5/32 (15.6%)
History of blood transfusion	4/32 (12.5%)
Japanese descent	1/32 (3.1%)
Birth or sex/c in Japan	1/32 (3.1%)
Other risk	1/32 (3.1%)

TABLE 5. *Univariate analysis—risk factors associated with HTLV-I/II infection*

	Random donors*	HTLV-I/II+ donors	Odds ratio	95% confidence interval
History of intravenous drug use				
Males	est. <0.1%	9/105 (8.6%)	93.6	(12.0–259.9)**
Females	est. <0.1%	1/105 (1.0%)	9.6	(0.9–92.9)
Sexual contact with IV drug user				
Males	est. <1.0%	2/105 (1.9%)	1.9	(0.2–20.8)
Females	est. <1.0%	25/105 (23.8%)	30.9	(13.6–71.7)**
Birth in Caribbean (Puerto Rico excluded)	est. <0.5%	11/99 (11.1%)	24.8	(7.7–84.2)**
Sexual contact with Caribbean native				
Males	est. <1.0%	12/32 (37.5%)	59.4	(7.2–225.0)**
Females	est. <1.0%	22/73 (30.1%)	42.7	(5.8–160.4)**
Birth in Japan or sexual contact with Japanese native	est. <0.5%	4/105 (3.8%)	7.8	(1.7–34.4)**
Japanese descent	est. <0.2%	2/105 (1.9%)	9.6	(1.9–48.4)**
History of blood transfusion	est. 6.0%	23/105 (21.9%)	4.4	(1.5–12.7)**
History of sexually transmitted disease				
Males	est. 20%	11/32 (34.3%)	2.1	(0.8–5.5)
Females	est. 10%	11/73 (15.1%)	1.5	(0.5–4.4)

* Note that the random donor comparative data are estimates. Odds ratio estimates have been based on the upper prevalence limit indicated. It will be necessary to collect control interview data so as to provide accurate odds ratio estimates for the individual donor risk characteristics.

** Denotes significance ($p < 0.05$).

Rico excluded). When IVDUs were excluded, infection was also strongly associated with female sex (OR = 3.8).

Although these data represent only the earliest phases of HTLV-I screening and should be interpreted as preliminary and nonrepresentative, they provide continuing evidence for the clustering of HTLV-I/II infection in individuals with a definable risk history.

Contacts of HTLV-I/II Seropositive Donors

In addition to the standardized assessment of seropositive blood donors, the American Red Cross encourages seropositive donors to refer family members and contacts for testing and has established mechanisms to capture data related to the epidemiology and virology of HTLV-I/II infection in these contacts. Of 148 donors notified of HTLV-I/II seropositivity, 41 donors have referred 32 sexual and 20 nonsexual family contacts for HTLV-I testing. Serological evidence of infection was identified in 31% of donor sexual contacts, but not in 12 children or 8 other family members of donors. Nine of the 10 infected donor-partner pairs were married and had a higher mean age (44 yr) than all seropositive donors (41 yr). The hierarchical risk of infection was greater in the male for five couples (four with prior transfusion, one with prostitute contact in Japan), and higher in the female for four couples (three women with prior contact with an IVDU, one woman of Japanese descent). One couple had no identifiable risk.

Although nonsexual family contacts of blood donors are not at high risk of infection, these data indicate that approximately one-third of donor sexual contacts referred for testing are likely to be infected. In contrast to reports from Japan (35), it appears that female-to-male sexual transmission is a factor in HTLV-I/II transmission dynamics in the U.S.

HTLV-I/II Lookback

The American Red Cross has also undertaken a systemwide HTLV-I/II lookback program, which serves as the framework for recipient notification and counseling, as well as establishing mechanisms for the centralized capture of recipient blood samples and seropositivity data. We estimate that approximately 6000 living recipients will have been exposed to transfusions of blood previously donated by HTLV-I–seropositive blood donors detected during the first year of HTLV-I testing.[4]

---

[4] HTLV-I estimated seroprevalence [0.025%] × an average of 4 prior donations per donor in the past five years × 2.5 products per unit × 40% living recipients = 100 living recipients per 100 000 collections. For six million collections per year, this is approximately 6000 recipients.

REFERENCES

1. Gallo RC, Streicher HZ. Human T-lymphotropic retroviruses (HTLV-I, II, and III): the biological basis of adult T-cell leukemia/lymphoma and AIDS. In: Broder S, ed. *AIDS: modern concepts and therapeutic challenges.* New York: Marcel Dekker, 1987:1–22.
2. Tajima K, Kuroishi T. Estimation of rate of incidence of ATL among ATLV (HTLV-I) carriers in Kyushu, Japan. *Jpn J Clin Oncol* 1985;15:423–430.
3. Kondo T, Nonaka H, Miyamoto N, et al. Incidence of adult T-cell leukemia-lymphoma and its familial clustering. *Int J Cancer* 1985;35:749–751.
4. Murphy EL, Hanchard B, Figueroa B, et al. Modeling the risk of adult T-cell leukemia/lymphoma in persons infected with human T-lymphotropic virus type I. *Int J Cancer* 1989;43(2):250–253.
5. Tajima K, Tominaga S, Kuroishi T, Shimizu H, Suchi T. Geographical features and epidemiological approach to endemic T-cell leukemia/lymphoma in Japan. *Jpn J Clin Oncol* 1979;9(suppl): 495–504.
6. Jaffe ES, Blayney DW, Blattner WA, et al. HTLV and associated T-cell malignancies in the United States. *Am J Surg Pathol* 1984;8:263–275.
7. Bove JR, Rigney PR, Kehoe PM, Campbell J. Look-back: preliminary experience of AABB members. *Transfusion* 1987;27(2):201–202.
8. Osame M, Kaplan J, Igata A, Kubota H, Nishitani H, Janssen R. Risk of development of HTLV-I–associated myelopathy (HAM/TSP) among persons infected with HTLV-I. Presented at V International Conference on AIDS, Montreal, Canada, 1989.
9. Roman GC. Retrovirus-associated myelopathies. *Arch Neurol* 1987;44:659–663.
10. Gessain A, Barin F, Vernant JC, et al. Antibodies to human T-lymphotropic virus type-I in patients with tropical spastic paraparesis. *Lancet* 1985;2:407–410.
11. Osame M, Usuku K, Izumo S, et al. HTLV-I associated myelopathy, a new clinical entity. *Lancet* 1986;1:1031–1032.
12. Brew BJ, Hardy W, Poiesz B, et al. Transfusion-related HTLV-I neurological disease: a putative marker of disease activity and response to therapy. Presented at the V International Conference on AIDS, Montreal, Canada, 1989.
13. Okochi K, Sato H, Himuma Y. A retrospective study on transmission of adult T cell leukemia virus by blood transfusion: seroconversion in recipients. *Vox Sang* 1984;46:245–253.
14. Inaba S, Sato H, Okochi K, et al. Prevention of human T-lymphotropic virus type 1 (HTLV-I) through transfusion by donor screening with antibody to the virus. One year experience. *Transfusion* 1989;29(1):7–11.
15. Okochi K, Sato H. Adult T-cell leukemia virus, blood donors and transfusion: experience in Japan. In: Dodd RY, Barker LF, eds. *Infection, immunity, and blood transfusion.* New York: Alan R. Liss, 1984;245–248.
16. Kamihira S, Nakasima S, Oyakawa Y, et al. Transmission of human T cell lymphotropic virus type I by blood transfusion before and after mass screening of sera from seropositive donors. *Vox Sang* 1987;52:43–44.
17. Nishimura Y, Yamaguchi K, Kiyokawa T, Takatsuki K, Inamura Y, Fujiwara H. Prevention of transmission of human T-cell lymphotropic virus type-I by blood transfusion by screening of donors. *Transfusion* 1989;29(4):372.
18. Manns A, Murphy E, Wilks R, et al. Detection of early HTLV-I seroconversion. Submitted to V International Conference on AIDS, Montreal, Canada, 1989.
19. Jason JM, McDougal JS, Cabradilla C, Kalyanaraman VS, Evatt BL. Human T-cell leukemia virus (HTLV-I) p24 antibody in New York City blood product recipients. *Am J Hematol* 1985;20: 129–137.
20. Minamoto G, Gold JWM, Scheinberg DA, et al. Infections with Human T-cell leukemia virus type I in patients with leukemia. *N Engl J Med* 1988;318:219–222.
21. Cohen ND, Munoz A, Reitz B, et al. Transmission of retroviruses by transfusion of screened blood in patients undergoing cardiac surgery. *N Engl J Med* 1989;320(18):1172–1176.
22. Williams AE, Sullivan MT, D'Aquila RT, Williams AB, McNelly A, Peterson LR. Prevalence of HTLV-I/II antibodies among intravenous drug users in Connecticut. (*submitted*)
23. Robert-Guroff M, Weiss SH, Giron JA, et al. Prevalence of antibodies to HTLV-I, -II, and -III in intravenous drug abusers from an AIDS endemic region. *JAMA* 1986;255:3133–3137.
24. Weiss SH, Ginzberg HM, Saxinger WC, et al. Emerging high rates of human T-cell lymphotropic

virus type 1 (HTLV-I) and HIV infection among U.S. drug abusers. In: *Abstracts of the Third International Conference on AIDS,* Washington, D.C. 1987;211.

25. Williams AE, Fang CT, Slamon DJ, et al. Seroprevalence and epidemiological correlates of HTLV-I infection in U.S. blood donors. *Science* 1988;240:643–646.

26. Fang CT, Williams AE, Sandler SG, Slamon DJ, Poiesz BJ. Detection of antibodies to human T-lymphotropic virus type 1 (HTLV-I). *Transfusion* 1988;28(2):179–183.

27. Blattner WA, Saxinger CW, Gallo RC. HTLV-I, the prototype human retrovirus: epidemiologic features. In: Dodd RY, Barker LF, eds. *Infection, immunity, and blood transfusion: Proceedings of the 16th Annual Scientific Symposium of the American Red Cross,* Washington D.C., May 9–11, 1984.

28. Anderson DW, Lee T-H, Lairmore M, et al. Serological confirmation of human T-lymphotrophic virus type I infection in healthy blood and plasma donors. *Blood* 1989;74:2585–2591.

29. Sullivan MT, Williams AE, Slamon DJ. Comparison of single and double-labeling of SLB1 and MT2 cells in the detection of HTLV-I antibodies by radioimmunoprecipitation assay. Submitted to V International Conference on AIDS, Montreal, Canada, 1989.

30. Lee H, Swanson P, Shorty VS, Zack JA, Rosenblatt JD, Chen ISY. High rate of HTLV-II infection in seropositive IV drug abusers in New Orleans. *Science* 1989;244:471–475.

31. Imagawa DT, Moon HL, Wolinsky SM, et al. Human immunodeficiency virus type-1 infection in homosexual men who remain seronegative for prolonged periods. *N Engl J Med* 1989;320:1458–1462.

32. Blattner WA, Nomura A, Clark JW, et al. Modes of transmission and evidence for viral latency from studies of human T-cell lymphotropic virus type 1 in Japanese migrant populations in Hawaii. *Proc Natl Acad Sci USA* 1986;83:4895–4898.

33. Burczak J, Fang C, Leete J, Schrode J, Laitila V, Lee H. HTLV-I and HTLV-II infection in U.S. blood donors. Presented at V International Conference on AIDS, Montreal, Canada, 1989.

34. U.S. Dept. of Health and Human Services. Licensure of screening tests for antibody to human T-lymphotropic virus type I. MMWR 1988;37(48):736–747.

35. Kajiyama W, Kashiwagi S, Ikematsu H, et al. Intrafamilial transmission of adult T cell leukemia virus. *J Infect Dis* 1986;154:852–857.

36. Fang CT, Akins R, Le P, et al. Relative specificity of two EIAs for antibodies to HTLV-I. *Abstracts, V International Conference on AIDS.* Montreal, Canada, 1989.

DISCUSSION

One discussant asked whether the blood bank population of another discussant is voluntary or paid. It was explained that in the American Red Cross population, all blood donors are voluntary donors. Eighty-five percent of donors have donated previously and all accepted donors have negative tests for hepatitis B surface antigen, hepatitis core antibody, and HIV antibody. Each donor signs a form stating that the donor has no risk factors for HIV infection at the time of interview.

One discussant commented that it is extremely interesting to note the decision for blood screening for HTLV-I in the United States. There has been considerable discussion in Europe whether HTLV-I should be screened there. So far, the answer has been negative for economic reasons, but the discussant felt that it is better to screen blood than to transmit infection.

Another discussant was queried as to the reason for screening blood in the United States: Was it for reasons of litigation, for medical indications, or for medical research? The discussant replied that it was the right thing to do.

A discussant wondered whether it was worthwhile to spend millions of dollars testing for HTLV-I if it would only prevent a few cases of ATL, as well as HAM. This discussant wondered whether the money could not be better spent elsewhere in the health care system. The previous discussant responded that considering the cost of open heart surgery and other surgery, the total cost of the procedure for testing blood is relatively inexpensive.

The additional cost of the screening appears to be a very small part of the entire cost of delivering health care to the recipient. This discussant further commented that there was no question that screening for this particular virus is the right thing to do. The long-term consequences of exposing the American population to this virus are not clear. No one could say at this time that either HTLV-I or HTLV-II will not significantly affect the health of the United States in the next decade or even in the next century. It is clear that once people are infected with these viruses they remain infected for their entire lives.

One speaker commented that there are two cases of HAM/TSP associated with transfusions in the United States at this time. One case is from New York, and the other case is from Los Angeles. The case from Los Angeles is that of a 65-year-old Jewish Caucasian who had not traveled outside the United States. The case from New York is that of a 60-year-old who had no previous risk factors for HAM/TSP but had received 45 units of blood when he had been extremely ill. There is no documented evidence of seroconversion, but this patient is believed to have HAM/TSP associated with blood transfusion.

Reiterating on the theme of myopathy, it was felt that the interval from transfusion to development of disease was six months in the case reported. Of the 100 cases of transfusion-associated HAM/TSP in Japan, the average interval for the development of the disease was approximately six months. Intervals of between six months and three years were noted.

Considerable discussion ensued concerning the presence of HTLV-I and HTLV-II in the drug addict population. One participant was asked what results had been found in the drug addict population with regard to HTLV-I and HTLV-II. The participant replied that assessment of the reasons for HTLV-I and HTLV-II risk among the donor population because of drug addiction was in progress.

A discussant queried the group as to the reasons for setting the sensitivity level of the ELISA screening and wondered whether true positive was being missed. Another discussant stated that a few HTLV-I– and HTLV-II–infected individuals may be missed by the tests, but this discussant felt that an epidemic of misinformation to individuals who were falsely positive needed to be avoided. This discussant wanted to have a test that was unequivocally positive, even though some false negatives would be possible.

Human Retrovirology: HTLV,
edited by William A. Blattner.
Raven Press, Ltd., New York © 1990.

Maternal-Infant Transmission of HTLV-I: Implication for Disease

Shigeo Hino

Department of Bacteriology, Nagasaki University
School of Medicine, Nagasaki 852, Japan

INTRODUCTION

Both adult T-cell leukemia (ATL) (1,2) and HTLV-I–associated myelopathy/ tropical spastic paraparesis (HAM/TSP) (3,4) develop specifically in people infected with a human retrovirus, human T-lymphotropic virus type I (HTLV-I) (5–8). Whereas ATL is associated with the malignant monoclonal growth of CD4-positive T-cells infected with HTLV-I, HAM/TSP is associated with poly- or oligo-clonal expansion of infected T-cells, resulting in post-infectional inflammatory and demyelinating processes.

The incidence of ATL among HTLV-I carriers is ∼1/1 000 per year regardless of whether HTLV-I is endemic or nonendemic in the area (9,10). This suggests that the life risk of ATL in HTLV-I carriers is ∼5%. In contrast, the incidence of HAM/TSP is at most one-tenth of ATL.

One of the distinct features of HTLV-I is a peculiar distribution of endemic areas in the world (7,8,11–17). Nagasaki, the southwesternmost prefecture of Japan (population: 1.5 million), is one of the hottest foci in the world, where we estimate the prevalence of carriers as high as 10% of the population over 40 yrs of age (10,18,19). We have 60–100 incidence cases of ATL per year, which corresponds to ∼0.5% of total death in the prefecture (10). Thus, HTLV-I is one of the most serious viruses for health care in the prefecture, and our duty is to provide an effective measure to stop the cycle of HTLV-I infection.

DISTRIBUTION OF HTLV-I CARRIERS IN NAGASAKI

Geographic Distribution of HTLV-I Carriers

We analyzed the geographic distribution of HTLV-I carriers in the southern half and remote islands of the Nagasaki prefecture. The prevalence of carriers varied strikingly from town to town, from 1% to over 20% in populations over 40 years of age (Fig. 1). HTLV-I carriers tend to be concentrated in rural areas,

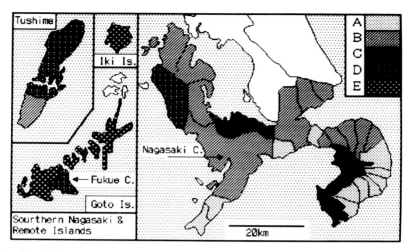

FIG. 1. Distribution of HTLV-I carriers in Southern Nagasaki Prefecture and remote islands; distances from Nagasaki City are 140 km north (Tsushima), 100 km north (Iki), and 100 km west (Goto). At least 50 blood donors, ages 40–59 yr, were tested in each town or city. Rating of prevalences: A, <4.0; B, 4.0–6.3; C, 6.4–9.6; D, 9.7–12.0; E, >12.0%.

including remote islands. The microdistribution of carriers in a small town in one of the remote islands again showed a scattered pattern (data not shown).

Socioeconomic classification of the population was insignificant for the scattered distribution, except for the lower prevalence rate in the eastern coast area of Shimabara peninsula (southeast peninsula in Fig. 1), where the majority of the population was replaced by immigrants from nonendemic areas after a local war in 1637. This suggests that HTLV-I has been endemic in restricted areas for centuries.

Most viruses confined to certain areas have specific vectors. However, infection of HTLV-I in nonvertebrate cells is unknown, and the amount of blood mechanically transferred by insects seems to be insufficient for infection. Furthermore, familial clustering of HTLV-I carriers even in a small homogeneous community (20) is hardly explained by vectors.

Age Distribution of HTLV-I Carriers

A cross-sectional study of carrier prevalences in a town in the Goto islands in 1981 showed age-dependent increase of carrier prevalences more predominant in the female population (Fig. 2), similar to other studies (5,6,9). Age-dependent increase of the seropositive cases in cross-sectional view usually suggests accumulation of population with experiences of viral infection by dominant horizontal

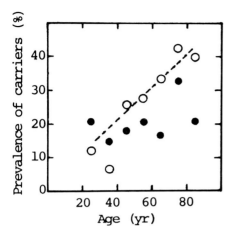

FIG. 2. Age-dependent prevalence of HTLV-I carriers in a small town of Goto, Nagasaki. Open symbols: females; closed symbols: males. Samples were graded by ages in 10-yr generations. Each point represents 15–205 samples.

transmissions. If that is the case, the prevalence in a cohort should show a significant increase during a 10-yr period, especially in females.

Three repeated surveys on the same town in 1971, 1976, and 1981 showed no such increase of carrier rates in each group adjusted by birth years (Fig. 3). The results suggested that apparent age-dependent increase of carriers is more dependent on the environmental factors at the time of birth than on age.

The results of the following studies were consistent with the conjecture that the prevalences are more dependent on birth year than aging. We surveyed the

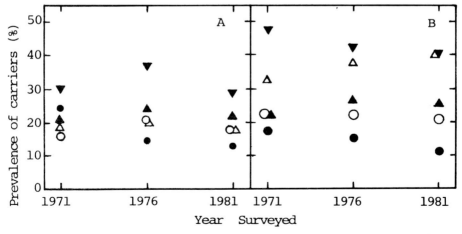

FIG. 3. Prevalence of HTLV-I carriers based on generations. The same town as Fig. 2 was surveyed on three occasions, 1971, 1976, and 1981. **A**: male; **B**: female. Symbols: people who were born in 1900s (▼), 1910s (△), 1920s (▲), 1930s (○), and 1940s (●).

TABLE 1. *Absence of increase in HTLV-I carrier rate with aging in children born to carrier mothers*

Age	No. positive/no. tested	% carriers
1	8/28	29
1–2	9/39	23
1–9	14/62	23

prevalence of carriers in children born to carrier mothers, but there was no significant increase of carrier rate by aging (Table 1) (21). Kusuhara et al. supported the conjectures that children seropositive at age 3 were consistently seropositive at age 18 and that there had been no incidence of seroconversion in children born to carrier mothers during the observation period (22).

We also undertook two surveys of school children of the town (Figs. 2 and 3) six years apart (Table 2) (21). The cross-sectional view on each survey revealed an apparent increase of the carrier prevalences by aging, for example, 1.7–1.6–3.9–8.7% and 0.4–0.0–1.3–2.1%, respectively. However, if we look at the data on the basis of birth years, both the cohorts born in 1968–1970 and in 1971–1973 did not show any significant increase in carrier rates during this 6-yr interval. The results are consistent with the conjecture that HTLV-I infections take place at the very early stage of life, and that carrier rate is dependent on the environments at the time of birth.

These results do not support the possibility of dormant infections with delayed appearance of antibody, because carrier rates did not increase with age. Moreover, we have no example of such a virus except for viruses causing prolonged viremia. The cohort effect can be explained by decreased infections either by disappearance of previously common infection routes or by decreased sensitivity of infection in recipients associated with change of lifestyle, nourishment, vaccination procedures, or with a decrease in mosquito bites. At present, it seems impossible to specify the single contributing factor.

TABLE 2. *Prevalence of HTLV-I carriers in schoolchildren in a town in Goto*

Birth year	Age			
	6–8	9–11	12–14	15–17
1956–1958				8.7 (264)
1959–1961			3.9 (512)	
1962–1964				
1965–1967		1.6 (370)		2.1 (243)
1968–1970	1.7 (290)		1.3 (298)	
1971–1973		0.0 (298)		
1974–1976	0.4 (273)			

Prevalence in percent (no. samples tested).

Male-to-Female Sexual Transmission

Male-to-female sexual transmission is indicated by higher prevalence of carriers in elderly females (9)—most elderly wives of carrier husbands being carriers—and the presence of HTLV-I–infected T-cells in carriers' seminal fluids (23). However, because the greater number of carrier females than males is usually significant only in the population over 60 yrs of age, the sexual transmission from male to female seems to be inefficient. An alternative explanation is the popular use of condoms in Japan for contraception. The average number of offspring decreased from 8–12 in the 1930s to 1–3 in the 1950s in Japan, not because of decreased sexual activity. This may be the reason why the young female group does not show significantly higher prevalence than males today.

HTLV-I Infections by Blood Transfusions

Approximately one-half of those patients who received transfusion with blood from a carrier became HTLV-I carriers themselves (19,24). The most striking feature of HTLV-I infection by blood transfusion is cell associated, as indicated by the fact that freshly frozen plasma without live cells is not infectious at all (24). This is consistent with the experiments with tissue culture systems, which show that the efficiency of HTLV-I infection by virions is far less than that of contact cell cultures (25,26).

In Nagasaki, 5% of donated blood (25 units a day) is from carriers. Therefore, blood transfusion is one of the pathways to an increase in the number of carriers, as indicated by relatively frequent cases of blood transfusion recipients among HAM/TSP patients (4) and carriers (19,27).

However, transfusion hardly explains the high incidence of ATL, because the development of ATL is dependent on infantile infections and transfusions for infants are rare. The high carrier rate in Nagasaki is hardly explained by transfusions, because only 0.3% of the population has experienced blood transfusions even as adults (Hino S, unpublished data). Nevertheless, screening of blood units since 1986 essentially terminated transfusion-mediated infections (28).

The high prevalence of HTLV-II carriers in drug abusers has been reported. This is probably because of the repeated use of unsterile needles, but predisposing factors, such as malnutrition, other infections, immunosuppression, and promiscuity, might explain the high incidence of infections. The number of infected cells in the blood remaining in a needle appear to be small for infection.

MATERNAL-INFANT TRANSMISSION OF HTLV-I

Maternal Transmission as the Major Pathway

Clustering of carriers in families (20) and epidemiological association of carriers in mothers and children (9) were consistent with the presence of maternal in-

TABLE 3. *Association of HTLV-I carriers in mothers and children in Nagasaki City area*

	Age	No. positive/no. tested	%	χ^2
Test group:				
Children born to carrier mothers	1–10	17/78	22	
Pregnant women		523/13,460	3.9	66
Control group:				
Pediatric patients	0–19	14/553	2.6	53
Blood donors	16–18	21/274	1.6	109
Nursing school students	18–19	1/192	0.5	40

$\chi^2 = 3.84$ ($p = 0.05$).

fections. Furthermore, as discussed earlier, most natural infections seem to occur at a very early stage of life. All of these observations were compatible with maternal-infant infections.

To confirm the significance of the maternal-infant infections of HTLV-I for endemic areas of Japan, we started to screen carriers from pregnant females in Nagasaki City area and looked for siblings of the target children (27). The prevalence of carriers in children born to carrier mothers (22%) was significantly higher than that of mothers (4%) (Table 3). Using these two numbers, we estimated the incidence of child carriers in Nagasaki to be roughly 1%, if the maternal-infant transmission is the main route in natural infections.

To estimate the prevalence of child carriers in Nagasaki, we tried to collect serum samples from age-matched controls, but we could obtain control groups with older children. However, if there is a bias in control prevalences because of higher age groups, the control data should be biased to higher age groups. Therefore, if the difference between the carrier and control groups is significant, the conjecture should also be applicable to younger age groups.

The first control group was hospitalized children with an age distribution of 0–19 yr. Of these, 2.6% were seropositive, but 12 of 14 seropositive patients suffered from leukemia and were treated with repeated transfusions. Therefore, we considered the profile did not represent the child population in Nagasaki. We next obtained other control groups, including blood donors (age 16–18) and nursing school students (age 18–19). They showed a prevalence of ~1%. This

TABLE 4. *Mothers of carriers are mostly carriers*

	No. positive/no. tested	%
Mothers of		
School children in a town of Goto	12/13	92
Carrier mothers	6/8	75*
Total	18/21	86

* Each of two mothers with seronegative grandmothers had received blood transfusion: one for open heart surgery, and the other for massive bleeding at previous labor.

was consistent with the value we estimated by the carrier prevalence of pregnant females and infection rate to children and also with the conjecture that maternal transmission is the major pathway of HTLV-I infection (27).

If most child carriers had been infected via their mothers, most mothers of carrier children should be carriers (27). We surveyed mothers of child carriers from two sources (Table 4). For the first group, we screened schoolchildren and adults separately in a town far away from Nagasaki City. Mothers of carrier children were traced from the town registry, and we looked for the seropositivity in screening data of the adult population. We found 13 matched pairs, and 12 out of 13 mothers were carriers. We could not determine a reason for the negative result in the remaining one case.

The second group consisted of mothers of the screened pregnant females. Because we have no positive care for them even if they turn out to be carriers, we tested these grandmothers only if they volunteered. Six of eight grandmothers were positive. Each of the remaining two pregnant cases had experienced blood transfusions, one for an open heart surgery and the other for massive bleeding during previous labor. Thus, most, if not all, mothers of carrier children were themselves carriers. This indicated not only the presence of maternal transmission but also that the major route of HTLV-I infection at a young stage is via carrier mothers.

Period of Maternal-Infant Transmission

Mother-to-child infection can be genetic, intrauterine, perinatal, or postnatal. Because HTLV-I is not endogenous to human species, the genetic infection was discarded first (8). Komuro et al. (29) reported the presence of HTLV-I–bearing cells in a cord blood sample of a baby born to a carrier mother, suggesting transplacental infections.

We surveyed cord blood samples of more than 227 babies born to carrier mothers (Table 5). Essentially every case had maternal antibody in IgG class, judged by close correlation in titers between mother and babies ($r = 0.92$), and by consistent decline of titers after birth (one-tenth-fold in 2 mo) (data not shown).

TABLE 5. *Absence of HTLV-I infection markers in cord bloods of babies born to carrier mothers*

	No. positive/no. tested	%
Anti-HTLV-I		
IgG class	222/227	98
IgM class	0/227	0
HTLV-I Ag positive cells after culture	0/227	0

Antibodies were tested by indirect immunofluorescent test using MT2 cells as targets. Second antibodies used were anti-human IgG or IgM rabbit IgG labeled with FITC.

HTLV-I antigen was tested by microcultures with RPMI1640 and IL-2, for 1–4 wk. At intervals, cells were harvested and tested for antigen expression by the indirect immunofluorescence.

We could not find any case which showed antibody activity in IgM class by indirect immunofluorescence assays or HTLV-I–bearing T-cells after culture up to 1 mo. Because we detected HTLV-I positive cells in two-thirds of carriers, we expected at least 25 positives if transplacental infection is the major route of maternal infections. These data suggested that intrauterine infections are less likely as the major route of maternal transmission of HTLV-I (27).

If perinatal transmission plays an important role in maternal transmission, we expect seroconversions within 2–3 mo, because the latency to develop antibodies is 1–2 mo in cases of infection by blood transfusions (24) and of oral infections in common marmosets (30,31). In contrast, seroconversions within 6 mo were very rare, and most seroconversion was observed at 12 mo after birth (27).

Although these negative results were not strong enough to preclude the possibility of perinatal infection, marked decrease of maternal infection in infants whose carrier mothers refrain from breast-feeding is a strong argument against the possibility, as shown later (32). These considerations prompted us to look for the possible mechanism of postnatal infection as the major pathway of maternal transmission.

Laboratory Evidences of Milk-Borne Infection

Because freshly frozen plasma is not infectious in blood transfusion, we looked for the transfer of infected cells rather than virions. We reached a working hypothesis that the potential candidate for the live-cell transfer, specific to the mother-and-child relationship, is the breast milk of carrier mothers. The first step was to prove the presence of significant numbers of infected cells in the breast milk of carrier mothers (30,33).

We quantitated the number of HTLV-I–carrying cells in the breast milk of carrier mothers. Although the numbers of infected cells varied from one carrier to another, total cells, T-cells (rosette-forming cells with sheep red blood cells), and HTLV-I–infected T-cells (viral antigen positive cells after culture) in breast milk samples were 10^6, 10^5, and 10^3 cells/ml in average, respectively (30). Because the concentration of infected cells did not change significantly during the breast-feeding period (34), we estimated that a baby can take as much as 10^8 cells infected with HTLV-I through breast-feeding by the time of weaning (30).

To strengthen the conjecture of milk-borne infection, it was necessary to study experimental infections in animals. We used common marmosets because Old World monkeys are known to have STLV-1, which cross-reacts with HTLV-I. To simulate the natural breast feeding, we dripped concentrated cell suspensions into mouth cavities, using plastic syringes without needles.

To inoculate known amounts of HTLV-I positive cells, we first used ATL cells from PBL of ATL patients in an amount of 7×10^7 cells (30). At 2.3 mo after the first inoculation, one of two marmosets was found to have seroconverted.

The specificity of the antibody was confirmed by radioimmunoprecipitation using MT2 cell lysates labeled with ^{3}H-leucine. The carrier state of the marmoset was confirmed by repeatedly positive culture of peripheral lymphocytes for HTLV-I positive cells. The results indicated three points: [1] we can use common marmosets as an experimental animal, [2] natural dose of infected cells by breast-feeding can be infectious, and [3] oral administration of live HTLV-I–bearing cells can cause infection.

Next, we inoculated fresh milk cells of carrier mothers (31). During a 2-mo period, we inoculated 26 milk samples obtained within 1 wk after delivery from 22 volunteer carrier mothers. The marmoset seroconverted at 2 mo after the first inoculation and was diagnosed as a carrier by repeated culture of peripheral blood lymphocytes. The total amount of milk we used in the experiment was ~200 ml. Thus, the HTLV-I was found transmissible to common marmosets through breast milk of carrier mothers within an amount of 200 ml.

These results strongly suggested that HTLV-I is transmitted from carrier mothers to children via breast milk. However, the port of entry is not known. I suspect infected T-cells may home into lymphatic tissues in oropharynx or Peyer's patches in the intestine. Higher susceptibility to HTLV-I infections in marmosets than in humans is also possible. Vertical milk-borne infection of HTLV-I in rabbits has been reported (35).

Retrospective Analysis of Children Born to Carrier Mothers in Relation to Feeding

To confirm the working hypothesis of milk-borne transmission as the major route of HTLV-I infection, we analyzed the children born to carrier mothers retrospectively in respect to feeding procedures. Of 83 mother-child pairs, none (0%) of 10 never-breast-fed children were infected, in contrast to 14 (38%) of 43 children mostly breast fed. Of 30 children who were partially breast fed, mixed fed from the beginning, or withdrawn prematurely from breast feeding, 3 (10%) were infected. The completely breast fed children had significantly higher risk of infection, and the data were consistent with the conjecture of milk-borne infection.

Prospective Analysis

The obvious extension of the conjecture is the intervention of breast-feeding. However, there were many sociomedical problems to solve before we started the intervention. The rationale for the intervention study was [1] the life risk for a carrier to develop ATL, as high as 5%, is more than the likelihood of being killed by a traffic accident, approximately 0.8%; [2] ATL and HAM are incurable by current medical standards; [3] compound milk feeding is as safe as breast feeding in developed countries, such as Japan; and [4] prevalence of pregnant carriers in Nagasaki is as high as 4%.

We started a small-scale intervention study in the area of Nagasaki City in 1986, and the program was expanded to cover the whole Nagasaki Prefecture in 1987 (ATL Prevention Program, Nagasaki; APP, Nagasaki). We gave hormones to minimize the distress due to milk production for volunteer mothers who decided to refrain from breast-feeding, and followed up the children at 6, 12, 18, 24, and 36 mo after birth.

The study, now in progress, was designed to analyze 200 formula-fed children born to carrier mothers with 3-yr follow-up data (32). Currently, we have ~150 cases with 1-yr follow-up data. We found four seropositives in formula-fed children. None of them had problems during or after the labor, and they are healthy. Of three cases with data at 6 mo, one was seropositive. The antibody titer of the child was one-half that of the mother, indicating he seroconverted before 6 mo. The possible infection route for these positive children awaits further study.

The frequency of seropositive children born to carrier mothers even by formula feeding, 3%, was <10% of those expected by breast-feeding. In other words, we believe 90% of HTLV-I infection through breast milk of carrier mothers could be interrupted by refraining from breast feeding. Another prospective study by Ando et al. (36) is consistent with our results.

We have 22 000 deliveries per year in Nagasaki; of these, 1000 are deliveries by carrier mothers. If the overall infection rate is 22%, 200 babies will be infected every year without the intervention. By the intervention, we think we could reduce 180 cases of infection every year, which reduces 9 cases of ATL every year in 2140s. The cost for the screening is approximately $15 (US) per head, $1500 per putative infection, and $30 000 per putative ATL case, which is probably less than the cost to treat an ATL case in a hospital. Because the cost is reciprocally proportional to the prevalence of carriers in the intervention area, this type of intervention may not be worthwhile in areas with <1% of carriers.

There are several important questions to be answered, but the intervention study made it impossible to answer them in Nagasaki. The first is the possible protective role of maternal antibodies for milk-borne infection. Although the data are not convincing, Takahashi et al. (37) proposed the possibility that most child carriers appeared to be those who received breast milk for >9 mo. Our data of decreased infection rate in partially breast-fed children may be explained this way. We experienced a seroconversion in a child who received breast feeding for only the first 2 wk, but we do not know whether he was infected by milk or not.

The second question concerns whether some carrier mothers are more infectious than others. We proposed that mothers are more infectious if they have high titer sera (38), detectable antigen in PBL (21), or detectable HTLV-I positive cells in the milk (34) (Table 6). Data have been presented consistent with the hypothesis that mothers with high titer antibody are at high risk for infection.

TABLE 6. *Incidence of mothers with carrier children*

Parameters to break carrier mothers	High risk	Low risk	χ^2	Ref.
High titer antibody in sera	yes (tested) 65% (17)	no (tested) 11% (54)	20.4	38
Antigen expression	yes (tested)	no (tested)		
In PBL culture	47% (19)	11% (19)	6.3	21
In milk cells	40% (10)	7% (15)	4.2	34

$\chi^2 = 3.84$ ($p = 0.05$).

This is very important practically, because we might select mothers who should refrain from breast feeding.

Although this type of intervention is rather simple and safe in developed countries, we cannot suggest it in developing countries where contaminated drinking water may cause infantile diarrhea and appropriate milk concentration is not maintained because of economical and educational problems, because in these areas more serious problems than ATL will ensue.

ACKNOWLEDGMENTS

The author thanks a number of people who collaborated in this study, including medical practitioners, health care officials, laboratory workers, and professionals in the Medical School. Part of this study was supported by Grants-in-Aid for Cancer Research by the Ministry of Education and Welfare.

REFERENCES

1. Uchiyama Y, Yodoi J, Sagawa K, Takatsuki K, Uchino H. Adult T cell leukemia: clinical and hematological features of 16 cases. *Blood* 1977;50:481–491.
2. Takatsuki K, Uchiyama T, Ueshima Y, et al. Adult T cell leukemia: proposal as a new disease and cytogentic, phenotypic, and functional studies of leukemic cells. *Gann Monogr Cancer Res* 1982;28:13–22.
3. Gessain A, Barin F, Vernant JC, et al. Antibodies to human T-lymphotropic virus type-I in patients with tropical spastic paraparesis. *Lancet* 1985;2:407–410.
4. Osame M, Usuku K, Izumo S, et al. HTLV-I associated myelopathy, a new clinical entity. *Lancet* 1986;1:1031–1032.
5. Poiesz BJ, Rescetti FW, Gazdar AF, et al. Detection and isolation of type-C retrovirus particles from fresh and cultured lymphocytes of a patient with cutaneous T-cell lymphoma. *Proc Natl Acad Sci USA* 1980;77:7415–7419.
6. Hinuma Y, Nagata K, Hanaoka M, et al. Adult T cell leukemia: antigen in an ATL cell line and detection of antibodies to the antigen in human sera. *Proc Natl Acad Sci USA* 1981;78:6476–6480.
7. Hinuma Y, Komoda H, Chosa T, et al. Antibodies to adult T-cell leukemia-virus-associated antigen (ATLA) in sera from patients with ATL and control in Japan: a nationwide seroepidemiologic study. *Int J Cancer* 1982;29:631–635.

8. Yoshida M, Miyoshi I, Hinuma Y. Isolation and characterization of retrovirus from cell lines of human adult T-cell leukemia and its implication in the disease. *Proc Natl Acad Sci USA* 1982;79: 2031.

9. Tajima K, Tominaga S, Suchi T, et al. Epidemiological analysis of the distribution of antibody to adult T-cell leukemia-virus-associated antigen: possible horizontal transmission of adult T-cell leukemia virus. *Gann* 1982;73:893–901.

10. Kinoshita K, Kamihira S, Yamada Y, et al. Adult T cell leukemia-lymphoma in the Nagasaki district. *Gann Monogr Cancer Res* 1982;28:167–184.

11. Blattner W, Kalyanaraman V, Robert-Guroff M, et al. The human type-C retrovirus, HTLV, in blacks from the Caribbean region and relationship to adult T-cell leukemia/lymphoma. *Int J Cancer* 1982;30:257–264.

12. Fleming AF, Yamamoto N, Bhusnurmath SR, Maharajan R, Schneider J, Hunsmann G. Antibodies to ATLV (HTLV) in Nigerian blood donors and patients with chronic lymphatic leukemia or lymphoma. *Lancet* 1982;2:962–963.

13. Merino F, Robert-Guroff M, Clark J, et al. Natural antibodies to human T-cell leukemia/lymphoma virus in healthy Venezuelan populations. *Int J Cancer* 1984;34:501–506.

14. Ishida T, Yamamoto K, Omoto K, et al. Prevalence of a human retrovirus in native Japanese: evidence for a possible ancient origin. *J Infection* 1985;11:153–157.

15. Robert-Guroff M, Clark J, Lanier AP, et al. Prevalence of HTLV-I in Arctic regions. *Int J Cancer* 1985;36:651–655.

16. Kazura JW, Saxinger WC, Forsyth K, et al. Epidemiology of human T-cell leukemia virus type I infection in East Sepik Province, Papua New Guinea. *J Infect Dis* 1987;155:1100–1107.

17. Liddle J, Bartlett B, May JT, Stent F, Schnagl RD. Antibody to HTLV-I in Australian aborigines. *Med J Aust* 1988;149:336.

18. Kinoshita K, Ikeda S, Suzuyama J, et al. Annual incidence of adult T-cell leukemia-lymphoma (ATL-L) from ATL virus-carriers in Nagasaki Prefecture. *Nagasaki Med J* 1985;60:56–60.

19. Hino S, Kawamichi T, Funakosi M, Kanamura M, Kitamura T, Miyamoto Y. Transfusion-medicated spread of the human T-cell leukemia virus in chronic hemodialysis patients in a heavily endemic area, Nagasaki. *Gann* 1984;75:1070–1075.

20. Ichimaru M, Kinoshita K, Kamihira S, Ikeda S, Yamada Y, Amagasaki T. T cell malignant lymphoma in Nagasaki district and its problems. *Jpn J Clin Oncol* 1979;9:337–346.

21. Sugiyama H, Doi H, Yamaguchi K, Tsuji Y, Miyamoto T, Hino S. Significance of postnatal mother-to-child transmission of human T-lymphotropic virus type-I on the development of adult T-cell leukemia/lymphoma. *J Med Virol* 1986;20:253–260.

22. Kusuhara K, Sonoda S, Takahashi K, Tokugawa K, Fukushige J, Ueda K. Mother-to-child transmission of human T-cell leukemia virus type I (HTLV-I): a fifteen-year follow-up in Okinawa, Japan. *Int J Cancer* 1987;40:755–757.

23. Nakano S, Ando Y, Ichijo M, et al. Search for possible routes of vertical and horizontal transmission of adult T-cell leukemia virus. *Gann* 1984;75:1044–1045.

24. Okochi K, Sato H, Hinuma Y. A retrospective study on transmission of adult T cell leukemia virus by blood transfusion; sero-conversion in recipients. *Vox Sang* 1985;46:245–253.

25. Miyoshi I, Kubonishi I, Yoshimoto S, et al. Type C particles in a cord T cell derived by cocultivating normal human cord leukocytes and human leukemic T cells. *Nature* 1981;294:770–771.

26. Yamamoto N, Okada M, Koyanagi Y, Kannagi M, Hinuma Y. Transformation of leukocytes by cocultivation with an adult T-cell leukemia virus producer cell line. *Science* 1983;217:737–739.

27. Hino S, Yamaguchi K, Katamine S, et al. Mother-to-child transmission of human T-cell leukemia virus type-I. *Jpn J Cancer Res (Gann)* 1985;76:474–480.

28. Kamihira S, Nakashima S, Oyakawa Y, et al. Transmission of human T-cell lymphotropic virus type I by blood transfusion before and after mass screening of sera from seropositive donors. *Vox Sang* 1987;52:43–44.

29. Komuro A, Hayami M, Fujii H, Miyahara S, Hirayama M. Vertical transmission of adult T-cell leukemia virus. *Lancet* 1983;1:240.

30. Yamanouchi K, Kinoshita K, Moriuchi R, et al. Oral transmission of human T-cell leukemia virus type-I into a common marmoset (*Callithrix jacchus*) as an experimental model for milk-borne transmission. *Jpn J Cancer Res (Gann)* 1985;76:481–487.

31. Kinoshita K, Yamanouchi K, Ikeda S, et al. Oral infection of a common marmoset with human T-cell leukemia virus type-I (HTLV-I) by inoculating fresh human milk of HTLV-I carrier mothers. *Jpn J Cancer Res (Gann)* 1985;76:1147–1153.

32. Hino S, Sugiyama H, Doi H, et al. Breaking the cycle of HTLV-I transmission via carrier mothers' milk. *Lancet* 1987;2:158–159.
33. Kinoshita K, Hino S, Amagasaki T, et al. Demonstration of adult T-cell leukemia virus antigen in milk from three seropositive mothers. *Gann* 1984;75:103–105.
34. Kinoshita K, Amagasaki T, Hino S, et al. Milk-borne transmission of HTLV-I from carrier mothers to their children. *Jpn J Cancer Res (Gann)* 1987;78:674–680.
35. Uemura Y, Kotani S, Yoshimoto S, Fujishita M, Yano S, Ohtsuki Y, Miyoshi I. Oral transmission of human T-cell leukemia virus type I in the rabbit. *Jpn J Cancer Res (Gann)* 1986;77:970–973.
36. Ando Y, Nakano S, Saito K, Shimamoto I, Ichijo M, Toyama T, Hinuma Y. Transmission of adult T-cell leukemia retrovirus (HTLV-I) from mother to child: comparison of bottle- with breast-fed babies. *Jpn J Cancer Res (Gann)* 1987;78:322–324.
37. Takahashi K, Usuku I, Osame M, Yashiki S, Sonada S. Factors on mother-to-child infection of HTLV-I: duration of breast feeding and maternal antibody. *Proc Jpn Assoc Virol* 1987;35:1040.
38. Hino S, Doi H, Yoshikuni H, et al. HTLV-I Carrier mothers with high-titer antibody are at high risk as a source of infection. *Jpn J Cancer Res (Gann)* 1986;78:1156–1158.

DISCUSSION

A speaker asked whether PCR studies have been performed to further assess mother-to-child transmission of HTLV-I and was told that such studies are in progress. The speaker further noted that data supporting transmission by breast-feeding appeared inconsistent with the data presented earlier. It was explained that the lack of seroconversions in 12 breast-fed children in the earlier study were probably due to an insufficient follow-up period. A participant asked whether the seropositive children have any clinical abnormalities associated with HTLV-I infection, such as meningitis or other infectious complications, and was informed that they do not.

Human Retrovirology: HTLV,
edited by William A. Blattner.
Raven Press, Ltd., New York © 1990.

Tropical Spastic Paraparesis-Associated Risk Factors in Tumaco, Colombia

*Cesar Arango, †Mauricio Concha, †Jorge M. Trujillo, and ‡Robin Biojo

*Department of Internal Medicine, Faculty of Health, Universidad del Valle, Cali, Colombia; †Tropical Spastic Paraparesis Research Unit, Tumaco, Colombia; ‡General Surgery Department, San Andres Hospital, Tumaco, Colombia

INTRODUCTION

A chronic, idiopathic, nonhereditary spastic paraparesis considered an isolated form of Tropical Spastic Paraparesis (TSP) was reported in 1981 in the southern Pacific lowlands of Colombia (1). Most of the patients were localized in the seaport of Tumaco. This port is on a sedimentary island located 2°N and 79°W, with a mean temperature of 28°C and an annual precipitation of 2500 mm (2). The island is surrounded by dense tropical rain forest with hundreds of rivers running through. The majority of the inhabitants are black people, descendants of the slave population imported from Africa to work in the gold mines three centuries ago (3).

The association of the human T-cell lymphotropic virus, type I (HTLV-I) with TSP in Colombia was published a few weeks after the initial report by Gessain and associates (4,5). Today there exists conclusive data implicating HTLV-I as an etiological agent in TSP (6–9). This retrovirus was discovered and subsequently shown to be the etiology of adult T-cell leukemia in 1980–1981 (10–12). The virus is transmitted by several routes (13). Although transplacental transmission is rare, the Japanese give evidence of mother-to-child transmission through breast milk (14). Sexual transmission is also an important route; in Japan, over a period of 10 yr, the possibility of HTLV-I transmission from husband to wife was 60.8% and only 0.4% from wife to husband (15). The virus is also transmitted through blood transfusions (16). Except for the latter, natural routes of transmission probably have maintained the virus in mankind for a long time before the appearance of new mechanisms introduced by modern medicine.

A very important feature in Tumaco is its high prevalence of TSP, 98/100 000 (17). To evaluate the possible risk factors associated to TSP and HTLV-I in Tumaco, a case-control study was performed. The study population was distributed as follows: A) 28 cases (14 males and 14 females), mean age of 50.5 yr, and 28 matched controls without TSP. These controls were matched by age ± 10 yr, sex, race, and a minimum of 5 yr living in the same neighborhood. B) A total of 204 household contacts who lived in the cases' and controls' dwellings were included. This group was divided in three subgroups: a) 28 sexual contacts (13 of cases and 15 of controls), b) 91 offspring (50 of cases and 41 of controls), and c) 85 other household contacts (54 of cases and 31 of controls).

After finishing the case-control study, the prevalence of HTLV-I in the general population of urban Tumaco was established.

RISK FACTORS

Family Clustering

All the cases but only seven controls had HTLV-I antibodies in their sera. An important family clustering for HTLV-I infection was suggested by the high seropositivity among the spouses and offspring of the cases, 85% and 25%, respectively, as opposed to 0% among the same control groups ($p < 0.001$). Only two spouses of the seven positive controls could be tested. Although 9% seropositivity was observed among the other household contacts of cases and no positivity in the equivalent control group, this difference was not statistically significant. Moreover, the mean age of the cases' positive household contacts was 40 yr and, if one considers the age-specific prevalence of HTLV-I in the whole population (see below), this age suggests that the chances of their having acquired the infection outside of their households are high.

In two couples from our study, both individuals had TSP. In one of these, one of their children, a 13-yr-old boy, presented increased patellar and Achilles reflexes, bilateral Babinski sign and complaints of occasional cramps in the lower limbs during the last two years. HTLV-I was isolated from his serum and cerebrospinal fluid (CSF) (6).

Sexual Behavior

In this natural setting it is important to remember that the family structure is a remnant of their African ancestors' polygamy and that today more than one-half of the couples cohabit without a formal marriage, a practice known as "union libre." Frequently these marriages last only few years, and an individual may have three, four, or more different spouses during his life. In general, males are more promiscuous than females. In our study, the mean number of intercourses per month, as well as the mean number of different sexual partners per month in either sex, did not associate with TSP. However, when promiscuity was defined as the number of sexual partners that lived for ≥ 1 yr with the studied case or control, it was found that four or more partners of this kind associated strongly with TSP ($p < 0.001$). This significance remained in males. The above results suggest that sexual transmission requires a repeated and/or prolonged contact with an infected individual for an amount of time not yet determined, not only a few or single casual intercourses. The larger the number of sexual partners, the greater the chance of finding an infected one.

Homosexuality and anal and oral intercourse were not commonly practiced by the study population, between 8% and 14%, 20% and 25%, and 8% and 14%, respectively. The few practices reported were evenly distributed between cases and controls. However, a homosexual cohort studied in Trinidad and Tobago showed a high prevalence of HTLV-I antibodies in their sera (19). In addition, the use of condoms in the study population was also infrequent, 13%.

Past history of urethritis or penile ulcers among men in this study tended to associate with TSP, although the difference was not statistically significant. However, an association between the history of genital ulcers and HTLV-I was found in a Jamaican study performed on a population from sexually transmitted disease clinics (18).

Blood Transfusions and Other Risk Factors

In Japan, 26% of 420 cases of HTLV-I–associated myelopathy (HAM) have been associated with a positive history of previous blood transfusion (20). However, this was not the case in Tumaco. So far, the number of previous blood transfusions is similarly distributed between cases and controls. Two facts can contribute to such an observation: a) in Tumaco this procedure is practiced to a lesser extent than in Japan; b) a larger sample size might be needed. Nevertheless, out of 12 randomly selected blood donors from Tumaco recently tested for HTLV-I, 3 (25%) were positive (unpublished data).

Another important risk factor for the transmission of retroviruses, specifically HTLV-I, is intravenous drug abuse (21). In both the case-control study and the prevalence study, none of the participants admitted to ever having been intravenous drug abusers. Because the use of nondisposable syringes is still a very common practice in Tumaco, this factor was also evaluated, but no association was found with TSP. Neither was any association found between TSP and the number of previous admissions to the hospital or a past history of yaws.

Transmission to Offspring and Other Household Contacts

Further in the analysis, the offspring and other household contact groups were subdivided into seropositive and -negative individuals. The mean number of years that these individuals shared the same house, bedroom, or bed with a TSP case was very similar. It is likely, then, that the risk factor for HTLV-I transmission is independent of such type of intimate contact. In addition, it was observed that almost every offspring received breast milk during an average period of 12 mo. Breast-feeding has been found to be the most important risk factor for mother-to-child transmission of HTLV-I (13).

Obstetric Impact

Neither the number of pregnancies nor the pregnancy outcome was associated with HTLV-I seropositivity or TSP in the female study population.

TSP and Other Infections

To assess a possible association between fluorescent treponemal antibody-absorption (FTA-ABS) serology with both HTLV-I infection and TSP, only cases and controls with a negative history for yaws were considered. It was observed that 50% of 14 cases and only 18% of 16 controls were FTA-ABS positive (odds ratio, OR = 4.3, $0.05 < p < 0.10$). This analysis was further adjusted by a negative history of previous blood transfusions, and although 50% association remained among the cases, 23% of 13 controls were positive for FTA-ABS (not significant). Moreover, the total population of 160 individuals with a negative history for previous blood transfusions were analyzed, and a strong association between HTLV-I and FTA-ABS was observed ($p < 0.001$). In the latter group, adjustments for a negative history of yaws could not be made. Yaws was eradicated from Tumaco during the early 1950s (17).

FTA-ABS serology not related to yaws or blood transfusion may reflect the presence of sexually acquired syphilis. The painless and many times unnoticed ulcer of syphilis could favor the transmission of HTLV-I. Other possibilities that may explain the association include a reactivation of an HTLV-I latent infection by the *Treponema sp.* infection or the fact that both microorganisms share the sexual route for transmission. FTA-ABS, however, has rarely been detected in the CSF of TSP patients from Jamaica and Colombia, making unlikely a direct interaction of both microorganisms in the central nervous system (22). Prospective studies are mandatory to clarify what type of association exists between these two microorganisms.

Hepatitis B surface antigen (HBsAg) was not associated with HTLV-I seropositivity or TSP. It can be inferred that, although both viruses share some routes for transmission, chronic hepatitis B carriage is not favored by the retroviral infection, TSP disease or vice versa.

With regard to *P. falciparum* and dengue virus antibodies, it was observed that neither one associated with HTLV-I infection or TSP. Thirty-three percent of 21 cases and 40% of 25 controls were positive for antibodies against *P. fal-*

ciparum. In addition, the prevalence among the 47 HTLV-I positive and 175 HTLV-I negative individuals was 36% and 31%, respectively. More than 90% of these individuals were positive for antibodies against the four serotypes of dengue virus. These results suggest that malaria and dengue, and their respective vectors, *Anopheles sp.* and *Aedes sp.,* are not favoring HTLV-I infection in this area. Nonconfirmed positive HTLV-I antibodies correlated with positive malaria serology in Africa in the past (23).

Sanitation

In general, Tumaco lacks appropriate sewage and aqueduct systems, and electric light runs for 16–20 h a day in an intermittent manner. Some of these aspects were evaluated as possible risk factors, but none of them differ. The use of different sewage systems was very similar in the TSP and control houses. Appropriate sewage systems were rarely used. A privy, shallow transitory holes in the back yard, or a common bucket with daily disposal into the sea were three alternative methods used by two-thirds of both groups. Public restrooms were used by less than one-third of the cases and controls. Both groups were probably equally and heavily exposed to human excretions during the use or manipulation of the buckets or other rudimentary restrooms.

SEROPREVALENCE STUDY

Tumaco has an urban population of 45 594 individuals and 7682 dwellings. The population is equally distributed between both sexes. In 1988, 1077 inhabitants living in 233 houses were randomly selected and later evaluated for HTLV-I antibodies. Thirty individuals (2.79%) were positive for HTLV-I antibodies in serum. This prevalence increased significantly with age (trend test for age $p <$ 0.001), following a similar pattern observed in different studies in Japan and the Caribbean Basin (24,25). No positive sera were detected in children under 5 yr of age. Among children between ages 5 and 9, only 1 of 281 subjects, a girl, was positive. After the population reaches 25 yr of age, the prevalence in women rises steadily, with a highest peak, 20.5% in the 50–59 age group. On the other hand, the prevalence in males begins to increase two decades later and a peak of 17.6% is observed during the sixth decade. The higher female than male prevalence becomes statistically significant after both reach 50 yr of age. This prevalence seems to be the result of an early acquired infection through breast milk, with an age-dependent activation plus a late acquisition of the virus by sexual exposure. It has been suggested that the female-predominant rates may represent a more efficient male-to-female than female-to-male transmission (14). However, the data shown above support the concept that in a subgroup of males with a high degree of promiscuity, female-to-male sexual transmission of HTLV-

I also occurs. In addition, blood transfusions might be involved in HTLV-I transmission during any period in life.

REFERENCES

1. Zaninovic V, Biojo R, Barreto P. Paraparesia espastica del pacifico. *Colombia Med* 1981;12: 111–117.
2. Roman G. The neuroepidemiology of tropical spastic paraparesis. *Ann Neurol* 1988;23(suppl): S113–S120.
3. Gutierrez I. *Historia del Negro en Colombia,* 2nd ed. Bogotá, Colombia: Editorial Nueva America, 1986:15–18.
4. Rodgers-Johnson P, Gajdusek DC, Morgan OStC, et al. HTLV-I and HTLV-III antibodies and tropical spastic paraparesis. *Lancet* 1985;2:1247–1248.
5. Gessain A, Barin F, Vernant JC, et al. Antibodies to human T-lymphotropic virus type-I in patients with tropical spastic paraparesis. *Lancet* 1985;2:407–410.
6. Jacobson S, Raine CS, Mingioli ES, McFarlin DE. Isolation of an HTLV-I like retrovirus from patients with tropical spastic paraparesis. *Nature* 1988;331:540–543.
7. Liberski PP, Rodgers-Johnson P, Char G, Piccardo P, Gibbs JC Jr., Gajdusek DC. HTLV-I–like viral particles in spinal cord cells in Jamaican tropical spastic paraparesis. *Ann Neurol* 1988;23(suppl):S185–S187.
8. Ceroni M, Piccardo P, Rodgers-Johnson P, et al. Intratechal synthesis of IgG antibodies to HTLV-I support an aetiological role for HTLV-I in tropical spastic paraparesis. *Ann Neurol* 1988;23(suppl): S188–S191.
9. Bhagavati S, Ehrlich G, Kula RW, et al. Detection of human T-cell lymphoma/leukemia virus type I DNA and antigen in spinal fluid and blood of patients with chronic progressive myelopathy. *N Engl J Med* 1988;318:1141–1147.
10. Poiesz BJ, Ruscetti FW, Gazdar AF, et al. Detection and isolation of type-C retrovirus particles from fresh and cultured lymphocytes of patients with cutaneous T-cell lymphoma. *Proc Natl Acad Sci USA* 1980;77:7415.
11. Hinuma Y, Nagata K, Hanaoka M, et al. Adult T-cell leukemia: antigen in an ATL cell line and detection of antibodies to the antigen in human sera. *Proc Natl Acad Sci USA* 1981;78:6476–6480.
12. Yoshida M, Miyoshi I, Hinuma Y. Isolation and characterization of retrovirus (ATLV) from cell lines of human adult T-cell leukemia and its implication in the disease. *Proc Natl Acad Sci USA* 1981;79:2031–2035.
13. Clark JW, Blattner WA, Gallo RC. Human T-cell leukemia virus: update VII, oncology. In: Petersdorff RG, Adams RD, Braunwald E, Isselbacher KI, Wilson, eds. *Principles of Internal Medicine,* 11th ed. New York: McGraw, 1986:29–39.
14. Sugiyama H, Doi H, Yamaguchi K, Tsuji Y, Miyamoto T, Hino S. Significance of postnatal mother to child transmission of human T-lymphotropic virus type-1 on the development of adult T-cell leukemia/lymphoma. *J Med Virol* 1986;20:253–260.
15. Kajiyama W, Kashiwagi S, Ikematsu H, et al. Intrafamilial transmission of adult T-cell leukemia virus. *J Infect Dis* 1986;154:851–857.
16. Okochi K, Sato H, Hinura Y. A retrospective study on transmission of adult T cell leukemia virus by blood transfusion: seroconversion in recipients. *Vox Sax* 1984;46:245–253.
17. Roman GC, Roman LN, Spencer PS, Schoemberg BS. Tropical spastic paraparesis. A neuroepidemiological study in Colombia. *Ann Neurol* 1985;17:361–365.
18. Murphy EL, Figueroa JP, Gibbs WN, et al. Sexual transmission of human T-lymphotropic virus type I (HTLV-I). *Ann Intern Med* 1989;555–560.
19. Bartholomew C, Saxinger WC, Clark JW, et al. Transmission of HTLV-I and HIV among homosexual men in Trinidad. *JAMA* 1987;257:2602–2604.
20. Osame M, Igata A, Matsumoto M, Izumo S, Kubota H. Blood transfusion and HTLV-I associated myelopathy in Japan. In: Roman GC, Vernant JC, Osame M, eds. *Neurology and Neurobiology. HTLV-I and the Nervous System,* vol. 51. New York: Alan R. Liss, 1989:547–549.
21. Guroff RM, Weiss SH, Giron JA, et al. Prevalence of antibodies to HTLV-I, -II, -III intravenous drug abusers from an AIDS endemic region. *JAMA* 1986;255:3133–3137.

22. Rodgers-Johnson P, Morgan OStC, Zaninovic V, et al. Treponematoses and tropical spastic paraparesis. *Lancet* 1986;i:809.
23. Biggar RJ, Saxinger C, Gardiner C, et al. Type-1 HTLV antibody in urban and rural Ghana, West Africa. *Int J Cancer* 1984;34(2):215–219.
24. Hinuma Y, Komoda H, Chosa T, et al. Antibodies to adult T-cell leukemia-virus–associated antigen (ATLA) in sera from patients with ATL and controls in Japan. A national seroepidemiologic study. *Int J Cancer* 1982;29:631–635.
25. Clark JW, Saxinger C, Gibbs WN, et al. Seroepidemiologic studies of human T-cell leukemia/ lymphoma virus type I in Jamaica. *Int J Cancer* 1985;36:37–41.

Human Retrovirology: HTLV,
edited by William A. Blattner.
Raven Press, Ltd., New York © 1990.

The Epidemiology of Human T-Lymphotropic Virus Type I in Panama

William C. Reeves

Division of Epidemiology, Gorgas Memorial Laboratory, Republic of Panama

INTRODUCTION

The human T-lymphotropic virus type I (HTLV-I) is an exogenous human retrovirus that was first reported in 1981. HTLV-I infection is highly cell-associated but does not cause T helper lymphocyte depletion or immunosuppression. Several chapters in this volume concern HTLV-I seroepidemiology so I will only briefly summarize key issues. Infection is endemic in southwestern Japan (overall rates as high as 15%), the Caribbean (crude seroprevalence ~5%), and parts of Africa. Antibody rates increase with age, reaching 15–30% in adults. HTLV-I transmission mechanisms within populations are not well understood. Transmission by transfusion is well documented in Japan and presumably occurs in the United States. Mother-to-child transmission occurs with breast-feeding and possibly by intrauterine or perinatal mechanisms. Finally, there is circumstantial evidence for venereal transmission (in particular male to female).

HTLV-I is etiologically associated with adult T-cell leukemia/lymphoma (ATL), and ATL incidence appears to covary with HTLV-I endemicity throughout Japan and the Caribbean. HTLV-I has also been associated with a chronic progressive degenerative neurologic disorder which presents as spastic paraparesis, currently referred to as HAM/TSP [a combination of the earlier names tropical spastic paraparesis (TSP) and HTLV-associated myelopathy (HAM)]. Several chapters also deal with HTLV-associated diseases and have complete reference lists. Unfortunately, most studies of HTLV-I–associated disease have been observational and have neither sampled defined populations of cases nor enrolled appropriate controls. Basic issues, such as the relationship between prevalence of HTLV-I infection and incidence of associated diseases and identification of risk factors for disease, remain unknown.

Descriptive information concerning HTLV has accrued extremely rapidly in the short time since its discovery. Many pioneering studies on which major current hypotheses are based used first-generation assays to test samples of convenience. This initial approach was logical (indeed necessary) and has provided important clues to be followed up by appropriately designed studies. Unfortunately, it is tempting to continue in this manner. Future HTLV-I protocols must be carefully designed to address key questions in such a manner that data from different areas are comparable. Defined normal populations must be sampled to compare geographic differences in seroprevalence, to understand the background against which disease occurs, and to test hypotheses concerning transmission. Clinical studies must carefully define both disease syndromes and patient populations and must include appropriate controls. It is no longer sufficient to present data from undocumented sources merely because serum samples are available or to continue with uncontrolled, descriptive studies of isolated or prevalent cases.

Laboratory studies of HTLV and other human retroviruses have progressed much more rapidly than our understanding of how these agents affect populations, and serologic methods exemplify this. The first studies used whole-virus preparations in enzyme-linked immunosorbent assay (ELISA) or IFAT to screen sera and confirmed seropositivity by blocking assays. It is now known that these tests embraced a wide range of nonspecificity, and standards now exist to define seropositivity by use of Western blot, radioimmunoprecipitation, and other tests for individual epitopes (1). Future studies must be designed taking this into account and utilize well-characterized test systems and standard definitions of seropositivity.

The overall objective of the Gorgas Memorial Laboratory (GML) retrovirus program is to describe the epidemiology of HTLV-I infection and disease in Panama and then conduct analytic studies to define factors associated with infection and disease development. The Program began in 1983 as a collaborative endeavor with Dr. William Blattner's group at the National Cancer Institute (NCI), National Institutes of Health (NIH). From the beginning, a key goal has been to assure standardized data collection and laboratory methodology so that results could be compared with those from other studies. The program includes two components, seroepidemiologic studies of representative (high/low risk) Panamanian populations and clinical epidemiology studies of incident ATL and HAM/TSP in defined populations. This paper will summarize our overall strategy, study design, and findings.

HTLV-I SEROEPIDEMIOLOGY

Basic descriptive studies of HTLV-I infection in a population must be accomplished to identify subpopulations with different infection rates in which to implement studies to determine risk factors for infection and disease. Three major findings have arisen from seroepidemiologic studies in Panama: overall HTLV-I antibody prevalence is similar to that in other Caribbean basin countries; infection rates are extremely high in Guaymi Indians from Bocas del Toro Province; and aberrant responses occur when different antibody assay systems are compared.

National Serologic Surveys

We estimated HTLV-I infection rates in the Panamanian population by testing sera from various surveys conducted, using three general sampling strategies, between 1978 and 1981. A serologic survey was conducted in 1978 to assess immunity to vaccinable diseases. The survey sampled the entire Republic with the exception of metropolitan Panama City and Colon and inaccessible areas in Darien, Chiriqui, Bocas del Toro, and Veraguas provinces. The survey enrolled all residents of randomly selected households within randomly selected boroughs from every county. We intended to sample 1% of the population and succeeded in obtaining sera from 0.8%. To estimate HTLV-I infection rates in Panama City and Colon, we used sera collected in 1981, during studies of ECHO 4 aseptic meningitis and enterovirus 70 acute hemorrhagic conjunctivitis. These studies sampled all members of houses randomly selected according to place of residence. Finally, to estimate seroprevalence in inaccessible areas, we used sera collected during various rural seroepidemiology studies in Darien, Veraguas, Chiriqui, and Bocas del Toro. These studies were conducted by specially trained (Indiana-Jones style) field teams visiting all villages in a selected area and attempting to map and census the village and obtain questionnaire data, sera, and other specimens from all residents.

Because HTLV-I prevalence is generally higher in adults than children, we only tested sera from subjects 15 yr or older. Sera were tested, independently and under code, in four HTLV-I–antibody assays. A first-generation ELISA screening test (2,3) used HTLV-I whole-virus antigen purified by zonal ultracentrifugation and then detergent disrupted. Sera with binding ratios greater than four were considered "screen-positive" and were retested by other methods. A first-generation confirmation assay (4) retested screen-positive sera after competition with heterologous HTLV-I antiserum; this will be referred to as the "blocking" test. Sera that were screen-positive were tested using the DuPont ELISA kit, according to manufacturer's instructions (DuPont/New England Nuclear). This highly standardized assay (which will be referred to as "ELISA") uses whole detergent-disrupted virus obtained from HUT-102 cells and is based on the first-generation screening test. Sera found positive in ELISA were tested by a radioimmunoassay (RIA) to measure antibody against HTLV-I p24, a major core protein (5); this will be referred to as the p24 RIA. The p24 RIA is less sensitive but more specific than ELISA. Sera with discordant results in ELISA and p24 RIA were further tested by Western blot (WB). HTLV-I for WB was obtained from HUT-102 cells and prepared by standard methods (5). Sera were considered WB-positive if both p19 and p24 bands could be seen, regardless of intensity or the presence of other bands. Sera with no bands were classified negative, and all other WB patterns were classified as indeterminate (for analysis indeterminate was considered negative).

One hundred thirty-five (5.4%) of 2496 sera from general population surveys had antibody by screening and blocking assays. Twenty-four (18% of screen-positives, 1% of the total) were not available for further testing (and have been eliminated from further analyses). Nineteen (20% of screen-positives, 0.8% of the total) were "confirmed" positive (positive by blocking or ELISA and p24 RIA or WB). The minimum estimated seroprevalence in subjects 15 yr or older varied from 0.2 to 2.0% throughout the Republic.

The small proportion of screen-positive sera which confirmed positive exemplifies the importance of uniform criteria for positivity. Although basic interpretations appear to be valid, our previous estimates of HTLV-I prevalence in Panama City and Colon were too high. Presumably similar problems haunt other serologic surveys that used only first-generation assays. Sera positive by first-generation screening and blocking tests but negative by newer methods (ELISA, p24 RIA, WB) had significantly lower initial binding ratios (mean 4.9; range 4.0–9.13; SD 1.2) than those which confirmed as positive (mean 11.3; range 5.2–17.5; SD 4.3).

No major geographic or urban/rural differences of HTLV-I antibody rates were observed, and male and female rates were similar, as were race-specific rates (with the exception of Guaymi Indians from Bocas del Toro). Specifically, HTLV-I infection rates were not elevated in black Panamanians of either Caribbean or mainland Latin American ancestry. HTLV-I prevalence rates were highest in Bocas del Toro Province (2%), but only 148 sera were collected in the national serologic survey. We conducted a separate analysis of 539 sera from isolated rural parts of Bocas del Toro; 21 (3.9%) were confirmed seropositive; 16 of these positive sera were from 161 Guaymi Indians (seroprevalence 9.9%).

Seroepidemiology in Guaymi Indians

Guaymi Indians have lived in extremely rugged inaccessible parts of Bocas del Toro for at least 300 yr. They remain largely unadmixed with people of European or African descent and still retain their traditional social organization. However, acculturation is occurring and is most obvious as migration to more urban areas to seek salaried employment, usually on banana plantations in Changuinola. In general, entire nuclear families migrate and establish permanent residence on plantations. Housing, medical care, and other amenities are provided during employment on the plantation.

We conducted a cross-sectional clinical seroepidemiology study of Guaymi Indians, 1 yr or older, from randomly selected households on banana plantations. We enrolled 366 of 395 (93%) eligible individuals and obtained 309 sera. Twenty-five (8.1%) subjects were seropositive by ELISA, but only those 15 yr or older had antibody (25/152, 16.5%), and male/female infection rates were similar. Most subjects had been born in Bocas del Toro Province and infection did not cluster according to birthplace. HTLV-I infection did not cluster by residence or family, and—because no one younger than 15 was seropositive—vertical transmission apparently did not occur. Similarly, there was no serologic evidence that transmission between spouses occurred. To date sera have been tested for HTLV-I antibodies by the qualitative DuPont HTLV-I ELISA kit, so these results must be considered tentative. Sera are being retested by ELISA, p24 RIA, RIPA, and WB assays.

HEMATOLOGIC MALIGNANCY

HTLV-I is highly associated with ATL in Japan, Jamaica, and Trinidad; in these areas ATL comprises a significant proportion of adult hematologic malignancy. GML maintains active country-wide surveillance of adult hematologic malignancy and we enroll all newly diagnosed (i.e., incident) patients into a clinical study which utilizes the same protocols as other NCI-funded HTLV-ATL studies. Our program has operated since 1984, and coverage encompasses 76% of hospital beds in the Republic. HTLV-associated ATL rarely occurs in Panama (6). Two hundred seventy-two hematologic malignancy patients have been enrolled through October 1988; only 23 (8%) had HTLV-I antibody. Only three patients (all seropositive) had clinical ATL and, as discussed by Dr. Levine at this meeting, ATL in Panama does not present the same clinical manifestations as in other areas. Other studies in Panama may help to determine how viral strain differences and host response interact to determine disease patterns after HTLV infection.

NEUROLOGIC DISEASE

GML maintains active neurologic disease surveillance at government-operated hospitals in Panama City. All newly diagnosed patients with selected neurologic diseases are enrolled into a standardized clinical protocol. Coverage includes ~90% of all neurologic disease patients from metropolitan Panama (7). Between 1985 and 1988 we enrolled 189 consecutive patients, and 12 (6%) had HTLV-I antibody. Eleven patients (6%) were diagnosed as spastic paraparesis, and 6 (55%) had serum and cerebrospinal fluid (CSF) antibody. Our clinical findings have been similar to those in other areas and indicated that HAM/TSP is more common in adult women and has a gradual onset and slowly progressive course, manifesting as a multifocal primarily demyelinating disorder that principally involves the spinal cord. Abnormalities have not been detected by myelography or computerized axial tomography scan and we have not found atypical lymphocytes in either blood or CSF. The other six seropositive patients represented a range of diseases (multiple sclerosis was not prominent), did not have detectable CSF antibody, and probably represent background infection. We have identified and enrolled seropositive cases' relatives, sexual, and household contacts and found no pattern of infection. We have sent specimens for virus isolation and diverse serologic studies to investigators at several NIH laboratories.

CONCLUSIONS

HTLV-I infection rates are similar in Panama as other parts of the Caribbean basin but there is no indication that other observed patterns apply. There are no major differences in male/female infection rates and no familial clustering; blacks do not have higher infection rates than other racial groups. The extremely high seroprevalence among Guaymi Indians needs to be further explored. A high proportion of screen-positive sera do not confirm by other assays, which may reflect lack of specificity or occurrence of other virus types.

In Panama, typical HTLV-associated ATL is extremely rare and the incidence of HAM/TSP is significantly greater. This may be related to occurrence of other viral types in the population or to different host responses after infection.

ACKNOWLEDGMENTS

The Gorgas Memorial Laboratory HTLV Epidemiology Program is a collaborative endeavor involving the following Panamanian investigators: Lics. M. Cuevas, S. Loo de Lao, M. Garcia, M. M. Brenes, Drs. F. Gracia, L. Castillo, E. Quiroz, J. R. Arosemena, C. Archbold, M. Altafulla, M. Larreategui, J. de Bernal, B. Rios, H. Xatruch, H. Espino, and E. Triana. The following Panamanian Institutions collaborate in the Program: the Ministry of Health, the Social Security Administration, Santo Tomas Hospital, the Metropolitan Medical Center—CSS, the National Oncology Institute, and the Integrated Health System of Chiriqui.

This research was supported in part by the National Cancer Institute, National Institutes of Health, under contract NCI-CP-31015 with the Gorgas Memorial Institute.

REFERENCES

1. Centers for Disease Control. Licensure of screening test for antibody to human T-lymphotropic virus type I. *MMWR* 1988;37:736–747.
2. Saxinger C, Gallo RC. Application of the indirect enzyme linked immunosorbent assay microtest to the detection and surveillance of human T-cell leukemia-lymphoma virus. *Lab Invest* 1983;49: 371–377.
3. Reeves WC, Saxinger C, Brenes MM, et al. Human T-cell lymphotropic virus type I (HTLV-I) seroepidemiology and risk factors in metropolitan Panama. *Am J Epidemiol* 1988;127:532–539.
4. Saxinger CW, Blattner WA, Levine PH, et al. Human T-cell leukemia virus (HTLV-I) antibodies in Africa. *Science* 1984;225:1473–1476.
5. Agius G, Biggar RJ, Alexander SS, et al. Human T lymphotropic virus type I antibody patterns: evidence of difference by age and risk group. *J Infect Dis* 1988;158:1235–1244.
6. Levine PH, Reeves WC, Cuevas M, et al. Human T-cell leukemia virus-I and hematologic malignancies in Panama. *Cancer* 1989;63:2186–2191.
7. Castillo L, Gracia F, Reeves WC. Spastic paraparesis and HTLV-I in the Republic of Panama. In: Roman GC, Vernant JC, Osame M, eds. *HTLV-I and the nervous system.* New York: Alan R. Liss, 1989:147–148.

Human Retrovirology: HTLV,
edited by William A. Blattner.
Raven Press, Ltd., New York 1990.

Development by the Public Health Service of Criteria for Serological Confirmation of HTLV-I/II Infections

*David W. Anderson, *Jay S. Epstein, *Lauren T. Pierik,
†Tun-Hou Lee, §Michael D. Lairmore, ‡Carl Saxinger,
‖V. S. Kalyanaraman, #Dennis Slamon, **Wade Parks,
††Bernard J. Poiesz, and ‡‡William Blattner

*Center for Biologics Evaluation and Review, FDA, Bethesda, Maryland 20892;
†Department of Cancer Cell Biology, Harvard School of Public Health,
Boston, Massachusetts; ‡Laboratory of Tumor Cell Biology, National Cancer
Institute, Bethesda, Maryland; §Retrovirus Diseases Branch, Centers for
Disease Control, Atlanta, Georgia; ‖Advanced Bioscience Laboratories, Inc.,
Kensington, Maryland; #Division of Hematology and Oncology, UCLA School of
Medicine, Los Angeles, California; **Department of Pediatrics/Immunology,
University of Miami, Miami, Florida; ††Division of Hematology/Oncology,
Department of Medicine, SUNY at Syracuse, Syracuse, New York; and ‡‡Viral
Epidemiology Section, National Cancer Institute, Bethesda, Maryland

INTRODUCTION

Soon after the discovery of HTLV-I in 1980 (1), epidemiologic studies in the United States identified the presence of this agent in certain populations (2); however, regard for the public health implications of wider infectious spread of HTLV-I was limited. Although the transmission of HTLV-I by cellular blood products was demonstrated in 1984 (3), it was not until a report of the association of HTLV-I–associated myelopathy (HAM/TSP) with transfusion (4) that alarm was raised by the American Red Cross (ARC) about the potential threat of HTLV-I and HTLV-II transmission to transfusion recipients (5). Prompted by these concerns, the ARC undertook a study of 39 898 random blood donors in eight U.S. cities, in which 10 donors (0.025%) were found to have antibodies against HTLV-I (6). A subsequent study confirmed that HTLV-I transmission from transfusion was occurring in this country (7). These events led to consideration by the Public Health Service (PHS) of the need to screen U.S. blood donors for antibodies to HTLV-I.

Despite the absence of any proven cases of HTLV-I–associated disease among transfusion recipients of HTLV-I–infected blood products in the U.S., a broad consensus was developed within the PHS and the larger medical and scientific community that donations of whole blood for transfusion (but not donations of plasma for further manufacture into derivatives) should be screened for antibodies to HTLV-I to prevent transmission of the infection to recipients of cellular blood products. This position, which was first taken by the Food and Drug Administration (FDA) in April 1987, was supported at a meeting of the FDA's Blood and Blood Products Advisory Committee in April 1988 and in an FDA-sponsored public workshop held at the National Institutes of Health on September 15, 1988. The rationale for blood screening was based on the following: 1) known disease associations of the virus, including adult T-cell leukemia/lymphoma (ATLL) and HAM/TSP; 2) proviral integration, implying lifelong risk of virus expression and possibly disease; and 3) known virally induced cell transformation, implying long-term cancer risk. Also in 1988, the unpublished reports from Japan of a short disease incubation period of several months to several years for HAM/TSP following transfusion was considered in favor of a decision to screen blood donations.

The FDA assumed that HTLV-I antibody screening would identify a high percentage of any healthy blood donors whose blood products could transmit HTLV-I infection. The task of validating the efficacy of such tests was therefore reduced to determining their sensitivity for antibody detection in asymptomatic carriers. In 1987, this effort was complicated by the fact that no reference procedure existed for determining the presence of HTLV-I antibodies and no reference sera were in existence. It therefore became necessary for the FDA to establish confirmatory criteria that would allow definition of seropositivity in order to develop reference material. Only then could a minimum standard be defined for sensitivity of HTLV-I antibody detection kits submitted for license by evaluation of their performance with reference sera and by the institution of a lot release panel for control testing.

DEVELOPMENT OF REFERENCE SERA FOR HTLV-I

It was initially assumed that serum from a person with an HTLV-I–associated disease could serve as a standard reference serum. However, it is well known that not all patients with ATLL or HAM/TSP have HTLV-I specific antibodies. Furthermore, even when antibodies are present in disease cases, it is not known whether they are the same antibodies that are present in asymptomatically infected persons. Thus, it became necessary to study the antibody profiles of asymptomatically infected persons. The approach which was taken was to obtain sera suspected of positivity for antibodies to HTLV-I based on available screening and confirmatory tests and to subject them to validation in expert laboratories. The presence or absence of antibodies to viral antigens was then determined by consensus among assays in different laboratories. The results were also analyzed to establish diagnostic criteria for confirming seropositivity that could be recommended to the medical community.

Procurement of Donor Sera. Between December 1987 and July 1988, with the assistance of blood and plasma establishments, pharmaceutical companies, the ARC, and U.S. governmental agencies, the FDA obtained 142 HTLV-I reactive sera from healthy Japanese (38), American (69), and Caribbean (35) blood donors. All donor samples were found nonreactive by a licensed HIV-1 EIA.

Collaborative Testing. Six laboratories with prior expertise in characterizing HTLV-I-specific antibodies and one in-house FDA laboratory were organized to participate in serological characterization of donor samples. Assays included four independent Western immunoblot (WIB) assays, including MT-2 (8), HUT 102 (9,10), and MJ cell-derived antigens (11); three radioimmunoprecipitation (RIP) assays using infected cell lysates from MT-2 (8,12) and SLB-1 cells (6); two p24 RIP using purified viral proteins from HUT 102 cells (13,14); one competetive enzyme-linked immunosorbent assay (EIA) based on antigen from HUT 102 cells (15,16); and one immunofluorescence (IFA) assay using whole infected MT-2 cells (17). In scoring the WIB and RIP assays, each laboratory chose to study antibodies to one or several of the *gag*-encoded core proteins p19, p24, and precursor p55; the *env*-encoded glycoproteins gp46 or gp61, gp63, gp65 or gp68, or the *tax*-encoded transcriptional activator protein $p40^x$. These particular viral proteins were selected for study because, with the exception of gp68 (referred to here as gp61), each represents the product of only one HTLV-I gene. IFA and competitive EIA could not be evaluated for gene product-specific antibodies. In addition to the above serological assays, follow-up studies were done on peripheral blood lymphocytes from seven of the U.S. donors to determine on a molecular genetic basis whether HTLV-I or HTLV-II was present by the use of culture, Southern blotting, and polymerase chain reaction (PCR).

To standardize the assays in each collaborating laboratory, FDA distributed eight positive reference sera from U.S. patients clinically diagnosed as having ATL (3 cases) or TSP (5 cases) (provided by Dr. Bernard Poiesz, State University of New York, Syracuse, New York). All cases had evidence of HTLV-I–specific gene sequences in peripheral lymphocyte DNA either by direct Southern blotting or after amplification by PCR (18). Negative reference sera from 20 normal U.S. blood donors were also provided. Each laboratory found strong seroreactivity to all analyzed viral proteins for the 8 positive sera and no detectable seroreactivity for any of the 20 negative reference sera. Identical aliquots of each HTLV-I study sample were then simultaneously distributed by the FDA under code to the seven collaborating laboratories.

Results of Comparative Serologic Testing. The results of this study have been published by Anderson et al. in *Blood* (19). Summarized briefly, it was determined that 1) WIB are more sensitive than RIP assays for detection of antibodies to p24; 2) RIP assays are more sensitive than WIB for detection of antibodies to gp46 and/or gp61; 3) antibodies to $p40^x$ are detected only in conjunction with antibodies to gp61 and at reduced frequency; 4) antibodies to p19 are found in association with antibodies to gp61 only in samples that also contain anti-p24; and 5) antibody patterns of HTLV-I–infected cases are not distinguishable from HTLV-II–infected cases.

Development of Criteria for HTLV-I/II Antibody Confirmation. Criteria for antibody confirmation were developed by the collaborating investigators at a meeting in July 1988. It was recognized that antibodies to the combination of two gene products—namely, *gag,* represented by p24, and *env,* represented by either gp46 or gp61—provided a sensitive and specific criterion within the limits of available assay systems, and that these assays did not distinguish HTLV-I from HTLV-II. This position became the basis for specific recommendations by the PHS. Among the 142 donor samples in the study, 137 were verified as containing antibodies to both p24 and gp61. These samples were defined as positive reference material for use by the FDA.

PHS RECOMMENDATIONS FOR CONFIRMATION OF HTLV-I/II ANTIBODIES

HTLV-I EIA tests were licensed by the FDA on November 29, 1988 and have been recommended by the FDA for screening of blood and cellular components donated for transfusion (20). They have also been approved as diagnostic tests, which may be useful in evaluating patients with clinical diagnoses of ATLL and TSP/HAM. All users of licensed HTLV-I screening tests are cautioned that additional, more specific tests are necessary to confirm that sera found repeatably reactive in screening tests are positive for HTLV-I antibodies.

Requirements of Additional, More Specific Serological Tests. Tests used to confirm the presence of HTLV-I-specific antibodies must be capable of identifying antibodies to the p24 and gp46 and/or gp61 proteins of HTLV-I. The PHS recommends examination of samples by WIB for p24 and gp46 and/or gp61 antibodies. A specimen must demonstrate seroreactivity to the *gag* gene product p24 and to an *env* gene product (gp46 and/or gp61) to be considered "positive." If antibodies to p24 are present but antibodies to gp46 and/or gp61 are lacking by WIB, a RIP assay for gp61 antibodies is indicated. IFA testing for HTLV-I has been used in some laboratories but it does not detect antibody to specific HTLV-I gene products. Specimens not satisfying these criteria but having antibodies to at least one suspected HTLV-I gene product (such as p19 only, p19 and p28, or p19 and *env*) are designated "indeterminate." Serum specimens with no antibodies to any HTLV-I gene products are designated "negative." Both WIB and RIP assays may be required to determine whether a serum specimen is positive, indeterminate, or negative.

Limitations of the Proposed Strategy for Confirmation of HTLV-I/II Antibodies. Infection with either HTLV-I or HTLV-II gives rise to a similar pattern of antibodies detected by WIB and RIP to the HTLV-I derived proteins p19, p24, p40^x, gp46, and gp61. In the absence of virus isolation or characterization of gene sequences by PCR, the virus type of seropositives in the U.S. is therefore best classified as HTLV-I/II; 2) our lack of molecular epidemiologic data as to the virus type of HTLV-I/II seropositive persons in the U.S. complicates accurate interpretation of antibody patterns by any serological test. Antibody patterns classified as "indeterminate" by the present confirmatory strategy (e.g., p19 antibodies only, p24 antibodies only, etc.) might reflect cross-reacting antibodies to other, as-yet-uncharacterized retroviruses; and 3) results of sero-conversion studies in persons with virologically defined infections as well as look-back studies to characterize antibody patterns of transfused blood linked to recipient seroconversion may justify the use of other serological markers as confirmatory criteria. In specific, antibodies to p19, p40^x, or other *env*-encoded proteins such as gp21^E, may be shown to contribute additional predictive power for identifying HTLV-I/II infected persons.

REFERENCES

1. Poiesz BJ, Ruscetti FW, Gazdar AF, et al. Detection and isolation of type-C retrovirus particles from fresh and cultured lymphocytes of a patient with cutaneous T-cell lymphoma. *Proc Natl Acad Sci USA* 1980;77:7415–7417.
2. Blayney DW, Blattner WA, Robert-Guroff M, et al. The human T-cell leukemia/lymphoma virus in the southeastern United States. *JAMA* 1983;250:1048–1052.
3. Okochi K, Sato H, Hinuma Y. A retrospective study on transmission of adult T-cell leukemia virus by blood transfusion: seroconversion in recipients. *Vox Sang* 1984;46:244–253.
4. Osame M, Izuma S, Matsumoto M, Matsumoto T, Sonada S, Tara M, Shibata Y. Blood transfusion and HTLV-I associated myelopathy. *Lancet* 1986;2:104–105.
5. Sandler SG. HTLV-I and II: new risks for recipients of blood transfusion? *JAMA* 1986;256:2245–2246.
6. Williams AE, Fang CT, Slamon DJ, et al. Seroprevalence and epidemiological correlates of HTLV-I infection in U.S. blood donors. *Science* 1988;240:643–646.
7. Minamoto GY, Gold JWM, Scheinberg DA, et al. Infection with human T-cell leukemia virus type I in patients with leukemia. *N Engl J Med* 1988;318:219–222.
8. Lairmore MD, Jason JM, Hartley TM, et al. Absence of HTLV-I Coinfection in HIV-infected hemophilic men. (*submitted*).
9. Saxinger WC, Levine PH, Dean AG, et al. Evidence for exposure to HTLV-III in Uganda prior to 1973. *Science* 1985;227:1036–1038.
10. Schupbach J, Kalyanaraman VS, Sarngadharan MG, et al. Antibodies against three purified proteins of the human type C retrovirus, human T-cell leukemia-lymphoma virus, in adult T-cell leukemia-lymphoma patients and healthy blacks from the Caribbean. *Cancer Res* 1983;43:886–891.
11. Tsang VC, Peralta JM, Simons AR. Enzyme-linked immunoelectrotransfer blot techniques for studying the specificities of antigens and antibodies separated by gel electrophoresis. *Methods Enzymol* 1983;92:377–391.
12. Lee TH, Coligan JE, Homma T, et al. Human T-cell leukemia virus-associated membrane antigens (HTLV-MA): identity of the major antigens recognized following virus infection. *Proc Natl Acad Sci USA* 1984;81:3856–3860.
13. Kanner SB, Cheng-Mayer C, Geffin RB, et al. Human retroviral *env* and *gag* polypeptides: serologic assays to measure infection. *J Immunol* 1986;137:674–678.
14. Kanner SB, Parks ES, Scott GB, et al. Simultaneous infection with HTLV-I and HIV. *J Infect Dis* 1987;155:617–625.
15. Saxinger WC, Gallo RC. Application of the indirect ELISA microtest to the detection and surveillance of human T-cell leukemia-lymphoma virus (HTLV). *Lab Invest* 1983;49:371–377.

16. Saxinger W, Blattner WA, Levine PH, et al. Human T-cell leukemia virus (HTLV-I) antibodies in Africa. *Science* 1984;225:1473–1476.
17. Yamamoto N, Hinuma Y. Antigens in an adult T-cell leukemia virus-producer cell line: reactivity with human serum antibodies. *Int J Cancer* 1982;30:289–293.
18. Kwok S, Ehrlich G, Poiesz B, Kalish R, Sninsky JJ. Enzymatic amplification of HTLV-I viral sequences from peripheral blood mononuclear cells and infected tissues. *Blood* 1988;72:1117–1123.
19. Anderson DW, Epstein JS, Lee TH, et al. Serological confirmation of human T-lymphotropic virus type I infection in healthy blood and plasma donors. *Blood* 1989;74:2585–2591.
20. Centers for Disease Control. Licensure of screening tests for antibody to human T-lymphotropic virus type I. *MMWR* 1988;37:736–747.

DISCUSSION

One discussant was asked why p19, p24 antibody presence cannot be used as sufficient criteria for HTLV-I seropositivity. The discussant replied that p19 antibodies alone are not predictive of envelope antibody presence. P24 antibodies are, however, always present in envelope antibody-positive sera. Another discussant commented that only 98 of 122 sera with both p19 and p24 antibodies in Western blot assays also had envelope antibodies in radioimmunoprecipitation assays, meaning that HTLV-I seropositivity could be over-estimated by 10%–20% if envelope antibody testing is not done. One participant stated that requiring p19, p24 antibody positive sera also to be envelope antibody-positive in order to be classed as HTLV-I seropositive makes the testing more specific rather than more sensitive.

Human Retrovirology: HTLV,
edited by William A. Blattner.
Raven Press, Ltd., New York © 1990.

Significance of Positive and Indeterminate HTLV-I Serologic Results in Blood Donors

*Rima F. Khabbaz, †Gerald Shulman, *Michael D. Lairmore,
*Trudie M. Hartley, *Barun De, and *Jonathan E. Kaplan

*Retrovirus Diseases Branch, Division of Viral Diseases, Center for Infectious Diseases,
Centers for Disease Control, Atlanta, Georgia 30333; and †American Red Cross
Blood Services, Atlanta Region, Atlanta, Georgia 30324*

INTRODUCTION

The human T-lymphotropic virus type I (HTLV-I), causative agent of the adult T-cell lymphoma (1) and HTLV-I–associated myelopathy/tropical spastic paraparesis (HAM/TSP) (2,3) can be transmitted by transfusion of HTLV-I–infected blood products (4). In November 1988, the Food and Drug Administration licensed screening tests for HTLV-I antibodies and recommended their use to screen the United States blood supply (5). The screening tests, as well as the supplementary serologic tests (Western blot and in some instances radioimmunoprecipitation assays) recommended to confirm the screening test results, cross-react with antibodies to the closely related retrovirus, HTLV-II. Interpretation of the supplementary serologic tests leads to three categories of results: HTLV-I "seropositive," defined by the presence of antibodies to the *gag* gene product (p24) and *env* (gp46 and/or gp61/68); "indeterminate" HTLV-I serologic results, defined by reactivity to at least one suspected HTLV-I gene product but failing to meet the recommended criteria for seropositivity; and "negative," for sera with no immunoreactivity to any HTLV-I gene products.

The significance of positive and indeterminate HTLV-I serologic results in U.S. blood donors is not well known. This chapter summarizes data that are emerging from our evaluation of Atlanta blood donors tested for HTLV-I.

SEROPREVALENCE

In the summer of 1988 and from December through the end of March 1989, 89 717 volunteer blood donors at the American Red Cross, Atlanta Region, were tested for HTLV-I antibodies; 14 were positive. The seroprevalence of HTLV-I in Atlanta blood donors was therefore 0.016%. Another 39 blood donors (0.043%) had "indeterminate" HTLV-I serologic results. Of these, 23 had antibodies to p19 only on Western blot and 7 had antibodies to p24 (and not to p19); 9 had antibodies to both p19 and p24.

DEMOGRAPHIC CHARACTERISTICS

HTLV-I–seropositive donors were more likely to be women (12/14) than HTLV-I–seronegative controls (systematically chosen to be the donors before and after each donor with a reactive HTLV-I screening test) and than donors with indeterminate HTLV-I serologic results; HTLV-I–seropositive donors also tended to be older than HTLV-I–negative controls and than donors with indeterminate HTLV-I serologic results (Table 1). Four of six HTLV-I seropositive donors interviewed to date were black; three of them were from the southeastern United States. One HTLV-I–seropositive donor was from southern Japan. In contrast, 3/5 interviewed donors with indeterminate serologic results were white, one was black, and one was Hispanic, and all three HTLV-I seronegative donors interviewed were white.

OTHER SEROMARKERS

Although none of the 14 HTLV-I–seropositive donors was seropositive for HIV, syphilis, or hepatitis B surface antigen, a significantly higher percentage (5/14, 35.7%) were positive for hepatitis B core antibody than HTLV-I–negative controls (2/106, 1.9%, $p < 0.01$) and than donors with indeterminate HTLV-I serologic results (1/39, 2.6%, $p < 0.01$ Fisher's exact).

EPIDEMIOLOGIC RISK FACTORS

Five of six seropositive donors interviewed gave histories of exposures that might have put them at risk for HTLV-I infection. These included sex with a person with a history of intravenous drug use (two donors), sex with a person

TABLE 1. *Demographic data, HTLV-I seropositive, indeterminate, and a sample of HTLV-I negative blood donors, Atlanta*

	HTLV-I serologic status		
	Positive ($N = 14$)	Negative ($N = 106$)	Indeterminate ($N = 39$)
Male	2	55	23
Female	12*	51	16
Mean age	40†	36	34

* $p < 0.01$, Chi-square.
† $p = 0.09$, *t* test for trend.

from a known HTLV-I endemic area (three donors), and blood transfusion (one donor).

GENE AMPLIFICATION FOR HTLV-I AND HTLV-II

To detect HTLV-I– and HTLV-II–specific sequences, DNA from peripheral blood lymphocytes from each of six HTLV-I–seropositive donors, eight donors with indeterminate HTLV-I serologic results, and three HTLV-I–seronegative donors was examined by polymerase chain reaction (PCR). Amplification was done in parallel using HTLV-I specific *pol* and *env* sequences and HTLV-II specific *pol* and *env* sequences as primer pairs. Amplified *env* and *pol,* and an amplified HLA sequence, were detected after Southern blotting by hybridization with HTLV-I, HTLV-II, and HLA oligo probes contained in the amplified regions, respectively. Cells were counted before lysis and all PCR reactions were carried on equivalent DNA from 150 000 cells. Dilutions of HTLV-I–infected cells, MT-2, or HTLV-II–infected cells, MoB, were run as controls with each gel, and the sensitivity of the PCR was close to one infected cell per reaction. PCR was positive for HTLV-I (negative for HTLV-II) in 3/6 HTLV-I–seropositive donors and was positive for HTLV-II (negative for HTLV-I) in one seropositive donor. In two seropositive donors, neither HTLV-I nor HTLV-II sequences were detected; in addition, 1/8 donors with an indeterminate HTLV-I serologic result was positive for HTLV-II; the Western blot pattern of serum from this donor showed reactivity to p19 and p24.

CONCLUSIONS

The seroprevalence of HTLV-I in Atlanta blood donors is 0.016%. There appears to be a preponderance of black women among HTLV-I–seropositive Atlanta donors. HTLV-I seropositivity in these donors also appears to be associated with hepatitis B core antibody positivity. Risk factors for seropositivity in Atlanta blood donors include sex with a person who has used intravenous drugs and sex with a person from a known HTLV-I–endemic area such as the Caribbean or southern Japan (6). PCR on cells from these donors shows that HTLV-I seropositivity can be due to HTLV-I (3/6 donors) or HTLV-II (one donor). Our inability to pick up a signal for either HTLV-I or HTLV-II in 2/6 seropositive donors suggests that in these donors, the number of infected circulating lymphocytes may be less than 1 in 10^5 cells. This contrasts with PCR done simultaneously on DNA from four HAM/TSP patients, which shows HTLV-I sequences in 4/4 patients. Our PCR finding of HTLV-II specific sequences in a donor with an indeterminate HTLV-I serologic result suggests that some of these indeterminate serologic results may be caused by HTLV-II infection.

REFERENCES

1. Kuefler PR, Bunn PA Jr. Adult T cell leukaemia/lymphoma. *Clin Haematol* 1986;15:695–726.
2. Osame M, Usuku J, Izumo S, et al. HTLV-I associated myelopathy: a new clinical entity. *Lancet* 1986;1:1031–1032.
3. Gessain A, Barin F, Vernant JC, et al. Antibodies to human T-lymphotropic virus type-I in patients with tropical spastic paraparesis. *Lancet* 1985;2:407–410.
4. Okochi K, Sato H, Hinuma Y. A retrospective study on transmission of adult T-cell leukemia virus by blood transfusion: seroconversion in recipients. *Vox Sang* 1984;46:245–253.
5. Public Health Service Working Group. Licensure of screening tests for antibody to human T-lymphotropic virus type I. *MMWR* 1988;37:736–747.
6. Blattner WA. Retroviruses. In: AS Evans, ed. *Viral infections of humans: epidemiology and control, 3rd ed.* New York: Plenum, 1989.

DISCUSSION

One speaker was asked what the risk factors were for indeterminants on the p19 assay. No major risk factors were identified for the indeterminants. The speaker commented that blood donors in Atlanta are predominantly white and that assessing race is very difficult in view of the sensitivity of that issue in the region. A discussant asked whether the serologic testing was to be continued on the indeterminants. The speaker responded that a longitudinal study on this particular group would be done. Another discussant commented that look-back studies on indeterminants would be most critical in determining their significance. Another participant had found that 0.25% are positive for the p19 and non-p24 *gag* antibodies who were all negative on PCR for HTLV-I and HTLV-II. One participant commented that many of the p19 alone would be found to have false positivity. This participant felt that the true positives were likely to be p19 positive and also positive on RIP with the presence of envelope antibody.

A discussant asked whether the HTLV-I assay was sensitive to IgG or IgM. Another discussant responded that there is cross reactivity with IgG and IgM. The previous discussant stated that there are some patients who have only IgM antibodies against HTLV-I, thus screening involving only IgG would not identify such individuals.

A speaker raised the issue that PCR primers in the p19 region may help to elucidate the significance of indeterminants. Another speaker commented that there is a normal cellular sequence closely related to the p19. It was also noted that in one study monoclonal antibody reacted with normal human tissue, suggesting that the p19 reactivity may be antibodies against host determinants.

Human Retrovirology: HTLV,
edited by William A. Blattner.
Raven Press, Ltd., New York © 1990.

The Particle Agglutination (PA) Assay and Its Use in Detecting Lower Titer HTLV-I Antibodies

*†Akihiko Okayama, *Junzo Ishizaki, *Nobuyoshi Tachibana, *Kazunori Tsuda, †Myron Essex, and ‡Nancy Mueller

*Second Department of Internal Medicine, Miyazaki Medical College, Miyazaki 889-16, Japan; and Departments of †Cancer Biology and ‡Epidemiology, Harvard School of Public Health, Boston, Massachusetts 02115

INTRODUCTION

The human T-cell leukemia virus type 1 (HTLV-I) can exist in a latent form, integrated as a provirus in the genome of lymphocytes in its human host. Some infected people eventually develop adult T-cell leukemia (ATL) or other HTLV-I–associated diseases, but the majority of others remain persistently infected for an entire lifetime without apparent disease development. Nevertheless, during the viral latent period, individuals remain asymptomatic carriers of the virus, and presumably produce antibodies against specific viral antigens. Because of the asymptomatic nature of HTLV-I infection, epidemiological data concerning its prevalence has had to be gleaned from serological studies (1). Here the detection of serum antibodies against HTLV-I antigens indicates infection with the virus (2). The fact that latent infections can be detected by serological methods became especially significant when it was recognized that HTLV-I could be transmitted through blood products via transfusion. To limit the spread of this virus through this mode of transmission, it became imperative to screen donated blood supplies, especially in countries such as Japan where the virus is endemic and thus highly prevalent in certain areas. For this reason numerous mass-screening tests have been developed, and, in Japan, nationwide screening of donated blood has already been initiated.

METHODS OF DETECTION

Various methods for detecting anti-HTLV-I antibodies in human serum samples have been used successfully. One of the first assays which was developed is the indirect immunofluorescence (IF) assay. This assay, which uses fixed HTLV-

I–infected cells as antigen, was developed by Hinuma et al. (3) and has been used extensively in studying HTLV-I seroprevalence. Radioimmunoassays (RIA) to detect antibodies to HTLV-I *gag* proteins such as p24 and p19 have also been used (4), as have enzyme-linked immunosorbent assays (ELISA) with antigens from either viral lysates or cellular lysates from HTLV-I–producing cells (5,6). Other blood screening techniques include indirect immunofluorescence assays using cellular membrane antigen (HTLV-MA) from HTLV-I–producing cells (7,8), and the particle agglutination assay (PA) in which specific viral antigens are adsorbed onto gelatin beads before being reacted with serum samples (9). Radioimmunoprecipitation assays (RIP) (10,11) and Western blotting (WB) (12) have also been used as serological confirmatory tests for identifying HTLV-I carriers.

PARTICLE AGGLUTINATION ASSAY

The Japanese Red Cross has been using the PA method to screen its blood supply since screening was instituted in 1986. This method has been demonstrated to have both high specificity and high sensitivity (9). It has also been shown, however, that more serum samples are positive by PA than by IF (13). This observation raises the question as to whether the higher number of positive PA tests occur because of nonspecific reactivity (that is, they are due to falsely positive reactions) or because the PA test is more sensitive than the IF test.

To address this question, we tested PA-positive, IF-negative blood samples for their reactivity in ELISA, HTLV-MA, RIP, and WB assays and analyzed the results we obtained (14). We initially screened 6915 donated blood samples by PA. Of these, 389 (5%) showed a positive result. These PA-positive samples were then tested by IF; 29 of the 389 samples (7.5%) tested negative. We were able to further examine 20 of these 29 samples using the four detection assays mentioned above (see Table 1). When we diluted our test sera to determine antibody titers, we found that all the PA+, IF− samples had relatively low antibody titers by PA (<1:256). At least 90% of the samples with titers ≥ 1:64 tested positive by HTLV-MA, RIP, and WB. There were nine samples with titers < 1:64; of these, one tested positive only by RIP, and one only by WB, but two others did not test positive by either of these techniques (RIP and WB).

Table 2 compares the reactivity of the 20 PA+, IF− serum samples to RIP and WB. In general, the envelope proteins are readily detectable in RIP assays; here 17/20 samples were able to detect gp61, the precursor envelope protein. WB, on the other hand, is more sensitive to the *gag* gene products; that is, the core proteins p24 and p19. With this method, 17/20 samples could detect HTLV-I core proteins. The differential results obtained by these two methods are most apparent in the samples with the lowest antibody titers.

Another hypothesis that may partially explain the high rates of PA reactivity as compared with IF reactivity suggests that PA is capable of detecting serum

TABLE 1. Reactivities of PA-positive, IF-negative serum samples by PA, HTLV-MA, ELISA, RIP, and WB

Serum	1	2	3	4	5	6	7	8	9	10	11	12	13	14	15	16	17	18	19	20
PA titer	16	16	16	32	32	32	32	32	32	64	64	64	64	64	128	128	128	128	256	256
HTLV-MA	+	+	−	+	−	+	−	−	+	+	+	+	+	+	−	+	+	+	+	+
ELISA	−	−	−	−	−	+	−	−	+	−	−	−	−	+	−	+	+	−	−	+
RIP	+	−	+	+	−	+	+	−	+	+	+	+	+	+	+	+	+	+	+	+
WB	+	−	−	+	+	+	+	−	+	+	+	+	+	+	+	+	+	+	+	+

PA, Particle agglutination assay (Serodia ATLA, Fuji Rebio, Tokyo); HTLV-MA, Indirect immunofluorescent assay using living HUT102 cell membrane antigen; ELISA, Enzyme linked immunosorbent assay (Eitest-ATL, Eisai, Tokyo); RIP, Radioimmunoprecipitation assay using [35-S] cysteine labelled HUT102 cell lysate as antigen; WB, Western blot assay using MT-2 cell lysate as antigen.

TABLE 2. *PA-positive, IF-negative serum antibodies to specific HTLV-I proteins by RIP and WB*

Serum	1	2	3	4	5	6	7	8	9	10	11	12	13	14	15	16	17	18	19	20
PA titer	16	16	16	32	32	32	32	32	32	64	64	64	64	64	128	128	128	128	256	256
RIP gp61	+	–	+	+	–	+	+	–	+	+	+	+	+	+	+	+	+	+	+	+
p24	–	–	–	–	–	–	–	–	–	–	–	–	+	+	–	–	–	–	–	–
p19	–	–	–	–	–	–	–	–	–	–	–	–	+	–	–	–	–	–	–	–
WB gp68	–	–	–	–	–	+	+	–	–	–	–	+	+	–	–	+	+	–	–	–
(IgG) p28	+	–	–	+	+	+	+	–	+	+	–	+	+	+	+	+	+	+	+	+
p24	+	–	–	+	+	+	–	–	+	+	–	+	+	+	+	+	+	+	–	+
p19	–	–	–	–	–	+	+	–	+	–	+	+	+	+	+	+	+	+	+	+
WB gp68	–	–	–	–	–	–	–	–	–	–	–	–	–	–	–	–	+	–	+	–
(IgM) p28	–	–	–	–	–	–	–	–	–	–	–	–	–	–	–	–	+	–	+	–
p24	–	–	–	–	–	–	–	–	–	–	–	–	–	–	–	–	–	–	–	–
p19	–	–	–	–	–	–	–	–	–	–	–	–	–	–	–	–	+	–	+	–

WB(IgG), IgG antibodies to HTLV-I by Western blot assay; WB(IgM), IgM antibodies to HTLV-I by Western blot assay.

IgM in addition to serum IgG, whereas IF usually only detects IgG. To examine this hypothesis further, we used WB assays that were also capable of detecting antigen reaction to serum IgM. In 2 of our 20 test samples we were able to detect anti-HTLV-I IgM (see Table 2). With both of these sera, however, we could also detect IgG antibodies against the same viral antigens that reacted with the IgM. Thus, at least in these cases, the PA positive assay could not be attributed to immunoreactivity with IgM that would not be detected by the IF method.

Overall, 18/20 (90%) of the PA+, IF− serum samples were confirmed as positive by RIP and/or WB. All of these samples had relatively low HTLV-I antibody titers. Because we did not test all 6915 of the donated blood samples, we cannot conclude that the PA assay is the most sensitive or the most specific of all the possible HTLV-I blood screening tests. Indeed, it is known that the specificity of the PA assay often depends on the source of the viral antigen preparation. Nonetheless, PA does seem to be especially effective in detecting latent HTLV-I infections in low-titer serum samples. This is important especially when screening blood from younger donors, for as we show in Table 3 (15), younger carriers tend to have a lower anti-HTLV-I antibody titer than do carriers of middle or older ages (15,16).

ISSUES IN DETECTING HTLV

Of course there are still other factors that impair our ability both to obtain accurate HTLV-I prevalence data and to screen donor blood supplies with 100% efficiency. It has recently been reported that HTLV-II, a second strain of HTLV virus, is relatively common within the intravenous drug–abusing population in the southern United States (17). This virus is virtually indistinguishable from HTLV-I by serologic methods (18). However, the newly developed and very

TABLE 3. *Age distribution of HTLV-I antibody titers determined by the HTLV-MA assay*

| | Antibody titers | | | | | | | | |
Age‡	1:4	1:10	1:20	1:40	1:80	1:160	1:320	Total	GMT†
15–19	7*	2	3	2				14	17.8
20–29	1	1	2	1				5	18.8
30–39	2	4	3	1	1		1	12	45.7
40–49	5	1	5	7	4	1		23	38.7
50–59	5	4	6	27	7	7		56	52.5
60–69	3	6	9	14	6	2	4	44	65.7
70+	1	2	10	12	8	6		39	59.1
Total	24	20	38	64	26	16	5	193	51.4

* Number of the HTLV-MA positive samples.

† Geometric Mean Titer.

‡ There is a statistical difference in GMT between age groups of above and below thirty years ($p < 0.05$).

sensitive retroviral detection technique called the polymerase chain reaction method (PCR) has been used successfully to distinguish between HTLV-I and HTLV-II infections. A second confounding factor is the possible existence of seronegative virus carriers. Already PCR has detected several cases of HTLV-I infection among seronegative multiple sclerosis patients (19). Seronegativity has also been noted in several cases of ATL in Japan (20); this disease is one of the classic manifestations of active HTLV-I infection. However, even more sensitive and highly specific detection techniques using various recombinant HTLV-I proteins are now being proposed (21). Perhaps between these new methods and a continual refinement of the PCR technique, we will be able to resolve the problems that have heretofore faced us in detecting latent HTLV infections.

REFERENCES

1. Hinuma Y, Komoda H, Chosa T, et al. Antibodies to adult T-cell leukemia-associated antigen (ATLA) in sera from patients with ATL and controls in Japan: a nation-wide sero-epidemiologic study. *Int J Cancer* 1982;29:631–635.
2. Gotoh Y, Sugamura K, Hinuma Y. Healthy carriers of a human retrovirus, adult T-cell leukemia virus (ATLV): demonstration by clonal culture of ATLV-carrying T-cells from peripheral blood. *Proc Natl Acad Sci USA* 1982;79:4780–4782.
3. Hinuma Y, Nagata K, and Hanaoka M. Adult T-cell leukemia antigen in an ATL cell line and detection of antibodies to the antigen in human sera. *Proc Natl Acad Sci USA* 1981;78:6476–6480.
4. Kalyanaraman VS, Sarngadharan MG, Bunn PA et al. Antibodies in human sera reactive against an internal structural protein (p24) of the human T-cell lymphoma virus. *Nature* 1981;294:271–273.
5. Taguchi H, Sawada T, Fujishita M et al. Enzyme-linked immunosorbent assay of antibodies to adult T-cell leukemia associated antigen. *Gann* 1983;74:185–187.
6. Saxinger WC, Gallo RC. Application of indirect enzyme-linked immunosorbent assay microtest to the detection and surveillance of human T cell leukemia-lymphoma virus. *Lab Invest* 1983;49:371–377.
7. Essex M, McLane MF, Lee TH et al. Antibodies to cell membrane antigens associated with human T-cell leukemia virus in patients with AIDS. *Science* 1983;220:859–862.
8. Essex M, McLane MF, Lee TH et al. Antibodies to human T-cell leukemia virus: membrane antigens (HTLV-MA) in hemophiliacs. *Science* 1983;221:1061–1064.
9. Ikeda M, Fujino R, Matsui T, et al. A new agglutination test for serum antibodies to adult T-cell leukemia virus. *Gann* 1984;75:845–848.
10. Yamamoto N, Schneider, Koyanagi Y, et al. Adult T-cell leukemia (ATL) virus specific antibodies in ATL patients and healthy virus carriers. *Int J Cancer* 1983;32:281–287.
11. Lee TH, Coligan JE, Homma T, et al. Human T-cell leukemia virus associated membrane antigens (HTLV-MA): identity of the major antigens recognized following virus infection. *Proc Natl Acad Sci USA* 1984;81:3856–3860.
12. Towbin, Staehelin T, Gordon J. Electrophoretic transfer of proteins from polyacrylamide gels to nitrocellulose sheets: procedure and applications. *Proc Natl Acad Sci USA* 1979;76:4350–4353.
13. Kobayashi S, Yoshida T, Hiroshige Y, et al. Evaluation of the passive particle agglutination method and enzyme-immunoassay for the antibodies to adult T-cell leukemia virus in the sera from healthy adults. *Acta Haematol Jpn* 1987;50:999–1006.
14. Ishizaki J, Okayama A, Tachibana N, et al. Comparative diagnostic assay results for detecting antibody to HTLV-I in Japanese blood donor samples: higher positive rates by particle agglutination assay. *J AIDS* 1988;1:340–345.
15. Okayama A, Tachibana N, Ishizaki J, et al. An immunological study of human T-cell leukemia

virus type-1 (HTLV-I): detection of low titer anti-HTLV-I antibodies by HTLV-I associated membrane antigen (HTLV-MA) method. *J Jpn Assoc Infect Dis* 1987;61:1363–1368.

16. Agius G, Bigger RJ, Alexander SS, et al. Human T-lymphotropic virus type-1 antibody patterns: evidence of difference by age and risk group. *J Infect Dis* 1988;158:1235–1244.
17. Lee H, Swanson P, Shorty VS et al. High rate of HTLV-2 infection in seropositive IV drug abusers in New Orleans. *Science* 1989;243:471–475.
18. Lee TH, Coligan JE, McLane MF, et al. Serological cross-reactivity between envelope gene products of type-1 and type-2 human T-cell leukemia virus. *Proc Natl Acad Sci USA* 1984;81:7579–7583.
19. Reddy EP, Sandberg-Woliheim M, Mettus RV et al. Amplification and molecular cloning of HTLV-I sequences from DNA of multiple sclerosis patients. *Science* 1989;243:529–533.
20. Shimoyama M, Kagami Y, Shimotohno K, et al. Adult T-cell leukemia/lymphoma not associated with human T-cell leukemia virus type-1. *Proc Natl Acad Sci USA* 1986;83:4524–4528.
21. Kuga T, Yamsaki M, Sekine S, et al. A *gag-env* hybrid protein of human T-cell leukemia virus type-1 and its application to serum diagnosis. *Gann* 1988;79:1168–1173.

Human Retrovirology: HTLV,
edited by William A. Blattner.
Raven Press, Ltd., New York © 1990.

Enhancing the Sensitivity of HTLV-I Immunoassays

James J. Lipka, Judy L. Parker, and Steven K. H. Foung

Department of Pathology, Stanford University Medical Center, and Stanford University Blood Bank, Palo Alto, California 94304

INTRODUCTION

Human T-cell lymphotropic virus type I, HTLV-I, is associated with adult T-cell leukemia (ATL) and a neurological disease known as either tropical spastic paraparesis (TSP) or HTLV-I–associated myelopathy (HAM) (1–4). The virus is endemic in Japan (3) and the Caribbean (4) and has been shown to be present in the United States, particularly among intravenous drug abusers (IVDA) (5). One mode of transmission of HTLV-I is by blood transfusion. Widespread antibody screening for seropositive HTLV-I blood donors has recently begun in the United States. Initial studies in the United States have indicated that over 50% of seropositive HTLV-I blood donors have some association with IVDA (6). Therefore, the effectiveness of blood donor screening to prevent transfusion transmission of HTLV-I should be dependent, in large part, on the ability of screening assays to detect antibody-seropositive HTLV-I individuals from the IVDA community.

We will discuss the sensitivity of three commercially available, Food and Drug Administration (FDA)-approved HTLV-I screening immunoassays in a San Francisco Bay area IVDA population. Additionally, we will discuss how these enzyme-linked immunosorbent assays (ELISA) may be interpreted in a manner that allows for the detection of all seropositive HTLV-I IVDAs. We will consider inherent serological differences in IVDA versus normal blood donor populations, and reproducibility errors in the ELISA and how these differences may affect the interpretations of ELISA results. Although our interest of establishing the sensitivity of HTLV-I ELISAs on IVDAs is based on the observation that a significant portion of seropositive HTLV-I blood donors have had some association with IVDA (6), it must be noted that serological assays are currently not able to distinguish HTLV-I from HTLV-II infection. Additionally, a significant proportion of our local IVDA population may well be infected with HTLV-II instead of HTLV-I, as has been reported elsewhere in the U.S. (Drs. I. Chen and W. Blattner, UCLA Symposium on Retroviruses, 1989; I. Chen, this conference;

7). Therefore, it is likely that our interpretation of these assays will permit the detection of a significant proportion of HTLV-II–, as well as HTLV-I–, infected blood donors.

REAGENTS

HTLV-I Immunoassays

ELISAs were obtained from DuPont Medical Products Department (DuPont), Cellular Products, Inc. (CPI), and Abbott Laboratories (Abbott). Viral antigens for these assays are from isolated whole virus propagated in the HUT-102.B2 human T-cell line. These three assays were recently licensed by the FDA for blood bank screening use and for use as diagnostic kits. All assays were performed as recommended by the manufacturer, and the assay results were interpreted as described below.

Confirmation Assays

HTLV-I Western blot (WB) kits were provided by DuPont Medical Products Department and were performed as recommended by the manufacturer. Western blots were considered negative with no bands, HTLV-I seropositive with either p19 and p24 (both *gag* gene products) or p24 (*gag*), p40 (*tax*), and gp46 (*env*) bands, and indeterminant with any other nonpositive banding pattern. Radioimmunoprecipitation assays (RIPA) were performed by Steven Alexander of DuPont/Biotech. Samples were considered HTLV-I seropositive by a combination of WB and RIPA (WB-RIPA) if there was a minimum of p24 (*gag*) reactivity by WB and gp68 (*env*) reactivity by RIPA. None of these tests (WB, RIPAs, and/or ELISAs) unambiguously distinguishes immune response to HTLV-I versus HTLV-II infections, and therefore confirmed seropositive HTLV-I samples may be referred to as HTLV-I/II positive.

Sera Samples

A panel of sera from 269 IVDAs was provided by Dr. Anthony J. Puentes, County of Santa Clara Health Department, San Jose, CA. Of 269 IVDA sera, 52 sera were HTLV-I seropositive by either WB alone (37 samples) or WB-RIPA (15 samples). Nineteen of the WB nonpositive HTLV-I/II IVDA samples were either p24 only or p24 and p53 WB indeterminant. Twelve of these were subsequently confirmed HTLV-I seropositive by WB-RIPA, whereas the other 7 were negative by RIPA. This group of 12 WB p24/RIPA gp68 sera represented the lowest signals of the HTLV-I seropositive samples on the three assays. Five additional IVDA sera were WB or WB-RIPA indeterminant (two p19 only, two p19 and p28, and one p19, p28, and p32) by WB, all *env* gp68 negative by RIPA. Normal and HTLV-I/II–positive blood donors were all from the Stanford University Blood Bank, Palo Alto, CA.

DATA ANALYSIS AND CUTOFF ADJUSTMENTS

Normalization of ELISA Results

The sample signal for each assay was initially recorded as the ratio of signal (s) to cutoff (c) (i.e., the observed absorbance of the sample divided by the absorbance of the manufacturer's recommended cutoff, s/c), then converted to the logarithm of the ratio, log(s/c). A signal-to-cutoff ratio (s/c) of 1.000 [i.e., log(s/c) of 0.000] or larger represents a signal equal to or exceeding the recommended cutoff and is considered reactive by the manufacturer. The sensitivity of each of the three HTLV-I ELISAs toward the 52 seropositive HTLV-I IVDAs is summarized in Table 1.

The distribution of values for the ratio of signal to cutoff in seronegative blood donor populations for each assay was invariably skewed toward higher values, and a logarithmic transformation served to symmetrize the donor distributions. The statistics of the logarithm distribution of normal blood donors are summarized in Table 2 for the three ELISAs. The distribution of sample values for each assay was converted to a standard normal curve by subtraction of the mean value of the donor population and dividing the result by the standard deviation of the donor population for each individual sample value in the logarithm distribution. This normalization procedure results in a distribution of responses from the seronegative blood donor population for each assay, such that the mean is 0.00 and the standard deviation is 1.00 for each assay. The magnitude of the separation of signals from seropositive HTLV-I IVDAs relative to the average

TABLE 1. *Sensitivity of ELISAs to detect seropositive HTLV-I IVDAs*

Assay	ELISA RR	False negatives	Lowest positive log(s/c) (s/c)
DuPont	47	5	−0.105
			0.786
CPI	52	0	0.023
			1.055
Abbott	50	2	−0.068
			0.856

All 52 seropositive HTLV-I IVDAs were assayed in triplicate on each of the three HTLV-I ELISAs and the average value was determined. The number of repeat reactive (RR), false negatives (52 seropositives minus the number of RR) and the log of the signal (s)-to-cutoff (c) ratio, i.e., log (s/c), of the lowest signal from a seropositive HTLV-I IVDA are listed for each assay. The original signal-to-cutoff ratio (s/c) of the lowest seropositive HTLV-I IVDA is given below the log (s/c).

TABLE 2. *Logarithm distribution of normal blood donors on HTLV-I ELISAs*

Test	n	Mean (m)	SD
DuPont	500	−0.891	0.280
CPI	289	−0.700	0.255
Abbott	500	−0.881	0.110

Mean (m) and standard deviation (SD) are from a population of size *n* where the sample signals are expressed as logarithm of the signal-to-cutoff ratio, i.e., log (s/c), and where the cutoff is the manufacturer's recommended cutoff C.O.$_m$ for each assay. The manufacturer's recommended cutoff is by definition at (s/c) = 1.000 and log (s/c) = 0.000 for each assay.

blood donor can now be readily measured on the same scale for each assay, i.e., the number of standard deviations away from the mean value of the normal blood donor population.

Data Display

Figures 1–3 display the log normal distribution of a limited number of randomly drawn normal blood donors and all the seropositive IVDAs from the sensitivity panel. The histograms display the log normal of the sample signal versus the probability of a given score for each of the two groups.

Manufacturer's Recommended Cutoff, C.O.$_m$

The seropositive HTLV-I IVDA panel exhibited a wide range of antibody responses to HTLV-I/II infection as measured by ELISA, WB, and WB-RIPA. The results of the IVDA sensitivity panel studies were initially interpreted with the manufacturer's recommended cutoff, C.O.$_m$ (Table 1). The cutoff values C.O.$_m$ are summarized in Table 3 for each assay, in units of standard deviation of the average blood donor relative to the mean blood donor, i.e., in the log normal convention. The CPI assay has a C.O.$_m$ which is adequate to produce repeat reactive (RR) signal-to-cutoff ratios for all seropositive HTLV-I IVDA samples. The C.O.$_m$ of the DuPont and the Abbott assays are inadequate to capture all of the seropositive HTLV-I IVDA samples in our panel and result in five and two false-negative seropositive HTLV-I IVDA samples, respectively, as noted in Table 1.

Lowest Positive Cutoff, C.O.$_{lp}$

The sensitivity of each HTLV-I ELISA screening assay for seropositive HTLV-I IVDAs was initially adjusted to allow the detection as RR all the seropositive

HTLV-I samples in the panel of IVDA sera. Results of the IVDA sensitivity panel studies can be interpreted to determine a new cutoff which is minimally adequate to detect all HTLV-I/II positive sera. To eliminate false-negative samples for the DuPont and Abbott assays, redefined cutoff values lower than $C.O._m$ are necessary, whereas for the CPI assay the $C.O._m$ can be raised slightly. The choice of a cutoff that is adequate to detect all true seropositive HTLV-I IVDAs can be accomplished by an empirical or a mathematically modeled approach (see K. Kafadar, J. J. Lipka, and S. K. H. Foung, manuscript in preparation). Here we will consider an empirical approach whereby the lowest value of the ratio of signal to cutoff of a seropositive IVDA will be arbitrarily considered minimally adequate to eliminate all false negatives. Therefore this new cutoff determined by the lowest positive IVDA, $C.O._{lp}$, is defined as the lowest value of all the confirmed seropositive HTLV-I IVDA samples for each assay. The $C.O._{lp}$ for each assay is noted in Table 3, expressed in the log normal convention.

IVDA "Shifted" Cutoff, $C.O._{iv}$

The distribution of sample signals for the seronegative normal blood donor population and the seronegative IVDA population are qualitatively similar for each of the three HTLV-I ELISAs. There is a notable increase in the mean signal value for a seronegative IVDA, m_{IV}, versus the mean signal value for a seronegative blood donor, m_{DN}, which is probably due to an increased nonspecific antibody binding secondary to hypergammaglobulinemia in the IVDA population. The "IVDA-Shifted" seronegative population versus the normal seronegative blood donor population is demonstrated in Figure 4 for the Abbott HTLV-I ELISA. We will assume that the increase in m_{IV} relative to m_{DN} is representative of a simultaneous increase between seropositive IVDAs and seropositive blood donors (i.e., the average seropositive blood donor is not hypergammaglobulinemic, whereas the average seropositive IVDA is). Therefore, the cutoff again must be modified to measure as RR a blood donor with an equally weak serologic response specific to HTLV infection equal to the weakest seropositive, assumed hypergammaglobulinemic IVDA (i.e., the sera determining $C.O._{lp}$). The difference $m_{IV} - m_{DN}$ and new "IVDA-Shifted" cutoff, $C.O._{iv}$, are listed in Table 3 for each ELISA.

Reproducibility Errors and $C.O._{gz}$

The repeated measurement of any one sample on any of the three HTLV-I ELISAs reveals that there is some error in the measurement of the mean sample signal. When we measured this reproducibility error by triplicate measurements of individual samples over the entire dynamic range of the assays, we determined that the standard deviation of any sample value was ~5% of that value (data not shown). Because the usual experimental procedure involves a single measurement to determine if a sample is ELISA reactive or nonreactive, it is important that the reproducibility errors be considered. To be 95% confident that a single measurement of a sample with a mean value exactly equal to the cutoff (for example, the lowest seropositive IVDA adjusted to a blood donor population) will be detected as reactive in the initial screening measurement, a "gray zone" (or a lowering of the cutoff) of about two times the standard deviation of the

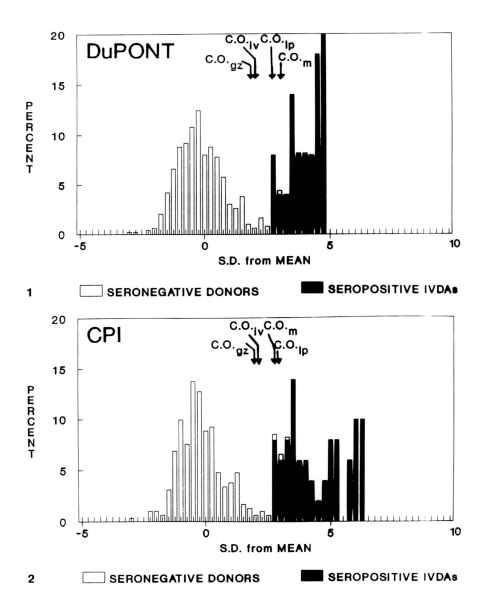

TABLE 3. *Cutoff adjustments for HTLV-I ELISAs*

Test	$C.O._m$	$C.O._{lp}$	$m_{IV} - m_{DN}$	$C.O._{iv}$	$C.O._{gz}$
DuPont	3.19	2.81	0.80	2.0	1.96
CPI	2.75	2.84	0.70	2.1	2.09
Abbott	7.99	7.38	2.19	5.2	5.14

The various cutoff values and $m_{IV} - m_{DN}$ are expressed as the log normal convention in units of standard deviation of the normal blood donor population relative to the mean of the normal blood donor population.

reproducibility error is required. This amounts to a lowering of the cutoff to 90% of the value of the cutoff for each assay, or a lowering of the cutoff by 0.046 in the log normal distribution convention [since log (0.90) = 0.046]. This new cutoff, $C.O._{gz}$, which incorporates a "gray zone" to account for reproducibility errors in each HTLV-I ELISA is listed in Table 3 and is a cumulative adjusted cutoff for all of the effects discussed above.

We have used the cutoffs as described above to interpret the results of our

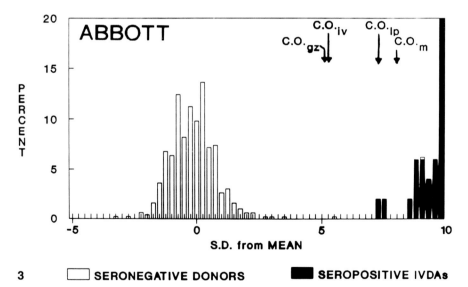

3 ☐ SERONEGATIVE DONORS ■ SEROPOSITIVE IVDAs

FIGS. 1–3. Log normal distribution of ELISA scores for both seronegative blood donors (open bars) and HTLV-I seropositive IVDAs (closed bars) displayed as the log normal of the sample signal versus probability. Log normal of the signal is as described in text and probability is expressed as the percent of the total for each group. Each bar is 0.25 standard deviations wide; the manufacturer's recommended cutoff $C.O._m$, the lowest positive cutoff $C.O._{lp}$, the "IVDA-Shifted" cutoff $C.O._{iv}$, and the gray zone cutoff $C.O._{gz}$ are denoted by appropriate arrows. The number of seropositive HTLV-I IVDAs is 52 and the number of blood donors is from Table 2. The cutoff values are listed in Table 3. The figures represent the probability distributions for DuPont, Figure 1; CPI, Figure 2; and Abbott, Figure 3.

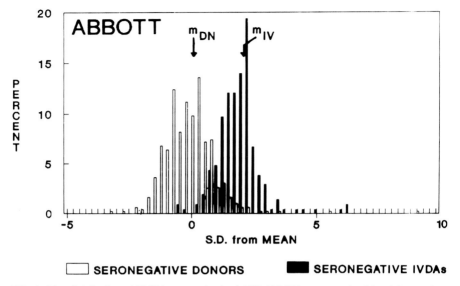

FIG. 4. The distribution of ELISA scores for both 500 HTLV-I seronegative blood donors (open bars) and 206 HTLV-I–seronegative, WB-negative IVDAs (closed bars) is displayed for the Abbott ELISA as a log normal probability, as described in Figures 1–3. The arrows represent m_{DN} and M_{IV}. The difference between M_{IV} and m_{DN} is about 2.2 standard deviations in the log normal convention.

blood donor population at the Stanford University Blood Bank. These procedures have allowed us to detect as reactive two seropositive HTLV-I blood donors who would otherwise have been ELISA false negative by an FDA-licensed HTLV-I ELISA. It is interesting to note that at least one of these two donors was infected with HTLV-II, as determined by the polymerase chain reaction technique (8).

ACKNOWLEDGMENTS

The authors wish to thank Dr. Anthony Puentes for providing the necessary IVDA sera, the technical support staff of the Stanford University Blood Bank, Drs. Michael McGrath and Valerie Ng for providing invaluable support in preliminary studies, and Drs. Karen Kafadar and Ed Engleman for helpful discussions. This work was supported in part by Grant HL33811 from the National Institutes of Health.

REFERENCES

1. Poiesz BJ, Ruscetti FW, Gazdar AF, Bunn PA, Minna JD, Gallo RC. Detection and isolation of type C retrovirus particles from fresh and cultured lymphocytes of a patient with cutaneous T-cell lymphoma. *Proc Natl Acad Sci USA* 1980;77:7415–7419.

2. Popovic M, Reitz MS Jr, Sarngadharan MG, et al. The virus of Japanese adult T-cell leukaemia is a member of the human T-cell leukaemia virus group. *Nature* 1982;300:63–66.
3. Gessain A, Vernant JC, Maurs L, et al. Antibodies to human T-lymphotropic virus type-I in patients with tropical spastic paraparesis. *Lancet* 1985;2:407–409.
4. Osame M, Usuku K, Izumo S, et al. HTLV-I associated myelopathy: a new clinical entity. *Lancet* 1986;1:1031–1032.
5. Robert-Guroff M, Weiss SH, Giron JA, et al. Prevalence of antibodies to HTLV-I, -II, and -III in intravenous drug abusers from an AIDS endemic region. *JAMA* 1986;255:3133–3137.
6. Williams AE, Fang CT, Slamon DF, et al. Seroprevalence and epidemiological correlates of HTLV-I infection in U.S. blood donors. *Science* 1988;240:643–646.
7. Lee H, Swanson P, Shorty VS, Zack JA, Rosenblatt JD, Chen IY. High rate of HTLV-II infection in seropositive IV drug users in New Orleans. *Science* 1989;244:471–475.
8. Kwok S, Ehrlich G, Poiesz B, Kalish R, Sninsky JJ. Enzymatic amplification of HTLV-I viral sequences from peripheral blood mononuclear cells and infected tissues. *Blood* 1988;72:1117–1123.

DISCUSSION

It was asked whether some of the sera classified by another speaker as HTLV-I sero-positive lacked envelope antibodies. The speaker replied that two of the sera lacked envelope antibodies, but that these two sera tested positive in all three HTLV-I antibody assays and therefore did not affect the revised cut-offs that had been suggested. One discussant objected to the use of sera from high-risk individuals as gold standards for use in determining an antibody assay's sensitivity. This discussant felt that only sera from individuals in whom HTLV-I presence had been demonstrated should be used to determine assay sensitivity.

Human Retrovirology: HTLV,
edited by William A. Blattner.
Raven Press, Ltd., New York © 1990.

Initial Evaluation of a Recombinant HTLV-I gp21 Enzyme Immunoassay

*Richard M. Thorn, Virginia M. Braman, Gerald A. Beltz, Annelie Wilde, Chung-Ho Hung, Jonathan Seals, and Dante J. Marciani

Cambridge BioScience Corporation, Biotechnology Research Park, Worcester, Massachusetts 01605; and *Baxter Healthcare Corporation, Miami, Florida 33121

HTLV-I/II infection induces antibodies reactive with *env* and *gag* viral proteins (1). The major *env* proteins of HTLV-I are gp21, gp46, and the precursor gp61 (2). The *gag* proteins p19 and p24 are the major mature polypeptides usually identified, but there are numerous intermediates with molecular weights up to that of the full length precursor, p55. Commercially available viral lysate blots are acknowledged to detect the *gag*-directed antibodies very sensitively, but only poorly detect *env* antibodies. Radioimmunoprecipitation (RIP) is more sensitive than blots for the *env* gp46 and gp61 antibodies, but neither method detects gp21 *env* antibody with good sensitivity.

Because sensitive detection of *env* antibody is important for screening and confirmation and because RIP technology is too demanding for routine use, there is a need for more suitable methods. We report here our progress in developing a recombinant gp21 enzyme immunoassay (EIA) which is practical for screening blood samples.

CHARACTERISTICS OF THE gp21 CLONE AND ANTIGEN POLYPEPTIDE

The CBC recombinant gp21-expressing clone is composed of nucleotides which code for amino acids 307 through 440 of the HTLV-I *env* protein, similar to that described by Samuels et al. (3). The clone has a bacteriophage lambda cII leader sequence and linkers which are translated as the N-terminal 14 amino acids. A region of nucleotides representing amino acids 314 through 330 has been deleted to improve expression levels in *Escherichia coli*. Expression is controlled by a temperature shift.

As shown in Fig. 1, the expressed polypeptide is easily visualized on Coomassie-stained, sodium dodecyl sulfate polyacrylamide gels. It migrated as a 14-kd pro-

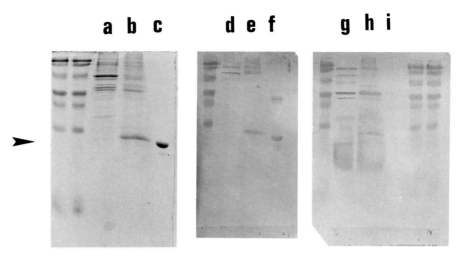

FIG. 1. SDS-PAGE and Western Blot Analysis of Recombinant gp21. Samples of *E. coli* not expressing recombinant gp21 (a, d, g); *E. coli* expressing recombinant gp21 (b, e, h); and purified recombinant gp21 (c, f, i) were analyzed by SDS-PAGE. Gels were stained with Coomassie blue (a, b, c) or transferred to nitrocellulose and developed with serum from an HTLV-I–positive patient (d, e, f) or with serum from a rabbit immunized with host *E. coli* (g, h, i). Molecular weights of standards (unmarked lanes) are 130k, 75k, 50k, 39k, 27k, and 17k.

tein, consistent with the molecular weight predicted from the plasmid construct. The expressed peptide was purified on ion-exchange resins after solubilization by alkylation and reversible acylation (4). The purified antigen is also shown in Fig. 1. The small amount of slightly shorter polypeptide (12 kd) is due to enzymatic cleavage which removes the N-terminal 18 amino acids.

Western blots of induced *E. coli* and purified antigen are also shown in Fig. 1. Human anti-HTLV-I/II sera reacted with the induced and purified proteins. The high-molecular-weight band in the purified antigen lane is a dimer of the antigen. Rabbit anti-*E. coli* sera did not react with the purified polypeptide. The anti-*E. coli* serum was generated against extracts of the host bacteria most likely to be contaminates of the purified antigen. The identity of the expressed polypeptide was confirmed by N-terminal amino acid sequencing.

INITIAL PERFORMANCE OF CBC RECOMBINANT gp21 EIA

The purified antigen preparations were used to develop a solid phase indirect immunoassay (Cambridge BioScience, Cat. #6017, for research use only). The assay was tested at Cambridge BioScience on a set of 1177 samples which were classified as shown in Table 1. The HTLV-I Western blot was from Biotech/ DuPont, and a conservative definition for an HTLV-I/II seropositive sample

TABLE 1. CBC recombinant gp21 EIA: in-house testing

| Samples | Number tested | CBC-EIA | |
		Pos	Neg
Positive[a]	101	101	0
Indeterminate[b]	2	1	1
Negative	1074	7[c]	1067[d]

Sensitivity = 100%; specificity = 99.3%.
[a] Blot p19 and p24.
[b] Blot bands were present, but not both p19 & p24.
[c] Blot negative (no bands).
[d] Licensed EIA negative, blot not done.

was used (presence of at least both p19 and p24 antibody). The 101 positive and 2 indeterminate samples were from intravenous drug abuser (IVDA), Japanese, and Caribbean persons living in the San Francisco, northeast, and southeast areas of the U.S. About 1000 seronegative samples were from prescreened (licensed HIV and HTLV EIA tests) blood donors in the northeast U.S. Approximately 70 samples were from individuals with no known HTLV risk, but with diseases or conditions which might cause false-positive reactions in EIA tests such as ANA, myeloma, or rheumatoid factor. The two indeterminate samples were tested by RIP, but the band patterns on repeated testing were inconsistent.

From preliminary testing with a small panel of known positive and negative samples, we selected a provisional cut-off of 0.3 times the positive control, which is usually about 0.35 OD units. In Table 1, all OD values for positive samples were greater than two times the cut-off. All but seven negative samples had OD values less than the provisional cut-off, and three of the seven were near the cut-off.

Further testing was done at four sites using kits provided by Cambridge BioScience. The results are combined and shown in Table 2. All data are classified

TABLE 2. CBC recombinant gp21 EIA: four non-CBC studies combined

| Samples | Number tested | CBC-EIA | |
		Pos	Neg
Positive[a]	177[b]	177	0
Indeterminate[c]	78	36	42
Negative	2156	22[d]	2134[e]

Sensitivity = 100%; specificity = 99.0%.
[a] Blot p24 and p19 or RIA p24 positive.
[b] Three samples were excluded because testing was not complete.
[c] Blot bands were present, but not both p19 & p24.
[d] Blot negative (no bands).
[e] Licensed EIA negative, blot not done.

TABLE 3. *CBC recombinant gp21 EIA: all data combined*

Samples	Number tested	CBC-EIA	
		Pos	Neg
Positive[a]	278[b]	278	0
Indeterminate[c]	80	37	43
Negative	3230	29[d]	3201[e]

Sensitivity = 100%; specificity = 99.1%.
[a] Blot p24 and p19 or RIA p24 positive.
[b] Three samples were excluded because testing was not complete.
[c] Blot bands were present, but not both p19 & p24.
[d] Blot negative (no bands).
[e] Licensed EIA negative, blot not done.

positive, indeterminate, or negative, as in Table 1. Samples primarily included transfusion-induced seroconversions, Caribbean blood donors, and IVDA sera. Table 3 shows the combined in-house and external testing.

CONCLUSIONS

These results show that the CBC recombinant HTLV-I gp21 in an EIA format performed with excellent sensitivity and adequate specificity. The status of the indeterminate samples remains to be determined and is an important factor in setting the proper cut-off. Many of the low positive samples may be HTLV-II. If so, the cut-off may have been lower than necessary to assure reaction with HTLV-I sera. However, it is also possible that HTLV-I induces antibody which does not react very strongly with this recombinant antigen. It may be necessary to develop a new antigen or combination of antigens which give better cross-reactivity with HTLV-II or reactivity with HTLV-I antibodies.

ACKNOWLEDGMENTS

The authors thank James Drummond, James Lipka, Angela Manns, James Damato, and William Blattner for permission to summarize their data.

REFERENCES

1. Anonymous. Licensure of screening tests for antibody to human T-lymphotropic virus type I. *MMWR* 1988;37:736–747.
2. Lee TH, Coligan JE, Homma T, McLane MF, Tachibana N, Essex M. Human T-cell leukemia virus-associated membrane antigens: identity of the major antigens recognized after virus infection. *Proc Natl Acad Sci USA* 1984;81:3856–3860.
3. Samuel KP, Lautenberger JA, Jorcyk CL, Josephs S, Wong-Staal F, Papas TS. Diagnostic potential for human malignancies of bacterially produced HTLV-I envelope protein. *Science* 1984;226: 1094–1097.
4. Hung CH, Thorn R, Riggin C, Marciani DJ. Process of purifying recombinant proteins, and products thereof. U.S. Patent No. 4,732,362;1988.

Human Retrovirology: HTLV,
edited by William A. Blattner.
Raven Press, Ltd., New York © 1990.

Use of Recombinant HTLV-I Proteins in the Confirmation of HTLV-I/II Infection

Mark Carle Connelly

*E. I. Du Pont de Nemours & Company, Inc., Medical Products Department,
Glasgow Research Laboratory, Box 713, Glasgow, Delaware 19702*

WESTERN BLOT SENSITIVITY AND SPECIFICITY

Infection by either of the human T-cell lymphotropic viruses I or II (HTLV-I/II) may cause the host to mount an immune response to a variety of viral specific proteins. Enzyme-linked immunosorbent assay (ELISA) is an efficient and inexpensive way to screen populations, and blood product donors, for the presence of antibodies to HTLV-I/II. A repeatedly positive reactivity in ELISA suggests, but does not prove, that antibodies are present to HTLV-I/II. Occasionally, a positive ELISA result may be obtained for reasons unrelated to HTLV-I/II viral infection (1,2). Thus, additional tests must be conducted on ELISA-positive sera to provide more specific information regarding the nature of the antigen-antibody interaction being detected. The Western blot assay (WB) employing proteins from disrupted virions of purified HTLV-I virus is one such more-specific test. WB may be used to determine whether the antibodies present are specific for any of the viral gene products, and if so, for which gene products.

Antibody reactivity to any gene product of HTLV-I/II is not necessarily diagnostic of HTLV-I/II infection. To determine the minimum WB band pattern diagnostic of HTLV-I/II infection, sera and plasma from patients with adult T-cell leukemia and/or tropical spastic paraparesis, from asymptomatic individuals from HTLV-I/II endemic regions, and from uninfected seronegative individuals were tested by ELISA, WB, and radioimmunoprecipitation (RIPA) assays. Patient information, geographic location, and test results were entered into a DBASE III PLUS™ program customized for this purpose. Computer analysis of 1475 samples demonstrated that antibody reactivity to a minimum of two gene products must be present to confirm HTLV-I/II infection serologically. Antibody must be present to *gag* (e.g., p24) and envelope (e.g., gp46) gene products; reactivity to *gag* protein(s) alone gave a poor correlation with other markers of HTLV-I/II infection.

WB was a very sensitive means of detecting antibody to p24, correctly identifying 99.5% of all p24 RIPA positive samples. In contrast, the RIPA assay detected anti-p24 in only 73.2% of WB p24-positive sera. The advantage of WB over RIPA in detecting anti-*gag* antibody was offset by its diminished sensitivity, with respect to RIPA, for detecting anti-envelope. Table 1 shows the 95% confidence intervals for sensitivity and specificity of anti-envelope antibody detection by WB as compared with RIPA. Although the specificity of the WB gp46 reactivity was excellent, only 60.7% of samples having antibody to envelope, as measured by RIPA, were scored positive for gp46 in WB. It may be concluded from these data that a positive band at gp46 in WB is highly predictive of a positive RIPA result. However, the reverse is not true; the absence of gp46 reactivity in WB is not predictive of a negative RIPA result. Although serologic confirmation of HTLV-I/II infection may be achieved by the use of either WB or RIPA, in some cases both technologies are required to detect antibody to two gene products. Having to do both WB and RIPA is time-consuming, technically difficult, and expensive. We report here our efforts to improve the process of serologic confirmation through the use of HTLV-I recombinant proteins.

RECOMBINANT PROTEINS

Six HTLV-I recombinant proteins were examined for their purity, homogeneity, and serologic reactivity by WB. Four of the recombinant proteins were to the gp46 or gp21 envelope proteins. The fifth was a *gag* protein construct (3), and the sixth was a full-length p40X or TAX gene recombinant (3). The recombinant proteins were run on a 4%–20% gradient gel and transferred to nitrocellulose by the use of electrophoretic transfer (4,5). Figure 1 shows the reactivity of each recombinant protein in WB using a strong HTLV-I seropositive sample. The reactivity with TBE in the experiment shown was unusually weak for reasons unknown; however, TBE, like most of the recombinants, gave a single immunoreactive band in WB. The p40X recombinant showed a minor amount of protein degradation, and TR24 gave evidence of a weak, diffuse, high-molecular-weight species. Both the TR24 and p40X preparations were 98%–99% pure. Each recombinant was tested using a large number of seronegative normal sera and found to be free of contaminating protein and nonspecific immunoreactivity. Because some recombinants were of similar size, it was not possible to combine them in a single WB strip. However, after extensive individual WB analysis,

TABLE 1. *Anti-envelope detection by Western blot versus RIPA[a]*

WB specificity	99.7 ± 0.4[b]
WB sensitivity	60.7 ± 3.5
Positive predictive value of WB result	99.6 ± 0.6
Negative predictive value of WB result	71.3 ± 2.8

[a] Comparison of WB reactivity for gp46 versus RIPA reactivity for gp61.
[b] Results are expressed as percent ± a 95% confidence interval (n = 1475).

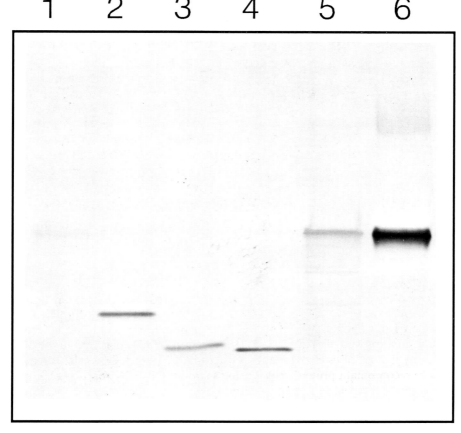

FIG. 1. Gradient gel WB analysis of HTLV-I recombinant proteins. Lane 1, 22 ng of TBE, obtained from Triton Biosciences Inc. (3), containing the C-terminus of gp46 and most, but not all, of the gp21 transmembrane region of the envelope protein. Lane 2, 59 ng of r1, containing the C-terminus of gp46, but no gp21 sequences. Lane 3, 63 ng of r2, a C-terminus gp46 recombinant identical to r1 except r2 lacks 35 amino acids at its N-terminus that are present in r1. Lane 4, 61 ng of r3, most but not all of the gp21 transmembrane envelope protein. Lane 5, 34 ng of p40X, obtained from Triton Biosciences Inc. (3), representing a full length p40X or Tax gene product. Lane 6, 63 ng of TR24, obtained from Triton Biosciences Inc. (3), representing all of p24, and a portion of p19.

each was judged to be sufficiently pure, homogeneous, and free of nonspecific reactivity to make it possible to slot blot these proteins. Recombinant protein–based slot blots (rSB) were simpler and faster to prepare, and provided greater flexibility in reagent format, than WB. Proteins were bound to nitrocellulose using a slot blot device of our own making. Figure 2 shows a rSB of the four envelope proteins and their reactivity with monoclonal antibody reagents. Positive control sera from an HTLV-I/II–infected individual reacted with all four proteins.

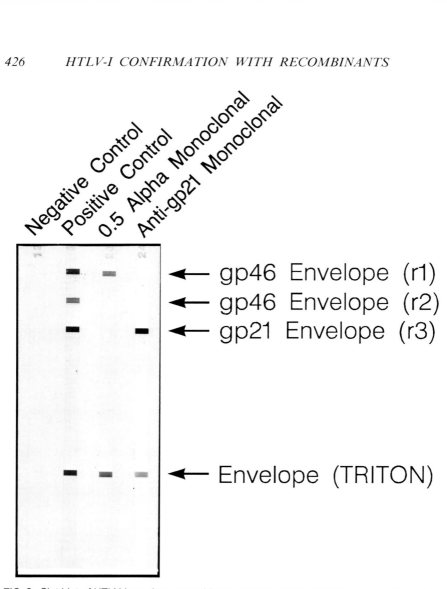

FIG. 2. Slot blot of HTLV-I envelope recombinant proteins. Proteins were bound to nitrocellulose using a slot blot device. Strips were incubated with samples overnight at room temperature with constant rocking. Negative control serum was from an HTLV-I–seronegative individual. Positive control was from a HTLV-I–seropositive individual. 0.5 α was a human monoclonal antibody to gp46, and the anti-gp21 monoclonal antibody was a mouse monoclonal (M7G7) developed at DuPont.

The 0.5 α monoclonal, kindly provided by Dr. Sam Broder, is a human monoclonal antibody reactive with gp46 of HTLV-I (6). Its failure to react with r2, in light of the positive reactivity with r1, suggests 0.5 α may bind to an epitope located between amino acids 165 and 200 of gp46. TBE also contained the 0.5

α epitope. A mouse monoclonal antibody (M7G7, Du Pont), raised against recombinant gp21 protein, reacted with r3 and TBE, but failed to react with r1 and r2. The reactivity of this monoclonal antibody with a single 21 000 molecular weight band in viral lysate WB (Fig. 3), suggests the epitope for this monoclonal antibody is also present and accessible in gp21 of viral origin. Thus, the nature and immunoreactivity of the four envelope recombinant proteins being consid-

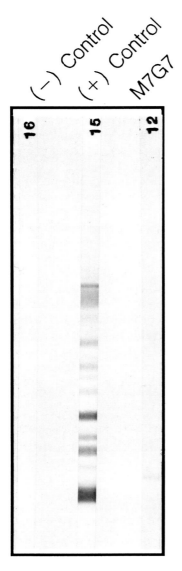

FIG. 3. Envelope Reactivity (p21 Transmembrane Region, HUT-102): Commercial viral lysate Western blot strips (DuPont/Biotech) were incubated with kit positive control, negative control, and mouse monoclonal antibody to recombinant gp21 protein (M7G7).

TABLE 2. *Western blot positive samples*

		Recombinant	
Reactivity in Western blot	RIPA *env* reactivity	*gag*	Envelope
Multiple bands[a]	556/559[b]	214/214	189/191

[a] Antibody reactivity to any two of three gene products; p19 or p24 for *gag,* p36 (HUT-102) for *tax,* and gp46 for envelope.
[b] Number positive/number tested.

ered were validated, at least in part, by reaction with a reagent of known specificity (0.5 α) and by their ability to generate antibody to appropriate viral gene products.

CONFIRMATION OF HTLV-I/II REACTIVITY BY RECOMBINANTS

If recombinant proteins are to be used to improve the HTLV-I/II confirmation process, they must 1) not react with seronegative samples; 2) react with known seropositive samples; and 3) show greater sensitivity for the detection of antibody to two gene products then viral lysate WB. The six recombinants under consideration were nonreactive with seronegative samples. Known positive samples were tested for reactivity with the *gag* and envelope recombinant proteins and by RIPA (Table 2). Ninety-nine percent of WB positive sera reacted with envelope protein, and 100% reacted with recombinant *gag.* Ninety-nine percent of WB positive sera were RIPA envelope positive. All sera in Table 2 were among the 60.7% of anti-envelope positive sera that scored positive for gp46 in viral lysate WB. It was concluded that WB using recombinant proteins was as sensitive as viral lysate WB and RIPA for the detection of antibody to two gene products with this class of sera. However, the data in Table 2 do not address whether recombinant proteins would be more sensitive than viral lysate WB for the detection of anti-envelope antibody.

Table 3 summarizes the results of ~400 WB indeterminate samples, none of which gave evidence of anti-envelope antibody in WB. The WB reactivity was

TABLE 3. *Western blot* gag *only reactive samples*[a]

		Recombinant	
Reactivity in Western blot	RIPA *env* reactivity	*gag*	Envelope
p19 only	18/86[b]	33/47	13/28
p24 only	70/144	11/11	7/11
p19 and p24	98/122	47/47	36/44

[a] Samples showing reactivity to only *gag* and *gag*-related proteins, no envelope.
[b] Number positive/number tested.

divided into three groups, depending on the nature and extent of anti-*gag* antibody detected. Eighteen sera initially classified as p19 only on the basis of WB were found to contain anti-envelope antibody when examined by the more sensitive RIPA assay. Of the 13 p19-only samples positive on envelope recombinant proteins, all were *gag*-antibody positive. Five of these 13 sera were RIPA positive but RIPA data were not available on the other 8. The *gag* recombinant used (TR24) was primarily p24 with only a small portion of p19 (3). WB p19-only sera that failed to react with the *gag* clone (14 of 47) may have been reacting with a region of p19 not expressed in TR24. WB p19-only sera positive on TR24 may have been reacting to the portion of p19 contained in TR24, or there may have been improved detection of anti-p24 antibody in these sera, or both. All of the WB p24-only sera reacted with TR24. The positive results obtained with TR24 confirm the viral specificity of WB p19- and p24-only sera. In addition, antibody to envelope was detected in many p24-only sera when examined by RIPA or by recombinant WB. Of the 11 WB p24-only sera tested on recombinant envelope proteins, only 4 were negative. RIPA data were not available on these 4 sera. Of seven p24-only recombinant envelope–positive sera, five were RIPA positive. The two p24-only RIPA negative sera were collected from intravenous drug users (IVDU) from New Orleans and gave several bands on HTLV-II WB. The majority of sera having antibody to both p19 and p24 in WB were RIPA and recombinant envelope–positive, and all were reactive with TR24.

Antibody to two gene products could be demonstrated in many WB indeterminate samples by either the combination of WB and RIPA or the use of a recombinant-based WB or rSB. This point is illustrated further in Fig. 4. Five

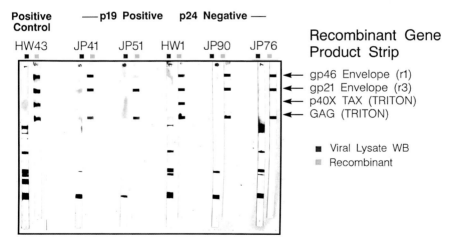

FIG. 4. HTLV-I Western Blot Indeterminate RIPA Positive Samples: Reactivity of WB indeterminate samples with viral lysate WB and rSB. Four recombinant proteins (r1, r2, p40X, and TR24), representing three HTLV-I gene products, were bound to nitrocellulose using a slot blot device. Samples were incubated overnight at room temperature with constant rocking.

sera that gave indeterminate results in WB were tested by RIPA and rSB. Three samples from Japan (JP41, JP51, and JP90) were p19-only, whereas two samples (HW1 and JP76) were p19 with faint or questionable bands at p24. All five sera were RIPA positive for envelope and reacted with TR24 in addition to the gp46 (r1) and gp21 (r3) envelope proteins in rSB. One sample (HW1) also reacted with the TAX protein. In contrast to the indeterminate serologic result obtained with viral lysate WB, antibody to at least two gene products was clearly demonstrated by use of the rSB. Thus, rSB was capable of confirming as positive sera which would otherwise have been considered indeterminate. It is worth noting, however, that it is rare to encounter a WB p19-only serum that can be confirmed by either RIPA or reactivity with recombinants. A few such samples have been observed only because a large number of sera from HTLV-I endemic regions, HTLV-I associated disease patients, and high-risk individuals have been examined. The existence of a serologically confirmable WB p19-only sample in the US population has not been established.

CONFIRMATION OF HTLV-II

The HTLV-I and HTLV-II viruses are so closely related immunologically that it is difficult to distinguish them with the use of current serologic techniques. This provides a challenge for those who seek to determine which virus is present for the proper counseling of infected individuals. Often, when the same serum is tested using a HTLV-I and a HTLV-II WB, the serum shows multiple bands on both blots. Other techniques, such as viral culture or polymerase chain reaction, must be used to establish whether this represents dual infection or results from a high titer of crossreacting antibodies. In some cases, however, it is possible to make a presumptive differentiation. Table 4 shows the results of HTLV-II WB analysis on sera from 16 IVDUs from New Orleans. Antibody was present in these sera that bound to numerous HTLV-II viral specific proteins. The samples 1057, 1066, 1077, and 1109 clearly had antibody against most of the HTLV-II proteins (Table 4). Table 5 shows the HTLV-I WB and RIPA results on all the IVDU samples. In contrast to their strong reactivity with numerous HTLV-II proteins, strong reactivity was observed only to the p24 of HTLV-I. Reactivity to p19 and other HTLV-I *gag*-related bands was weak or absent for these IVDU sera. These individuals were determined to be HTLV-II infected on the basis of their relative reactivities to the different viruses and their risk factor of having been drawn from IVDUs (7,8). All sera were classified as HTLV-I WB indeterminate. Follow-up RIPA analysis demonstrated anti-envelope reactivity in some but not all sera. The samples 1057, 1066, 1077, and 1109, which showed strong reactivity to HTLV-II (Table 4), were HTLV-I RIPA negative (Table 5). Thus, even the combination of both WB and RIPA failed to demonstrate antibody reactivity to two gene products in four samples.

All of the putative HTLV-II sera reacted with at least one envelope recombinant

TABLE 4. *Putative HTLV-II samples*

Sample ID	HTLV-II Western blot[a]								
	p15	p22	p24	p26	p32	p40	p42	gp46	p53
1043	0	1	3	1	2	1	2	0	2
1050	1	3	3	3	2	2	2	2	3
1057	0	3	3	1	2	2	3	1	2
1062	0	2	2	1	2	1	2	2	1
1066	0	3	3	2	2	2	3	1	3
1067	0	2	2	0	0	0	1	0	0
1070	1	2	3	2	2	2	2	2	2
1073	0	3	3	2	2	2	2	0	2
1077	0	1	2	1	0	1	2	1	2
1079	0	3	3	2	1	2	2	1	2
1101	2	2	2	2	2	1	3	1	3
1104	1	3	3	3	3	3	3	1	3
1108	1	2	3	2	2	2	2	1	2
1109	0	1	2	1	0	1	2	0	1
1110	1	2	3	2	2	2	3	1	3
1529	0	3	2	2	0	1	0	1	1

[a] Band intensity was scored on a scale of 0 for negative to 3+. HTLV-II WB data from Dr. S. Alexander of Biotech Research Laboratory.

protein (Table 5). The ability of a rSB to detect antibody to two gene products is illustrated in Fig. 5. All 16 sera were reactive to both the *gag* and gp21 recombinant proteins. Thus, serologic confirmation of HTLV-I/II infection was achieved in each case, including the four samples missed by RIPA. The reactivity to gp46, as measured by r1, was weak or absent. Reactivity with *gag* coupled with a strong gp21 and a weak or absent gp46 band was suggestive of HTLV-II infection (Fig. 5 and S Alexander, personal communication). Such a pattern of reactivity was not usually seen with confirmed HTLV-I seropositive samples. However, it remains to be seen whether this pattern is characteristic of all HTLV-II sera or whether it is limited to those HTLV-II sera having generally poor crossreactivity to HTLV-I.

CONCLUSIONS

The serologic response to HTLV-I/II infection is complex and not fully understood. A variety of sophisticated technologies are being used at Du Pont and elsewhere to better understand the biology and immunology of HTLV-I/II. A sensitive and specific confirmatory assay is essential to the study of HTLV-I and HTLV-II epidemiology, as well as for making an accurate and informed diagnosis of infection based on serology. Recombinant proteins may be formatted in a way that makes confirmation as sensitive and specific as RIPA, but easier, less expensive, and less time-consuming to perform. Many samples that give indeterminate WBs may be unequivocally characterized as HTLV-I/II negative or

TABLE 5. *Putative HTLV-II samples*

Sample ID	HTLV-I Western blot[a]										HTLV-I RIPA[b]		HTLV-I envelope recombinants[a]		
	p15	p19	p24	p26	p28	p32	p36	p42	gp46	p53	gag	env	r1	r2	r3
1043	0	0	3	0	0	0	0	0	0	0	0	1	0	0	1
1050	0	2	2	0	0	0	0	0	0	0	1	1	1	0	1
1057	0	1	3	0	0	0	0	0	0	0	0	0	1	0	2
1062	0	0	2	0	0	0	0	1	0	0	0	1	1	1	1
1066	0	2	3	1	1	0	0	0	0	2	1	0	3	1	3
1067	0	0	2	0	0	0	0	1	0	0	0	0	0	0	1
1070	0	0	3	0	1	0	0	0	0	1	0	1	1	1	1
1073	0	0	2	0	0	0	0	0	0	0	0	1	1	0	1
1077	0	0	2	0	0	0	0	0	0	0	0	0	0	1	2
1079	0	1	2	0	0	0	0	0	0	0	0	1	0	0	1
1101	0	1	3	0	0	0	0	0	0	0	1	1	1	1	1
1104	0	1	3	0	0	2	0	2	0	2	2	1	0	1	1
1108	0	1	3	0	0	1	0	1	0	1	1	1	1	1	1
1109	0	1	2	0	0	0	0	0	0	0	1	0	0	0	1
1110	0	1	3	0	0	0	0	2	0	0	1	1	1	1	1
1529	0	0	2	0	0	0	0	0	0	0	1	2	0	0	1

[a] Band intensity was scored on a scale of 0 for negative to 3+.
[b] Band intensity was scored on a scale of 0 for negative to 2+.

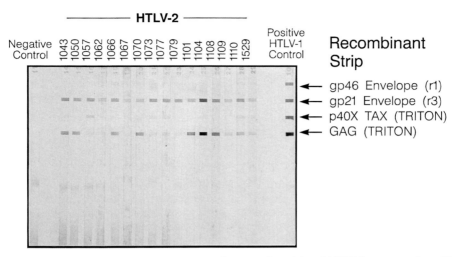

FIG. 5. HTLV-II Western Blot Indeterminate Samples: Reactivity of HTLV-II sera samples with rSB. Four recombinant proteins (r1, r2, p40X, and TR24), representing three HTLV-I gene products, were bound to nitrocellulose using a slot blot device. Samples were incubated overnight at room temperature with constant rocking.

positive by virtue of their reactivity with recombinant proteins. Furthermore, in the case of some HTLV-II sera, recombinant protein–based assays may be a more sensitive confirmatory test than the combination of viral lysate WB and RIPA.

ACKNOWLEDGMENTS

I thank Cathleen O'Connell for her excellent technical assistance and Monica Tadler and Ray Ryan for help in the development of the rSB. Special thanks to Dr. Jim Stave, Dr. Conrad Heilman, and Denise Cervelli for their efforts in the development of the gp21 specific monoclonal antibody. I also thank Dr. Ann Bodner and Dr. Steve Alexander of Biotech Research Laboratories, Inc. for their contribution of time and talent in the development of WB and RIPA data.

REFERENCES

1. Constantine NT, Fox E. Need to confirm HTLV-I screening assays. *Lancet* 1989;1:108–109.
2. Phair JP, Wolinsky S. Diagnosis of infection with the human immunodeficiency virus. *J Infect Dis* 1989;159:320–323.
3. Coats S, Harris A, Keitelman E, et al. Serological reactivity of recombinant HTLV-I proteins. *J Cell Biochem Suppl* 1989;13B:G206.
4. Towbin H, Staehelin T, Gordon J. Electrophoretic transfer of proteins from polyacrylamide gels

to nitrocellulose sheets: procedure and some applications. *Proc Natl Acad Sci USA* 1979;76:4350–4354.

5. Gershoni JM, Palade GE. Protein blotting: principles and applications. *Anal Biochem* 1983;131:1–15.

6. Matsushita S, Robert-Guroff M, Trepel J, Cossman J, Mitsuya H, Broder S. Human monoclonal antibody directed against an envelope glycoprotein of human T-cell leukemia virus type I. *Proc Natl Acad Sci USA* 1986;83:2672–2676.

7. Robert-Guroff M, Weiss SH, Giron JA, et al. Prevalence of antibodies to HTLV-I, -II, and -III in intravenous drug abusers from an AIDS endemic region. *JAMA* 1986;255:3133–3137.

8. Lee H, Swanson P, Shorty VS, Zack JA, Rosenblatt JD, Chen ISY. High rate of HTLV-II infection in seropositive IV drug abusers in New Orleans. *Science* 1989;244:471–475.

DISCUSSION

A speaker found the p19-only Western-blot group interesting. In a similar cross-sectional study of 26 000 serum samples, this speaker found 178 Western-blot indeterminants, of which 57 were p19-only and RIPA-negative. PCR had been done on a few of those samples, with no evidence of virus. This speaker asked another participant if virus in cells from those individuals had been looked for. In those particular cases, the answer was no. The participant explained that many of these serum samples are from Japan. They were from unlinked studies, so it was not possible to go back and obtain cells. The participant was now trying to identify more of those in linked studies, and thought the point was well taken. There clearly are some individuals, mostly in high risk groups, that do exhibit this reactivity. Another discussant cautioned that investigators must be careful in interpreting the data, because the work was done using different techniques in different laboratories. One speaker questioned if the reactivity of the 21E peptide had been examined with a large number of negative, well-tracked normal blood donors. This speaker had found a lot of false positives with the 21E recombinant when a similar group of normal blood donors was examined and asked if another speaker had found any false positives. The other speaker replied that it was a problem initially, but the assay was formatted in such a way that the specificity was under control. The relative hydrophobicity of the particular 21E clone sequences may be an important factor in eliminating false positives. There were problems with the *tax* gene clone, because if there is a normal that is *tax*-positive and nothing else, is it one of the sera that was referred to earlier, or is it a false positive? Unless one can show that it is false, one is "kind of stuck."

One discussant brought up the family data where "this sort of thing" was shown clustering in families. This discussant felt that, since there are MHC determinants of immune response, there might be some way of looking at the relationship of MHC to this type of reactivity. Another discussant agreed and stated that another possibility is that it is a result of viral infection, since more *tax* and anti-*tax* reactivity can be seen in ATL patients than in HTLV-I antibody-positive blood donors. So, basically, some type of continuing viral antigen stimulation may be playing another role. One participant made the point that perhaps in the earlier reactivity, with the age of the group, the age of the donors was matched to the family members. It wasn't that the family members were recently infected. The previous discussant agreed and stated that the age preclusion of the family members is from 20 to 40. This discussant's study did not have very young family members. It was thought that some Japanese investigators had similar group collections that do contain young carriers. It was hoped that they would do a similar study and confirm these results.

Human Retrovirology: HTLV,
edited by William A. Blattner.
Raven Press, Ltd., New York © 1990.

Mapping of Epitopes on Human T-Cell Leukemia Virus Type I Envelope Glycoprotein

Thomas J. Palker

Division of Rheumatology and Immunology, Duke University Medical Center, Durham, North Carolina 27710

INTRODUCTION

The discovery of two human retroviruses—human T-cell leukemia virus (HTLV) types I and II—at the beginning of the past decade marked the first time that retroviruses were etiologically linked to human disease (1–6). After a prolonged latent period, HTLV-I can cause adult T-cell leukemia/lymphoma in some infected individuals (1–5) and can also give rise to a chronic neurological degenerative syndrome, tropical spastic paraparesis (TSP) (7–9). In contrast, an etiologic link between HTLV-II and human disease is less well established, although an association between HTLV-II infection and a rare variant of T hairy cell leukemia has been reported (6,10).

There is evidence that an appropriate host immune response to HTLV-I can inhibit viral infection. Antibodies from HTLV-I seropositive individuals inhibit HTLV-I–induced syncytium formation (11,12) and inhibit plaque formation obtained with vesicular stomatitis virus (VSV)/HTLV-I pseudotype particles (13,14). Further, cynomolgus monkeys can be protected from a challenge with HTLV-I by prior immunization with HTLV-I envelope gene products produced in bacteria (15). Additional studies performed with recombinant HTLV-I envelope products have confirmed that these proteins are recognized by antibodies from HTLV-I–infected individuals and can evoke virus-neutralizing antibody responses in animals (16–18). However, little is known regarding the locations of epitopes on HTLV-I envelope glycoproteins that are either immunogenic in HTLV-I+ individuals or that confer protective immunity to HTLV-I. A major focus of our work has been to use viral *env*-gene–encoded synthetic peptides and monoclonal antibodies to map immunogenic and functionally important regions of retroviral envelope glycoproteins (19–21). We describe here recent work on the epitope mapping of regions of HTLV-I envelope gp46 and gp21 glycoproteins.

PURIFICATION OF HTLV-I ENVELOPE GLYCOPROTEINS
FROM LYSATES OF HTLV-I–INFECTED CELLS

The mature gp46 and gp21 envelope glycoproteins of HTLV-I result from proteolytic cleavage of a gp61–65 envelope precursor encoded by the viral *env* gene (22,23). Based on a proposed structure for animal retroviruses (24), the gp46 molecule of HTLV-I is thought to be bound to the outside of the virion by a nonconvalent association with the p21 transmembrane protein that is inserted into the lipid bilayer surrounding the viral core. A major problem in obtaining sufficient native gp46 from viral preparations has been that gp46 detaches from the surface of virions during sucrose density purification and is thereby depleted in these samples (25). Although supernatants from cell cultures of HTLV-I–infected cells contain shed viral glycoproteins, substantially greater amounts appear to be associated with infected cells (23). To obtain sufficient amounts of envelope glycoprotein for study, we extracted 50 ml cell pellets of HTLV-I–infected HUT-102 cells with nonionic detergent (26). Supernatants were then passed over an affinity column containing IgG from an HTLV-I seropositive individual followed by chromatography of the bound fraction over a lentil lectin column (Fig. 1). A 46-kd molecule was precipitated with HTLV-I+ patient serum (Fig. 2) and human monoclonal antibody 0.5α (27) (not shown) to HTLV-I gp46 in radioimmunoprecipitation (RIP) assays with affinity-purified, radiolabeled sample. In sequential RIP assays, it was determined that anti-HTLV-I patient antibodies and 0.5α recognized the same 46-kd molecule (21). To evaluate further the purification of gp46, HTLV-I radiolabeled sample was subjected to sodium dodecyl sulfate polyacrylamide gel electrophoresis and autoradiography. A major band of 46K was observed (Fig. 2). After treatment of ^{125}I-labeled sample with endoglycosidase F, a 34-kd molecule was seen (26), in agreement with the reported molecular weight of deglycosylated HTLV-I gp46 (28).

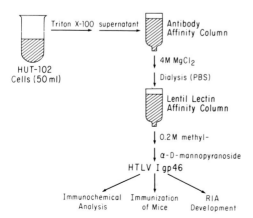

FIG. 1. Procedure for sequential affinity purification of HTLV-I envelope glycoproteins (26).

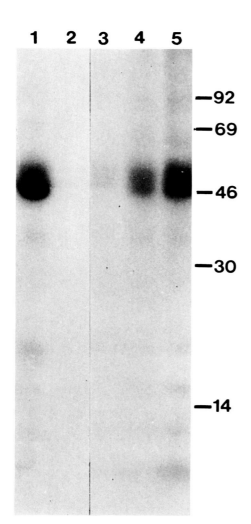

FIG. 2. Analysis of affinity-purified material from lysates of HTLV-I+ HUT-102 cells. Purified glycoproteins (see Fig. 1) were labeled with ^{125}I and tested in radioimmunoprecipitation (RIP) assay with HTLV-I+ patient serum (lane 1) or normal human serum (lane 2). A 46- to 54-kilodalton (kd) band was precipitated with HTLV-I+ patient serum. The molecular weight range of this band is consistent with the molecular weight of HTLV-I gp46 envelope glycoprotein. To evaluate the affinity-purified sample, increasing amounts (lanes 3–5) of ^{125}I-labeled sample were subjected to SDS-PAGE and autoradiography. A major band at 46–54 kd was observed.

Results of these studies were consistent with the conclusion that HTLV-I gp46 had been purified by sequential affinity chromatography.

PRODUCTION OF A MURINE MONOCLONAL ANTIBODY (MoAb) TO HTLV-I gp46

Balb/c mice were immunized with affinity purified gp46 for the production of MoAb with which to perform epitope mapping studies. One MoAb, designated

1C11, was obtained with a binding specificity consistent with that of an anti-HTLV-I gp46 antibody (21). In Western blot assays, MoAb 1C11 reacted with a 46- to 54-kd molecule in affinity-purified lysates of HUT-102 cells (Fig. 3) and also with a 63-kd molecule (21), presumably the HTLV-I envelope precursor. A similar pattern of reactivity was observed with HTLV-I patient sera and human monoclonal antibody 0.5α. MoAb 1C11 and human anti-HTLV antibodies recognized the same 46-kd molecule from envelope-enriched samples used in sequential RIP assays. In immunofluorescence (IF) assays, antibody 1C11 reacted with >90% of cells from four different HTLV-I+ T-cell lines but did not recognize either acetone-fixed or viable HTLV-II or HIV-1 (HTLV-III$_B$) cell lines. MoAb 1C11 was nonreactive when tested in IF assay against a wide panel of HTLV-I–negative cell lines and normal tissues, thus demonstrating a high degree of specificity for HTLV-I (Table 1). A panel of 12 synthetic peptides (21, Table 2) containing hydrophilic amino acid sequences of HTLV-I envelope was used to

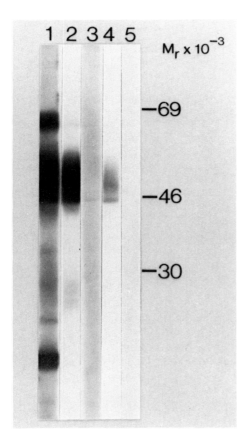

FIG. 3. Antibody reactivity to affinity purified HTLV-I gp46/gp63 in Western blot assay. Lane 1, HTLV-I+ patient serum; Lane 2, human monoclonal anti-gp46 antibody 0.5α (27); Lane 3, normal human serum; Lane 4, murine monoclonal antibody 1C11 raised to HTLV-I gp46; Lane 5, P3X63 ascites fluid (negative control). HTLV-I patient antibodies, 0.5α, and 1C11 all reacted with a 46- to 54-kd molecule. Patient antibodies also reacted with a 63-kd molecule also recognized by 0.5α and 1C11 in another study (26). The pattern of reactivity obtained with 0.5α and 1C11 in this and another study (26) indicates that these antibodies react with HTLV-I gp46 and gp63 envelope glycoproteins.

TABLE 1. *Reactivity of anti-HTLV-I gp46 antibody 1C11*

Assay	Reactivity
Western blot	gp46, gp63
Radioimmunoprecipitation	gp46; same molecule recognized by human gp46 antibodies
Immunofluorescence	
Acetone-fixed cells	Positive on HTLV-I+ cell lines: HUT-102, MT-2, MJ, C10MJ Negative on HTLV-I T-cell lines: HSB-2, H9, HUT-78, Jurkat; HTLV-I− B-cell lines SB, EB-3; HTLV-II+ C3-44/ MO cells, HIV-1 (HTLV-III$_B$)+ CEM cells, normal human tissues: thymus (2), lymph node (3), spleen (2), liver (3), skin (2), kidney (4), brain (1)
Viable cells	Positive on the surface of HTLV-I+ cell lines HUT-102, MT-2; negative on HTLV-I− cell lines CEM and Jurkat
Competitive RIA with HTLV-I *env*-encoded synthetic peptides	Reacts with synthetic peptide SP-4A (amino acids 190–209)
Neutralization of HTLV-I	Syncytium formation—no inhibition VSV/HTLV-I plaque formation—no inhibition

map the region of gp46 that was recognized by 1C11. In competitive radioimmunoassay (RIA), 1C11 was inhibited from binding to gp46 in the presence of synthetic peptide SP-4A containing amino acids 190–209 of HTLV-I gp46. When tested for the ability either to neutralize the infectivity of HTLV-I in syncytium inhibition assay or to inhibit gp46-receptor interaction in an assay utilizing VSV/HTLV-I pseudotype particles (13), MoAb 1C11 showed no inhibitory effect. These data suggested that the epitope of gp46 identified by 1C11 between amino acids 190 and 209 was not a critical determinant for virus neutralization. The binding characteristics of MoAb 1C11 are summarized in Table 1.

TABLE 2. *HTLV-I envelope synthetic peptides used in the study*

Synthetic peptide	Amino acid number[a]	Amino acid sequence[b]
1	33–47	VSSYHSKPCNPAQPV
2	86–107	(C)PHWTKKPNRNGGGYYSASYSDP
3	176–189	(C)LNTEPSQLPPTAPP(Y)
4	129–149	SSPYWKFQHDVNFTQEVSRLN(C)
4A	190–209	(C)LLPHSNLDHILEPSIPWKSK(Y)
5	269–280	(Y)LPFNWTHCFDPQ(C)
6	296–312	(C)PPFSLSPVPTLGSRSRR
7	374–392	YAAQNRRGLDLLFWEQGGL(C)
8	400–415	CRFPNITNSHVPILQE
9	411–422	(C)PILQERPPLENR
10	462–480	CILRQLRHLPSRVRYPHYS
11	475–488	(C)RYPHYSLIKPESSL

[a] Initiation methonine = 1.
[b] From Seiki et al. (28).

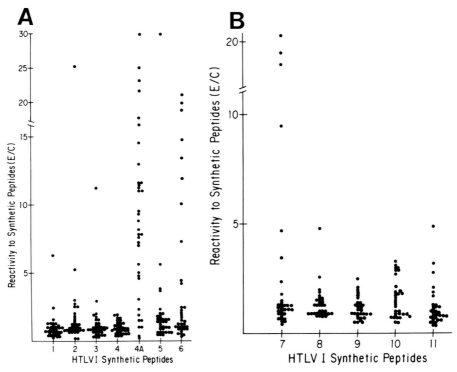

FIG. 4. HTLV-I+ patient antibody reactivity to epitopes on synthetic peptides (see Table 2) containing amino acid sequences of HTLV-I gp46 envelope glycoprotein. Synthetic peptides SP1-11 (25 μg/well) were bound to microtiter wells and used as antigen in radioimmunoassay to evaluate human antibody binding to epitopes on HTLV-I envelope. Patient antibodies reacted predominantly with synthetic peptides SP-4A and SP-6 containing amino acid sequences of gp46 (**A**) and with SP-7 from the N-terminus of p21 transmembrane protein (**B**). Data are expressed as the ratio (E/C) of mean cpm values obtained in RIA with HTLV-I+ patient (experimental, E) sera and normal human (control, C) sera. From Palker et al. (21) used with permission.

EPITOPE MAPPING WITH HTLV-I *env*-GENE–ENCODED SYNTHETIC PEPTIDES

To map immunogenic regions of HTLV-I envelope and to evaluate the topography of gp46 with epitope-specific antibodies, we synthesized peptides (Table 2) containing hydrophilic regions (29) of HTLV-I gp63 envelope glycoprotein. Peptides were either bound to microtiter wells and used as antigens in RIA or coupled to tetanus toxoid (19) and used to immunize rabbits. When synthetic peptides were tested in RIA for the ability to bind antibodies from HTLV-I+ individuals, synthetic peptide SP-4A (amino acids 190–209) bound antibodies from 75% of samples tested (Fig. 4). Peptides SP-6 (29% positive) from the C-terminus of gp46 and SP-7 (18% positive) from the N-terminus of gp21 transmembrane protein also bound antibodies in this assay. These data indicated that the C-terminal portion of HTLV-I gp46 contains epitopes that elicit major antibody responses in HTLV-I–infected persons. Next, anti-peptide antisera were evaluated for the ability to bind gp46 antigenic determinants in an epitope-specific manner and to neutralize HTLV-I in syncytium assay and VSV/HTLV-I pseudotype assay (Palker T, Haynes B, Clapham P, Weiss R, manuscript in preparation). Specificity of antibody binding relative to preimmune serum controls was determined in RIA, Western blot, and RIP assays. In RIA, antibodies raised to peptides 1–6 (Table 2) containing sequences of gp46 selectively bound to the immunizing synthetic peptide with signal-to-noise ratios ranging between 21 (SP-4) and 284 (SP-5). Antisera to peptides 1, 3, 4, 4A, and 6 bound to reduced HTLV-I gp46 and gp63 in Western blot assay and to gp46 in RIP assay. Antiserum to SP-2 precipitated ^{125}I-labeled gp46 but did not react with gp46/63 in Western blot assay. Thus, with the exception of antiserum raised to SP-5, all antisera raised to peptides SP-1–6 containing sequences from HTLV-I gp46 reacted with either denatured or native HTLV-I envelope glycoprotein in Western blot or RIP assay, respectively. In contrast, only one (anti-SP-11 antiserum) of five antisera raised against synthetic peptides SP7–11 bearing sequences of HTLV-I p21 transmembrane protein reacted with HTLV-I gp63 in Western blot assays. Also, in Western blot assays, antisera to peptides 7–9 recognized a recombinant HTLV-I p21 transmembrane protein produced in bacteria (17).

We next evaluated the ability of epitope-specific antibodies to neutralize HTLV-I either in syncytium inhibition assay or in an assay using VSV/HTLV-I pseudotype particles. None of the antisera raised to peptides SP1–11 could inhibit HTLV-I–induced syncytium formation, whereas 5 μg of purified IgG from sera of two HTLV-I+ subjects (but not normal human IgG) could block syncytium formation by >50% (21). However, when tested for the ability to inhibit plaque formation induced by VSV/HTLV-I pseudotype particles, anti-peptide antisera raised to peptides with sequences from N-terminal and central regions of HTLV-I gp46 could block pseudotype plaque formation by >80% (Weiss R, Clapham P, Haynes B, Palker T, manuscript in preparation). These data suggested that at least two sites on HTLV-I gp46 are implicated in virus neutralization.

MAPPING OF A CYTOTOXIC T-CELL EPITOPE ON HTLV-I gp46

HTLV-I *env*-gene–encoded synthetic peptides SP1–11 (Table 1) were tested for the ability to act as targets for MHC-restricted cytotoxic T-cell killing (Jacobson S, McFarlin D, Palker TJ, manuscript in preparation). A cytotoxic T-cell line and an Epstein-Barr virus-transformed B-cell line were both cloned from the same HTLV-I patient with TSP. B-cells were then preincubated with synthetic peptides SP1–11 for 6 hr, labeled with ^{51}Cr and used as autologous targets in cytotoxicity assays with the cytotoxic T (Tc) cell line. Up to 90% of B cells coated with synthetic peptide SP-4A were killed when incubated with the autologous Tc cells; none of the remaining peptides could act as targets when presented to Tc cells on the surface of B cells. In other studies, only SP-4A–coated B cells matched with the Tc cell line at the DR-2 locus could be killed, indicating that killing was MHC class II restricted and that the Tc cell line was CD4$^+$.

SUMMARY AND CONCLUSIONS

Synthetic peptides, anti-peptide antisera, and monoclonal antibodies have been valuable probes for the mapping of immunogenic and functionally important determinants of human retroviral envelope glycoproteins (19–21,27,30–36). A monoclonal antibody, 1C11, raised to affinity-purified HTLV-I envelope glycoproteins was selectively reactive with the surface and cytoplasm of HTLV-I–infected cell lines (21). Despite amino acid sequence homology between the envelope glycoproteins of HTLV-I and HTLV-II (37), 1C11 did not react with HTLV-II infected cells. The epitope of gp46 that bound 1C11 was mapped with synthetic peptides to a site between amino acids 190 and 209. Substantial amino acid sequence divergence between HTLV-I and -II envelope glycoproteins in this region (37) could account for the selective reactivity of 1C11 to HTLV-I gp46. Synthetic peptides containing hydrophilic amino acid sequences of HTLV-I envelope were useful in mapping the human anti-envelope antibody response to determinants within the C-terminal half of gp46. A major immunogenic region of gp46 between amino acids 190 and 209 was found to contain epitopes that bound antibodies from 75% of HTLV-I+ individuals and that, as well, were targets for cytotoxic T cell killing. Studies are underway to map further the T- and B-cell determinants within the region defined by peptide SP-4A.

Antisera raised to HTLV-I *env*-gene–encoded synthetic peptides reacted specifically with the immunizing peptide in RIA. Five of seven anti-peptide antisera raised to determinants of gp46 were found to react with either denatured or native gp46. Antisera to peptides with amino acid sequences from N-terminal and central regions of gp46 could inhibit plaque formation induced by VSV/HTLV-I pseudotype particles; in contrast, anti-peptide antisera had no inhibitory effect on HTLV-I–induced syncytium formation. In a similar manner, antisera raised to synthetic peptides containing amino acid sequences of bovine leukemia virus (BLV) gp51 neutralized VSV/BLV pseudotype particles but did not inhibit BLV-induced syncytium formation (38). It is unclear why anti-peptide antisera neutralized HTLV-I or BLV pseudotype particles but did not inhibit syncytial cell formation; it may be that the pseudotype assay is more sensitive than the syncytial cell assay for detecting an inhibition of receptor-envelope interaction.

A comprehensive understanding of the human immune response to HTLV-I envelope and of the epitopes that evoke virus-neutralizing antibody responses should facilitate the design of vaccines capable of eliciting protective immunity to HTLV-I in man. Testing of synthetic peptide inocula in primate or rabbit models of HTLV-I infection will undoubtedly provide valuable insights into the feasibility of this approach.

ACKNOWLEDGMENTS

The authors gratefully acknowledge Richard Scearce, Robert Streilein, and Mary Tanner for expert technical support; and Ms. Kim R. McClammy for expert secretarial assistance. This research was supported by NIH Grant CA40660 to TJP. Thomas J. Palker is a Scholar of the Leukemia Society of America.

REFERENCES

1. Poiesz BJ, Ruscetti FW, Gazdar AF, Bunn PA, Minna JD, Gallo RC. Detection and isolation of type-C retrovirus particles from fresh and cultured lymphocytes of a patient with cutaneous T-cell lymphoma. *Proc Natl Acad Sci USA* 1980;77:7415–7419.
2. Poiesz BJ, Ruscetti FW, Reitz MS, Kalyanaraman VS, Gallo, RC. Isolation of a new type-C retrovirus (HTLV) in primary uncultured cells of a patient with Sézary T-cell leukemia. *Nature* 1981;294:268–271.
3. Popovic M, Reitz MS, Sarngadharan MG, et al. The virus of Japanese adult T-cell leukemia is a member of the human T-cell leukemia virus group. *Nature* 1982;300:63–66.
4. Yoshida M, Miyoshi I, Hinuma Y. Isolation and characterization of retrovirus from cell lines of human adult T-cell leukemia and its implication in the disease. *Proc Natl Acad Sci USA* 1982;79: 2031–2035.
5. Haynes BF, Miller SE, Palker TJ, et al. Identification of human T-cell leukemia virus in a Japanese patient with adult T-cell leukemia and cutaneous lymphomatous vasculitis. *Proc Natl Acad Sci USA* 1983;80:2054–2058.
6. Kalyanaraman VS, Sarngadharan MG, Robert-Guroff M, et al. A new subtype of human T-cell leukemia virus (HTLV-II) associated with a T-cell variant of hairy cell leukemia. *Science* 1982;218: 571–573.
7. Gessain A, Francis H, Sonan T, et al. HTLV-I and tropical spastic paraparesis in Africa. *Lancet* 1986;2:698.
8. Osame M, Usuku K, Izumo S, et al. HTLV-I associated myelopathy, a new clinical entity. *Lancet* 1986;1:1031–1032.
9. Roman GC, Spencer PS, Schoenberg BS et al. Tropical spastic paraparesis: HTLV-I antibodies in patients from the Seychelles. *N Engl J Med* 1987;316:51.
10. Rosenblatt JD, Golde DW, Wachsman W et al. A second isolate of HTLV-II associated with atypical hairy-cell leukemia. *N Engl J Med* 1986;315:372–377.
11. Nagy K, Clapham P, Chengsong-Popov R, Weiss R. Human T-cell leukemia virus type 1: induction of syncytia and inhibition by patient's sera. *Int J Cancer* 1983;32:321–328.
12. Hoshimo H, Shinoyama M, Miwa M, Sugimura T. Detection of lymphocytes producing a human retrovirus associated with adult T-cell leukemia by syncytium induction assay. *Proc Natl Acad Sci USA* 1983;80:7337–7341.
13. Clapham P, Nagy K, Weiss RA. Pseudotypes of human T-cell leukemia virus types 1 and 2: neutralization by patient's sera. *Proc Natl Acad Sci USA* 1984;81:2886–2890.
14. Hoshino H, Clapham PR, Weiss RA, Miyoshi I, Yoshida M, Miwa M. Human T-cell leukemia virus type I: pseudotype neutralization of Japanese and American isolates with human and rabbit sera. *Int J Cancer* 1985;36:671–675.

15. Nakamura H, Hayami M, Ohta Y, et al. Protection of cynomolgus monkeys against infection by human T-cell leukemia virus type-I by immunization with viral *env* gene products produced in *Escherichia coli. Int J Cancer* 1987;40:403–407.

16. Kiyokawa T, Yoshikura H, Hattori S, Seiki M, Yoshida M. Envelope proteins of human T-cell leukemia virus. Expression in *Escherichia coli* and its application to studies of *env*-gene functions. *Proc Natl Acad Sci USA* 1984;81:6202–6206.

17. Samuel KP, Lautenberger JA, Jorcyk CL, Joseph S, Wong-Staal F, Papas TS. Diagnostic potential for human malignancies of bacterially produced HTLV-I envelope protein. *Science* 1984;226: 1094–1097.

18. Kanner SB, Cheng-Mayer C, Geffen RB, et al. Human retroviral *env* and *gag* polypeptides: serologic assays to measure infection. *J Immunol* 1986;137:674–678.

19. Palker TJ, Matthews TJ, Clark ME, et al. A conserved region at the COOH-terminus of human immunodeficiency virus gp120 envelope protein contains an immunodominant epitope. *Proc Natl Acad Sci USA* 1987;84:2479–2483.

20. Palker TJ, Clark ME, Langlois AJ, et al. Type-specific neutralization of the human immunodeficiency virus with antibodies to *env*-encoded synthetic peptides. *Proc Natl Acad Sci USA* 1988;85:1932–1936.

21. Palker TJ, Tanner ME, Scearce RM, Streilein RD, Clark ME, Haynes BF. Mapping of immunogenic regions of human T cell leukemia virus type I (HTLV-I) gp46 and gp21 envelope glycoproteins with *env*-encoded synthetic peptides and a monoclonal antibody to gp46. *J Immunol* 1989;142:971–978.

22. Lee TH, Coligan JE, Homma T, McLane MF, Tachibana N, Essex M. Human T-cell leukemia virus-associated membrane antigens: identity of the major antigens recognized after virus infection. *Proc Natl Acad Sci USA,* 1984;81:3856–3860.

23. Schneider J, Yamamoto N, Hinuma Y, Hunsmann G. Sera from adult T-cell leukemia patients react with envelope and core polypeptides of adult T-cell leukemia virus. *Virology* 1984;132:1–11.

24. Bolognesi DP, Montelaro RC, Frank H, Shäfer W. Assembly of type C oncornaviruses: a model. *Science* 1978;199:183–186.

25. Copeland TD, Tsai WP, Kim YD, Oroszlan S. Envelope proteins of human T-cell leukemia virus type I: characterization of antisera to synthetic peptides and identification of a natural epitope. *J Immunol* 1986;137:2945–2951.

26. Palker TJ, Clark ME, Sarngadharan MG, Matthews TJ. Purification of envelope glycoproteins of human T-cell lymphotropic virus type (HTLV-I) by affinity chromatography. *J Virol Methods* 1987;18:243–256.

27. Matsushita S, Robert-Guroff M, Trepel J, Cossman M, Mitsuya H, Broder S. Human monoclonal antibody directed against an envelope glycoprotein of human T-cell leukemia virus type I. *Proc Natl Acad Sci USA* 1986;83:2672–2676.

28. Seiki M, Hattori S, Hirayama Y, Yoshida M. Human adult T-cell leukemia virus: complete nucleotide sequence of the provirus genome integrated in leukemia cell DNA. *Proc Natl Acad Sci USA* 1983;80:3618–3623.

29. Kyte J, Doolittle RF. A simple method for displaying the hydrophilic character of a protein. *J Mol Biol* 1982;157:105–132.

30. Hattori S, Imagawa K, Shimizu F, Hashimura E, Seiki M, Yoshida M. Identification of envelope glycoprotein encoded by *env* gene of human T-cell leukemia virus. *Gann* 1983;74:790–793.

31. Hattori S, Kiyokama T, Imagawa K, et al. Identification of *gag* and *env* gene products of human T-cell leukemia virus (HTLV). *Virology* 1984;136:338–347.

32. Sugamura K, Fujii M, Ueda S, Hunuma Y. Identification of a glycoprotein, gp21, of adult T cell leukemia virus by monoclonal antibody. *J Immunol* 1984;132:3180–3184.

33. Newman MJ, Baker IT, Reitz MS, et al. Serological characterization of human T-cell leukemia (lymphotropic) virus, type I (HTLV-I) small envelope protein. *Virology* 1986;150:106–116.

34. Kinney Thomas E, Weber JN, McClure J, et al. Neutralizing monoclonal antibodies to the AIDS virus. *AIDS* 1988;2:25–29.

35. Rusche JR, Javaherian K, McDanal C, et al. Antibodies that inhibit fusion of human immunodeficiency virus-infected cells bind to a 24-amino acid sequence of the viral envelope, gp120. *Proc Natl Acad Sci USA* 1988;85:3198–3202.

36. Kennedy RC, Chanh TC, Allan JS, et al. Overview on the use of synthetic peptides in human immunodeficiency virus infection. In: Mizrahi A, ed. *Synthetic peptides in biotechnology.* New York: Alan R. Liss, 1988;149–172.

37. Sodroski J, Patarca R, Perkins D, et al. Sequence of the envelope glycoprotein gene of type II human T lymphotropic virus. *Science* 1984;225:421–424.
38. Portetelle D, Dandoy C, Burny A, et al. Synthetic peptides approach to identification of epitopes on bovine leukemia virus envelope glycoprotein gp51. *Virology* 1989;169:34–41.

DISCUSSION

A discussant wondered about peptides that are useful to discriminate HTLV-II sera and asked if the monoclonal of HTLV-I is specific to 4-A. Another discussant replied that that is correct and stated that there is at least one epitope on the 4A peptide that is HTLV-I specific, but there may be other epitopes that are shared by both the HTLV-I and HTLV-II envelope proteins. A participant inquired about the four antisera that show some neutralizing activity and wondered if the corresponding peptides could be used to absorb out the neutralizing activity in human sera. The previous discussant replied that those experiments were in progress. Another participant asked if this discussant had seen virus enhancement with polyclonal rabbit antisera to any of these peptides. The discussant had not looked for virus enhancement but stated that there were one or two antisera that gave a few more syncytia than other antisera, but it was variable.

Human Retrovirology: HTLV,
edited by William A. Blattner.
Raven Press, Ltd., New York © 1990.

Antibody Responses to Different Regions of HTLV-I gp61 in Asymptomatic Carriers, ATL Patients, and Individuals with TSP

Yi-ming A. Chen and Myron Essex

*Department of Cancer Biology, Harvard School of Public Health,
Boston, Massachusetts 02115*

INTRODUCTION

The human T-cell leukemia virus type-I (HTLV-I) has been shown to be etiologically associated with human adult T-cell leukemia and lymphoma (ATL) (1–3) and with a rare neurological disorder called tropical spastic paraparesis (TSP) and HTLV-I–associated myelopathy (HAM) (4,5). The primary protein product of the HTLV-I *env* gene, gp61, is cleaved to produce both the exterior (gp46) and the transmembrane (gp21) portions of the HTLV-I envelope protein. It is known that gp61 is essential in the early stages of viral infection and it is thought to be the most immunogenic protein of all the viral antigens (6). Several lines of evidence suggest that the HTLV-I *env* glycoprotein may be a critical viral antigen for targeting by the host immune system. First, it has been demonstrated that the HTLV-I *env* protein is the predominant viral antigen recognized by antibodies in the sera of HTLV-I–infected subjects (6). Second, the infectivity of pseudotypes of vesicular stomatitis virus bearing the HTLV-I *env* protein on their surfaces can be neutralized by antibodies specific to this protein (7). And finally, in other retroviral infections, including human immunodeficiency virus type 1 (HIV-1) infections, most neutralizing antibodies are directed against envelope proteins (8).

Previous studies on the HTLV-I *env* protein have revealed interesting observations about its antibody reactivity (9–13). Experiments have shown dramatic variations in the reactivity of different regions of this protein by testing seroreactivities to three synthetic peptides, each representing 11–15 amino acids of gp61 (9). Further experiments have studied a *Bam*HI-*Xho* I fragment of the HTLV-I *env* gene which was cloned into plasmid pkS400 which, on induction, expressed a recombinant protein containing the final four amino acids at the C-terminus of gp46 and two-thirds of the length of gp21 (12). This recombinant protein was able to detect 11 of 11 HTLV-I positive test sera. Nevertheless, information that allows a comprehensive understanding of the various rates of

447

human antibody reactivity to specific major regions of gp61 is not yet available. Such information might be useful for clinical diagnostic procedures, blood screening, and perhaps even vaccine development.

In this report we systematically compare the serologic reactivity of specific regions of the HTLV-I gp61 protein. We will first describe how we cloned different regions of the HTLV-I *env* gene into two plasmid vectors, which then expressed recombinant proteins (RPs) in *Escherichia coli.* Each RP represents a specific region of the HTLV-I gp61 protein, and when taken together, they span its entire length. We will then compare the antibody reactivity of three groups of HTLV-I–seropositive subjects to each RP: asymptomatic carriers, ATL patients, and individuals with TSP. Finally, we will discuss the clinical implications and the significance of this study.

RECOMBINANT HTLV-I ENVELOPE PROTEINS

Construction of Recombinant Plasmids Using Two Vectors

Our initial approach to studying the relative immunoreactivity of different regions of HTLV-I gp61 was to construct recombinant plasmids and then to induce them to express RP's containing different regions of the HTLV-I gp61 protein. In this study, the HTLV-I proviral DNA sequences reported by M. Seiki et al. (14) were used as the reference for the restriction enzyme sites for gene cloning and screening. As shown in Fig. 1, a 4.2-kb *sph* I-*sph* I fragment, containing both the HTLV-I *env* gene and *x* region, was excised from pMT2 and subcloned into plasmid p806. The plasmid p806, a derivative of pUC18, contains an *Eco*RI-*Nco* I fragment from plasmid pXVR (15). To ensure that the inserted *env* gene was in the same reading frame as the preceding *v-ras*^{H} sequence, the DNA sequence in the junction region of plasmid p*sph* I-*env* 1 was analyzed by the dideoxynucleotide sequencing method.

To generate plasmid pS3 containing the distal two-thirds of the HTLV-I *env* gene (nucleotides 5694–6665), the p*sph* I-*env* 1 DNA was digested with *Sal* I, which cuts at nucleotide 5694 of the *env* gene and at the Sal I sequence in the polylinker site of the plasmid. The sticky ends of the linearized plasmid DNA were repaired with Klenow, and then fused by blunt-end ligation (see Fig. 1b). Plasmid pD (Fig. 1c), which contains the gene corresponding to the HTLV-I transmembrane protein gp21 (nucleotides 6140–6665), was similarly generated by digesting the p*sph* I-*env* 1 DNA with *Kpn* I which cuts at nucleotide 6140 of the *env* gene and at the *Kpn* I sequence of the polylinker site. The plasmid DNA was then re-ligated.

To construct plasmids expressing the middle regions of gp46, we took plasmid pS3 DNA, which had been isolated from DNA methylase-free cells (strain JM110), digested it with either *Cla* I or *Xho* I, repaired the ends with Klenow, and then re-fused them by blunt-end ligation with an *Nhe* I nonsense codon

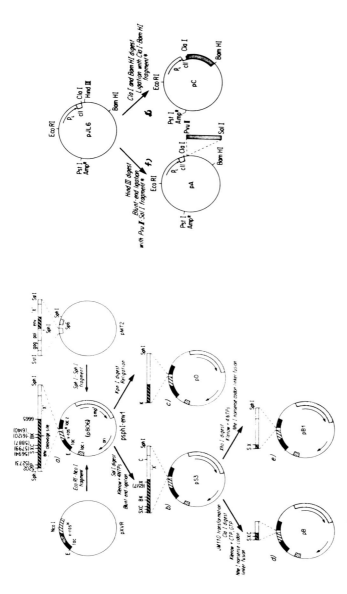

FIG. 1. Construction of plasmids containing specific regions of the HTLV-I *env* gene (B, *Bam*HI; C, *Cla* I; E, *Eco*RI; K, *Kpn* I; S, *Sal* I; X, *Xho* I, x, x region of the HTLV-I provirus; *, fragment from the HTLV-I provirus clone pMT2).

449

linker: d(CTAGCTAGCTAG) (see Fig. 1, d and e). These new plasmids, called pB and pB1, contained HTLV-I *env* gene nucleotide sequences 5694–5887 and 5694–5799, respectively.

Another expression vector used in this study was pJL6 (16). It contains a λ P_L promoter and the N-terminal fragment of the λ cII gene, including the ribosomal binding site and an ATG start codon. As shown in Fig. 1f, a 422-bp *Pvu* II-*Sal* I fragment (nucleotides 5273–5694), obtained from pMT2 and encoding the N-terminal half of the HTLV-I envelope exterior glycoprotein gp46, was fused in-frame by blunt-end ligation to the repaired *Hind*III cloning site of pJL6 (plasmid pA). The plasmid pC (Fig. 1g), which encodes the C-terminal region of gp46, was constructed by force-cloning a 234-bp *Cla* I-*Bam*HI fragment (nucleotides 5887–6120) from plasmid pMT2 into plasmid pJL6 which had been linearized by *Cla* I and *Bam*HI digestion.

Recombinant Protein Production and Identification

RPs containing specific regions (see Fig. 2) of HTLV-I gp61 were expressed either by vector p806 (pB, pB1, and pD) or by vector pJL6 (pA and pC) in two different bacterial culture systems. *E. coli* X-90 cells carrying plasmids derived from p806 were induced to express RPs by treatment with IPTG (isopropyl-b-D-thiogalactopyranoside). *E. coli* strain DC1148 was used to carry recombinant plasmids derived from pJL6. This strain is a lysogen that contains a mutant temperature-sensitive repressor encoded by the phage λ gene c1857, and can be induced to express RPs when exposed to a temperature shift from 32°C to 42°C.

The characteristics of the RPs expressed by these plasmids are summarized in Table 1. In general, the RPs were initially identified by staining sodium dodecyl sulfate–polyacrylamide gels containing both controls and the induced clones with Coomassie blue. Reactivity of the RPs was then confirmed by Western blot using high titer HTLV-I–positive sera for all of the RPs and goat anti-*v-ras*[H] antiserum for the fusion proteins expressed by pB, pB1, and pD. As shown in Fig. 3, the observed molecular weights of the RPs correspond to the predicted values.

In the case of RP-D, there are two immunoreactive bands, one with a molecular weight (Mol wt) of 32 Kd, the other with a Mol wt of 30 Kd, which react by Western blot with most of the HTLV-I positive sera (Fig. 3) and with a rabbit anti-gp21 serum.

We initially tried and failed to induce either plasmid p*sph* I-*env* 1 to express the full length, or plasmid pS3 to express two-thirds of the length, of gp61. Several possibilities may be considered to explain this phenomenon; e.g., message instability, codon usage preference, a toxic effect of the expressed proteins on the bacterial cells (17), and metabolic instability of the protein itself (18). Nevertheless, after shortening the gene fragments for cloning, all the RPs were expressed at levels high enough to allow direct detection by Coomassie blue staining.

PLOTSTRUCTURE OF HTLV-1 gp61

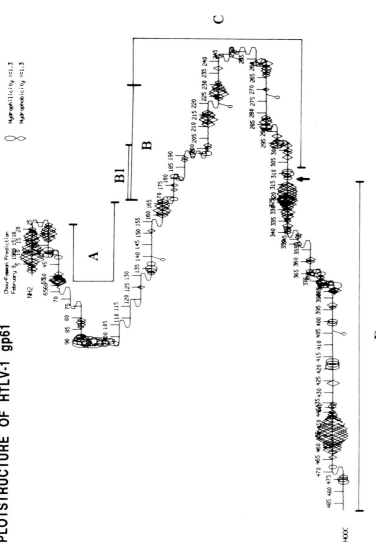

FIG. 2. Chou-Fasman prediction of the HTLV-I envelope protein gp61 derived from the sequence of λ ATK-1 and regions contained in the recombinant proteins A, B, B1, C, and D. The probability of there occurring α-helices, B-pleated sheets, random coils, and B-turn regions was evaluated with stringent conditions. The parameters for hydrophilicity, flexibility, and surface probability were averaged over 5 amino acid residues, with a limit of 0.7 for hydrophilicity, 1.040 for flexibility, and 5.0 for the probability of a given recombinant protein occurring on the cell surface. Arrows, cleavage sites; —, regions contained in individual recombinant proteins.

TABLE 1. *Characteristics of HTLV-I envelope recombinant proteins*

Plasmids	Expression vector	gp61 A.A. sequence	Regions of HTLV-I envelope gp61*	Observed Mol Wt (Kd)†	~Amt protein synthesized (%)‡
pA	pJL6	26–165	N-gp46	18	10
pB	p806	166–229	M	25	2
pB1	p806	166–201	M	23	5
pC	pJL6	229–308	C-gp46	15	8
pD	p806	313–488	gp21	32 and 30	5

* N: N-terminus; M: middle region; and C: C-terminus, all of gp46, the exterior domain of gp61. gp21: the transmembrane domain of gp61.

† In most instances, the observed Mol Wt deviates slightly from the predicted Mol Wt which is an intrinsic property of the input gene sequence.

‡ % values are rough estimates of the relative quantity of recombinant proteins expressed against a background of total bacterial proteins as observed in Coomassie-blue-stained SDS-PAGE.

The RPs, which have Mol wt ranging from 15 to 32 Kd, comprise ~98% of the HTLV-I *env* precursor complex, as shown in Fig. 2. Although both bacterial systems produced adequate quantities of the recombinant protein products, the pJL6 system produced them in greater quantities than did the p806 system. The reason for this difference is not clear, although the two systems themselves are not identical; protein expression with the pJL6 vector is induced by temperature shift, whereas that of the p806 system is induced by chemical treatment with IPTG.

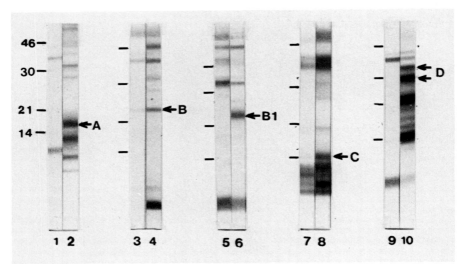

FIG. 3. Western blot data from the five recombinant proteins (arrow; A, B, B1, C, D) detected by HTLV-I–positive sera (lanes 2, 4, 6, 8, and 10). Control sera: lanes 1, 3, 5, 7, and 9. Mol wt markers are labeled at the left side of each of the reaction pairs.

However, data from the hydrophobicity plots and from the secondary structural analyses of the RPs suggest another explanation for this observation. It is clear that the N-terminal portions of RP-A and RP-C come from the phage λ cII gene product, while those of RP-B and RP-D are derived from the *v-ras* oncogene product. It is also clear that these gene products differ in their hydrophobicity such that the cII gene product is much more hydrophilic than the *v-ras* product. It is possible, therefore, that the hydrophilicity of the N-termini of RP-A and RP-C was able to exert a stabilizing effect on these RPs, allowing them to be synthesized in relative abundance.

Seroepidemiological Studies of the RPs

The Most Immunoreactive Region of gp61 is Located in the C-terminal Half of the HTLV-I gp46

Sera from two groups of HTLV-I positive subjects, healthy carriers and ATL patients, were used to study the prevalence of antibodies to each of the specific RPs. Initial screening for HTLV-I seropositivity was standardized using both the particle agglutination test and the immunofluorescence test (19). All samples were then confirmed as positive by both a Western blot assay using MJ cell lysates, and a radioimmunoprecipitation assay using ^{35}S-cysteine-labeled MT-2 antigens. The results from the seroreactivity studies, as shown in Table 2, show the following: 22.2% (10/45) of the carriers and 16.9% (13/77) of the ATL patients were reactive to RP-A; 80% (36/45) of the carriers and 95% (76/80) of the ATL patients were reactive to RP-B; 72.7% (32/44) of the carriers and 68.8% (53/77) of the ATL patients were reactive to RP-B1; and 91.1% (41/45) of the carriers and 93.8% (76/81) of the ATL patients were reactive to RP-C.

TABLE 2. *Rates of antibody positivity for various groups of HTLV-I–positive subjects to recombinant proteins of the HTLV-I envelope protein*

RPs	A	B	B1	C	D*	B + C*	A + B + C
HTLV-I carriers	10/45	36/45	32/44	41/45	25/45	43/45	44/45
	(22.2)§	(80.0)	(72.7)	(91.1)	(55.6)	(95.6)	(97.8)
ATL†	13/77	76/80	53/77	76/81	59/69	80/81	81/81
	(16.9)	(95.0)	(68.8)	(93.8)	(85.5)	(98.8)	(100)
Total	23/122	112/125	85/121	117/126	84/114	123/126	125/126
	(18.9)	(89.6)	(70.2)	(92.9)	(73.7)	(97.6)	(99.2)
TSP‡	4/15	12/15	12/15	13/15	12/15	13/15	13/15
	(26.7)	(80.0)	(80.0)	(86.7)	(80.0)	(86.7)	(86.7)

* In the three groups comprised of (1) HTLV-I carriers, (2) ATL patients and (3) their combined numbers, there was statistical significance between the rates of reactivity to [RP-D] and those to [RP-B + RP-C] ($p < .005$).

† ATL: Adult T-Cell Leukemia/Lymphoma patients.

‡ TSP: Tropical spastic paraparesis patients.

§ Numbers in parentheses indicate percentages.

Taken together, the seroreactivity rates for all the serum samples, as assayed by Western blot, are as follows: 18.9% (23/122) for RP-A, 89.6% (112/125) for RP-B, 70.2% (85/121) for RP-B1, and 92.9% (117/126) for RP-C. These results indicate that the C-terminal half of gp46 (RP-B plus RP-C) can detect 97.6% (123/126) of the HTLV-I–positive sera, a percentage which is significantly higher ($p < 0.005$) than that detected by the N-terminal half of gp46 (RP-A). RPs-A, B, and C, which together span the entire length of gp46 except the first five amino acids at the N-terminus and the last four amino acids at the C-terminus, can detect 99.2% (125/126) of the HTLV-I–positive subjects. More significantly, this same combination of RPs can detect every ATL patient in this study. In contrast, RP-D, which contains the transmembrane envelope protein gp21 minus the first amino acid at the N-terminus, has only a 73.7% (84/114) rate of seroreactivity. This difference in the rates of seroreactivity between gp46 (RP-A, plus -B, plus -C) and gp21 (RP-D) is statistically significant ($p < 0.005$). The difference in the rate of reactivity with RP-D is statistically significant ($p < 0.01$) when HTLV-I carriers (55.6%) and ATL patients (85.5%) are compared.

The results of the Western blot assay shown in Fig. 3 and the serological data summarized in Table 2 indicate that many antigenic determinants of the gp61 protein retain their antibody-binding specificities in the denatured form of the RPs. Indeed, all of the RPs, with the exception of RP-A, are able to react with over 70% of the HTLV-I–positive sera.

Low Seroreactivity to the N-terminal Half of gp46 in HTLV-I-infected People

RP-A, on the other hand, detects a much smaller percentage of the HTLV-I–positive sera, although it does contain many potentially reactive epitopes (see antigenic index of HTLV-I envelope protein in Fig. 4c). It is possible that many epitopes residing on the N-terminus of gp46 may be hindered by interactions between gp46 and other molecules such as gp21. Such a scenario was in fact considered in an analysis of the HIV-1 external glycoprotein (20). For this study, Kowalski et al. constructed a panel of HIV-1 envelope mutants and analyzed the phenotypic changes corresponding to each genetic mutation. Their results indicated that among the regions tested, those responsible for the association of the HIV-1 gp120 to the HIV-1 gp41 (analogous to the HTLV-I gp46 and gp21, respectively), are clustered at the N-terminal half of the gp120 molecule.

On the other hand, conformational epitopes may also play an important role in determining the relative immunogenicity of the N-terminal half of gp46. Thus, when the proteins are not in their natural configurations, they may not be recognizable by the same antibodies which react with the native protein. As shown in Fig. 4d, there are 10 cysteine residues in the N-terminal half of gp46, but only 4 in the C-terminal half. It is therefore possible that disulfide bonds between these cysteine residues might be especially significant to the tertiary structure of

HTLV-1 gp61

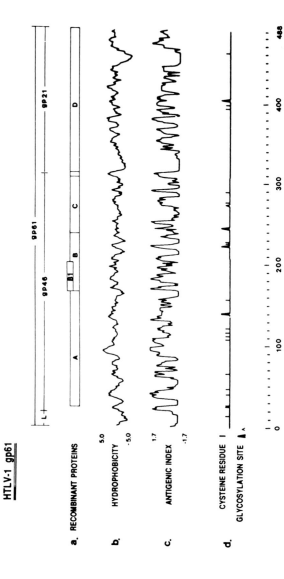

FIG. 4. Schematic diagram of the hydrophobicity plot, predicted antigenic index, and the positions of predicted cysteine residues and glycosylation sites of regions contained within specific recombinant proteins along the HTLV-I envelope protein (gp61). The primary envelope protein gp61 is proteolytically cleaved at the N-terminus [to remove a short leader (L) sequence] and within the gp61 precursor to generate the mature envelope glycoprotein comprised of the gp46 (exterior) and gp21 (transmembrane) proteins. The antigenic index was calculated by summing several weighted measures of secondary structure, including the hydrophobicity, surface probability, flexibility, Chou-Fasman structure, and the Robson-Garnier structure.

the gp46 protein at its N-terminus. Because all the RPs used for the Western blot assays were solubilized in either a 3M or an 8M urea/Tris-HCl buffer solution, their conformational epitopes were denatured and could not be studied by the assay.

A Middle Region of gp46 Contains Highly Immunogenic Epitopes

Nevertheless, when comparing the relative antibody reactivities of the various regions of HTLV-I gp61 by using denatured recombinant proteins in Western blot assays (see Table 2), we found that the C-terminal half of gp46 (RP-B plus RP-C) is the most immunogenic region of the protein. This region reacted with over 95% of the HTLV-I positive sera, whereas the N-terminal half of gp46 (RP-A) had a much lower seropositive rate. Perhaps more interesting still is the fact that RP-B1, which contains only 36 amino acids from the middle region of the HTLV-I gp46, can react with 70% of the HTLV-I positive sera. If one examines antibody reactivity as a function of relative polypeptide length, RP-B1 may be considered a highly reactive domain in the HTLV-I envelope protein. As illustrated in Fig. 4c, RP-B1 does contain three major antigenic peaks in its polypeptide sequence.

Seroreactivity to the HTLV-I Transmembrane Protein gp21

Seropositivity rates for RP-D (HTLV-I gp21) were significantly lower than those for RP-B or RP-C (C-terminus of gp46) in the HTLV-I carrier and ATL patient groups, and in the combined numbers of the two groups ($p < 0.005$). An earlier report described a higher degree of reactivity for a recombinant polypeptide representing HTLV-I gp21 (12), but that study was based on the analysis of a relatively small panel of HTLV-I–positive serum samples.

Of further interest to the present study was the difference in rates of RP-D seroreactivity between HTLV-I carriers (55.6%) and ATL patients (85%) ($p < 0.01$). Whether this difference bears any direct relationship to disease development could not be determined from the current studies.

Seroreactivity of TSP Patients to the RPs

In addition to samples from the above two groups of HTLV-I–positive subjects, serum samples from TSP patients were also tested for reactivity to the same panel of RPs. RPs B, B1, C, and D react with 80–86.7% of the TSP serum samples; reactivity with RP-A, however, was only 26.7%. Two of 15 (13.3%) TSP patients did not react with any of the RPs. This nonreactivity rate is significantly higher than that shown by both the HTLV-I carriers and the ATL patients ($p < 0.030$).

As controls, 50 HTLV-I seronegative Japanese sera from the same HTLV-I–endemic region, 3 HIV-1 positive sera, and 4 sera from systemic lupus erythematosis patients were tested for their reactivities to the various RPs. None of the control sera reacted with any of the HTLV-I *env* RPs.

Patterns of Seroreactivity to Different Regions of the HTLV-I gp61 in Infected People

Table 3 provides an analysis of seroreactivity patterns found in sera from HTLV-I carriers and ATL patients. The regions of HTLV-I gp61 as expressed in *E. coli* are defined as RP-A for the N-terminal half of gp46 (N-gp46), RP-B plus RP-C for the C-terminal half of gp46 (C-gp46), and RP-D for the gp21 transmembrane protein. The most common serological pattern, Pattern II, occurs such that there is a negative reaction to N-gp46 and a positive reaction to both C-gp46 and gp21. There are, however, 24.2%–28.9% of HTLV-I–positive sera that have antibodies to C-gp46 only (Pattern I). Eleven of 78 (14.1%) of the HTLV-I–positive serum samples had antibodies to all regions of gp61 represented by the RPs (Pattern IV). Surprisingly, two cases (one an ATL patient, and the other a healthy carrier) had antibodies only to N-gp46, which, according to the results of this study, is the least reactive region of the entire HTLV-I envelope protein.

The serum of one HTLV-I carrier was also found to show no antibody reactivity to any of the RPs tested. However, both Western blot assay and radioimmunoprecipitation showed that this patient's serum had antibodies to HTLV-I p19 and p24 *gag* gene proteins (group-specific antigens). It is possible that this person was actually infected with an HTLV-I–related virus or that antibodies to gp46, gp61, and/or the RPs may be detected by techniques other than Western blotting. It may even be possible to observe different seropositivity rates when the same RPs are tested by different immunoassay procedures.

TABLE 3. *Patterns of seroreactivity to different regions of the HTLV-I envelope protein in HTLV-I carriers and ATL patients*

Pattern	RP-A (N-gp46)	RP-[B + C] (C-gp46)	RP-D (gp21)	ATL	HTLV-I carriers	Total
I	−	+	−	8 (24.2%)	13 (28.9%)	21 (26.9%)
II	−	+	+	20 (60.6%)	21 (46.7%)	41 (52.6%)
III	+	+	−	0	2 (4.4%)	2 (2.6%)
IV	+	+	+	4 (12.1%)	7 (15.6%)	11 (14.1%)
V	+	−	−	1 (3.0%)	1 (2.2%)	2 (2.6%)
VI	−	−	−	0	1 (2.2%)	1 (1.3%)
			Total	33	45	78

No Correlation Between the anti-tax-1 Antibody and the Patterns of Seroreactivity to the HTLV-I gp61 Protein

In further analysis, as shown in Table 4, we used a recombinant protein containing the complete p42 HTLV-I *tax* protein to study the correlation between the pattern of seroreactivity to the HTLV-I gp61 and the status of antibody response to the *tax*-1 protein. As shown in Table 3, 19 of 45 (42.2%) HTLV-I carriers had antibodies to RP-*tax*-1, and the cases were distributed evenly among the HTLV-I envelope reactivity patterns. This suggests that there is no correlation between anti-p42-*tax* antibodies and any pattern of seroreactivity to the HTLV-I envelope protein.

In conclusion, we constructed five recombinant plasmids and induced them to express recombinant proteins spanning close to the entire length of the HTLV-I envelope protein gp61. After partially purifying these RPs, we used them as the antigens for Western blot assays to conduct a cross-sectional epidemiological study on two groups of HTLV-I–positive subjects—healthy HTLV-I carriers and ATL patients. Our results show that a region defined by RP-B1 (amino acid sequence 166–201) and located in the center of the HTLV-I envelope gp46 has a relatively high immunogenicity in terms of both its length and its rate of reactivity with seropositive sera. Furthermore, the C-terminal half of gp46 is much more immunogenic than is the full length of gp21, whereas the N-terminal half of gp46 is the least immunogenic portion of the entire gp61 molecule. This final observation may be related to the hindering effects of the gp46-gp21 association.

Because RP-A, B, and C, together spanning the length of gp46, can detect over 99% of the HTLV-I–positive subjects, a recombinant protein, or a cocktail containing various proteins representing this region, may be a logical candidate for use in blood screening and epidemiological studies. Such an antigen preparation would probably provide greater specificity than current diagnostic assays based on disrupted whole-virus and/or infected cell antigen preparations. Furthermore, because the rate of in vitro production for HTLV-I is quite low, the use of bacterially expressed RPs as antigen may in fact provide a more economical

TABLE 4. *Association of anti-tax-1 antibody to different patterns of seroreactivity to HTLV-I gp61*

	Antibody reactivity to			HTLV-I carriers	Anti-*tax*-1 antibody
Pattern	N-gp46	C-gp46	gp21		
I	−	+	−	13	7/13 (53.8%)
II	−	+	+	21	9/21 (42.9%)
III	+	+	−	2	0/2 (0%)
IV	+	+	+	7	3/7 (42.9%)
V	+	−	−	1	0/1 (0%)
VI	−	−		1	0/1 (0%)
			Total	45	19/45 (42.2%)

and abundant source of well-defined envelope proteins for use in the assay systems currently in use.

ACKNOWLEDGMENTS

This work was supported in part by grants 1 ROI CA 38450 from the NIH, I-P50-HL-33774 from the Center for Blood Research, and the Outstanding Investigator Award CA39805 from the NCI.

The authors thank S. Finkelstein for manuscript preparation; G. W. Smythers of NCI Advanced Scientific Computing Laboratory for computer analysis; A. Okayama for Anti-*tax*-1 antibody analysis; and K. Samuel, T. S. Papas, N. Tachibana, I. Miyoshi, T. H. Lee, M. F. McLane, X. F. Yu, Z. Matsuda, M. Matsuda, M. Chen, B. Du, W.-J. Syu, and G. C. DuBois for technical assistance and helpful discussions.

REFERENCES

1. Hinuma Y, Nagata K, Hanaoka M, et al. Adult T-cell leukemia: antigen in an ATL cell line and detection of antibodies to the antigen in human sera. *Proc Natl Acad Sci USA* 1981;78:6476–6480.
2. Poiesz BJ, Ruscetti FW, Gazdar AF, Bunn PA, Minna JD, Gallo RC. Detection and isolation of type C retrovirus particles from fresh and cultured lymphocytes of a patient with cutaneous T-cell lymphoma. *Proc Natl Acad Sci USA* 1980;77:7415–7419.
3. Yoshida M, Seiki M, Yamaguchi K, Takatsuki K. Monoclonal integration of human T-cell leukemia provirus in all primary tumors of adult T-cell leukemia suggests causative role of human T-cell leukemia virus in the disease. *Proc Natl Acad Sci USA* 1984;81:2534–2537.
4. Jacobson S, Raine CS, Mingiolio ES, McFarlin DE. Isolation of an HTLV-I–like retrovirus from patients with tropical spastic paraparesis. *Nature* 1988;331:540–543.
5. Osame M, Usuku K, Izumo S, et al. HTLV-I associated myelopathy, a new clinical entity. *Lancet* 1986;1:1031–1032.
6. Lee TH, Coligan JE, Homma T, McLane MF, Tachibana N, Essex M. Human T-cell leukemia virus associated membrane antigens (HTLV-MA): identity of the major antigens recognized following virus infection. *Proc Natl Acad Sci USA* 1984;81:3856–3860.
7. Clapham P, Nagy K, Weiss RA. Pseudotypes of human T-cell leukemia virus type 1 and 2: neutralization by patients' sera. *Proc Natl Acad Sci USA* 1984;81:2886–2889.
8. Siliciano RF, Lawton T, Knall C, et al. Analysis of host-virus interactions in AIDS with anti-gp120 T cell clones: effect of HIV sequence variation and a mechanism for $CD4^+$ cell depletion. *Cell* 1988;54:561–575.
9. Copeland TD, Tsai W, Kim YD, Oroszlan S. Envelope proteins of human T-cell leukemia virus type 1: characterization by antisera to synthetic peptides and identification of a natural epitope. *J Immunol* 1986;137:2945–2951.
10. Eiden M, Newman M, Fisher AG, Mann DL, Howley PM, Reitz MS. Type 1 human T-cell leukemia virus small envelope protein expressed in mouse cells by using a bovine papilloma virus–derived shuttle vector. *Mol Cell Biol* 1985;5:3320–3324.
11. Kiyokawa T, Yoshikura H, Hattori S, Seiki M, Yoshida M. Envelope proteins of human T-cell leukemia virus: expression in *Escherichia coli* and its application to studies of *env* gene functions. *Proc Natl Acad Sci USA* 1984;81:6202–6206.
12. Samuel KP, Lautenberger JA, Jorcyk CL, Josephs S, Wong-Staal F, Papas TS. Diagnostic potential for human malignancies of bacterially produced HTLV-I envelope protein. *Science* 1984;226:1094–1097.

13. Shida H, Tochikura T, Sato T, et al. Effect of the recombinant vaccinia viruses that express HTLV-I envelope gene on HTLV-I infection. *EMBO J* 1987;6:3379–3384.
14. Seiki M, Hattori S, Hirayama Y, Yoshida M. Human adult T-cell leukemia virus: complete nucleotide sequence of the provirus genome integrated in leukemia cell DNA. *Proc Natl Acad Sci USA* 1983;80:3618–3622.
15. Feig LA, Pan BT, Roberts TM, Cooper GM. Isolation of *ras* GTP-binding mutants using an in situ colony-binding assay. *Proc Natl Acad Sci USA* 1986;83:4607–4611.
16. Lautenberger JA, Seth A, Jorcyk C, Papas TS. Useful modifications of *Escherichia coli* expression plasmid pJL6. *Gene Anal Technol* 1984;42:49–57.
17. Amann E, Broker M, Wurm F. Expression of herpes simplex virus type 1 glycoprotein C antigens in *Escherichia coli. Gene* 1984;32:203–215.
18. Bachmair A, Finley D, Varshavsky A. In vitro half life of a protein is a function of its amino-terminal residue. *Science* 1986;234:179–186.
19. Ishizaki J, Okayama A, Tacibana N, et al. Comparative diagnostic assay results for detecting antibody to HTLV-I in Japanese blood donor samples: higher positive rates by particle agglutination assay. *J AIDS* 1988;1:340–345.
20. Kowalski M, Potz J, Basiripour L, et al. Functional regions of the envelope glycoprotein of human immunodeficiency virus type 1. *Science* 1987;237:1351–1355.

Human Retrovirology: HTLV,
edited by William A. Blattner.
Raven Press, Ltd., New York © 1990.

Epitope Profiles of HTLV-I *env* and *gag* Gene-Encoded Proteins

Peter Horal, Stig Jeansson, *Lars Rymo,
Bo Svennerholm, and Anders Vahlne

*Departments of Clinical Virology and *Medical Biochemistry,
University of Göteborg, Sweden*

INTRODUCTION

After infection with the human T-cell leukemia virus type I (HTLV-I), the humoral immune response to the virus structural proteins encoded by the *env* and *gag* genes may readily be demonstrated in Western blot or radioimmunoprecipitation assays (1–5). The antigenic determinants on these proteins may either be linear sequences of the amino acid backbone, so-called continuous epitopes, or be discontinuous, in which the amino acids that form the epitope are brought together by the three-dimensional folding of the protein. We will here present the locations on the protein amino acid sequences for continuous determinants of the *env*- and *gag*-encoded products leading to a humoral response in man after infection.

SOME CONSIDERATIONS ON THE SIZES OF PEPTIDES EMPLOYED FOR IDENTIFICATION OF IMMUNOREACTIVE DOMAINS OF PROTEINS

With the use of small synthetic peptides (usually 6–25 amino acids of length) as antigens in antibody assays, only antibodies to linear or continuous epitopes will be detected. Such epitopes may be found in hydrophilic sequences, probably corresponding to surface structures of the protein. However, because proteins are processed in the antigen-presenting cells, antibodies might be induced also to amino acid sequences not located at the surface of the native protein. Although antibodies to such "internal" epitopes probably are of little value, if any, in the combat of the infection, they might prove useful for diagnostic purposes.

Recombinant polypeptides, usually 100 amino acids long, or more, will undertake a tertiary structure, allowing antibodies also to conformational (discontinuous) epitopes to be detected. However, linear internal epitopes of diagnostic value might be hidden within the molecule and inaccessible for antibody binding.

EPITOPES OF HTLV-I *env*-GENE ENCODED PRODUCTS

Glycoprotein 46

By using a recombinant polypeptide (6) as well as synthetic peptides (7,8) as antigens in antibody assays, the C-terminal half of gp46 has been shown to contain domains that are immunogenic in man. In Fig. 1 the location of tested synthetic peptides on the gp46/gp21 transcript is depicted. Of the three peptides reported by Copeland et al. (7), only SP70 was found to react with HTLV-I antibody positive sera. Of the gp46 peptides reported on by Palker et al. (8), peptides 4A and 6 were predominantly reactive.

We have synthesized 35 different 18-to-26-amino-acids-long synthetic peptides, overlapping one another by five amino acids or more and covering the entire *env*-gene transcript (21 gp46 peptides and 14 gp21 peptides) according to the predicted amino acid sequence (9). The peptides were bound to the wells of microtiter plates and screened for reactivity with antibodies in six sera from HTLV-I–positive human subjects in an enzyme-linked immunosorbent assay (EIA). The 11 peptides (7 gp46 peptides and 4 gp21 peptides) that gave a positive reaction with 1 or more of the 6 HTLV-I–positive sera were then further tested with an additional 65 HTLV-I–positive sera. The location on the *env* glycoprotein amino acid sequence of these 11 peptides, as well as the percentage of the 71 sera that reacted with respective peptide, is shown in Fig. 1. As shown, the carboxyl terminal half of gp46 is highly immunogenic, containing at least three different linear epitopes. The three immunogenic regions are located between amino acids 176 and 212 (defined by two peptides, 176–199 and 190–212), 223 and 261 (defined by two peptides, 223–243 and 239–261), and 273 and 313 (defined by two peptides, 273–295 and 291–313), respectively (the initiation methionine = 1).

These results corroborate the results by Copeland et al. (7) and Palker et al. (8), identifying the C-terminus of gp46 (Fig. 1, peptides SP70 and 6, respectively) to be highly immunogenic, as well as the region around amino acid 200 (peptide 4A of Palker et al.). We have identified one additional region (amino acids 223–261) that is recognized by antibodies from HTLV-I–positive individuals. Only one peptide from the N-terminal half of gp46 (amino acids 89–110) gave a positive reaction with HTLV-I–positive sera. This peptide overlaps peptide 2 (amino acids 86–107) of Palker et al. (8).

The C-terminal half of gp46 contains three potential N-glycosylation sites, two of which are predicted to be glycosylated (10). Interestingly, peptides covering these latter two sites were not reactive with sera from HTLV-I positive human subjects, whereas a peptide (amino acids 239–261) covering the third

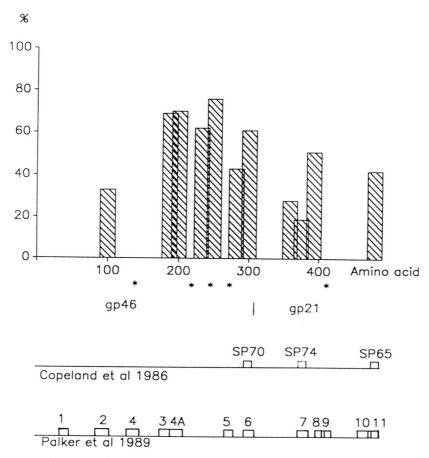

FIG. 1. HTLV-I *env* peptide reactivity. The peptides assayed were synthesized on an Applied Biosystems 430A. Removal of the peptides from the solid phase support and of protective side chains was accomplished by use of hydrogen fluoride. The peptides were then coupled to bovine serum albumin (BSA) via SPDP (Pharmacia, Sweden). To the original peptide sequence a C-terminal cysteine had been added to facilitate coupling. The peptide-BSA complex was subsequently coated to microtiter plates (NUNC, Denmark). The sera evaluated for peptide reactivity were assayed at a dilution 1/50 in a standard indirect EIA. Sera that gave an absorbance value greater than or equal to mean of five negative controls + 6 SD were considered positively reactive to the peptide. The relative position of the peptides along the amino acid sequence of the glycoproteins, as well as the percentage of HTLV-I samples (*n* = 71) being reactive to respective peptide, is depicted. Only the seven gp46 peptides and four reactive gp21 peptides that bound antibodies in HTLV-I positive sera are shown. Potential glycosylation sites are marked (∗). The location of the envelope peptides previously described by Copeland et al. (1986) and Palker et al. (1989) are also depicted.

Asn-X-Ser/Thr site gave a positive signal with 53 of the 71 HTLV-I positive sera.

Glycoprotein 21

Palker et al. (8) recently reported that a peptide corresponding to amino acids 374 to 392 (numbering gp46/gp21 in consecutive sequence) of gp21 (peptide 7 in Fig. 1) reacted with 18% of HTLV-I positive human sera. With a few such sera they found some antibody reactivity also directed to peptides 8, 10, and 11 (8). Copeland et al. (7) could not find any antibody reactivity to peptides SP74 (amino acids 373–385) or SP65, corresponding to the 12 C-terminal amino acids of gp21 (Fig. 1). We found two regions in gp21 containing epitopes being immunogenic in man (Fig. 1): one region in the N-terminal half of gp21 corresponding to amino acids 345–404 as defined by three peptides (amino acids 345–367, 363–385, and 381–404) and another region in the C-terminus as defined by a peptide corresponding to amino acids 466–488 (Fig. 1). A peptide (amino acids 400–422) covering the only potential N-glycosylation site in gp21 at the position 404–406, was not reactive to any of the HTLV-I–positive human sera tested. This peptide corresponds exactly to the region covered by peptides 8 and 9 of Palker et al. (8).

EPITOPES OF HTLV-I *gag*-GENE ENCODED PRODUCTS

For mapping epitopes of the *gag*-gene encoded transcripts, we used a different strategy from the one above. Thus, hexamer peptides overlapping one another by five amino acids and covering the entire predicted *gag*-gene transcript were synthesized on polyethylene rods as described by Geysen et al. (11). An EIA was then performed with the peptides deprotected but still bound to the rods. Two human sera reacting strongly to all three HTLV-I *gag* proteins (p19, p24, and p15) in Western blot assay were pooled and tested in an EIA at a dilution of 1/50, i.e., the final dilution for each serum was 1/100. The absorbance values of each peptide are shown in Fig. 2. Guided by these results, five peptides (17–21 amino acids in length) were synthesized and coated onto microtiter plates, as was done with the *env*-gene encoded peptides above. When tested to HTLV-I–antibody-positive human sera, one peptide corresponding to the C-terminal 19 amino acids of p19 reacted with 20 of 21 sera, confirming the results by Palker et al. (12) that the C-terminus of p19 is highly immunogenic in man. The four others, including one (amino acids 206–226) encompassing the sequence of the most reactive hexamer peptide in Fig. 2, did not react with any of the 21 HTLV-I–positive human sera tested.

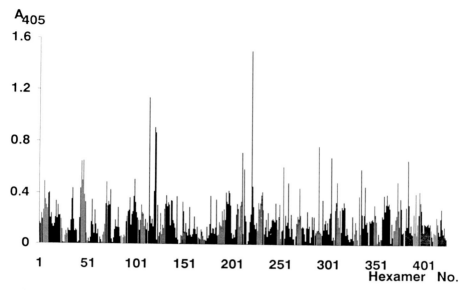

FIG. 2. Epitope mapping of HTLV-I *gag* gene–encoded proteins. Hexamer peptides corresponding to the entire HTLV-I *gag* (p24, p19, and p15) encoded transcript were synthesized on solid phase rods, as described by Geysen et al. (11), the peptide frame being shifted one amino acid C-terminally per peptide (one copy of each individual peptide). Reagents and synthesis protocol were according the manufacturer (Cambridge Research Biochemicals, UK). The peptide-coated rods were then assayed in a standard EIA with two HTLV-I pooled HTLV-I positive sera. Each serum had a final concentration of 1/100. The numbers on the X-axis represent amino acid number of the first (N-terminal) amino acid in respective hexapeptide. Y-axis shows the corresponding optical density reading for each peptide obtained with the two pooled HTLV-I–positive sera.

CONCLUSIONS

The *env*-gene of HTLV-I encodes for a 54 kilodalton (kD) polypeptide that subsequently is glycosylated and cleaved into two structural proteins, one surface glycoprotein of an apparent molecular weight of 46 kD (gp46), and a transmembrane glycoprotein having an apparent molecular weight of 21 kD (gp21) (7). Data collected using as antigens for detection of antibodies in HTLV-I–infected human sera either a recombinant polypeptide (6) or synthetic peptides (7,8, present study) from the gp46-encoded sequence indicate that the C-terminal half of the HTLV-I surface glycoprotein is highly immunogenic in man, and entails at least three strong linear epitopes [present study], situated N-terminal and C-terminal, respectively, of the two Asn-X-Ser/Thr sites predicted to be glycosylated (10) as well as in the hydrophobic region between these two glycosylation sites. A less predominant linear epitope is also found in the N-terminal half of the protein, around amino acid 100 (8, present study). There are two immunogenic domains containing continuous epitopes in the transmembrane gp21, one in the N-terminal half of the protein (12, present study) and one at the C-terminus [present study]. The *gag*-gene of HTLV-I encodes three proteins of 19, 24, and 15 kD, respectively (10). With the epitope-mapping technique described by Geysen (11)—i.e., using as antigens hexamer peptides synthesized on rods and covering the entire *gag*-encoded amino acid sequence and differing by only one amino acid—a number of potential humoral epitopes were obtained with two pooled HTLV-I positive human sera. However, when the most promising sequences were incorporated in larger (17–21 amino acids) synthetic peptides and tested against a collection of human sera from HTLV-I–infected subjects, we were only able to demonstrate one strong epitope at the C-terminus of p19. This linear epitope, also described by Palker et al. (12), is highly immunogenic in man and will give a humoral immune response in >90% of HTLV-I–infected humans (12, present study). One obvious reason for our relative failure to pick strong *gag*-gene encoded epitopes using the Geysen technique could be that the two sera used were not representative, i.e., they did not contain antibodies directed to the major epitopes of p24 and p15. However, when testing 19-to-25-amino-acids-long overlapping synthetic peptides covering the entire *gag*-gene encoded sequence of human immunodeficiency virus type 1 (HIV-1) with a collection of human sera containing strong antibody reactivities to the core proteins of HIV-1, as demonstrated in Western blot assay, <50% of the sera were reactive to any of the peptides (our unpublished observations). Thus, it might be that most of the epitopes of retrovirus *gag*-encoded proteins are discontinuous, i.e., dependent on the conformational structure of the protein.

ACKNOWLEDGMENTS

The skillful technical assistance of Ms. Susanne Wennerström, Ms. Annika Hallgren, Ms. Tua Karlsson, and Ms. Eva Lundin is gratefully acknowledged. This research was supported financially by Syntello AB, Göteborg, Sweden, and Virovahl SA, Zug, Switzerland.

REFERENCES

1. Schneider J, Yamamoto N, Hinuma Y, Hunsmann G. Sera from adult T-cell leukemia patients react with envelope and core polypeptides of adult T-cell leukemia virus. *Virology* 1984;132:1–11.
2. Gessain A, Vernant JC, Maurs L, et al. Antibodies to human T-lymphotropic virus type-I in patients with tropical spastic paraparesis. *Lancet* 1985;2:407–410.
3. Asher DM, Goudsmit J, Pomeroy KL, et al. Antibodies to HTLV-I in populations of the Southwestern Pacific. *J Med Virol* 1988;26:339–351.
4. Dalgleish A, Richardson J, Matutes E, et al. HTLV-I infection in tropical spastic paraparesis: lymphocyte culture and serologic response. *AIDS Res Hum Retroviruses* 1988;4:475–485.
5. Williams AE, Fang CT, Slamon DJ, et al. Seroprevalence and epidemiological correlates of HTLV-I infection in U.S. blood donors. *Science* 1988;240:643–646.
6. Samuel KP, Lautenberger JA, Jorcyk CL, Joseph S, Wong-Staal F, Papas TS. Diagnostic potential for human malignancies of bacterially produced HTLV-I envelope protein. *Science* 1984;226:194–197.

7. Copeland TD, Tsai W-P, Kim YD, Oroszlan S. Envelope proteins of human T cell leukemia virus type I: characterization by antisera to synthetic peptides and identification of a natural epitope. *J Immunol* 1986;137:2945–2951.
8. Palker TJ, Tanner ME, Scearce RM, Streilein RD, Clark ME, Haynes BF. Mapping of immunogenic regions of human T cell leukemia virus type I (HTLV-I) gp46 and gp21 envelope glycoproteins with *env*-encoded synthetic peptides and a monoclonal antibody to gp46. *J Immunol* 1989;142:971–978.
9. Seiki M, Hattori S, Hirayama Y, Yoshida M. Human adult T-cell leukemia virus: complete nucleotide sequence of the provirus genome integrated in leukemia cell DNA. *Proc Natl Acad Sci USA* 1983;80:3618–3622.
10. Oroszland S, Copeland TD. Primary structure and processing of *gag* and *env* gene products of human T-cell leukemia viruses HTLV-I$_{cr}$ and HTLV-I$_{ATK}$. In: Vogt PK, ed. *Current topics in microbiology and immunology.* Berlin: Springer-Verlag, 1986:221–233.
11. Geysen HM, Meloen RH, Barteling SJ. Use of peptide synthesis to probe viral antigens for epitopes to a resolution of single amino acid. *Proc Natl Acad Sci USA* 1984;81:3998–4002.
12. Palker TJ, Scearce RM, Copeland TD, Oroszlan S, Haynes BF. C-terminal region of human T cell lymphotropic virus type I (HTLV-I) p core protein is immunogenic in humans and contains and HTLV$_I$-specific epitope. *J Immunol* 1986;136:2393–2397.

Human Retrovirology: HTLV,
edited by William A. Blattner.
Raven Press, Ltd., New York 1990.

The American T-Cell Leukemia/ Lymphoma Registry (ATLR): An Update

Paul H. Levine

Viral Epidemiology Section, Environmental Epidemiology Branch, Division of Cancer Etiology, National Cancer Institute, National Institutes of Health, Bethesda, Maryland 20892

INTRODUCTION

The American T-cell Leukemia/Lymphoma Registry (ATLR) was established in 1988 (1) with several purposes intended:

1. To identify sufficient cases to allow clinical, pathologic, and laboratory evaluation to provide an improved case definition and a clearer picture of the disease in the United States;
2. Assuming most cases of adult T-cell leukemia/lymphoma (ATL) were human T-cell lymphotropic virus (HTLV-I)–associated, to identify geographic areas in the United States where the virus is in the highest concentration;
3. To evaluate family members and provide counseling as to the prevention of virus spread;
4. To identify possible risk factors for ATL in HTLV-I–infected individuals; and
5. To assist in identifying patients for active therapeutic trials.

In our first report (1) we evaluated 74 cases either referred to the National Cancer Institute (NCI) for evaluation (71 cases) or reported in the literature (3 cases). We carefully identified each individual by name to avoid duplication because patients were frequently reported in more than one series. Noting that the criteria for diagnosis of ATL varied widely, our next step was to develop a set of criteria that we hoped could standardize case ascertainment and allow a series more comparable with those collected in endemic areas. We initially developed an algorithm (Table 1) that emphasized the clinical and laboratory features distinguishing ATL from other non-Hodgkin's lymphomas and that utilized the experience of the Registry pathologist, Dr. Elaine Jaffe.

As of March 1, 1989, the ATLR has identified 118 cases diagnosed in the Western Hemisphere outside of the Caribbean Islands, 107 of them diagnosed in the United States. Eighty-eight of the Western Hemisphere cases were referred to the National Cancer Institute for evaluation and 30 were identified initially

TABLE 1. *Proposed ATLL criteria*

Definition of ATLL
Clinical criteria
Hypercalcemia = 1 pt.
Skin lesions = 1 pt.
Leukemic phase = 1 pt.
Pathologic criteria
Characteristic ATLL = 2 pts.
Compatible with ATLL = 1 pt.
Inconsistent with ATLL = 0 pts.
Laboratory criteria
T-cell lymphoma = 1 pt.
HTLV-I antibody = $\frac{1}{2}$ pt.
Titer in ATLL range = $\frac{1}{2}$ pt.
TAC positive = 1 pt.
HTLV-I *pos* tumor cells = 2 pts.
Classical = 6 pts.
Probable = 4 or 5 pts.
Possible = 3 pts.
Unconfirmed < 3 pts. (missing data)
Inconsistent with ATLL = 0 pts.
Exclusion data
B-cell positivity

in the literature. ATL occurred in patients of various racial/ethnic groups. Of the 90 diagnosed in the United States where racial/ethnic data were available, there were: 63 black; 15 Caucasian; 5 Hispanic; 4 Japanese; 2 other Asian-Pacific; and 1 Mestizo. The places of birth of those patients where these data were available are indicated in Fig. 1. The places of diagnosis, however, clustered primarily in cities with large black populations and alert clinicians.

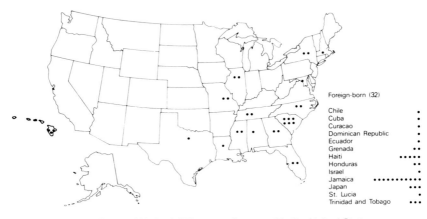

FIG. 1. Place of birth of ATL cases diagnosed in the United States.

Because one of the goals of the ATLR has been to identify geographic areas in the United States where HTLV-I is in high concentration, we initiated a serosurvey with Dr. Harvey Dosik and his colleagues in Brooklyn, who identified 15 cases of ATL occurring in a 22-mo period (2). The initial results from this serosurvey have confirmed the high prevalence of HTLV-I antibody in the Bedford-Stuyvesant area (3), and studies are currently in progress to investigate various racial/ethnic groups in this area.

The algorithm developed thus far emphasizes our concept that the diagnosis of ATL is based not only on virologic features, but also on clinical, pathologic, and immunologic features. Therefore, details are provided below for two cases reported to the registry—cases that we would not consider to be ATL—because of their epidemiologic and clinical significance.

ATLR #035

A 75-yr-old, Alaska native male presented in 1979 with a mass in the area of the left breast. Within the next 5 mo, the tumor was identified in multiple locations, including the nasopharynx, bone marrow, and subcutaneous tissue of the right breast and shoulder area. A biopsy was classified at NCI as malignant lymphoma, diffuse large cell, immunoblastic type compatible with ATL. HTLV-I antibody, tested in 1980 by radioimmunoprecipitation (RIP) and radioimmunoassay (RIA), was reported as positive (4).

Comment

The original diagnosis of ATL was based primarily on pathologic and serologic findings. Clinically, there was no leukemic phase, hypercalcemia, or involvement of the skin by tumor in the course of the patient's illness. However, the pathology is not specific for ATL and the titer of HTLV-I antibody was not higher than that reported in healthy donors living in Alaska (5). The disease had not been well defined in the United States at the time of patient diagnosis, and an atypical presentation for ATL in the polar region was a distinct possibility. The clinical features are not typical of ATL and, more important, Dr. Elaine Jaffe, NCI, was recently able to phenotype slides from a paraffin block obtained by Dr. Anne Lanier, Centers for Disease Control, documenting the tumor to be a B-cell lymphoma. Therefore, as of this report, there is no documented case of ATL in the polar region.

ATLR #109 (from reference 6)

A 34-yr-old black woman, the sister of an ATL patient, was reported to have had an intermittent maculopapular rash responsive to steroids. No signs of hypercalcemia, lymphadenopathy, or other clinical abnormalities were observed. No biopsies were taken, but the peripheral blood revealed 1% highly convoluted lymphocytes. HTLV-I *gag* antigen was identified in peripheral blood mononuclear cells after 20 d in culture. The patient was diagnosed as having "chronic ATL."

Comment

The diagnosis of chronic ATL was made based on the history of skin rash, abnormal "ATL-like" cells and the integration of HTLV-I, findings resembling "preleukemic" or "smoldering" ATL as previously described (7). Because similar cells are observed in HTLV-I–associated neurologic disease (8) [now referred to as HAM/TSP (9)] and do not necessarily indicate leukemia, in our opinion the diagnosis cannot be considered to be definitive without biopsy confirmation of a lesion. We recognize that this is a controversial area. The ATLR intends to include such cases under a separate category tentatively designated as "HTLV-I–associated leukocyte dysplasia," because leukemia is not an inevitable sequela and reversion to normal status has been observed (10).

Our perspective is based on two decades of experience with Epstein-Barr virus (EBV), the first human tumor-associated virus, which is etiologically implicated in Burkitt's lymphoma (BL), nasopharyngeal carcinoma (NPC), and other fatal malignant and nonmalignant diseases (11). The findings of abnormal circulating HTLV-I–associated leukocytes may well herald the subsequent appearance of ATL, but, because precursor lesions for EBV-associated NPC (12) and papilloma virus–associated cervical cancer (13) frequently regress, the condition as described in case #109 above should be considered as a potential *pre*malignant lesion at the present time.

Regarding case collection and laboratory evaluation, the ATLR has the potential of providing important biologic information comparable with that resulting in large part from the American Burkitt's Lymphoma Registry (ABLR). Although most cases of ATL are HTLV-I associated, at the present time it appears that many are not (14). This situation is comparable with that of BL and EBV, where a significant proportion (estimated at 5%–10%) of African cases are not EBV-related (do not contain EBV genome) even though they may occur in EBV-infected individuals (15). Recently, in a collaborative study with Dr. Ian Magrath's group at NCI, we have identified at least two distinct patterns of BL based on the relationship with EBV and independent of the geographic origin of the patient (16). The better prognosis of EBV-associated BL, suggested earlier in a previous series (17) based solely on serology, now is associated with specific chromosomal, biochemical, and immunologic markers that may lead to an improved understanding of the pathogenesis of one virus-associated B-cell lymphoma.

Perhaps the ATLR will be able to provide the same support role for molecular studies in this important T-cell lymphoma.

ACKNOWLEDGMENTS

The author thanks Drs. Jeffrey Clark, Angela Manns, and William Blattner, who helped develop the ATLR; and Dr. Peter Johnstone and Ms. Ruth Maloof for assistance in collating the data for this report.

REFERENCES

1. Levine PH, Jaffe ES, Manns A, Murphy EL, Clark J, Blattner WA. Human T-cell lymphotropic virus type I and adult T-cell leukemia/lymphoma outside Japan and the United States. *Yale J Biol Med* 1988;61:215–222.
2. Dosik H, Denic S, Patel N, Krishnamurthy M, Levine PH, Clark JW. Adult T-cell leukemia/lymphoma in Brooklyn. *JAMA* 1988;259:2255–2257.
3. Dosik H, Williams L, Goldstein M, Silberman E, Levine P, Maloof R. Seroprevalence of HTLV-I in a black Brooklyn community. *Proceedings of the Vth International Conference on AIDS.* 1989, 145.
4. Gallo RC, Kalyanaranam VS, Sarngadharan MG, et al. Association of the human type-C retrovirus with a subset of adult T-cell cancers. *Cancer Res* 1983;43:3892–3899.
5. Robert-Guroff M, Clark J, Lanier AP, et al. Prevalence of HTLV-I in Arctic regions. *Int J Cancer* 1985;36:651–655.
6. Ratner L, Poiesz B. Leukemias associated with Human T-cell Lymphotropic Virus Type I in a non-endemic region. *Medicine* 1988;401–422.
7. Yamaguchi K, Nishimura H, Kohrogi H, Jono M, Miyamoto Y, Takatsuki K. A proposal for smoldering adult T-cell leukemia: a clinicopathologic study of five cases. *Blood* 1983;62:758–766.
8. Dalgleish A, et al. *AIDS Res Hum Retroviruses* 1988;4:475–477.
9. Scientific Group on HTLV-I Infections and Its Associated Disease. Regional Office for the Western Pacific of WHO, Kagoshima, Japan, Dec. 10–15, 1988.
10. Kinoshita K, Amagasaki T, Ikeda S, et al. Preleukemic state of adult T-cell leukemia: abnormal T lymphocytosis induced by human adult T-cell leukemia-lymphoma virus. *Blood* 1985;66:120–127.
11. Levine PH, Ablashi DV, Nonoyama M, Pearson GR, Glaser R, eds. *Epstein-Barr virus and human disease.* Clifton: Humana Press, 1987.
12. Li ZQ, Chen JJ, Li WJ. Early detection of nasopharyngeal carcinoma (NPC) and nasopharyngeal mucosal hyperplastic lesion (NPHL) with its relationship to carcinomatous change. In: Prasad U, Ablashi DV, Levine PH, Pearson GR, eds. *Nasopharyngeal carcinoma—current concepts.* Kuala Lumpur: University of Malaya Press, 1983;11–23.
13. Stern E, Nelly PM. Carcinoma and dysplasia of the cervix. *Acta Cytol* 1963;7:357–361.
14. The T- and B-cell Malignancy Study Group. Statistical analysis of clinico-pathological, virological and epidemiological data on lymphoid malignancies with special reference to adult T-cell leukemia/lymphoma: a report of the second nationwide study of Japan. *Jpn J Clin Oncol* 1985;15:517–535.
15. de The G, Geser A, Day NE, et al. Epidemiological evidence for a causal relationship between Epstein-Barr Virus and Burkitt's lymphoma: results of the Ugandan prospective study. *Nature* 1978;274:756–761.
16. Barriga F, Kiwanuka J, Alvarez-Mon M, et al. Significance of Chromosome 8 breakpoint location in Burkitt's lymphoma: correlation with geographical origin and association with Epstein-Barr Virus. *Current Top Microbiol Immunol* 1988;141:128–137.
17. Levine PH, Kamaraju LS, Connelly RR, et al. The American Burkitt's Lymphoma Registry: eight years' experience. *Cancer* 1982;49:1016–1022.

Subject Index

T-lymphocytes (*contd.*)
  cytotoxic. *See* Cytotoxic T-lymphocytes
  "double negative", 67
  in HAM/TSP, 66–67, 82–83
  HTLV-I-infected, TNFβ secretion,
     108–109
  proliferation. *See* Lymphoproliferative re-
     sponse
  viruses infecting, 32
T-lymphotropic virus
  human. *See* HTLV
  simian (STLV), 8–9
TNFα. *See* Tumor necrosis factor
TNFβ. *See* Lymphotoxin
T-non-Hodgkin's lymphoma. *See* ATL, lym-
     phoma type
Tobago
  ATL in, 185–190, 238, 239, 243
  HTLV-I infections in, 237–242
  HTLV-I seropositivity, 255
  HTLV-I seroprevalence, 238
Trinidad
  ATL in, 185–190, 238, 239, 243
  HTLV-I infections in, 237–242
  HTLV-I seropositivity, 237, 255
Tropical ataxic neuropathy (TAN), 199
Tropical neuromyelopathies, 338–342
Tropical spastic paraparesis (TSP). *See also*
     HAM/TSP
  age and sex incidence, 200
  in Caribbean migrants in Britain, 221–223
  clinical features, 199–204
  clinical presentation, 200–201
  concomitant ATL and TSP, 239
  course and prognosis, 203
  defective provirus in, 21, 22–23, 25
  defining criteria, 200
  demographic features, 200
  diagnosis, 202–203
  discussion, 203–204
  in equatorial Africa, 339–340
  familial, in Trinidad and Tobago, 239–240
  HTLV-I role in, 47–48
  laboratory investigations, 202–203
  neurologic examination, 201–202
  neuropathology, 10
    visna vs, 17

non-HTLV-I-associated, in South India,
     247–249
seroreactivity to recombinant HTLV-I en-
     velope proteins, 453, 456–457
in Trinidad and Tobago, 238–240
in Tumaco, Colombia, 377–382
Tumor necrosis factor (TNFα), 105, 106, 108

V
Vesicular stomatitis virus (VSV) pseudotype
     assay, HTLV-I receptor, 45, 46
Viral persistence, 27–34, 33–34
Virology
  ATL in France, 13–134, 131
  HAM/TSP, 65, 67–75
  tropical spastic paraparesis, 202
Visna
  causation, 17–19
  neuropathology, tropical spastic paraparesis
     vs, 17
  neurotropism, 20–21
  parallels with HAM/TSP, 15–16

W
Weibull model, age-dependent occurrence of
     ATL, 308–312
Western blot, 410
  confirmation of HTLV reactivity. *See also*
     Serologic testing, confirmation assays
  HTLV-I recombinant proteins in,
     428–430
  confirmation of HTLV-I, 353, 394
  confirmation of HTLV-II, HTLV-I recom-
     binant proteins in, 430–431, 432, 433
  HTLV-I gp61 recombinant proteins, 452
  HTLV-I recombinant proteins, 424–425,
     428
  recombinant HTLV-I gp21, 420
  sensitivity, 393
  sensitivity and specificity, 423–424
World Health Organization (WHO), Kago-
     shima meeting of Scientific Group on
     HTLV-I Infections and Associated
     Diseases, 191–197